Lecture Notes in Artificial Intelligence   2680

Edited by J. G. Carbonell and J. Siekmann

Subseries of Lecture Notes in Computer Science

Springer
*Berlin
Heidelberg
New York
Hong Kong
London
Milan
Paris
Tokyo*

Patrick Blackburn Chiara Ghidini
Roy M. Turner Fausto Giunchiglia (Eds.)

Modeling and Using Context

4th International and Interdisciplinary Conference
CONTEXT 2003
Stanford, CA, USA, June 23-25, 2003
Proceedings

 Springer

Series Editors

Jaime G. Carbonell, Carnegie Mellon University, Pittsburgh, PA, USA
Jörg Siekmann, University of Saarland, Saarbrücken, Germany

Volume Editors

Patrick Blackburn
INRIA Lorraine
615, rue du Jardin Botanique, 54602 Villers les Nancy Cedex, France
E-mail: patrick@aplog.org

Chiara Ghidini
University of Liverpool, Department of Computer Science
Chadwick Building, Peach Street, Liverpool L69 7ZF, UK
E-mail: chiara@csc.liv.ac.uk

Roy M. Turner
University of Maine, Department of Computer Science
5752 Neville Hall, Orono, ME 04469-5752, USA
rmt@umcs.maine.edu

Fausto Giunchiglia
University of Trento, Department of Information and Communication Technology
38050 Povo, Trento, Italy
E-mail: Fausto.Giunchiglia@unitn.it

Cataloging-in-Publication Data applied for

A catalog record for this book is available from the Library of Congress.

Bibliographic information published by Die Deutsche Bibliothek
Die Deutsche Bibliothek lists this publication in the Deutsche Nationalbibliografie;
detailed bibliographic data is available in the Internet at <http://dnb.ddb.de>.

CR Subject Classification (1998): I.2, F.4.1, J.3, J.4

ISSN 0302-9743
ISBN 3-540-40380-9 Springer-Verlag Berlin Heidelberg New York

This work is subject to copyright. All rights are reserved, whether the whole or part of the material is
concerned, specifically the rights of translation, reprinting, re-use of illustrations, recitation, broadcasting,
reproduction on microfilms or in any other way, and storage in data banks. Duplication of this publication
or parts thereof is permitted only under the provisions of the German Copyright Law of September 9, 1965,
in its current version, and permission for use must always be obtained from Springer-Verlag. Violations are
liable for prosecution under the German Copyright Law.

Springer-Verlag Berlin Heidelberg New York
a member of BertelsmannSpringer Science+Business Media GmbH

http://www.springer.de

© Springer-Verlag Berlin Heidelberg 2003
Printed in Germany

Typesetting: Camera-ready by author, data conversion by PTP-Berlin GmbH, Heidelberg
Printed on acid-free paper   SPIN: 10927588   06/3142   5 4 3 2 1 0

Preface

Whether you are a computer scientist, a logician, a philosopher, or a psychologist, it is crucial to understand the role that context and contextual information plays in reasoning and representation. The conference at which the papers in this volume were presented was the fourth in an international series devoted to research on context, and was held in Stanford (USA) on June 23–25, 2003. The first conference in the series was held in Rio de Janeiro (Brazil) in 1997, the second was held in Trento (Italy) in 1999, and the third was held in Dundee (Scotland, UK) in 2001.

CONTEXT 2003 brought together representative work from many different fields: in this volume you will find philosophical theorizing, logical formalization, computational modelling — and, indeed, computational applications — together with work that approaches context from a more cognitive orientation. While we don't believe that this volume can capture the lively flavor of discussion of the conference itself, we do hope that researchers interested in context (in any of its many manifestations) will find something of interest here, perhaps something that will inspire new lines of work.

We are very grateful to our invited speakers: Patrick Brézillon (University of Paris VI, France), Keith Devlin (CSLI, Stanford), and David Leake (Indiana University, USA) for presenting three important contemporary perspectives on the study on context.

Special thanks are due to the program committee and the additional reviewers (listed below) who shouldered a heavy load and stuck to an exacting reviewing schedule. Reviewing for an interdisciplinary conference is never easy, and the CONTEXT series, because of the extraordinarily broad nature of the topic, imposes especially heavy demands. The program committee and additional reviewers rose to the challenge splendidly, and we are extremely grateful for their efforts.

Last but not least, we would like to thank all those people who made CONTEXT 2003 happen at Stanford itself. So thank you to Dikran Karagueuzian (who chaired the local arrangements committee) and to Keith Devlin, Michele King, John Perry, and Elisabetta Zibetti. Finally, special thanks to Roberta Ferrario, who handled the CONTEXT 2003 publicity.

June 2003

Patrick Blackburn
Chiara Ghidini
Roy Turner
Fausto Giunchiglia

Organization

CONTEXT 2003 took place in the Gates Building, Computer Science Department, Stanford University.

Organizing Committee

Conference Chair: Fausto Giunchiglia (Università degli Studi di Trento, Italy)
Program Co-chairs: Patrick Blackburn (INRIA Lorraine, France)
Chiara Ghidini (University of Liverpool, UK)
Roy Turner (University of Maine, USA)
Local Arrangements: Dikran Karagueuzian (Chair)
Keith Devlin
Michele King
John Perry
Elisabetta Zibetti
Publicity Chair: Roberta Ferrario (Università degli Studi di Trento, Italy)

Program Committee

Horacio Arló Costa
Carla Bazzanella
Jose Luis Bermudez
Anind K. Dey
Bruce Edmonds
Tim Fernando
Christopher Gauker
Avelino Gonzalez
Lucja Iwanska
David Leake
Bernard Moulin
Jean-Charles Pomerol
Carles Sierra
Steffen Staab
Peter Turney
Robert J. van den Bosch
Terry Winograd

John Barnden
John Bell
Matteo Bonifacio
Christo Dichev
Paul Feltovich
Anita Fetzer
Alain Giboin
Jerry Hobbs
Ruth Kempson
Mark Maybury
Rolf Nossum
Marina Sbisà
Munindar Singh
Elise Turner
Johan van Benthem
Teun A. van Dijk

Additional Reviewers

Josep Lluis Arcos
Massimo Benerecetti
Paolo Bouquet
Paolo Busetta
Eric de la Clergerie
Enrico Franconi
Fabien Gandoni

Jorge Gonzalez Olalla
Floriana Grasso
Stefan Kaufmann
Saturnino Luz
Pablo Noriega
Josep Puyol-Gruart
Sergi Robles

Juan Rodríguez-Aguilar
York Sure
Valentina Tamma
Roger Young
Stefano Zanobini

Sponsoring Institutions

We would like to thank the University of Trento (Trento, Italy), the Stanford University Center for the Study of Language and Information (CSLI), the Linguistics Department of Stanford University, the Philosophy Department of Stanford University, and the Psychology Department of Stanford University, for supporting CONTEXT 2003.

Table of Contents

Full Papers

Presupposition Incorporation in Adverbial Quantification 1
  David Ahn

A Theory of Contextual Propositions for Indicatives 15
  Horacio Arló Costa

Context-Sensitive Weights for a Neural Network 29
  Robert P. Arritt, Roy M. Turner

A Common Sense Theory of Causation 40
  John Bell

How to Refer: Objective Context vs. Intentional Context.............. 54
  Claudia Bianchi

A SAT-Based Algorithm for Context Matching 66
  Paolo Bouquet, Bernardo Magnini, Luciano Serafini, Stefano Zanobini

On the Difference between Bridge Rules and Lifting Axioms 80
  Paolo Bouquet, Luciano Serafini

Context Dynamic and Explanation in Contextual Graphs 94
  Patrick Brézillon

A Deduction Theorem for Normal Modal Propositional Logic 107
  Saša Buvač

Natural Deduction and Context as (Constructive) Modality 116
  Valeria de Paiva

Communicative Contributions and Communicative Genres: Language
Production and Language Understanding in Context 130
  Anita Fetzer

Explanation as Contextual Categorization........................... 142
  Leslie Ganet, Patrick Brézillon, Charles Tijus

Effects of Context on the Description of Olfactory Properties 154
  Agnès Giboreau, Isabel Urdapilleta, Jean-François Richard

Varieties of Contexts 164
  R. Guha, John McCarthy

Ubi-UCAM: A Unified Context-Aware Application Model 178
  Seiie Jang, Woontack Woo

Contextual Effects on Word Order: Information Structure and
Information Theory 190
  Nobo Komagata

A Generic Framework for Context-Based Distributed Authorizations 204
  Ghita Kouadri Mostéfaoui, Patrick Brézillon

Unpacking Meaning from Words: A Context-Centered Approach to
Computational Lexicon Design 218
  Hugo Liu

A Contextual Approach to the Logic of Fiction 233
  Rolf Nossum

Predictive Visual Context in Object Detection 245
  Lucas Paletta

Copular Questions and the Common Ground 259
  Orin Percus

Contextual Coherence in Natural Language Processing 272
  Robert Porzel, Iryna Gurevych

Local Relational Model: A Logical Formalization of
Database Coordination 286
  Luciano Serafini, Fausto Giunchiglia, John Mylopoulos,
  Philip Bernstein

What to Say on What Is Said 300
  Isidora Stojanovic

Modelling "but" in Task-Oriented Dialogue 314
  Kavita E. Thomas

Dynamic Contextual Intensional Logic: Logical Foundations and
an Application 328
  Richmond H. Thomason

Comparatively True Types: A Set-Free Ontological Model of
Interpretation and Evaluation Contexts 342
  Martin Trautwein

Discourse Context and Indexicality 356
  Mark Whitsey

A Mathematical Model for Context and Word-Meaning 369
  Dominic Widdows

Demonstratives, Reference, and Perception 383
  R.A. Young

Perceiving Action from Static Images: The Role of Spatial Context 397
  Elisabetta Zibetti, Charles Tijus

Short Papers

How to Define the Communication Situation: Determining Context
Cues in Mobile Telephony 411
  Louise Barkhuus

How to Use Enriched Browsing Context to Personalize Web Site Access .. 419
  Cécile Bothorel, Karine Chevalier

Modular Partial Models: A Formalism for Context Representation 427
  Harry Bunt

Contextual Modeling Using Context-Dependent Feedforward
Neural Nets 435
  Piotr Ciskowski

Context-Based Commonsense Reasoning in the DALI Logic
Programming Language 443
  Stefania Costantini, Arianna Tocchio

An Ontology for Mobile Device Sensor-Based Context Awareness 451
  Panu Korpipää, Jani Mäntyjärvi

The Use of Contextual Information in a Proactivity Model for
Conversational Agents 459
  Marcello L'Abbate, Ulrich Thiel

GloBuddy, a Dynamic Broad Context Phrase Book 467
  Rami Musa, Madleina Scheidegger, Andrea Kulas, Yoan Anguilet

Exploiting Dynamicity for the Definition and Parsing of Context
Sensitive Grammars 475
  Emanuele Panizzi

Co-text Loss in Textual Chat Tools 483
  Mariano Gomes Pimentel, Hugo Fuks, Carlos José Pereira de Lucena

Context Proceduralization in Decision Making 491
  Jean-Charles Pomerol, Patrick Brézillon

GRAVA: An Architecture Supporting Automatic Context Transitions
and Its Application to Robust Computer Vision 499
  Paul Robertson, Robert Laddaga

Speaking One's Mind 507
    Alice G.B. ter Meulen

Connecting Route Segments Given in Route Descriptions 517
    Ladina Tschander

Author Index 525

Presupposition Incorporation in Adverbial Quantification

David Ahn

University of Rochester, Rochester, NY 14627, USA,
http://www.cs.rochester.edu/~davidahn

Abstract. In this paper, we present a critique of the uniform treatment of nominal and adverbial quantification in van der Sandt's binding theory of presuppositions. We develop an alternative account of adverbial quantification framed in terms of Beaver's reconstruction of Karttunen's satisfaction theory. This account provides a simpler characterization of adverbial quantification that does not require recourse to accommodation to explain the basic facts regarding presupposition incorporation.

1 Introduction

Natural language quantifiers presuppose their domains of quantification. Or, alternatively, natural language quantifier domains are anaphoric. Whichever way you look at it, it is uncontroversial that the domain of a natural language quantifier depends on the context in which it is used. Consider the discourse (1).

(1) There are fifteen boys and fifteen girls at a boarding school. Five girls and two boys are day-pupils, though, so only ten girls and thirteen boys live in the dormitory. As the dormitory has ten rooms on each of two floors, every girl has a room to herself, but some boys have to share.

The domains of the quantified noun phrases *every girl* and *some boys* are restricted by the context to those girls and boys living in the dormitory, even though there are explicitly other girls and boys in our domain of discourse.

Van der Sandt's binding theory of presuppositions [1], together with the assumption that quantifiers presuppose their domains, provides a straightforward account of the contextual restriction of quantifier domains. However, Beaver [2] points out that the binding theory incorrectly predicts that a quantifier domain can be restricted through accommodation of presuppositions in the scope of the quantifier. Thus, the following sentences are predicted to be equivalent.

(2) a. #Every German loves his kangaroo.
    b. Every German who has a kangaroo loves his kangaroo.

Geurts and van der Sandt [3] present a revised version of the binding theory which accounts for the difference between these two sentences by giving up on the possibility of determining the domain of a quantifier through intermediate accommodation. However, as Ahn [4,5] points out, accommodation of scopal presuppositions appears to be crucial in the interpretation of quantificational adverbs (qadverbs). For example, in the discourse (3), it is the presupposition of the verb *beat* that determines the domain of the qadverbs *usually* and *always*.

(3) Marvin and John play various games together. Marvin has spent a lifetime practicing racket sports, but John is simply the superior athlete. John usually beats Marvin at badminton. He always beats him at tennis.

In this paper, we present an alternative account of the semantics of qadverbs which provides for incorporation of presuppositions into qadverb domains without accommodation. Our account is based on Beaver's reconstruction [6] of the satisfaction theory of Karttunen [7] and relies on the treatment of contexts as sets of possible worlds and possible worlds, in turn, as composed of situations.

2 The Binding Theory and Quantification

The presuppositions of a sentence are those propositions that the sentence seems to take for granted. Presuppostitions are triggered by a wide variety of linguistic phenomena, most notably definite descriptions (which presuppose the existence and uniqueness of an entity satisfying the description) and factive predicates (which presuppose their propositional complements), but also clefts, selectional restrictions, adverbial clauses, aspectual verbs, iteratives, quantifiers, and so on.

Presuppositions are more robust than other entailments of a sentence. They are usually preserved when the sentence that carries them is embedded under negation or modality, for example. One of the most perplexing problems regarding presuppositions is the projection problem — the problem of accounting for their behavior under various embeddings. In this section, we present one widely accepted theory that attempts to solve the projection problem.

2.1 The Standard Binding Account

Van der Sandt [1] frames his account in Discourse Representation Theory (DRT) [8]. A discourse is represented by a Discourse Representation Structure (DRS) — a pair of a universe (a set of discourse referents) and a set of conditions (atomic predications or complex formulas built out of other DRSs). We notate a DRS by enclosing it in square brackets with a vertical line separating the universe from the conditions. A presupposition is represented as an underlined DRS which is initially placed in the DRS in which its trigger is represented. For example, *John's son* in the discourse (4a) triggers the presupposition that there is an individual who is John's son. This presupposition is represented by the innermost DRS in (4b).

(4) a. John does not have a daughter. If John has a son, John's son is bald.

b. $\left[J \; \middle| \; \begin{array}{l} \neg \left[D \; \middle| \; \text{daughter}(D, J) \right], \\ \left[S \; \middle| \; \text{son}(S, J) \right] \to \left[\; \middle| \; \text{bald}(X), \underline{\left[X \; \middle| \; \text{son}(X, J) \right]} \right] \end{array} \right]$.

Van der Sandt's basic insight is that presuppositions behave like discourse anaphors (i.e. pronouns), in that they must be bound to an antecedent. DRT provides a structural notion of accessibility that constrains a pronoun's search for an antecedent. On van der Sandt's account, the same notion of accessibility constrains presuppositions, as well. Basically, a potential antecedent is accessible to an anaphor if it is introduced in a DRS that contains the anaphor or that is the antecedent of a conditional whose consequent contains the anaphor. Thus, in the DRS (4b), referents J and S, but not D, are accessible to the presupposition.

A pronoun or presupposition is bound to an antecedent by replacing the discourse marker corresponding to the pronoun or presupposition with the antecedent discourse marker and, in the case of a presupposition, adding the conditions of the presuppositional DRS to the DRS in which the antecedent is introduced (and then deleting the presuppositional DRS). For example, the result of binding the presupposition in (4b) to the referent S would be (5).

(5) $\left[J \;\middle|\; \begin{array}{l} \neg\, [D \,|\, \text{daughter}(D, J)], \\ [S \,|\, \text{son}(S, J)] \rightarrow [\,|\, \text{bald}(S)\,] \end{array} \right]$.

There is an important difference between presuppositions and pronouns. If a suitable antecedent for a presupposition cannot be found, one can simply be created in an accessible DRS using the descriptive content of the presupposition. The creation of an antecedent for a presupposition that otherwise would not have one is the realization in this theory of Lewis's notion of accommodation [9]. For example, if the antecedent of the conditional in (4a) were *John is bald* rather *John has a son*, then, assuming that John cannot be his own son, there would be no suitable antecedent for the presupposition in (4b). Thus, we would simply create one, which, in the case of global accommodation, would result in (6).

(6) $\left[J\,X \;\middle|\; \begin{array}{l} \neg\, [D \,|\, \text{daughter}(D, J)], \text{son}(X, J), \\ [\,|\, \text{bald}(J)\,] \rightarrow [\,|\, \text{bald}(X)\,] \end{array} \right]$.

Van der Sandt's mechanisms of binding and accommodation provide an account of the projection problem — global resolution results in projection; local or intermediate resolution, in cancellation. There are a variety of preferences and constraints that govern the resolution of presuppositions. In short, resolution must not result in a DRS that is uninterpretable, inconsistent, or redundant. Also, binding is preferred to accommodation, and more local binding and more global accommodation are preferred.

2.2 Quantification in the Standard Account

In DRT, the conditional is given a strong semantics (every assignment that verifies its antecedent must verify its consequent), so that (7a) and (7b) may be treated equivalently as (7c).[1]

---

[1] In our binding theory examples, we omit the domain set presupposition necessary to capture the context-dependence of a quantifier.

(7) a. If a farmer owns a donkey, he beats it.
   b. Every farmer who owns a donkey beats it.
   c. $[\,|\,[X\,Y\,|\,\text{farmer}(X), \text{donkey}(Y), \text{owns}(X,Y)] \to [\,|\,\text{beats}(X,Y)]\,]$.

As a consequence, the restrictor of a quantifier is predicted to be accessible for the resolution of pronouns and presuppositions. It seems that as far as binding goes, this prediction is correct. For example, the correct interpretation of (8a) is the one in which the complex presupposition triggered by *his donkey* is bound to the discourse referents F and D introduced in the restrictor of the quantifier.

(8) a. Every farmer who owns a donkey beats his donkey.
   b. $\left[\,\Big|\,\left[\,\Big|\,[F\,D\,|\,\text{farmer}(F), \text{donkey}(D), \text{owns}(F,D)] \to \left[\,\Big|\,\text{beats}(F,X), \underline{[X\,|\,\text{donkey}(X), \text{owns}(Y,X), \underline{[Y\,|\,\text{male}(Y)]}]}\right]\right]\right]$.

Unfortunately, as Beaver [2] points out, the prediction that presuppositions may also be accommodated by the restrictor seems to be wrong. Consider sentences (2a) and (2b). The initial DRS for sentence (2a) is depicted in (9).

(9) $\left[\,\Big|\,\left[\,\Big|\,[G\,|\,\text{german}(G)] \to \left[\,\Big|\,\text{loves}(G,X), \underline{[X\,|\,\text{kangaroo}(X), \text{owns}(Y,X), \underline{[Y\,|\,\text{male}(Y)]}]}\right]\right]\right]$.

The standard binding account predicts that the presupposition is accommodated by the restrictor, resulting in the DRS (10). This DRS is also predicted to be the final interpretation of (2b), after the presupposition triggered by *his kangaroo* is resolved by the restrictor (we omit the intial representation for brevity).

(10) $[\,|\,[G\,K\,|\,\text{german}(G), \text{kangaroo}(K), \text{owns}(G,K)] \to [\,|\,\text{loves}(G,K)]\,]$.

Of course, sentence (2a) is clearly not as felicitous as sentence (2b). It does not seem to be the case that the presupposition triggered by *his kangaroo* can restrict the domain of the quantificational NP *every German* to just those Germans who own kangaroos. The binding theory simply makes the wrong predictions regarding intermediate accommodation and quantification.

2.3 Adverbial Quantification in the Standard Account

The situation is somewhat different when we turn to adverbial quantification. Unlike a quantificational determiner, which is syntactically associated with a constituent that provides its restrictor argument (namely, the N'), a qadverb has no syntactic relationship with its restrictor argument. Several authors have suggested that the presuppositions of an adverbially quantified sentence determine the restrictor of a qadverb [10,11]. Thus, for example, in the discourse (3), it is the presupposition of the verb *beat* — *x beats y in some game* presupposes that *x* plays *y* in that game — that provides the description of the domains of quantification for the two adverbially quantified sentences.

Intermediate accommodation seems to be exactly the mechanism which is at work here. Ahn [4,5] presents an account which relies heavily on intermediate accommodation and on the explicit representation of situations. Adverbial quantification is treated as quantification over situations. Presuppositions are associated with resource situations, following Cooper [12]. Initially, an adverbially quantified sentence is represented with a vacuous restrictor that provides a situation variable with no restrictions, as in (11b), a simplified DRS for (11a)[2].

(11) a. John usually beats Marvin at badminton.
  b. $\left[JM \mid [S \mid S : [\mid]] \stackrel{\text{USU}}{\Longrightarrow} \left[\mid S : \left[\mid \begin{array}{c} \text{win}(J), \\ \underline{[S' \mid S' : [\mid \text{play\_bdtn}(J, M)]]} \end{array} \right] \right] \right]$.

Accommodation of scopal presuppositions in the restrictor results in binding their resource situations to the situation variable. This then provides the restriction, and a presupposition of a set of situations based on this restriction is triggered. Resolving this presupposition then accounts for the context-dependence of the qadverb domain. Thus, in (11b), the presupposition associated with *beat* is accommodated by the restrictor of *usually*, resulting in a restrictor that specifies situations of badminton-playing, as in (12). A presupposition of a set of badminton-playing situations would then be generated and resolved by global accommodation (there is a set of situations available in the discourse, but since the available set is of situations of playing various games, it doesn't match the presupposition).

(12) $\left[JM \mid [S \mid S : [\mid \text{play\_bdtn}(J, M)]] \stackrel{\text{USU}}{\Longrightarrow} [\mid S : [\mid \text{win}(J)]] \right]$.

While this account provides more or less the correct interpretations for adverbially quantified sentences, it suffers from several faults. First of all, it invokes accommodation, which, in the binding theory, is intended as a repair strategy, to account for all adverbially quantified sentences in all contexts. Secondly, it imposes a strict interleaved order on steps of presupposition resolution and computation; in particular, it requires that first, presuppositions in the scope be computed; second, these presuppositions be resolved; and third, a presupposition corresponding to the restrictor be computed. Finally, it is parasitic on the standard account of nominal quantification, even though nominal and adverbial quantification are syntactically quite different and behave differently precisely with respect to intermediate accommodation of scopal presuppositions.

2.4 The Revised Account

In response to Beaver's criticisms, Geurts and van der Sandt [3] present a revised account of quantification in the binding theory. The most striking departure from the standard account is that DRSs are reified, so that there is a sort of discourse

---

[2] The colon operator is intended to indicate a support relation between a situation and a formula.

referent that is associated with a DRS and that refers to the set of assignments which embeds the DRS. By making such a referent anaphoric, it is possible for local operations, such as intermediate accommodation, to have non-local effects.

In particular, the restrictor of a quantifier is treated as such an anaphoric discourse referent. Since, on this account, presuppositions are resolved from left to right, this restrictor referent is bound to an antecedent before the presuppositions of the scope are resolved. These scopal presuppositions may then be accommodated by the restrictor DRS, but this results in the direct imposition of conditions on the antecedent set. If these conditions are not compatible with the antecedent set, resolution fails. Thus, the scopal presuppositions cannot be used as the basis for accommodating a new domain set — the identity of the domain set is decided before the scopal presuppositions are resolved.

With this revised account, Geurts and van der Sandt manage to make the correct predictions for (2a). The restrictor is resolved first by being to bound to a set of Germans. Out of the blue, this set of Germans is apt to be the set of all Germans. Accommodation of the scopal presupposition requires imposing an additional condition on this domain set, namely, that every German has a kangaroo. Since this additional condition is clearly in conflict with what we know about Germans, the resolution fails, resulting in infelicity. There is no option to accommodate a set which satisfies the scopal presuppositions.

Geurts and van der Sandt also present an account of adverbial quantification that is based on their account of nominal quantification. Like Ahn, they begin with a vacuous restrictor, and as with nominal quantifiers, this (vacuous) restrictor is resolved before the scopal presuppositions. Unfortunately, as we have seen in the example (3), the domain set must be accommodated on the basis of the scopal presuppositions, which is exactly the option that this revised mechanism makes impossible. This account of adverbial quantification, too, is plagued by an undue reliance on an analysis of nominal quantification.

3 Satisfaction

Van der Sandt's binding theory of presupposition, in which presuppositions are treated as entities that can be resolved (and thus, in a sense, cancelled) within a highly structured context, is only one of many ways to approach the problem of presuppositions. In the remainder of this paper, we adopt an alternative account of presuppositional phenomena — the satisfaction theory of Karttunen [7].

Karttunen's theory takes seriously the pretheoretic intuition that a presupposition is something that is taken for granted. The presuppositions of a sentence are taken to be constraints on the contexts in which the sentence can be uttered. Thus, a context admits a sentence if and only if it satisfies the presuppositions of the sentence. Crucially, the presuppositions of the components of a complex sentence do not necessarily constrain the context of the complex sentence itself. Instead, one part of a complex sentence may establish an updated context of evaluation for another part. For example, the first conjunct of a conjunction is evaluated with respect to the same context as the entire conjunction, but the second conjunct is evaluated with respect to the update of that context with the first conjunct. Thus, if the first conjunct entails the presuppositions of the second, those presuppositions do not project — the updated context with respect to which the second conjunct is evaluated will always entail them.

We build on Beaver's formalization of the satisfaction theory in terms of an update semantics [6], which, in turn, builds on Heim's dynamic account in terms of context-change potential [13]. By treating the meaning of a sentence dynamically, as a function from input contexts to output contexts, it is possible to tie the admittance conditions of a sentence directly to its semantics. Following Stalnaker [14], Beaver treats a context as a set of possible worlds. A context consists of the possible worlds compatible with what has already been expressed in a conversation and thus represents the "live options" left open by the conversation. We show that by treating possible worlds as structures that are decomposable into constituent situations, adverbial quantification can be given an analysis that accounts for both domain determination by presupposition incorporation and the anaphoricity of qadverb domains *without* recourse to intermediate accommodation. This analysis also avoids the inaccurate analogy with nominal quantification and is thus free to provide a one-place logical operator, which corresponds more closely to the syntactic type of a qadverb.

3.1 Context-Change Potential and Situation Theory

The theory of interpretation we adopt here identifies the meaning of a sentence not with its truth conditions but with its context-change potential. The denotation of a sentence is defined as a relation between an input context and an output context. As we stated earlier, we take a context to be a set of possible worlds, and we use an update semantics, along the lines of Veltman [15], to model context change. The set of possible worlds that constitute a context is intended to represent the possibilities left open by the discourse so far. To update a context with a sentence, the worlds which are not compatible with the sentence are thrown out. Thus, as a discourse progresses, possibilities are winnowed away and the context comes closer and closer to the speaker's view of the actual world.

We augment this notion of context by taking possible worlds to have situational substructure. We take a view of situations along the lines of Schubert et al. [16,17]. In the ontology of situations given in [17], situations are individual entities in the domain of discourse. The set of situations $\mathcal{S}$ within the domain of discourse is subject to a partial part-of ordering (notated $\sqsubseteq$) which induces a join semi-lattice structure (the join operation is notated with the infix operator $\sqcup$). Thus, any two situations may be joined to form a larger situation. We identify the maximal element of the semi-lattice as the world.

In order to introduce multiple possible worlds into an ontology such as this one, we weaken the structure of the set of situations. Instead of requiring that it be a join semi-lattice (in other words, that every pair of situations have a unique join), we merely require that it be a set of possibly overlapping join semi-lattices. In order to enforce this requirement, we designate a subset of the set of situations $\mathcal{S}$ as the set of worlds $\mathcal{W}$ and require that every situation stand in the part-of relation with at least one world. We further require that if two situations t, u are both part of the same world w, then there exists a situation $s \sqsubseteq w$ such that $t \sqsubseteq s$, $u \sqsubseteq s$, and for every situation $s' \sqsubseteq w$, if $t \sqsubseteq s'$ and $u \sqsubseteq s'$, then $s \sqsubseteq s'$. We call s the join of t and u with respect to w and write $s = t \sqcup_w u$.

The most distinctive characteristic of Schubert's situation theory is that there are two different relations in which a situation and a formula may find themselves. The *support* relation holds between a formula and a situation which is (at least) partially described by the formula; this more familiar relation is similar to the support relation in the Situation Semantics of Barwise and Perry [18]. The *characterization* relation, which is a generalization of the relation between a Davidsonian event predication and the event described by the predication, holds between a situation and a formula which provides a description that applies to the entire situation as a whole. Note that this description need not be fully specified or detailed, only that it must describe the entire situation and not merely a part of it. Interestingly, the characterization relation between a situation and a negated formula does not hold simply whenever the characterization relation fails to hold between the situation and the non-negated formula. Instead, there is a notion of anti-characterization — a formula $\neg \phi$ characterizes a situation s if and only if s is a situation of ϕ not holding.[3] Both the support and characterization relations are defined in terms of the basic Davidsonian relation between events and atomic event predicates and a corresponding anti-relation.

Following Ahn and Schubert [20], we adopt a reconstruction of the propositional fragment of Schubert's first-order situation logic FOL** as a modal propositional logic, in which the states are situations and there are modalities corresponding to the subsumption (the dual of part-of) and join relations. The satisfaction[4] relation of our logic corresponds to the characterization relation. Support can be defined as possibility with respect to the subsumption modality. Our update semantics, then, is exactly that of Beaver's, except that the atomic update condition applies to all modal formulas and is stated with respect to the modal satisfaction relation.

3.2 A Propositional Update Logic with Adverbial Quantification

A model for our logic $\mathcal{M}$ is a a 5-tuple $\langle \langle \mathcal{S}, \sqsubseteq \rangle, \mathcal{W}, \sqsupseteq, \sqcup, \langle \mathcal{I}^+, \mathcal{I}^- \rangle \rangle$. $\langle S, \sqsubseteq \rangle$ is the partially ordered set of situations; $\mathcal{W}$ is the subset of $\mathcal{S}$ that is the set of worlds. The binary relation $\sqsupseteq$ (subsumption) is the dual of the partial order $\sqsubseteq$. The ternary relation $\sqcup$ is the "is-the-join-of-with-respect-to-some-world" relation; thus, $\sqcup(s, t, u)$ iff s is the join of t and u with respect to some world.

---

[3] What constitutes a situation of some formula not holding is an interesting question. One possibility is a situation of some positive event that precludes the truth of the formula. See Schubert et al. [16,17], as well as Cooper [19], for further discussion.

[4] We use the term *satisfaction* for two different relations. One is Karttunen's relation between a context and a sentence; the other is the relation between a state in a model and a formula of our modal logic. Where there may be confusion, we use *context* or *modal* satisfaction.

Finally, $\mathcal{I}^+$ and $\mathcal{I}^-$ are the positive and negative interpretation functions — they correspond directly to atomic characterization and anti-characterization.

First, we give the modal satisfaction conditions that will form the atomic basis for our update semantics.

Definition 1 (Modal satisfaction conditions).
For a model $\mathcal{M}$ and a situation s:

$\mathcal{M}, s \Vdash p$ *iff* $s \in \mathcal{I}^+(p)$,
$\mathcal{M}, s \Vdash \Diamond \phi$ *iff* $\exists s'.(s \sqsupseteq s')$ *and* $(\mathcal{M}, s' \Vdash \phi)$,
$\mathcal{M}, s \Vdash \Box \phi$ *iff* $\forall s'.(s \sqsupseteq s')$ *only if* $(\mathcal{M}, s' \Vdash \phi)$,
$\mathcal{M}, s \Vdash \phi \circ \psi$ *iff* $\exists t, u.(\mathcal{M}, t \Vdash \phi)$ *and* $(\mathcal{M}, u \Vdash \psi)$ *and* $\sqcup(s, t, u)$,
$\mathcal{M}, s \Vdash \neg \phi$ *iff* $\mathcal{M}, s \Vdash^- \phi$,
$\mathcal{M}, s \Vdash \phi \vee \psi$ *iff* $\mathcal{M}, s \Vdash \phi$ *or* $\mathcal{M}, s \Vdash \psi$.

The conditions for anti-satisfaction ($\Vdash^-$) are the dual of these conditions (i.e. $\Vdash$ and $\Vdash^-$ are interchanged, as are $\circ$ and $\vee$, and universal and existential quantification, and so on).

There are several crucial things to observe here. The first is that the satisfaction relation between a situation and a formula corresponds to characterization. Another is that the binary modal operator $\circ$ corresponds to conjunction under characterization for FOL**. Since each characterization statement of FOL** is intended to correspond to a single tensed clause, $\circ$ is intended to be used to "conjoin" the atomic predications that correspond to the translation of a single tensed clause. Thus, we treat $\circ$ as the dual of $\vee$. Finally, we can think of the $\sqsupseteq$ modality as a sort of accessibility relation betweeen situations, which can be used to define Schubert's notions of support and inward persistence.

Definition 2 (Support and inward persistence).
A situation s supports a formula ϕ iff $s \Vdash \Diamond \phi$.
A situation s is inward persistent with respect to a formula ϕ iff $s \Vdash \Box \phi$.

A context is a set of worlds. The denotation of a formula of our logic is a relation between contexts, given by the following update semantics.

Definition 3 (Update conditions).
For contexts σ and τ, the denotation $[\![\cdot]\!]$ of a formula is given recursively, as follows:

$\sigma[\![\phi_{modal}]\!]\tau$ *iff* $\tau = \{w \in \sigma | w \Vdash \Diamond \phi_{modal}\}$,
$\sigma[\![\neg\phi]\!]\tau$ *iff* $\exists \nu.\sigma[\![\phi]\!]\nu$ *and* $\tau = \sigma \setminus \nu$,
$\sigma[\![\phi \wedge \psi]\!]\tau$ *iff* $\exists \nu.\sigma[\![\phi]\!]\nu$ *and* $\nu[\![\psi]\!]\tau$,
$\sigma[\![\phi \to \psi]\!]\tau$ *iff* $\sigma[\![\neg(\phi \wedge (\neg\psi))]\!]\tau$,
$\sigma[\![\phi \vee \psi]\!]\tau$ *iff* $\sigma[\![\neg(\neg\phi \wedge \neg\psi)]\!]\tau$,
$\sigma[\![\partial\phi]\!]\tau$ *iff* $\sigma \models \phi$ *and* $\tau = \sigma$.

In the first condition, ϕ_{modal} indicates any atomic formula or any formula whose top-level operator is one of $\Diamond, \Box$, or $\circ$. The condition states that the result of updating an input context with an atomic proposition is the subset of the input context containing just those possible worlds which support the proposition.

The remaining conditions are just as Beaver defines them. The second condition states that an input context is updated by a negation just in case it can be updated by the non-negated formula; the output context is the set of those worlds in the input context which are not present in update by the non-negated formula. The third condition states that an input context is updated by a conjunction just in case it can be updated by the first conjunct and the result of that update can be updated by the second conjunct; the output context is the result of this second update. The fourth and fifth clauses define disjunction and implication in terms of negation and conjunction. The sixth clause defines the semantics of the unary presupposition operator (∂) in terms of the notion of context satisfaction, which is defined next.

Definition 4 (Context satisfaction).
$\sigma \models \phi$ iff $\sigma[\![\phi]\!]\sigma$.

A context satisfies a formula just in case updating the context with the formula results in no new information being added to the context. Thus, an input context is updated by an elementary presupposition indicated by the ∂ operator just in case the context already satisfies the presupposition. If the input context does not satisfy the presupposition, the update fails.

Beaver formalizes Karttunen's notion of admittance as follows.

Definition 5 (Admittance).
$\sigma \triangleright \psi$ iff $\exists \tau . \sigma[\![\phi]\!]\tau$.

A context admits a formula if and only if it is possible to update the context with the formula. Beaver defines the relation of presupposition between complex formulas in terms of admittance and satisfaction.

Definition 6 (Presupposition).
$\phi \gg \psi$ iff $\forall \sigma . \sigma \triangleright \phi \Rightarrow \sigma \models \psi$.
Alternatively, $\phi \gg \psi$ iff $\exists \chi . [\![\phi]\!] = [\![\partial \psi \wedge \chi]\!]$.

There are two equivalent characterizations of presupposition given in this definition. The first characterization states that one formula presupposes another just in case every context that admits the first satisfies the second. The second characterization states that one formula presupposes another just in case the first is equivalent to the conjunction of the presupposition of the second and some other residual formula.

We now define a new notion which figures in our semantics for adverbial quantification. We would like to be able to talk about *the* presupposition of a formula, a notion which we formalize as maximal presupposition, as well as *the* asserted content, which is the non-presupposing residual.

Definition 7 (Maximal presupposition).
$\phi \gg_{max} \psi$ iff $\forall \sigma . \sigma \triangleright \phi \Leftrightarrow \sigma \models \psi$.
Alternatively, $\phi \gg_{max} \psi$ iff $\exists \chi . [\![\phi]\!] = [\![\partial \psi \wedge \chi]\!]$ and $\forall \sigma . \sigma \triangleright \chi$. We call χ the asserted content of ϕ.

The relationship between the definitions of presupposition and maximal presupposition should be clear. In order for a formula ψ to be the maximal presupposition of another formula ϕ, ψ must be a presupposition of ϕ, and further, it must be sufficient that a context satisfy ψ for it to admit ϕ. The alternative characterization is clearer and allows us to define our notion of asserted content: the residual formula χ must be non-presupposing.

We introduce one further auxiliary notion, which allows us to extract from a set of situations the subset of situations which are part of a particular world.

Definition 8 (Situation slicing).
For a set of situations S and a possible world w, $S_w = \{s \in S | s \sqsubseteq w\}$.

We now add an operator to our language that corresponds to adverbial quantification. Unlike the qadverb-like operators introduced in the accounts discussed above — two-place operators, analogous to quantificational determiners, but with an unspecified restrictor argument — our operator takes only a single sentential argument. This brings it closer to the syntactic type of a qadverb.

Definition 9 (Adverbial quantification).
$\sigma [\![Q(\phi)]\!] \tau$, where Q is a quantifier whose denotation is $\mathcal{Q}$, iff $\exists \psi, \chi, R, S$:

1. $\sigma \triangleright \phi$,
2. $\phi \gg_{max} \psi$,
3. $R = \{s | \exists w \in \sigma . s \sqsubseteq w \text{ and } s \Vdash \psi\}$,
4. χ is the asserted content of ϕ,
5. $S = \{s \in R | \exists s' . s \sqsubseteq s' \text{ and } s' \Vdash \psi \circ \chi\}$,
6. $\tau = \{w \in \sigma | \mathcal{Q}(R_w, S_w)\}$.

The first three conditions together have the effect creating a domain of quantification — the set R — out of the presuppositions of the scope ϕ and requiring that the input context satisfy the existence of this domain. Because the input context satisfies ψ, the maximal presupposition of the scope, each possible world in the input context must support ψ, which in turn means that each possible world must have sub-situations which are characterized by ψ. These sub-situations form the domain R. To form the nuclear scope set, we must find those situations that stand in the modal satisfaction relation with ϕ, which we reinterpret as the modal formula $\psi \circ \chi$. The nuclear scope set S is then set of those members of R which can be extended into a situation which satisfies $\psi \circ \chi$. The final condition outputs those worlds in the input for which the properly restricted domain set and scope set stand in the quantifier relation.

Returning to the example discourse (3), if we take p to be the specific proposition that John plays (i.e. is playing) Marvin at badminton and q to be the specific proposition that John wins, we might represent the adverbially quantified sentence (11a), as the formula USU($\partial p \wedge q$). By the first condition of the definition of adverbial quantification, this formula is admitted by a context σ just in case $\partial p \wedge q$ is admitted by σ. It should be clear that σ admits $\partial p \wedge q$ just in case every world in σ supports the proposition p. A world w, in turn, supports p just in case one (or more) of its constituent situations are situations of John playing Marvin at badminton. Thus, σ admits USU($\partial p \wedge q$) just in case every world in σ has constituent situations of John playing Marvin at badminton.

These situations of John playing Marvin at badminton constitute the domain of quantification for the qadverb. Thus, our formula both incorporates scopal presuppositions into its restriction and carries the presupposition that its domain exists. Updating σ with this formula results in a context containing just those worlds in which most situations of John playing Marvin at badminton can be extended to situations of John playing Marvin at badminton and winning.

3.3 Accommodation as Context Selection

Although we have emphasized the absence of intermediate accommodation as a feature of our analysis, we still need a general account of accommodation. Even in our example discourse (3), some notion of accommodation is needed to explain the felicity of the adverbially quantified sentences. The sentence we have been focusing on, (11a), presupposes that there are one or more situations of John playing Marvin at badminton. Unfortunately, the preceding discourse makes no claim regarding whether or not they have ever played badminton, and thus, the context could presumably include worlds in which they never have. Such a context would fail to admit our sentence.

Heim [13] provides a mechanism of accommodation much like van der Sandt's — simply add presuppositions that are not already present in the context. This would solve the problem of admittance, but it would fail to account for the anaphoric link between the presupposed situations of John and Marvin's badminton playing and the asserted situations of their game playing.

Beaver [6] suggests an alternative view of accommodation. Instead of identifying a discourse participant's information state with a context, he takes it to be a set of contexts, ordered according to plausibility. Updating such an information state with an utterance consists of updating each of the member contexts, throwing out contexts that cannot be updated. Contexts, then, are a way to encode commonsense knowledge.

For example, a commonsensical hearer should be aware that two people who played various games might include badminton among those games. Such a hearer begins with an information state which includes one or more contexts which satisfy a rule along the lines of *if two people play games together, some of those games may be badminton*. After updating with the first sentence of our discourse, every world in those contexts includes a set of situations of John and Marvin playing games, some of which are games of badminton. Of course, other (possibly more plausible) contexts would have rules involving other games, but when the hearer gets to sentence (11a), only those contexts with the badminton rule are admissible. The domain of the qadverb *usually*, then, is the set of those situations of John and Marvin playing badminton, which is a subset of those situations of John and Marvin playing games. This is an admittedly informal characterization, but we hope that it is at least suggestive of an approach to integrating our analysis into a general account of bridging and partial matches.

4 Conclusion

Our analysis of adverbial quantification has several advantages over an account framed in terms of van der Sandt's binding theory. Principally, no recourse to accommodation is required to explain the incorporation of scopal presuppositions into a qadverb's domain of quantification. On our account, a qadverb domain is determined directly through the satisfaction of scopal presuppositions — it is composed of the sub-situations of the satisfying worlds that are characterized by the presuppositions. Furthermore, no separate presupposition needs to be computed to account for the context-dependence of the qadverb domain. Determining the domain through contextual satisfaction of scopal presuppositions ensures that the context contains the domain set. Finally, the qadverb-like operator we introduce is more similar in its syntactic type to natural language qadverbs than the two-place operator normally proposed.

There are some obvious deficiencies with this account. One is that we have not formally related it to an account of accommodation, although we have informally outlined one possible avenue of exploration. Another important caveat is that we cannot seriously propose a formula like $\partial p \wedge q$ as a compositional semantic interpretation of the sentence *John beat Marvin at badminton*. Rather, we must develop a realistic compositional semantics that would assign to such a sentence an interpretation which both presupposed something like p and asserted something like q. Furthermore, we must hope that such an interpretation would render unnecessary our admittedly awkward reinterpretation of the scope of a qadverb as a modal formula.

Beaver, in other work [21], has demonstrated that it is possible to give a compositional treatment of predication and quantification over ordinary individuals that yields interpretations that are equivalent, at the propositional level, to formulas like the ones we have been working with in this paper. Of course, it is not entirely clear how to extend the static semantic notion that relates situations to formulas — satisfaction/characterization — to work with the dynamic variable binding required for an account of quantification. Ahn's dynamic extension of Schubert first-order models [22] may provide a starting point, but even as it is, the static conditions for characterization are uncomfortably unrelated to the update semantics for the rest of the language.

In at least one respect, there is a bright side to this last deficiency. We have provided an account of adverbial quantification that is not parasitic on an account of nominal quantification. Given that the two phenomena diverge significantly with respect to presupposition incorporation, it is a small victory to be able to give them naturally heterogeneous accounts.

Acknowledgements. This work was supported by the National Science Foundation under Grant No. IIS-0082928. The author would like to thank Len Schubert for detailed feedback on several drafts of this paper and the anonymous reviewers for their comments.

References

1. van der Sandt, R.: Presupposition projection as anaphora resolution. Journal of Semantics **9** (1992) 333–377
2. Beaver, D.: Accommodating topics. In van der Sandt, R., Bosch, P., eds.: Proceedings of the IBM/Journal of Semantics Conference on Focus. Volume 3. (1994)
3. Geurts, B., van der Sandt, R.: Domain restriction. In Bosch, P., van der Sandt, R., eds.: Focus: Linguistic, Cognitive, and Computational Perspectives. Cambridge University Press (1999)
4. Ahn, D.: Computing adverbial quantifier domains. In Striegnitz, K., ed.: Proceedings of the Sixth ESSLLI Student Session. (2001) 1–12
5. Ahn, D.: The role of presuppositions and situations in the interpretati on of adverbial quantifiers. Technical Report 793, University of Rochester Computer Science Department (2002)
6. Beaver, D.: Presupposition: A plea for common sense. In Moss, L., Ginzburg, J., de Rijke, M., eds.: Logic, Language, and Computation. Volume 2. CSLI (1999)
7. Karttunen, L.: Presuppositions and linguistic context. Theoretical Linguistics (1974)
8. Kamp, H., Reyle, U.: From Discourse to Logic. Kluwer Academic Publishers (1993)
9. Lewis, D.: Scorekeeping in a language game. In Bäuerle, R., Egli, U., von Stechow, A., eds.: Semantics from Different Points of View. Springer, Berlin (1979) 172–187
10. Schubert, L.K., Pelletier, F.J.: Problems in the representation of the logical form of generics, plurals, and mass nouns. In LePore, E., ed.: New Directions in Semantics. Academic Press, London (1987) 385–451
11. Berman, S.: Situation-based semantics for adverbs of quantification. In: University of Massachusetts Occasional Papers in Linguistics. Volume 12. Graduate Linguistic Student Association, University of Massachusetts, Amherst (1987) 45–68
12. Cooper, R.: The role of situations in generalized quantifiers. In Lappin, S., ed.: Handbook of Contemporary Semantic Theory. Blackwell (1995)
13. Heim, I.: On the projection problem for presuppositions. In: Proceedings of the Second Annual West Coast Conference on Formal Linguistics, Stanford Linguistics Association, Stanford University (1983) 114–125
14. Stalnaker, R.C.: Assertion. In Cole, P., ed.: Syntax and Semantics: Pragmatics. Volume 9. Academic Press, New York (1978)
15. Veltman, F.: Defaults in update semantics. Journal of Philosophical Logic (1996)
16. Schubert, L.K., Hwang, C.H.: Episodic Logic meets Little Red Riding Hood. In Iwanska, L., Shapiro, S.C., eds.: Natural Language Processing and Knowledge Representation. MIT/AAAI Press (2000) 111–174
17. Schubert, L.K.: The situations we talk about. In Minker, J., ed.: Logic-Based Artificial Intelligence. Kluwer Academic Publishers, Dordrecht (2000)
18. Barwise, J., Perry, J.: Situations and Attitudes. MIT Press (1983)
19. Cooper, R.: Austinian propositions, Davidsonian events and perception complements. In Ginzburg, J., Khasidashvili, Z., Vogel, C., Lévy, J.J., Vallduví, E., eds.: Tbilisi Symposium on Language, Logic and Computation. CSLI (1998)
20. Ahn, D., Schubert, L.: HLC** (hybrid logic of characterization). Technical report, University of Rochester Computer Science Department (in preparation)
21. Beaver, D.: Presupposition and Assertion in Dynamic Semantics. CSLI (2001)
22. Ahn, D.: A dynamic situation logic. In Bunt, H., van der Sluis, I., Morante, R., eds.: Proceedings of IWCS-5. (2003)

A Theory of Contextual Propositions for Indicatives

Horacio Arló Costa

Carnegie Mellon University, Pittsburgh PA 15213, USA
hcosta@andrew.cmu.edu

Abstract. A theory of contextual propositions for indicative conditionals is presented. The main challenge is to give a precise account of how the dynamics of possible worlds depends on epistemic context. Robert Stalnaker suggested in [Stalnaker 84] that even when selection functions for evaluating indicatives cannot be *defined* in terms of epistemic context, they can be importantly constrained by a principle of context dependency that we adopt here. In addition, we show how to define a gradation of possibilities for each point in an epistemic context by taking into account a proposal first introduced by Wolfgang Spohn in [Spohn 87] and later refined by Darwiche and Pearl in [Darwiche & Pearl 97]. The resulting theory of contextual propositions (unlike some alternative views) is shown to be compatible with basic qualitative consequences of the Bayesian principle of conditionalization (which is frequently used in probabilistic semantics for indicative conditionals).

1 Introduction

There is an area of pragmatics and semantics where the problems of context-dependency are particularly poignant. The area in question is concerned with the semantics of indicative conditionals. While many assume that other conditionals (subjunctives, for example) express propositions and carry probability, there is considerably less consensus about that regarding indicatives (see, for example, [Gibbard 80]). In this paper I propose to revisit the problem of the semantic of indicatives and to use it as a tool in order to think about context-dependency in semantics. I shall argue that a theory of contextual propositions for indicatives is possible, but that in order to build it up it is necessary to radicalize assumptions about context dependency common in semantics and pragmatics. In particular I shall argue that the right theory requires making the dynamic properties of possible words dependent on epistemic context. The picture that thus arises confirms the ideas that many have expressed before concerning the hidden indexicality in conditional statements [van Fraassen 76]. Indicatives do express contextual propositions, which are highly sensitive to epistemic context. But a theory of such propositions, which is also compatible with basic epistemological and semantic tenets, requires to think anew the role of possible worlds in semantic constructions and in formal epistemology.

16   H. Arló Costa

Frank Ramsey provided basic intuitions about the semantics of conditionals in a footnote of a paper on laws and causality [Ramsey 90]. Here is the passage that the followers of at least three orthogonal research programs in semantics of conditionals see at the root of their proposals (for an overview of both epistemic and ontic theories see [Cross & Nute 98]):

> ...the belief on which the man acts is that if he eats the cake he will be ill, taken according to our above account as a material implication. We cannot contradict this proposition either before or after the event, for it is true provided the man doesn't eat the cake, and before the event we have no reason to think he will eat it, and after the event we know he hasn't. Since he thinks nothing false, why do we dispute with him or condemn him?[1] Before the event we do differ from him in a quite clear way : it is not that he believes p, we $\bar{p}$; but he has a different degree of belief in q given p from ours ; and we can obviously try to convert him to our view. But after the event we both know that he did not eat the cake and that he was not ill ; the difference between us is that he thinks that if he had eaten it he would have been ill, whereas we think he would not. But this is *prima facie* not a difference of degrees of belief in any proposition, for we both agree as to all the facts.

The footnote (1) in the above text provides further clarification:

> If two people are arguing 'If p, then q?' and are both in doubt as to p, they are adding p hypothetically to their stock of knowledge and arguing on that basis about q ; so that in a sense 'If p, q' and 'If p, $\bar{q}$' are contradictories. We can say that they are fixing their degree of belief in q given p. If p turns out false, these degrees of belief are rendered *void*. If either party believes *not* p for certain, the question ceases to mean anything to him except as a question about what follows from certain laws or hypothesis.[1]

In order to illustrate Ramsey's example we can use a small universe of four points (or possible worlds) $\{w_1, w_2, w_3, w_4\}$. Say that in w_1 and w_2 it is true that the cake is good; and worlds w_1 and w_3 are situations where the man does eat the cake. In addition, world w_3, where the man eats the cake and the cake is bad is also a situation where the man is ill.

An agent can have a prior probability distribution according to which: $P(w_1)$ = .04, $P(w_2)$ = .01, $P(w_3)$ = .16, $P(w_4)$ = .79. The idea being that the agent puts a high probability (.8) on the event that the man will not eat the cake and most of this probability mass concentrates on the event that the man will not eat it when it is bad, etc. A second agent assessing the situation might diverge from him regarding his *degrees of belief*. For example, he might swap the probabilities attributed to w_1 and w_3. I.e. this second agent might have a probability function P' which coincides with P, except that $P'(w_1)$ = .16 and $P'(w_3)$ = .04. The

---
[1] See [Ramsey 90], pages 154-55.

conditional probabilities of these two agents differ in an important manner. In particular the probability that the man is ill conditional on his eating the cake is .8 for the first agent and .2 for the second agent. So, if the probability of a conditional is conditional probability,[2] these two agents assign very different probabilities to: 'if the man eats the cake, he will be ill'.

Of course, as Ramsey points out, before the fact neither agent can contradict the proposition that either the man does not eat the cake or he will be ill. In fact, before the fact either agent does not have a formed belief concerning whether the man eats or does not eat the cake. The corresponding material conditional cannot be contradicted after the fact either. But in addition after the fact the degrees of belief of the two agents converge. After the fact (i.e. after the man decides not to eat the cake) both agents assessing the situation assign zero probability to w_1 and to w_3. And they converge in assigning $P(w_2) = .0125$ and $P(w_4) = .9875$. Both agents are sure that the man did not eat the cake (this proposition receiving measure one). Of course, I am assuming here that both agents condition on the fact that the man did not eat the cake.

Ramsey points out in addition that even after this convergence in degrees of beliefs, it seems that the two agents in our story can still be differentiated. The difference being that the first agent will still accept that if the man had eaten the cake he would have been ill, while the second agent will reject this conditional. This new conditional is often classified as a subjunctive (as opposed to the indicatives considered above).

Notice that before the fact the crucial issue was the evaluation of conditional beliefs. In a nutshell the important thing is evaluating $P(w_3 \mid w_1 \cup w_3)$. For one agent this conditional probability is high (.8) for the other agent is low (.2). Why not to use the same strategy after the fact? The reason is that after the fact this conditional probability is not defined because $P(w_1 \cup w_3) = 0$. So, degrees of belief conditional on $w_1 \cup w_3$ are, as Ramsey says, void. In other words, in order to evaluate 'if the man had eaten the cake he would have been ill' one needs to make a supposition (that the man had eaten the cake) which contradicts an event of measure one. And this cannot be done with the apparatus of standard probability theory (for recent work on using non-standard measures in the semantics of conditionals see [McGee 94] and [Arlo-Costa 01]).

The analysis of indicatives is done with the help of two elements: a *context set* [Stalnaker 84] [Stalnaker 98] consisting of a space of possible worlds, and a probability distribution defined over them. The evaluation of conditionals is then sensitive to both qualitative and numerical differences in the representation. Before the fact differences in the attitudes towards conditionals are ultimately traced back to differences in degrees of beliefs, that are registered in terms of

---

[2] The status of this thesis (probability of conditionals are conditional probabilities), is quite problematic. There is a consensus to the extent that it can be saved for non-nested conditionals. In addition van Fraassen has offered an indexical interpretation of conditional propositions for which stronger variants of the thesis also hold [van Fraassen 76]. More robust versions of the thesis for non-standard probability are offered in [McGee 94] and [Arlo-Costa 01].

the probability function defined over the context set. Recent work extending Ramsey ideas has proposed to adopt a generalized notion of conditional probability (adopted as an epistemological primitive) which allows for conditioning on events of measure zero. Alternatively the idea is to supplement the standard representational tools of Bayesian theory with a metric that can make comparisons between events of measure zero. In our example one can put the problem as follows: even when both w_1 and w_3 carry zero measure (after the fact), which is the state that one judges more plausible from the point of view of the context set composed by $w_2 \cup w_4$? This is equivalent to supplement the context set and the probability function defined over it, by a *belief revision function*. In a companion paper I suggested that this should be done in order to have a reasonably powerful representational tool [Arlo-Costa forthcoming]. I shall offer here new reasons for extending context sets with belief revision functions (which are based on basic facts related to the semantics and pragmatics of indicatives).

One important aspect of the representational framework tacitly introduced above resides in the relationships between the underlying context set and the probability function defined over it. One can see the context set as associated with the probability function via a function ρ. So, in the above example, $\rho(P) = \{w_1, w_2, w_3, w_4\}$. The idea is to define $\rho(P)$ as the set of points that receive non-zero probability.[3] Now, in our example for any proposition A, such that $\rho(P) \cap A \neq \emptyset$, we have that the context set corresponding to the probability function P updated with A (P_A) is calculated as $\rho(P) \cap A$. We will use the notation $\rho(P)_A$ to denote the process of updating the context set corresponding to P, with the proposition A. So, as long as every state in the prior context set receives non-zero probability, we have that:

(Preservation) If $\rho(P) \cap A \neq \emptyset$, then $\rho(P)_A = \rho(P_A) = \rho(P) \cap A$.[4]

More in general for any epistemic state E (which can be different from a standard probability function), such that it is associated with a context set via the function ρ, we have:

(Preservation-E) If $\rho(E) \cap A \neq \emptyset$, then $\rho(E)_A = \rho(E_A) = \rho(E) \cap A$.[5]

---

[3] This idea is coherent with a probabilistic outlook. Regarding the epistemological interpretation of $\rho(P)$, in [Arlo-Costa 01] I offer an argument to the extent that it encodes the *expectations* of the agent described by P, rather than his certainties. The set $\rho(P)$ can also be assumed as a primitive without probabilistic interpretation. This yields a more sophisticated account of the relationships between probability and belief. Our arguments here can be stated in either case, but below we will follow the idea that $\rho(P)$ encodes the set of points carrying positive probability.

[4] See [Gardenfors 88] for the use of this terminology and for a basic introduction to the theory of belief revision. Here $P_A(X|Y) = P(X|Y \cap A)$.

[5] E here can be any representation of the epistemic state. For example, later on we will propose to use *rankings* or *ordinal conditional functions* as representations of epistemic states. Those rankings can be substituted by the variable E in the previous formualtion.

and when the epistemic state is identified with a *belief set* K, i.e. when E = K = $\rho(E)$:

(Preservation-K) If K ∩ A ≠ ∅, then K_A = K ∩ A

This is a very simple principle, ultimately justified in terms of the properties of conditionalization. Of course, in our example, we start with a probability function P and $\rho(P) = \{w_1, w_2, w_3, w_4\}$. Then we condition with A $\{w_2, w_4\}$. As a consequence P_A = P', and $\rho(P') = \{w_2, w_4\}$. In the following analysis we will assume Preservation.

2 Context Set and Indicatives: Stalnaker's Constraint and Preservation

Say that one of the agents in our example learns that the cake is in bad shape. Then his context set shrinks to $\{w_3, w_4\}$. From this point of view he accepts 'if the man eats the cake he will be ill'. As a matter of fact the corresponding conditional probability is one. Does this conditional express a proposition? If so, how to build it up with the elements we have?

Propositions are built up by worlds, so a natural manner of proceeding is to say that the conditional proposition will be built (if it exists at all) by a set of possible worlds obeying some natural constraint. An obvious constraint is to collect all the possible worlds where the conditional in question is true. Here is the analysis proposed by Robert Stalnaker in his article on indicative conditionals:

> We need a function which takes a proposition (the antecedent of the conditional) and a possible world (the world as it is) into a possible world (the world as it would be if the antecedent were true). Intuitively the *value* of the function should be that world in which the antecedent is true which is most similar, in relevant aspects, to the actual world (the world which is one of the arguments of the function). In terms of such function - call it f - the semantic rule for the conditional may be stated as follows: a conditional *if A, then B*, is true in a possible world *i* just in case B is true in possible world f(A, i).

So, now our representational framework has been amplified. Now, in addition to our context set and, eventually, a probability function P on it, we need a primitive selection function f defined for each world and argument A. Notice that this new primitive used in the construction is not *definable* from the initial context set and the probability function P. It is a new primitive, and if we follow Stalnaker's suggestions, the motivation for it is purely ontological. The similarity of worlds has not been used so far in our analysis. For the moment and , for the sake of the argument, I shall expand the basic context set with it. Later on we will have the opportunity of discussing its use.

Are there any useful constraints we can impose on f? Notice that so far the function is completely unconstrained. One useful constraint is presented by Stalnaker in his piece in indicative conditionals.

> I cannot *define* the selection function in terms of the context set, but the following constraint imposed on the context set on the selection function seems plausible: if the conditional is being evaluated at a world in the context set, then the world selected must, if possible, be within the context set as well (where C is the context set, if $i \in$ C, then f(A, i) $\in$ C). In other worlds, all worlds within the context set are closer to each other than any worlds outside it. The idea is that when a speaker says 'If A', then everything he is presupposing to hold in the actual situation is presupposed to hold in the hypothetical situation in which A is true.

This constraint seems sensible and useful. Even when the selection functions f are a new primitive, at least their behavior is constrained by the existing context, in a way that seems intuitive. So, since our goal here is to try to understand what type of propositions are expressed by indicatives (if any), it seems natural to add this constraint to the principle of Preservation (formulated in the previous section). Unfortunately this is not possible. I shall use a slightly modified form of our example in order to illustrate the problem and in order to begin to present a solution.

We can start with a context set C = $\{w_1, w_2, w_3, w_4\}$. Say that in addition we have a probability distribution on this context such that $P(w_1)$ = .16, $P(w_2)$ = .01, $P(w_3)$ = .04, $P(w_4)$ = .79. Say that in w_1 and w_2 it is true that Gore is elected; and worlds w_1 and w_3 are situations where Gore wins the popular vote. This could be the context set of an agent prior to the last general election in USA. The agent in question thinks that it is highly likely that Gore will not be elected and that he will not win the popular vote. Tiny probabilities are assigned to the cases where Gore wins the popular vote and is not elected; and to the situation where is elected even when he does not won the popular vote. We know today that this prior will be modified drastically after the election, but as a prior it is a perfectly possible one. States w_1 and w_4 can be considered as 'good' states, while their complement can be seen as bad outcomes. In the good states Gore is elected when he wins the popular vote, and Gore is not elected when he loses the popular vote. This partition of states into good and bad ones is the qualitative precursor of having a value function for outcomes.

Let's now consider two possible states of belief definable over the given context set C. One of them is K = $\{w_3, w_4\}$, which corresponds to the proposition that Gore has not been elected. Another is G = $\{w_1, w_4\}$, which corresponds, to the proposition that the outcome of the election is good. Of course, when the probability measure P is updated either with K or G the context set C shrinks either to K or to G. Shifting to G is tantamount to deny any credibility whatsoever to the marginal outcomes when Gore wins the popular vote but he is not elected and/or when Gore is elected and he loses the popular vote. There might be situations where shifting to such state might be justified - in the presence of appropriate legal or institutional arrangements (or statistical and pool data before the fact).

We need in addition information about the *f* function. Here is a possible distribution of *f*-values. Let A = $\{w_1$ and $w_3\}$. Then define: f(A, w_1) = $\{w_1\}$, f(A, w_2) = $\{w_1\}$, f(A, w_3) = $\{w_3\}$, f(A, w_4) = $\{w_3\}$. According to Stalnaker 'intuitively the *value* of the function should be that world in which the antecedent is true which is most similar, in relevant aspects, to the actual world (the world which is one of the arguments of the function).' The function proposed above implements the most elemental notion of closeness in terms of mereological similarity (by counting shared atoms under a constraint). There might be, of course, other relevant notions of similarity. The important issue is that most of the existing notions of ontological similarity are context-independent. They are not supposed to be sensible to changes in epistemic context, being determined absolutely in terms of certain objective features.

Consider then the conditional 'if Gore wins the popular vote this will be a good outcome' where the notion of goodness is determined as we explained above via the proposition G. We can abbreviate this conditional by $A > G$. Which is the proposition expressed by such conditional? In order to determine this it would be good to add a further degree of precision. We are here evaluating propositions relative to a model M, which consists of a prior context set C, a probability function P, and a function f for each possible world in the context set C.

$[A > G]^M = \{w \in C: f(A, w) \subseteq G\} = \{w_1, w_2\}$.

It seems intuitive that after shifting (from C) to the context set G ('the outcome is good' – where 'bad' outcomes have zero probability of occurring) one would be willing to assert 'if Gore wins the popular vote this is a good outcome'. The reason for this is that $G_A = \{w_1\}$, which is a good state of affairs (Gore is elected and he wins the popular vote). On the other hand, after shifting (from C) to K (Gore is not elected president) one would be willing to accept 'if Gore wins the popular vote this will not be a good outcome'. This is so, given that $K_A = \{w_3\}$, which is not a 'good' outcome (Gore wins the popular vote but is not elected president).

Notice that after conditioning with the proposition G ('the outcome is good') the acceptance of 'if Gore wins the popular vote this is a good outcome' is mandated probabilistically. In fact, after conditioning with G, all probability is distributed among the worlds w_1 (.168) and w_4 (.83). If P' is $P(.|G)$ it is clear that $P'(G|A)$ is one. So, 'if Gore wins the popular vote this is a good outcome' should be accepted with respect to P'.

Notice, nevertheless, that $[A > G]^M = \{w_1, w_2\}$. This means that $[A > G]^M$ *is not* entailed by G, contrary to intuition and the probabilistic account of acceptance. Notice also that in this situation Stalnaker's constraint on selection functions is violated. In fact, after shifting to G the constraint requires that '...if the conditional is being evaluated at a world in the context set, then the world selected must, *if possible*, be within the context set as well' - the emphasis is mine. This seems to indicate that when f(A, w_4) is evaluated in the context set G, the constraint requires to set f(A, w_4) to $\{w_1\}$, a world which is *not* mereologically similar to $\{w_4\}$. If one is guided by an objective criterion for determining the similarity of worlds (like the one we are using, mereological similarity) then the content of f(A, w_4) is fixed in a context-independent manner. For example, according to our simple implementation of mereological distance, f(A, w_4) = $\{w_3\}$.

Perhaps Stalnaker's criterion can be seen as an extra consideration to be taken into account when the initial criterion for determining similarity is not powerful enough to specify the content of all functions. Or it can be read as a complementary consideration, which can clash with an underlying criteria for establishing similarity, in such a way that one has to decide case by case. Finally, it can be seen as a proposal for creating selection functions which depend on three arguments, the actual world, a given proposition and the given context.[6] Perhaps the latter option is the most charitable manner of interpreting Stalnaker. It should be noted, nevertheless, that the aforementioned examples show that this is tantamount to not having a unified criterion for determining the content of selection functions. Selection functions will pick up the mereologically closest worlds, when this is permitted by the constraint for indicatives, and other convenient worlds in the context of evaluation otherwise. This is not a very satisfying situation. In fact, it seems that the result of applying the constraint is to abandon a principled way of understanding world selection in general.

Let me re-estate some of the issues just mentioned in a more formal manner. When we consider mereological similarity as the determining factor in constructing selection functions one has: $[A > G]^M = \{w_1, w_2\}$ and $[\neg(A > G)]^M = [A > \neg G]^M = \{w_3, w_4\}$. So, we do have $K \subseteq [\neg(A > G)]^M$, but G does not entail $[A > G]^M$, against intuition. More in general, when conditional propositions are rigidly determined for a given context across of possible epistemic states definable for the context, we do not have the following bridges that in the literature usually receive the name of Ramsey tests:

(RT) For every model M, $K_A \subseteq B$ iff $K \subseteq [A > B]^M$.

and the corresponding:

(NRT) For every model M, K_A is not included in B iff $K \subseteq [\neg(A > B)]^M$.

We are assuming, of course, that the operation K_A does obey preservation, for every K and A. As it is explained in [Arlo-Costa & Levi 96] (RT) is compatible with a different operation which sometimes receives the name of *imaging* in the literature [Lewis 76]. The idea is to use the underlying *f*-function in order to construct the revision operation used in the Ramsey test. Define for every

---

[6] The latter option was suggested by the comments of an anonymous referee.

context set K and proposition A (and for a fixed selection function f across contexts):

$K \# A = \cup \{f(A, w): w \in K\}$

Then we do have:

(RT) For every model M, $K \# A \subseteq B$ iff $K \subseteq [A > B]^M$.

The motivations for the adoption of $\#$ are unclear in the context of our analysis. It seems that one certainly wants to have Preservation, a notion which is tightly connected with the use of probabilistic tests in the acceptance of conditionals. We argued at length above for preservation, by showing how it is anchored in the use of Bayesian conditioning. In order to see that preservation is violated by uses of $\#$ notice that $G \# A = \{w_1, w_3\} \neq G \cap A$. Notice also that this is a violation of Slanaker's constraint. In fact, relative to context G, the criterion mandates that w_1 should be the 'closest' world to w_4 – a world that is not mereologically close to w_4.

Is there any way of constructing conditional propositions for indicatives that respect both (a strict version) Stalnaker's constraint and Preservation? In other words, is there an unified epistemic criterion for determining the content of selection functions, respecting both Preservation and Stalnaker's constraint? In the next section I shall argue that it is indeed possible, as long as the propositions in question are dependent on epistemic context.

3 Contextual Propositions for Indicatives

If the function f used in the model picks up worlds according to measures of overall similarity, then it makes perfect sense to have one and only one selection function for each world in the model. If the function encodes mereological similarity, then it should be a matter of fact that either w_4 is most similar (among A-worlds) either to w_1 or to w_3. Perhaps one just can count atoms and say that w_3 is the 'closest' world to w_4, period.

It seems that in order to accommodate the Ramsey tests, $f(A, w_4)$ should yield w_1, when evaluated at G, and w_3, when evaluated at K. But this, of course, cannot be articulated in terms of overall mereological similarity. We should remind the reader here that when the f-function was introduced we pointed out that we will adopt it for the sake of the argument in order to present Stalnaker's constraint. We pointed out, nevertheless, that the function in question seemed an ontological intrusion in which otherwise is an epistemological account. Here is a proposal for retaining some of the role of the f-function in constructing propositions, while its motivation is presented in epistemic terms.

In order to present the new proposal let's go back to our motivation for supplementing conditioning with rankings of worlds which receive probability zero after conditioning. Let's focus on the initial context C. After one conditions on the proposition K (Gore is not elected president) the worlds w_1 and w_2 receive zero measure. But we can still order these two worlds in different manners according to their degree of plausibility with respect to K. The formal tool needed here is what the philosopher W. Spohn calls an ordinal conditional function [Spohn 87]. Here we just call them rankings. A ranking is a function κ from the set of all interpretations of the underlying language (worlds) into the class of ordinals. A ranking is extended to propositions by requiring that the rank of a proposition be the smallest rank assigned to a world that satisfies:

$\kappa(A) = min_{w \models A} \kappa(w)$.

The set of models corresponding to the *belief set* $\rho(\kappa)$ associated with a ranking κ is the set {w: k(w) = 0}. There are various ways of understanding what a ranking is. Spohn's own interpretation is that $\kappa(A) \leq \kappa(B)$ means that A is less disbelieved than B. Spohn's formalism is, in turn, an elaboration of Shackle's notion of *potential surprise* [Shackle 61]. There are also various possible manners of updating rankings. Spohn's methods for updating rankings are in accordance with probability theory where probability takes infinitesimal values. Here we will appeal to a method of revision proposed in the Artificial Intelligence literature by A. Darwiche and J. Pearl. Here is a presentation of the update rule as presented in [Darwiche & Pearl 97]:

$$(\kappa \bullet (A))(w) = \begin{cases} \kappa(w) - \kappa(A) & \text{if } w \models A \\ \kappa(w) + 1 & \text{otherwise} \end{cases}$$

The selection of this dynamics is based on purely pragmatic considerations. The reader will see below that it offers a convenient encoding of change for our representational goals. There are, to be sure, other feasible dynamics that can be used instead. Of those perhaps this is the simplest possible. The central feature of the dynamics that interest us is that if K is a set of points in rank 0, and A $\cap$ K $\neq \emptyset$, after updating with any of the points in K, the closest A-points with respect to this point are also K-points.

One can see G and K as beliefs sets associated with richer representations of context κ_G and κ_K in such a way that $\rho(\kappa_G) = G$ and $\rho(\kappa_K) = K$.

κ_K	Possible worlds
0	w3, w4
1	w1
2	w2

Notice that if one updates the ranking in question with the proposition {w_4}, the updated rank has w_4 in rank zero and w_3 in rank 1. On the other hand, one can have the following ranking giving the dynamic properties associated with G.

A Theory of Contextual Propositions for Indicatives 25

κ_G	Possible worlds
0	w1, w4
1	w3
2	w2

And the result of updating this second ranking with w_4 preserves w_4 in rank zero, but locates w_1 in rank 1. The intuition here is that in order to determine the content of the *f*-functions f(A, w) for each world w in the context set K one should utilize the ranking that has $\rho(\kappa_K) = K$ and then one should shift to $\kappa_K \bullet (\{w\}) = \kappa_w$. Then this ranking is used to calculate $\kappa_w(A) = $ f(A, w). It is important to notice that under this definition the content of f(A, w) need not be a singleton. Under this point of view we are following the advice of David Lewis in [Lewis 73], rather than the account of selection functions offered by Stalnaker.

Since the act of indicative supposition typically requires entertaining propositions that are not belief contravening; we can start with a flat initial context C (determined by the points receiving non-zero measure in an initial measure) and use the Darwiche-Pearl algorithm in order to construct rankings naturally associated with each subset of C, including each point in C. So, if we proceed this way we would have:

κ'_K	Possible worlds
0	w3, w4
1	w1, w2

and,

κ'_G	Possible worlds
0	w1, w4
1	w3, w2

Even when these rankings have poorer information concerning belief contravening worlds, they can perfectly well be utilized in order to carry out our analysis. The underlying intuition is simple. What is most plausible from the point of view of a world w depends on epistemic context. If one is in κ_K (κ'_K), then the most plausible A-world from w_4's point of view is w_3. In other words, $\kappa'_K \bullet (\{w_4\})(A) = \{w_3\}$. On the other hand, if one is in κ_G (κ'_G), then the most plausible A-world from w_4's point of view is w_1. In other words, $\kappa'_G \bullet (\{w_4\})(A) = \{w_1\}$. So, here we are adopting an epistemic strategy all the way down to the possible worlds that are components of context. When we have not only the context set, but also the corresponding function κ for it, we can indeed *define* the *f*-functions in terms of them. In the previous paragraphs I argued that suitable functions κ can be determined from a flat initial prior by only assuming the general properties of the Darwiche-Pearl algorithm. Moreover these properties guarantee the satisfaction of Preservation. So, we can conclude that there is indeed an unified epistemic criterion for determining the content of selection functions, which respects both Preservation and Stalnaker's constraint.

We are taking advantage also here of a feature of Spohn's functions, namely that they are path-dependent. So, if w is both in K and in G, what is most plausible from the point of view of w depends on the plausibility orders for K and G. If one starts with K (and its plausibility order κ_K), one gets one estimation. If one starts with G (and its plausibility order κ_G), one might get another. And this seems quite natural. Path dependency is just another way of saying that the rankings retain memory of where they come from. One epistemic path from K to a ranking that has w_4 in the zero level might be different from an epistemic path from G to a ranking that also has w_4 at the zero level. This suggests the following recipe for defining contextual propositions.

$[A > G]^{K,C,\kappa} = \{w \in C: f(A, w) \subseteq G, \text{ where } \rho(\kappa) = K \text{ and } f(A, w) = \kappa \bullet (\{w\})(A)\}$

Now we can have:

(RT) For every context set C and $K \subseteq C$ and κ, such that $\rho(\kappa) = K$; $\rho(\kappa \bullet (A)) = K_A \subseteq B$ iff $K \subseteq [A > B]^{K,C,\kappa}$.

as well as a similar clause for negation of conditionals. This can be done even when Preservation is indeed obeyed. Rather than going into the details of the existence proof I shall focus on our example, showing how contextual propositions actually look like.

Let's first focus on G. If κ_G (κ'_G) is the ranking for G, then the f function is: $f(A, w_1) = \{w_1\}$, $f(A, w_2) = \{w_1\}$, $f(A, w_3) = \{w_3\}$, $f(A, w_4) = \{w_1\}$. Remember that this selection function seems difficult to motivate in terms of overall similarity (especially it is hard to motivate $f(A, w_4) = \{w_1\}$). Now this selection function emerges naturally by applying iterated changes to the initial ranking κ_G. Actually each selection function emerges quite naturally even if one starts with a flat state C, after calculating $C \bullet G \bullet (\{w\})$ for each w in G.

So, $[A > G]^{C,\kappa_G} = \{w_1, w_2, w_4\}$.[7] Therefore the proposition expressed by 'if Gore wins the popular vote this is a good outcome' relative to κ_G is indeed entailed by G. And this aligns perfectly well with the fact that updating G with A yields the 'good' state w_1.

By the same token $[A > G]^{C,\kappa_K} = \{w_1\}$. And $[\neg(A > G)]^{C,\kappa_K} = \{w_2, w_3, w_4\}$. So, K does entail $[\neg(A > G)]^{C,\kappa_K}$, which aligns with the fact that conditioning K with A, yields the 'bad' outcome w_3.

We can formulate Stalnaker's constraint more precisely in this setting as follows:

---

[7] Propositions are here indexed only by the initial context set C and the corresponding ranking, which in this case is κ_G. Adding G would be repetitious. As a matter of fact, the only parameters that matter in order to identify contextual propositions are the initial context set C, the current context set (in this case G) and the function $\bullet$. The ranking for G can be naturally constructed in terms of C and $\bullet$.

(Epistemic Dependency) If $\rho(\kappa) \cap A \neq \emptyset$, then for every $w \in \rho(\kappa) \cap A$, $\rho(\kappa \bullet (\{w\}) \bullet A) = \{w\}$ and for every $w \in \rho(\kappa) \,/\, \rho(\kappa) \cap A$, $\rho(\kappa \bullet (\{w\}) \bullet A) \subseteq \rho(\kappa) \cap A$.

The intuitive idea is simpler than what the notation suggests. We simply have that every world in a context set C is more plausible, from the point of view of C, than any other world outside C (without provisos). Indicative conditionals do express propositions, but they are delicately dependent on epistemic context. When an agent who is in G utters 'if Gore wins the popular vote, this is a good outcome' this utterance is ultimately related to the fact that when he conditions for the sake of the argument his view with A (the antecedent) he finds himself in an outcome he considers good (Gore wins the vote and the election). And when an agent in K utters the same conditional, this really means that from his point of view conditioning with A puts him in a 'bad' state (where Gore is not elected, but wins the popular vote). An important philosophical problem is to what extent the postulation of these propositions can be made compatible with a reasonable account of communication. Is this view too fine-grained in such a way that communication is made implausible? This requires a detailed analysis that I do not intend to offer here. My main goal was to show that a theory of contextual propositions for conditionals is possible. I wanted also to illustrate with example how these propositions are calculated in simple cases. With regard to the deeper philosophical problem of communication I shall say only in passing that a view that defends the contextual account has to show how we are capable of using utterances in order to elicit contexts of communication and to partially elicit the points of view of interlocutors. My own philosophical preferences lean towards the view that it is better to see conditionals as *cognitive carriers* (the expression is Ramsey's) than to see them as carrying propositions. But a purely epistemic theory of contextual propositions is possible. In this article my main goal was to show the kind of epistemological commitments that one should contract in order to build it up.

Acknowledgements. I would like to thank the useful comments of two anonymous referees.

References

[Arlo-Costa 01] Arlo-Costa, H. 2001. 'Bayesian Epistemology and Epistemic Conditionals: On the Status of the Export-Import Laws,' *Journal of Philosophy*, Vol. XCVIII, 11, 555–598.

[Arlo-Costa forthcoming] Arlo-Costa, H. forthcoming 'Epistemological foundations for the representation of discourse context,' forthcoming in *Studies on Language and Information* (CSLI), Stanford.

[Arlo-Costa & Levi 96] Arlo-Costa, H., and Levi, I. 1996. 'Two notions of epistemic validity'. *Synthese*, volume 109, No. 2, 217–262.

[Cross & Nute 98] Cross, C., and Nute, D. 1998 'Conditional Logic,' in *Handbook of Philosophical Logic, second edition*, Vol III Extensions of Classical Logic (eds.) D. Gabbay and F. Guenthner, Reidel, Dordrecht.

[Darwiche & Pearl 97] Darwiche, A., and Pearl, J. 1997. On the logic of iterated belief revision. *Artificial Intelligence* 89(1–2):1–31.

[Gardenfors 88] Gärdenfors, P. 1988 *Knowledge in Flux*, Bradford Book, MIT Press, Cambridge, Mass.

[Gibbard 80] Gibbard, A. 1980, 'Two recent theories of conditionals,' in *Ifs*, (eds.) W. Harper, R. Stalnaker, G. Pearce, Reidel Publishing Company, pp. 211–247.

[McGee 94] McGee, V. 1994. Learning the impossible. In Eells, E., and Skyrms, B., eds., *Probability and Conditionals: Belief Revision and Rational Decision*. Cambridge University Press.

[Lewis 73] Lewis, D. 1973. *Counterfactuals*, Basil Blackwell, Oxford.

[Lewis 76] Lewis, D. 1976. 'Probabilities of conditionals and conditional probabilities', *Philosophical Review* vol. no. 85, pp. 297–315.

[Ramsey 90] Ramsey, F.P. 1990. *Philosophical Papers*, (ed.) Mellor, H. A., Cambridge University Press, Cambridge.

[Shackle 61] Shackle, G.L.S. 1961 *Decision, Order and Time in Human Affairs*, (2nd edition), Cambridge, Mass., CUP.

[Spohn 87] Spohn, W. 1987. 'Ordinal conditional functions: a dynamic theory of epistemic states,' *Causation in Decision, Belief Change and Statistics*, W.L. Harper and B. Skyrms eds., Dordrecht: Reidel, vol 2, 105–134.

[Stalnaker 84] Stalnaker, R. 1984 newblock Indicative Conditionals, in *Conditionals*, (ed.) F. Jackson, OUP.

[Stalnaker 98] Stalnaker, R. 1998. 'On the representation of context,' *Journal of Logic, Language and Information* 7, 3–19.

[van Fraassen 76] van Fraassen, B. 1976 'Probabilities of Conditionals,' in Harper and Hooker (eds.) *Foundations of Probability Theory, Statistical Inference and Statistical Theories of Science* Vol. I, Reidel, Dordrecht, 261–308.

Context-Sensitive Weights for a Neural Network

Robert P. Arritt and Roy M. Turner*

Department of Computer Science
University of Maine
Orono, ME 04468–5752 USA
arritt@maine.edu, rmt@umcs.maine.edu

Abstract. This paper presents a technique for making neural networks context-sensitive by using a symbolic context-management system to manage their weights. Instead of having a very large network that itself must take context into account, our approach uses one or more small networks whose weights are associated with symbolic representations of contexts an agent may encounter. When the context-management system determines what the current context is, it sets the networks' weights appropriately for the context. This paper describes the approach and presents the results of experiments that show that our approach greatly reduces the training time of the networks as well as enhancing their performance.

1 Introduction

Neural networks are well known for their ability to separate continuous numerical data into a finite number of discrete classes. This functionality makes them a perfect candidate for converting real-valued data into symbolic values for a symbolic system. We will call such symbolic values "linguistic values", borrowing the term from fuzzy logic (e.g., [14]).

A problem arises though, in real-world situations where linguistic values are highly context-dependent. For example, an underwater agent may convert a depth of 5 meters into *TOO_DEEP* while in a harbor, but later, when it finds itself in the open ocean, 5 meters may be classified as *NOMINAL*. Thus, if we wanted a neural network to convert an agent's depth to a linguistic value we would need to encode context into the network, which would entail adding nodes and connections to the network. It is easy to see that as the number of contextual features and the number of possible contexts increases this would become unmanageable: in order to be context-sensitive, the network's size would make it impractical.

On the other hand, if a neural network could be constructed and trained for use within a single context, it could be much smaller, and it would be fast

---

* This work was supported in part by the United States Office of Naval Research through grants N0001-14-96-1-5009 and N0001-14-98-1-0648. The content does not necessarily reflect the position or the policy of the U.S. government, and no official endorsement should be inferred.

to train and highly accurate. A large number of these simpler networks could then do the work of the larger, more complex network. The problem would then become one of somehow identifying which network to use in which context.

We have developed an approach to making neural networks context-sensitive that takes this second approach. We solve the problem of deciding which network to use by making use of prior work on a context-management system that explicitly represents context an agent might reasonably be expected to encounter [13,12]. With each contextual representation, or *contextual schema*, is stored information useful for the agent in the context. For an agent that uses (or that is) a neural network, contextual schemas contain the weights appropriate for the network to use in the context represented by the schema. Rather than encoding all the features of the context into the neural network's weights, the context-management system handles the problem of diagnosing the current context. This frees the network to encode only those features having to do with categorizing a value within that context. This greatly decreases both the complexity and training time of the neural networks. In addition, the classification error rate is on par with the fuzzy rule-based system currently used by our system for classification.

There has been some prior work aimed at developing context-sensitive neural networks. For example, Henninger *et al.* [7] have developed context-sensitive neural networks that use context to determine which network is to be used. Their work, however, makes use of a very simple form of context, i.e., an agent's distance from a location. In contrast, our approach relies on a much richer notion of context that incorporates numerous environmental factors. This approach not only allows us to better tailor the networks to the situations in which they will be used, but also to remove most of the contextual information from the networks themselves, thus reducing their complexity.

In the remainder of this paper, we discuss this approach in detail. Section 2 discusses our domain, autonomous underwater vehicle (AUV) control, and the agent, Orca, within which our network resides and that provides the context management functionality. Section 3 presents two neural network designs, one using the context-management system and one that does not, that perform one classification task important for AUVs, depth categorization. Section 4 presents the experiments we used to compare these two approaches, including the results of the experiments. Section 5 describes how the neural network is integrated into the context-management system, and Section 6 concludes and discusses future work.

2 AUVs, Orca, and Depth Management

Autonomous underwater vehicles are untethered submersible robots that are capable of carrying out untended underwater missions. They are useful for a variety of tasks in oceanography, ocean engineering, aquaculture, industry, and defense [2]. One of the major advantages of AUVs is their ability to operate in an area without a human surface presence, which allows them to perform long-term missions without constant human supervision. AUVs can also operate under hazardous conditions, such as within minefields, under ice, and during inclement weather. They can also be combined into multi-AUV systems that can perform tasks such as autonomous oceanographic sampling and surveying [4]. Figure 1 shows the two EAVE (Experimental Autonomous VEhicles) AUVs that were used for developing AUV technology.

Fig. 1. Two EAVE vehicles on a support barge. Used by permission of the Autonomous Undersea Systems Institute (AUSI), Lee, NH.

The current state-of-the-art of AUV hardware technology is such that competent AUVs can be built and fielded. However, although there has been much recent work on intelligent control of AUVs (e.g., [9,3]), the competent control software needed to carry out, successfully, a complex, autonomous, possibly long-term mission is largely lacking. For this, an AUV must have sophisticated artificial intelligence (AI) control software that can not only autonomously plan and perform the missions, but that can also respond appropriately to the unexpected events that are sure to arise within the highly dynamic undersea environment.

The Orca project [10] has the goal of creating an intelligent mission controller for long-term, possibly multiagent, ocean science missions. Orca, which is currently in development at the University of Maine, is a context-sensitive agent that will be able to recognize the context it is in and behave appropriately.

Orca's context-sensitivity is conferred by its context-management module, ECHO (Embedded Context Handling Object) [12]. Orca represents all contextual knowledge as *contextual schemas* (c-schemas) [13]. Each c-schema explicitly represents a *context*, that is, a recognizable class of problem-solving situations.[1] A c-schema contains a wide variety of contextual knowledge useful for the agent

---

[1] See [13] for a more complete description of our definition of context.

as it operates within the context represented. For example, c-schemas contain knowledge about: the expected features of the context; context-specific semantic information (e.g., the meaning of fuzzy linguistic terms [11]); how to cope with unanticipated events (how to recognize them, diagnose them, evaluate their importance, and handle them); goal priorities in the current context; which actions are appropriate for which goals in the context; and the appropriate settings of behavioral and perceptual parameters for the context.

C-schemas are stored in an associative long-term memory [8] that, when presented with features of the current situation, returns the c-schemas that most closely match it. A process of diagnosis then occurs to determine which of these evoked c-schemas actually are germane [1]. These are then merged to form the context object, which is a coherent picture of the current context. The knowledge in this object can be used by the agent to quickly decide how to behave appropriately for the context.

As an example of the usefulness of this approach, consider how an agent might determine the appropriate response to a catastrophic event such as a leak. For such a thing, there will be very little time in which to decide on a response, so complex reasoning after the leak may be impossible. Yet the appropriate response is context-specific. In a harbor, the AUV should probably land on the bottom and release a buoy, since that will avoid collisions with surface traffic. In the open ocean, however, landing might be disastrous: the bottom may be below the crush depth of the vehicle. Instead, since there would be very little likelihood of a collision, the appropriate response would be to surface and radio for help. If the agent always maintains an idea of what its current context is, then it can automatically take the appropriate response.

For the purposes of this paper, we are concerned with context-sensitive perception, that is, appropriate classification of sensory inputs. As an example, consider the AUV's depth envelope, that is, the range of depths that are allowable. This is highly context-sensitive. If the agent finds itself in a harbor, it should tighten its envelope to keep it away from the surface, where it is in danger of being hit by traffic, and above the bottom, where it may encounter debris that could ensnare it. If, on the other hand, the agent finds itself in the open ocean, it can loosen its envelope; the probability of surface traffic is minimal, so the agent only has to worry about staying above its crush depth.

3 Neural Networks for Depth Classification

Elsewhere, we have proposed a solution to the problem of context-sensitive classification of sensory data that was based on a fuzzy rule-based system [11]. However, we are interested in a mechanism that can learn from the agent's own experience or by being presented with training examples, since the AUV's operating conditions may change, and it may be difficult or impossible to obtain the fuzzy rules from human experts. Consequently, we have begun investigating the feasibility of using neural networks for this task.

If the neural network is going to classify (e.g.) depth, context must be taken into account. This can be done in two ways: either a large network with a large number of inputs can be trained with data from instances of all the contexts the AUV may encounter, or numerous smaller networks can each be trained with data from instances of a single context. The former kind of network must not only classify the agent's current depth, but also implicitly determine its current context in the process. To achieve this, the network will have to be provided not only with the current depth, but also with a large amount of environmental data that will help it determine the current context. The idea behind the smaller networks is that they will each specialize in classifying depth within a specific context. It will be up to the context-management system to determine which network is the appropriate one for the current situation. Since these networks are so specialized, they will require a few inputs, chiefly the agent's current depth.

The larger network will be very large indeed, and although it may be adequate for all contexts, it is unlikely it will be especially good for any particular one. Training time will be long for the network. The smaller networks will each be simpler to train, and they will obviously be highly-tailored to depth categorization for the context in which they are used.

Our approach is conceptually the latter. However, instead of using numerous small networks, we have a single network whose weights are set based on the context. The weights for the network that are appropriate in a given context are stored in the c-schema representing that context. When that context is recognized, the weights are retrieved from the c-schema and given to the neural network. During the context, the network then behaves as a highly-specific network. When the context changes, any changes to the weights, e.g., from learning sessions within the context, are then stored in the c-schema for use the next time the agent is in that context. A new c-schema is found for the new context, and its weights are loaded. This approach is particularly useful in a system such as Orca, in which the agent's context is already being recognized and represented.

In the next section, we discuss experiments we performed to determine if a large number of small networks, such as is effectively used in our approach, is as good as or better than a large, multi-context network. In the following section, we discuss how this can be integrated into Orca.

4 Experiments and Analysis

There are several criteria for determining if the smaller networks are better suited for our task. First, the smaller networks would have to perform the classification task at least as well as the larger network. Second, since we would be training numerous smaller networks, they would have to train much faster than the larger network. Finally, we would have to show that the size of the larger network grew at an unacceptable rate as more contexts were added to the system. In this section we will show that the set of small networks classify better than the single large network. We will then show, in a less formal manner, that the smaller networks train faster and that the larger network grows unmanageably large.

For the sake of the following experiments we made a few decisions regarding our networks. All of our networks were two layer feed-forward networks [5] that had 5 output nodes. Each of these output nodes stood for one of our linguistic values: *TOO_SHALLOW, SHALLOW, NOMINAL, DEEP, TOO_DEEP*. The output node with the highest value was the classification that we chose. Our smaller networks each had five nodes in their hidden layer. The number of hidden nodes was determined by starting with a network with two hidden nodes this network was trained with two-thirds of the training data. Next, the performance of this network was evaluated using the final third of the training data. A node was added and the network was retrained and tested. This process was repeated until an acceptable level of performance was reached. This process is a form of cross-validation[6]. The number of nodes in the hidden layer of the larger network varied with the number of contexts, but was also chosen via cross-validation. Training was done with the Levenberg–Marquardt algorithm[5]. All data was collected via Matlab programs running on an 1.5 GHz Intel Xeon PC under the Linux operating system.

When we set out to test the classification performance of our networks we decided to keep the number of contexts small in order to simplify the training of the larger network. Thus, we chose to test the performance of the networks at classifying depths within a harbor, on a shoal, and in the open ocean. In order to implement the larger network we had to determine the most salient features of the contexts and use them as parameters. We chose the following seven parameters: depth, distance from shore, water column depth, density of fish, density of debris on the bottom, and the density of surface traffic. The training data for all of these parameters was then generated randomly via a normal distribution around the accepted values for each parameter. Since we wanted these networks to perform as well as our current fuzzy system, we used it to generate the correct classifications for our training data.

Next, the large network was trained with the entire training set while the smaller networks were trained with the data from there respective contexts. Upon completion of training, the networks were tested with more randomly generated data. The rate of classification error can be seen in table 1. Performing a t-test with $H_0 : \mu_{large\_net} = \mu_{small\_nets}$ and $H_1 : \mu_{large\_net} > \mu_{small\_nets}$ we get a t statistic of 2.218 and a critical value of 1.649 when $\alpha = 0.05$. Thus, we reject the null hypothesis and we can say that the smaller nets generate, on average, fewer errors.

Table 1. Rate of Classification Errors

	Harbor	Shoal	Ocean	μ	σ^2
Large Network	0.0216	0.0197	0.0202	0.0205	0.000008
Small Networks	0.0378	0.0118	0.001	0.0169	0.00008

Another important aspect of the networks is how long they take to train. Each of the smaller networks take about the same amount of time to train, thus as we increase the number of smaller networks, the training time increases linearly. Consequently, given the one-to-one correspondence between networks and contexts in this approach, training time overall is linear in the number of contexts. It is intuitive that the smaller networks should train faster, but we have to verify that as the number of contexts is increased that the training time for the larger network increases faster than the training time for the set of smaller networks. Table 2 and figure 2 show how the training times for the networks grow as the number of contexts increase from one context to six. It is apparent that the training time for the larger network is increasing much faster than that of the smaller networks. We attribute this to the fact that as the number of contexts increases, the larger network grows both in size and in the number of inputs. It also requires more training data to allow it to distinguish between contexts.

Table 2. Training times as the number of contexts is increased

# of Contexts	Large Network	Small Networks
1	12 seconds	11.5 seconds
2	1 minute 9 seconds	24 seconds
3	2 minutes 30 seconds	37 seconds
4	6 minutes 23 seconds	49 seconds
5	10 minutes 16 seconds	62 seconds
6	18 minutes 37 seconds	77 seconds

Our final test was to find out how fast the number of nodes in our larger network increased as the number of contexts increased. We wanted to know this because as this number increased, we would increase the runtime of the network and the training time of the network. The number of nodes needed was calculated through simple cross-validation[6], as discussed earlier. Table 3 shows the results of this experiment. It appears that the network is growing in an exponential fashion. It should be apparent that this will soon cause the larger network to become both inefficient and impossible to manage.

This set of experiments demonstrates several things. First it shows us that the smaller networks perform better at the classification task than the larger network. Next, we saw that the training time of the set of smaller networks increases linearly as the number of contexts increases and the training time of the larger network appears to increase at a much faster rate. Finally, we saw that the number of nodes in the larger network exploded as more contexts were added. Taking this all into account, the logical choice is to use the set of smaller networks for our classification tasks.

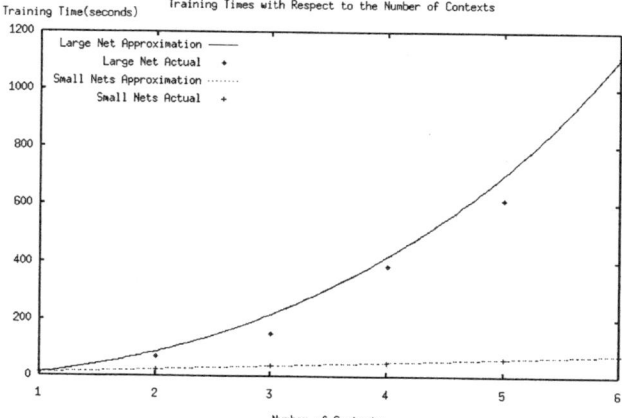

Fig. 2. Graph of training time as contexts are increased

Table 3. Number of nodes needed to classify depth

# of Contexts	# of nodes in hidden layer
1	5
2	8
3	20
4	32
5	63
6	112

5 Embedding the Network within a Schema

As previously discussed, Orca's context-management system explicitly represents contexts in the form of contextual schemas. A c-schema incorporates everything that is deemed important about a given context. This should include information about the operation of neural networks in the context represented. This can be done in one of two ways.

As mentioned earlier in this paper, the only difference between all of the smaller networks was the weights. This static architecture makes it very easy to store the weights in a c-schema. The simplest, although slightly short-sighted, way to make these networks context-sensitive would be to save the weights in a simple list. This makes it very easy for the context manager to manipulate the weights. When the weights are needed they can simply be read in order and slotted into the appropriate spot in the neural network. Likewise, if the context manager deems it necessary to retrain a network the new weights can be easily changed.

The problem with this method is that if in the future we decide to change the structure of the network we would have to rewrite all of our network handling functions. To solve this problem, we could encode the structure of the network into the list. By nesting lists we could generate any feed-forward network structure while maintaining a simple representation. Figure 3 shows how this could work. The idea behind this scheme is that each layer is represented as a list within the master list. Then the set of each node's input weights are stored within the layer list. We intend to use this method in our system, although we will have to be alert for performance degradation from the greater complexity it entails.

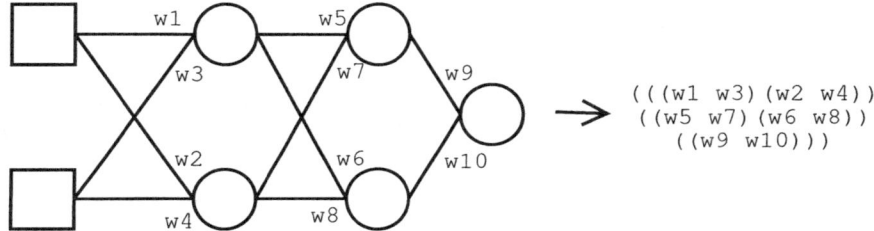

Fig. 3. Converting a network to a list

6 Conclusions

This paper presents a technique that allows a neural network to be context-sensitive without becoming unmanageable. This is achieved by allowing the contextual aspect of the problem to be handled by a symbolic context-management system. By using this technique we are able to to use the efficient classification and learning algorithms associated with neural networks without the explosion in network size that comes with the addition of more contextual data.

While we have only shown this technique to work for our classification task, we believe that it could be applied to many other neural network applications where the network's task and/or domain can be partitioned naturally into contexts in which the network must operate. In the future, we hope to show that this technique extends to other real world domains and also maps to other neural network architectures.

This project has raised many questions that we hope to look at in future work. First, we have to implement and test a method for retraining networks during a mission. One possibility is for the context manager to recognize that it is in a training context (by retrieving a corresponding c-schema) and to allow the weights to change while in that context. Another avenue that we are currently exploring, not only with respect to these networks but with our entire notion of context, is how to merge disparate contextual information. For example, an AUV may be in the context of a harbor and also a rescue mission. The harbor context may tell the AUV that the bottom of the harbor is too deep while the rescue context will want to eliminate the idea of a depth envelope entirely. Finally, another thing that we have to look at is whether we have lost anything by not using the larger network. This paper shows that it does not classify as well, but we may be able to draw other information from it such as when to merge contexts, or we may be able to use it as an indicator for when context is changing.

References

1. Robert P. Arritt and Roy M. Turner. A system for context diagnosis. Technical Report 2003-1, University of Maine Computer Science Department, 2003.
2. D. Richard Blidberg, Roy M. Turner, and Steven G. Chappell. Autonomous underwater vehicles: Current activities and research opportunities. *Robotics and Autonomous Systems*, 7:139–150, 1991.
3. Don Brutzman, Mike Burns, Mike Campbell, Duane Davis, Tony Healey, Mike Holden, Brad Leonhardt, Dave Marco, Dave McClarin, Bob McGhee, and Russ Whales. NPS Phoenix AUV software integration and in-water testing. In *Proceedings of the 1996 IEEE Symposium on Autonomous Underwater Vehicle Technology (AUV'96)*, pages 99–, Monterey, CA, USA, June 1996.
4. T.B. Curtin, J.G. Bellingham, J. Catipovic, and D. Webb. Autonomous oceanographic sampling networks. *Oceanography*, 6(3), 1993.
5. M. T. Hagan and M. Menha. Training feedforward networks with the Marquardt algorithm. *IEEE Transactions on Neural Networks*, 5(6):989–993, 1994.
6. Simon Haykin. *Neural Networks: A Comprehensive Foundation*. Prentice-Hall Inc., 1999.
7. Amy E. Henninger, Avelino J. Gonzalez, Michael Georgiopoulos, and Ronald F. DeMara. A connectionist–symbolic approach to modeling agent behavior: Neural networks grouped by context. In *Proceedings of the Third International and Interdisiplinary Conference on Modeling and Using Context*, pages 198–209, 2001.
8. J.H. Lawton, R.M. Turner, and E.H. Turner. A unified long-term memory system. In *Proceedings of the 1998 IEEE International Conference on Case-Based Reasoning*, 1998.
9. Henrik Schmidt, James G. Bellingham, Mark Johnson, David Herold, David M. Farmer, and Richard Pawlowicz. Real-time frontal mapping with AUVs in a coastal environment. In *Proceedings of the IEEE Oceanic Engineering Society Conference (OCEANS'96 MTS)*, pages 1094–1098, 1996.
10. Roy M. Turner. Intelligent control of autonomous underwater vehicles: The Orca project. In *Proceedings of the 1995 IEEE International Conference on Systems, Man, and Cybernetics*, 1995.
11. Roy M. Turner. Determining the context-dependent meaning of fuzzy subsets. In *Proceedings of the First International and Interdisiplinary Conference on Modeling and Using Context*, 1997.
12. Roy M. Turner. Context-mediated behavior for intelligent agents. *International Journal of Human–Computer Studies*, 48(3):307–330, March 1998.

13. Roy M. Turner. A model of explicit context representation and use for intelligent agents. In *Proceedings of the Second International and Interdisiplinary Conference on Modeling and Using Context*, 1999.
14. L. A. Zadeh. A theory of approximate reasoning. In J. Hayes, D. Michie, and L.I. Mikulich, editors, *Machine Intelligence 9*, pages 149–194. Halstead Press, 1979.

A Common Sense Theory of Causation

John Bell

Department of Computer Science
Queen Mary, University of London
London E1 4NS
jb@dcs.qmul.ac.uk

Abstract. Causes are defined informally to be events which are both contextually necessary and contextually sufficient for their effects. A formal, logico-pragmatic, definition is then given and discussed.

1 Introduction

We commonly talk of one event causing another, and think of the events as being connected; of there being a bond or nexus between them. However, Hume observed [4, Book I, Part III] that if we concentrate on a pair of such events in isolation from the rest of the universe, then the only relations that can be detected between them are succession and contiguity; the occurrence of the "cause" is followed by the occurrence of the "effect". Consequently he suggested that when we talk of an event e causing an event e', at least part of what we mean is that e is an event of type A and e' is an event of type B, and that events of type A are always followed by events of type B. The individual events are thus subsumed under an inductively learned regularity. Hume appears to have considered this sufficiency account to be equivalent to a necessity account:

> [W]e may define a cause to be *an object followed by another, and where all the objects, similar to the first, are followed by objects similar to the second. Or, in other words where, if the first object had not been, the second never had existed.* [3, §VII, Part II]

As Lewis [6] observes, Hume's necessity account is counterfactual in nature, as it involves implicit reference counterfactual reference to the context in which the purported cause occurs. Hume can thus be seen as proposing a pragmatic, or context-dependent, account of causation, according to which the concept has no independent reality, but rather should be seen as a useful abstraction from experience and particular contexts.

This paper proposes a formal theory of causation along these lines. The theory is based on a common sense theory of events [1] which takes their defeasible, context-dependent, nature seriously. Event types are defined by giving their preconditions and effects; the former being regarded as necessary conditions for the success of events of this type, and the latter being the effects which regularly follow them should they succeed. The idea is that events (tokens of the type) *normally* succeed if their preconditions obtain on occurrence; that their preconditions are, on occurrence, normally also sufficient for their effects. This idea is unpacked by considering the context in which an event occurs, its *context of occurrence*, which is typically incompletely specified or partial. The preconditions should be such that they are sufficient in most contexts, but will typically not be sufficient in all of them. The success of an event is then inferred by default whenever doing so is consistent with its context of occurrence. For example the preconditions of a blocks-world *Pickup* operator [2] are that the robot's hand is empty, and that the block is on the table and is clear (has nothing on top of it), and the effects are that the robot is holding the block, its hand is no longer empty, and the block is no longer on the table or clear. If a *Pickup* event occurs in a context in which its preconditions are all true and its success is consistent with the context, then the *Pickup* event should succeed, and its effects should follow. This view of events can be used to define causation as follows. An event e is said to be *contextually sufficient* for (a fact or other event) ϕ in context c iff e succeeds in c and ϕ is among e's effects. Event e is said to be *contextually necessary* for ϕ in c iff removing e from c would also remove ϕ from c. A (direct) cause can then be defined to be an event which is contextually necessary and contextually sufficient for its effect. Thus, for example, the *Pickup* event is said to be the cause of the block's being held in a given context iff the event succeeds in the context and, had it not occurred in the context, the block would have remained on the table.

The formal theory is expressed in a three-valued language called the Causal Temporal Calculus, which is presented in the next section. The theory of events is then summarized in Section 3, and the theory of causation is presented and discussed in Section 4.

2 The Causal Temporal Calculus

The Causal Temporal Calculus ($\mathcal{CTC}$) is a straightforward extension of the Temporal Calculus ($\mathcal{TC}$) [1], which in turn is based on Kleene's strong three-valued language [5]. This provides a means for reasoning demi-classically with partial information and classically with complete information. Accordingly, the truth conditions for the propositional operators return a Boolean truth value wherever possible. Thus the sentence $\neg \phi$ is true if ϕ is false, false if ϕ is true, and is undefined otherwise. And the sentence $\phi \wedge \psi$ is true if ϕ and ψ are both true, false if either is false, and is undefined otherwise. Further operators, such as inclusive and exclusive disjunction, can be defined as in classical logic, $\phi \vee \psi =_{\text{Df}} \neg(\neg \phi \wedge \neg \psi)$ and $\phi \mid \psi =_{\text{Df}} (\phi \vee \psi) \wedge \neg(\phi \wedge \psi)$, while a sentence of the form $\phi \equiv \psi$ is true if ϕ and ψ have the same truth value (true, false, undefined) and is false otherwise. This approach is extended to the first-order case. Atomic sentences may be true, false, or undefined. And a universal sentence $\forall x \phi$ is true if ϕ is true for all assignments to x, false if ϕ is false for one such assignment, and is undefined otherwise. The existential quantifier $\exists$ is then defined as in classical logic.

The *undefined* operator 'U' is added to Kleene's language in order to represent and reason about partiality. The sentence Uϕ is true if ϕ is undefined (is neither true nor false), and is false otherwise. This operator is used to define the classically-valued operators T, F, $\rightarrow$ and $\equiv$ as follows:

$$T\phi =_{Df} \neg(U\phi \vee \neg\phi), \quad F\phi =_{Df} \neg(U\phi \vee \phi), \quad \phi \rightarrow \psi =_{Df} \neg T\phi \vee T\psi,$$
$$\phi \equiv \psi =_{Df} (T\phi \wedge T\psi) \vee (F\phi \wedge F\psi) \vee (U\phi \wedge U\psi).$$

Thus, for sentences ϕ and ψ: Tϕ is true if ϕ is true, and is false otherwise; Fϕ is true if ϕ is false, and is false otherwise; and $\phi \rightarrow \psi$ is true if ψ is true or ϕ is not, and is false otherwise.

In order to represent time, and thus change, a temporal index is added to each atom of the underlying language. A *domain atom* is an atom of the form $r(u_1, \ldots, u_n)(t)$, where the u_i are terms denoting objects in the domain, and term t denotes a time point. Intuitively, a domain atom $r(u_1, \ldots, u_n)(t)$ states that the relation r holds between the objects $u_1, \ldots, u_n$ at time t, that the fact $r(u_1, \ldots, u_n)$ is true at t. For example, Axiom (16) in Table 5 states that object $O1$ is at location $L1$ at time 1, etc.

In order to represent change, events are added as a separate sort. Thus an *occurs atom* is an atom of the form $Occ(e)(t)$, stating that a token of event type e occurs at time t; for example, the occurs atom in Axiom (16) states that the event consisting of $O1$ moving from location $L1$ to $L2$ occurs at time 1.

In order to represent inertia, facts are added as a fourth sort. Formally, a *fact* is the atemporal component, α, of a domain atom $\alpha(t)$.

Finally, in order to define causation, $\mathcal{TC}$ is extended to $\mathcal{CTC}$ by adding a logical truth operator on $\mathcal{TC}$-sentences and allowing quantification over them. Thus if ϕ is a sentence of $\mathcal{TC}$, then $\Box\phi$ is a sentence of $\mathcal{CTC}$ which states that ϕ is a logical truth of $\mathcal{CTC}$. And a $\mathcal{TC}$-*formula atom* is an atom of the form $r_\Phi(\phi_1, \ldots, \phi_n)$, where the ϕ_i are $\mathcal{TC}$-formulas; for example $Cause(Occ(e)(t), Occ(e')(t+1))$ states that $Occ(e)(t)$ is the $Cause$ of $Occ(e')(t+1)$.

The five sorts of $\mathcal{CTC}$ are identified by the following letters: D for domain objects, T for time points, and E for events,

Definition 1 *The vocabulary of $\mathcal{CTC}$ consists of the symbols '$\neg$', 'U', '$\wedge$', '$\forall$', '<', '=', '(', ')', and the following, mutually disjoint, countable, sets of symbols:*

- C_D, C_T, C_E *(constants of sorts D, T and E)*,
- V_D, V_T, V_E, V_F, V_Φ *(variables of each sort)*,
- F_D, F_T, F_E *(function symbols of each arity $n \geq 1$ of sorts D, T, E), and*
- R_D, R_E, R_F, R_Φ *(relation symbols of each arity $n \geq 0$ of sorts D, E, F, and Φ).*

Definition 2 *The terms of each sort S are defined as follows:*

- *If S is of sort D or T then*
  $terms_S = C_S \cup V_S \cup \{f(u_1, \ldots, u_n) : n\text{-ary } f \in F_S, u_i \in terms_S\}.$

- $term_E = C_E \cup V_E \cup \{f(u_1,\ldots,u_n) : n\text{-ary } f \in F_E, u_i \in term_D\}$.
- $term_F = C_F \cup V_F$, where
  $C_F = \{r_D(u_1,\ldots,u_n) : n\text{-ary } r_D \in R_D, u_i \in term_D\}$.
- $term_\Phi = C_\Phi \cup V_\Phi$, where $C_\Phi = \mathcal{TC}$ is defined in Definition 3.

Definition 3 $\mathcal{TC}$ *is the minimal set which satisfies the following conditions.*

- If $t, t' \in term_T$ then $t < t' \in \mathcal{TC}$.
- If S is of sort D, E or F, $u_1,\ldots,u_n \in term_S$, r_S is an n-ary relation symbol in R_S, and $t \in term_T$, then $r_S(u_1,\ldots,u_n)(t) \in \mathcal{TC}$.
- If $v \in V_F$ and $t \in term_T$ then $v(t) \in \mathcal{TC}$.
- If S is of sort D, T, E or F, and $u, u' \in term_S$, then $u = u' \in \mathcal{TC}$.
- If $\phi, \psi \in \mathcal{TC}$, then $\neg \phi \in \mathcal{TC}$, $\mathsf{U}\phi \in \mathcal{TC}$, and $(\phi \wedge \psi) \in \mathcal{TC}$.
- If S is of sort D, T, E or F, $v \in V_S$ and $\phi \in \mathcal{TC}$, then $\forall v \phi \in \mathcal{TC}$.

The members of $\mathcal{TC}$ are called formulas *(of $\mathcal{TC}$). Those formulas in which no variable occurs free are called* sentences *(of $\mathcal{TC}$).*

Definition 4 $\mathcal{CTC}$ *is the minimal set which satisfies the following conditions.*

- $\mathcal{TC} \subseteq \mathcal{CTC}$.
- If $\phi_1,\ldots,\phi_n \in term_\Phi$ and r_Φ is an n-ary relation symbol in R_Φ, then $r_\Phi(\phi_1,\ldots,\phi_n) \in \mathcal{CTC}$.
- If $\phi \in \mathcal{TC}$ and ϕ' is the result of substituting zero or more variables in V_Φ for sub-formulas in ϕ, then $\Box\phi' \in \mathcal{CTC}$.
- If $\phi, \psi \in \mathcal{CTC}$, then $\neg \phi \in \mathcal{CTC}$, $\mathsf{U}\phi \in \mathcal{CTC}$, and $(\phi \wedge \psi) \in \mathcal{CTC}$.
- If S is any sort, $v \in V_S$ and $\phi \in \mathcal{CTC}$, then $\forall v \phi \in \mathcal{CTC}$.

The members of $\mathcal{CTC}$ are called formulas *(of $\mathcal{CTC}$). Those formulas in which no variable occurs free are called* sentences *(of $\mathcal{CTC}$).*

Models of $\mathcal{CTC}$ consist of a set $\mathcal{D}$ of domain objects, a set $\mathcal{E}$ of event types, a temporal frame $\langle \mathcal{T}, \mathcal{R}_\mathcal{T} \rangle$ (where $\mathcal{T}$ is a set of time points and $\mathcal{R}_\mathcal{T}$ is the before-after relation on $\mathcal{T}$), and interpretation functions for terms and relations. For simplicity, time is assumed to be discrete and linear. The denotations of terms are always defined and do not vary with time. By contrast relations are interpreted by time-dependent, partial, characteristic functions; thus the interpretation of relations may be partial and may vary with time.

Definition 5 *A model for $\mathcal{CTC}$ is a structure $M = \langle \mathcal{D}, \mathcal{E}, \langle \mathcal{T}, \mathcal{R}_\mathcal{T} \rangle, \mathcal{F}, \mathcal{R}, \mathcal{V} \rangle$, where:*

- $\mathcal{D}$, $\mathcal{E}$ *and* $\mathcal{T}$ *are mutually disjoint, countable, non-empty sets,*
- $\mathcal{R}_\mathcal{T}$ *is a binary relation on* $\mathcal{T}$ *which is discrete and linear,*
- $\mathcal{F} = \langle \mathcal{F}_D, \mathcal{F}_T, \mathcal{F}_E \rangle$, *where, for each pair* $\langle S, \mathcal{S} \rangle \in \{\langle \mathcal{D}, \mathcal{D} \rangle, \langle \mathcal{T}, \mathcal{T} \rangle, \langle \mathcal{E}, \mathcal{E} \rangle\}$, $\mathcal{F}_S$ *is a set of n-ary functions of type* $\mathcal{S}^n \to \mathcal{S}$ *for each* $n \geq 1$,

- $\mathcal{R} = \langle \mathcal{R}_D, \mathcal{R}_E, \mathcal{R}_F, \mathcal{R}_\Phi \rangle$, where for each pair $\langle S, \mathcal{S} \rangle \in \{\langle D, \mathcal{D} \rangle, \langle E, \mathcal{E} \rangle, \langle F, C_F \rangle, \langle \Phi, C_\Phi \rangle\}$, $\mathcal{R}_S$ is a set of partial n-ary functions of type $\mathcal{S}^n \to \{true, false\}$ for each $n \geq 0$,
- $\mathcal{V} = \langle \langle \mathcal{V}^C_D, \mathcal{V}^C_T, \mathcal{V}^C_E, \mathcal{V}^C_F, \mathcal{V}^C_\Phi \rangle, \langle \mathcal{V}^F_D, \mathcal{V}^F_T, \mathcal{V}^F_E \rangle, \langle \mathcal{V}^R_D, \mathcal{V}^R_E, \mathcal{V}^R_F \rangle \rangle$ is an interpretation function such that
  - $\mathcal{V}^C_S : C_S \to \mathcal{S}$ and $\mathcal{V}^F_S : F_S \to \mathcal{F}_S$ for $\langle S, \mathcal{S} \rangle \in \{\langle D, \mathcal{D} \rangle, \langle T, \mathcal{T} \rangle, \langle E, \mathcal{E} \rangle\}$,
  - $\mathcal{V}^C_F : C_F \to C_F$ and $\mathcal{V}^C_\Phi : C_\Phi \to C_\Phi$ are identity functions,
  - $\mathcal{V}^R_S : \mathcal{R}_S \times \mathcal{T} \to \mathcal{R}_S$.

Terms are interpreted in the standard way.

Definition 6 *A* variable assignment *for a $\mathcal{CTC}$-model is a function* $g = \langle g_D, g_T, g_E, g_F \rangle$, *where for* $\langle S, \mathcal{S} \rangle \in \{\langle D, \mathcal{D} \rangle, \langle T, \mathcal{T} \rangle, \langle E, \mathcal{E} \rangle, \langle F, C_F \rangle, \langle \Phi, C_\Phi \rangle\}$, $g_S : V_S \to \mathcal{S}$. *For $\mathcal{CTC}$-model M, interpretation function $\mathcal{V}$ and variable assignment g for M, the* term evaluation function $\mathcal{V}_g$ *is defined, for each $\mathcal{CTC}$-term u and sort S, as follows*

$$\mathcal{V}_g(u) = \begin{cases} \mathcal{V}_S(u) & \text{if } u \in C_S, \\ g_S(u) & \text{if } u \in V_S, \\ \mathcal{V}^F_S(f)(\mathcal{V}_g(u_1), \ldots, \mathcal{V}_g(u_n)) & \text{otherwise.} \end{cases}$$

The truth and falsity of sentences can now be defined by means of the intermediary notions of satisfaction and violation.

Definition 7 *Let $M = \langle \mathcal{D}, \mathcal{E}, \langle \mathcal{T}, \mathcal{R}_T \rangle, \mathcal{F}, \mathcal{R}, \mathcal{V} \rangle$ be a $\mathcal{CTC}$-model, g be a variable assignment for M, and ϕ be a $\mathcal{CTC}$-formula. Then g satisfies ϕ in M (written $M, g \models \phi$) or violates ϕ in M (written $M, g \dashv \phi$) according to the clauses given in Table 1; where the notation $g \stackrel{v}{\approx} g'$ is used to indicate that the variable assignments g and g' differ at most on the assignment to variable v. A formula ϕ is* true *in a $\mathcal{CTC}$-model M (written $M \models \phi$) if $M, g \models \phi$ for all variable assignments g. A formula ϕ is* false *in M (written $M \dashv \phi$) if $M, g \dashv \phi$ for all variable assignments g.*

It is straightforward to prove (by means of a parallel induction on the structure of $\mathcal{CTC}$-formulas) that, for any model M and sentence ϕ, either $M \models \phi$, or $M \models \neg\phi$, or $M \models \mathsf{U}\phi$. Consequently, as in classical logic, it is sufficient to consider the truth relation on sentences of $\mathcal{CTC}$.

Definition 8 *A $\mathcal{CTC}$-model M is said to be a* model *of a sentence ϕ if $M \models \phi$. Similarly M is said to be a model of a set of sentences Θ (written $M \models \Theta$) if $M \models \phi$ for every $\phi \in \Theta$. A set of sentences Θ* semantically entails *a sentence ϕ (written $\Theta \models \phi$) if all models of Θ are also models of ϕ.*

3 The Theory of Events

The theory of events begins with *primary events*, which can be thought of as defeasible STRIPS events [2]. Primary event types are defined by specifying their

Table 1. Satisfaction and violation conditions for $\mathcal{CTC}$ (see Definition 7)

$$M, g \models t < t' \text{ iff } \langle \mathcal{V}_g(t), \mathcal{V}_g(t') \rangle \in \mathcal{R}_\mathcal{T}$$
$$M, g \dashv t < t' \text{ iff } \langle \mathcal{V}_g(t), \mathcal{V}_g(t') \rangle \notin \mathcal{R}_\mathcal{T}$$
$$M, g \models u = u' \text{ iff } \mathcal{V}_g(u) \text{ is } \mathcal{V}_g(u')$$
$$M, g \dashv u = u' \text{ iff } \mathcal{V}_g(u) \text{ is not } \mathcal{V}_g(u')$$
$$M, g \models r_S(u_1, \ldots, u_n)(t) \text{ iff } \mathcal{V}_S^R(r_S, \mathcal{V}_g(t))(\mathcal{V}_g(u_1), \ldots, \mathcal{V}_g(u_n)) = \textit{true}$$
$$M, g \dashv r_S(u_1, \ldots, u_n)(t) \text{ iff } \mathcal{V}_S^R(r_S, \mathcal{V}_g(t))(\mathcal{V}_g(u_1), \ldots, \mathcal{V}_g(u_n)) = \textit{false}$$
$$M, g \models v(t) \text{ iff } M, g \models \mathcal{V}_g(v)(t)$$
$$M, g \dashv v(t) \text{ iff } M, g \dashv \mathcal{V}_g(v)(t)$$
$$M, g \models r_\Phi(u_1, \ldots, u_n) \text{ iff } \mathcal{R}_\Phi(r_\Phi)(\mathcal{V}_g(u_1), \ldots, \mathcal{V}_g(u_n)) = \textit{true}$$
$$M, g \dashv r_\Phi(u_1, \ldots, u_n) \text{ iff } \mathcal{R}_\Phi(r_\Phi)(\mathcal{V}_g(u_1), \ldots, \mathcal{V}_g(u_n)) = \textit{false}$$
$$M, g \models v \text{ iff } M, g \models \mathcal{V}_g(v)$$
$$M, g \dashv v \text{ iff } M, g \dashv \mathcal{V}_g(v)$$
$$M, g \models \Box \psi \text{ iff } M', g' \models \psi \text{ for every } M' \text{ and } g'$$
$$M, g \dashv \Box \psi \text{ iff } M', g' \dashv \psi \text{ for some } M' \text{ and } g'$$
$$M, g \models \neg \psi \text{ iff } M, g \dashv \psi$$
$$M, g \dashv \neg \psi \text{ iff } M, g \models \psi$$
$$M, g \models \mathsf{U}\psi \text{ iff neither } M, g \models \psi \text{ nor } M, g \dashv \psi$$
$$M, g \dashv \mathsf{U}\psi \text{ iff either } M, g \models \psi \text{ or } M, g \dashv \psi$$
$$M, g \models \psi \wedge \chi \text{ iff } M, g \models \psi \text{ and } M, g \models \chi$$
$$M, g \dashv \psi \wedge \chi \text{ iff } M, g \dashv \psi \text{ or } M, g \dashv \chi$$
$$M, g \models \forall v \psi \text{ iff } M, g' \models \psi \text{ for all } g' \text{ such that } g \stackrel{v}{\approx} g'$$
$$M, g \dashv \forall v \psi \text{ iff } M, g' \dashv \psi \text{ for some } g' \text{ such that } g \stackrel{v}{\approx} g'$$

preconditions and effects. The preconditions can be thought of as necessary conditions for the success of an event of this type, and the effects as the invariant effects of the event; examples are axioms (17) and (18) in Table 5. It is assumed that these definitions are *natural* in the sense that preconditions and effects are constructed entirely from fact and event atoms, that preconditions do not include posterior conditions, and that effects do not include prior effects. Thus it is assumed that in any definition instance $Pre(e)(t) \equiv \phi$ the sentence ϕ does not contain references to time points after t, and that in any instance $\textit{Eff}(e)(t) \equiv \phi$ the sentence ϕ does not contain references to time points before t. The preconditions of a primary event should normally be sufficient, on its occurrence, for its effects, but will typically not logically guarantee them. Call the context in which an event occurs its *context of occurrence*. Then the preconditions of a primary event should be such that they are sufficient in most contexts of occurrence, but need not be sufficient in all of them. In order to represent this, *success atoms* are introduced. Intuitively the *success atom*, $Succ(e)(t)$, states that event e succeeds at time t; that is, that e occurs at t, its preconditions are true on occurrence, and its effects are true at $t+1$. This is stated by the success axiom, Axiom (1) in

Table 2. The theory of events, Θ_E

$$\forall e, t(Succ(e)(t) \equiv \mathsf{T}(Occ(e)(t) \land Pre(e)(t) \land \mathit{Eff}(e)(t+1))) \qquad (1)$$

$$\forall e, t(\mathit{Fail}(e)(t) \equiv (Occ(e)(t) \land \neg Succ(e)(t))) \qquad (2)$$

$$\forall e, e', t(Inv(e, e')(t) \to \neg Inv(e', e)(t)) \qquad (3)$$

$$\forall e, e', t(Inv(e, e')(t) \to (Occ(e)(t) \land Occ(e')(t))) \qquad (4)$$

$$\forall e, e', t((Inv(e, e')(t) \land Succ(e')(t)) \to \exists e''(Inv(e'', e')(t) \land Succ(e'')(t))) \qquad (5)$$

$$\forall e, e', t(Inv^*(e, e')(t) \equiv (Inv(e, e')(t) \lor (Inv(e, e'')(t) \land Inv^*(e'', e')(t)))) \qquad (6)$$

$$\forall \alpha, t(Inert(\alpha)(t) \equiv (\alpha(t) \equiv \alpha(t+1))) \qquad (7)$$

$$\forall \alpha, t(Change(\alpha)(t) \equiv \neg Inert(\alpha)(t)) \qquad (8)$$

Table 2. Note that the presence of the truth operator in the axiom ensures that the *Succ* relation is bivalent; that is, the sentence $\forall e, t(Succ(e)(t) \lor \neg Succ(e)(t))$ is true in any model of the axiom. An event is said to fail if its preconditions do not hold on occurrence, or its effects do not result; Axiom (2). The success axiom is intended to be used in order to infer change. Given $Occ(e)(t)$ and $Pre(e)(t)$, the *success assumption*, $Succ(e)(t)$, should be made whenever it is consistent to do so (whenever it is consistent with *e*'s context of occurrence), and the axiom used to conclude $\mathit{Eff}(e)(t+1)$.

Primary events have the defeasibility of natural events, but are unlike natural events in that their effects, when successful, are invariant. But typically events also have context-dependent effects; for example, if block $B2$ is on block $B1$ when $B1$ is moved, then an additional effect of moving $B1$ is that $B2$ moves also. This limitation is overcome by introducing *secondary events*. Secondary events are defeasible STRIPS events which are *invoked* by other (primary or secondary) events in appropriate contexts, and their success depends on that of the events which invoke them. A common sense event can thus be thought of as a tree-structured object whose root is a primary event, and whose effects are the combined effects of all successful events in its invocation tree.

Invocations are represented *invocation atoms*, thus the atom $Inv(e, e')(t)$ states that event e invokes event e' at time t, and by invocation axioms of the form: $\forall e, e', t((Occ(e)(t) \land \Phi(e, e')(t)) \to Inv(e, e')(t))$; where $\Phi(e, e')(t)$ is a formula which distinguishes those contexts in which e invokes e' at t; examples are axioms (19) and (23) in Table 5.

In keeping with the suggested properties of secondary events, the invocation relation is required to satisfy axioms (3)-(5) in Table 2. Axiom (3) states that invocation is asymmetric. Axiom (4) requires that both the invoking and invoked events occur. Axiom (5) ensures that a secondary event succeeds only if it is directly invoked a successful event. Finally, Axiom (6) defines the transitive closure, Inv^*, of the invocation relation. Thus a primary event should be thought of as inheriting all of the effects of the events that it successfully invokes (either directly or indirectly).

It is also necessary to represent inertia, or what is not changed by events. Intuitively, the *inertia atom*, $Inert(\alpha)(t)$, states that the truth value of fact α does not change at time t; that is, that it persists to $t + 1$. This is stated by the inertia axiom, Axiom (7). Note that the nested equivalence operator makes the *Inert* relation bivalent. An atom changes truth value if it is not inert; Axiom (8). The intention is that the inertia axiom should be used to infer persistence of facts whenever possible. Given $\alpha(t)$, the *inertia assumption*, $Inert(\alpha)(t)$, should be made whenever it is consistent to do so (given the context of occurrence at t), and the axiom used to conclude $\alpha(t+1)$.

Definition 9 *The theory of events, Θ_E, consists of the axioms in Table 2; thus $\Theta_E = \{(1), \ldots, (8)\}$. An event theory is any theory $\Theta = \Theta_E \cup \Theta_B \subseteq \mathcal{TC}$, where the background theory Θ_B is natural; that is, all of its precondition and effects definitions are natural.*

The intended interpretation of event theories is obtained by defining an appropriate formal pragmatics for them. As noted, the intended interpretations of the success and inertia axioms are the positive ones. Given the preconditions of an event occurring at time t, its success at t should be assumed whenever possible, and the success axiom used to infer its effects at $t + 1$. Similarly, whenever possible it should be assumed that a fact is inert at t and the inertia axiom used to infer its persistence to $t + 1$. The temporal directedness of these interpretations suggests that the intended models of causal theories are among those in which they are interpreted *chronologically*. Moreover, in order to generate the intended success and inertia assumptions, the context of occurrence at each time point should be *minimal*; that is, it should be restricted to that which required by the previous pragmatic interpretation of the theory. These considerations suggest that the selected models of an event theory can be defined to be the *chronologically minimal* models of the theory [7].

However, a further refinement, *prioritization*, is necessary in order to establish the context of occurrence at each time point and to generate the appropriate success and inertia assumptions given it. Thus the selected models of an event theory should be those chronologically minimal models of the theory in which, at each time point, facts and events are minimized before invocations, invocations are minimized before maximizing success assumptions (by minimizing their negations), and success assumptions are maximized before maximizing inertia assumptions (by minimizing their negations). Minimizing facts, events and invocations at a time point has the effect of fixing the present context of occurrence before speculating about the future. Priority is given to the minimization of facts and events, as invocations depend on these. Maximizing success assumptions before maximizing inertia assumptions has the effect that, whenever possible, change is preferred to inertia. Thus, whenever possible, conflicts between the success axiom and the inertia axiom are resolved in favour of the former. An argument for doing so is as follows: maximizing inertia assumptions before success assumptions would have the effect that events always failed and nothing changed, while maximizing success and inertia assumptions with equal priority would result in the effects of events being much less predictable that we expect them to be. Finally, in order to keep models as small as possible, $\mathcal{TC}$-formula atoms are minimized.

Definition 10 *Let M and M' be $\mathcal{CTC}$ models which differ only on the interpretation of relations. Then M is E-preferred to M' (written $M \prec_E M'$) iff there is a time point t such that M and M' agree for any earlier time point and:*

- *at least one more domain atom or occurs atom is defined (is either true or false) in M' at t, or*
- *M and M' agree on the interpretation of domain and occurs atoms at t, and at least one more invocation atom is defined in M' at t, or*
- *M and M' agree on the interpretation of domain, occurs and invocation atoms at t, and at least one more success atom is false in M' at t, or*
- *M and M' agree on the interpretation of domain, occurs, invocation and success atoms at t, and at least one more inertia atom is false in M' at t, or*
- *M and M' agree on the interpretation of domain, occurs, invocation, success, and inertia atoms at t, and at least one more $\mathcal{TC}$-formula atom is defined in M' at t.*

A model M is an E-preferred model of a sentence ϕ iff $M \models \phi$ and there is no other model M' such that $M' \models \phi$ and $M' \prec_E M$. M is an E-preferred model of a set of sentences Θ iff $M \models \Theta$ and there is no other model M' such that $M' \models \Theta$ and $M' \prec_E M$.

An event theory Θ predicts a sentence ϕ, written $\Theta \models_E \phi$, iff ϕ is true in all E-preferred models of Θ. Event theory Θ is pragmatically consistent iff there is at least one E-preferred model of Θ.

The pragmatics can be made more concrete by considering model *schemas*. A single $\mathcal{CTC}$-model can be thought of as a schema representing the class of all of its classical completions. This idea can be pushed further by adopting a canonical interpretation of terms (as in Herbrand models), for then each pragmatic interpretation of a theory can be represented by a single model schema. Moreover, for theories of the kind considered in this paper, each such schema can be represented by the set of facts (domain atoms) and event structures (occurs and invocation atoms) which are defined in it; as the remaining relations, *Succ*, *Inert*, *Cause*, etc., are represented implicitly. For example, if the scenario of Example 1 below is simplified by removing block $B2$, then the model schema, M, for the resulting theory can be represented as follows:

$M/1 = \{Occ(\mathit{Init})(0), At(B1, L1)(1), Occ(\mathit{Move}(B1, L1, L2))(1),$
$\quad\quad\quad Inv(Occ(\mathit{Move}(B1, L1, L2), \mathit{Clear}(L1))(1), Occ(\mathit{Clear}(L1))(1)\}$
$M/2 = M/1 \cup \{At(B1, L2)(2), \neg At(B1, L1)(2), \mathit{Clear}(L1)(2)\}$

We can then take a dynamic view of the evolving context of occurrence in a preferred model schema M. The context of occurrence at time t in M arises from the earlier pragmatic interpretation of the theory Θ (for example, $M/1$ above represents the context of inference in M at time 1), then the axioms of Θ, especially the success and inertia axioms, are used to extend the context of occurrence in M to $t+1$ (for example, to $M/2$). This approach is adopted in the direct model-building implementation of the theory of primary events [8].

Table 3. The theory of causation, Θ_C

$$\forall e,t,\phi(PSCause(Occ(e)(t),\phi) \equiv (Succ(e)(t) \land \Box(\bigwedge \Theta_B \to (\mathit{Eff}(e)(t+1) \equiv \phi)))) \quad (9)$$

$$\forall e,e',t(CSCause(Occ(e)(t),Occ(e')(t)) \equiv$$
$$(\mathsf{T}Inv(e,e')(t) \land \neg\mathsf{T}\exists e''(Inv^*(e'',e)(t) \land Inv(e'',e')(t)))) \quad (10)$$

$$\forall e,t,\phi,\psi,\chi(SCause(Occ(e)(t),\phi) \equiv$$
$$(PSCause(Occ(e)(t),\phi) \lor CSCause(Occ(e)(t),\phi)$$
$$\lor (SCause(Occ(e)(t),\psi) \land \Box(\psi \to \phi))$$
$$\lor (SCause(Occ(e)(t),\psi) \land SCause(Occ(e)(t),\chi) \land \Box(\phi \equiv (\psi \land \chi))))) \quad (11)$$

$$\forall e,t,\phi(Cause(Occ(e)(t),\phi) \equiv$$
$$(SCause(Occ(e)(t),\phi) \land \neg\exists e'(\neg e' = e \land SCause(Occ(e')(t),\phi)))) \quad (12)$$

$$\forall e,t,\phi,\psi,\chi(Causes(Occ(e)(t),\phi) \equiv$$
$$(Cause(Occ(e)(t),\phi)$$
$$\lor \exists e',t'(Cause(Occ(e)(t),Occ(e')(t')) \land Causes(Occ(e')(t'),\phi))$$
$$\lor (Causes(Occ(e)(t),\psi) \land Causes(Occ(e)(t),\chi) \land \Box(\phi \equiv (\psi \land \chi))))) \quad (13)$$

$$\forall e,t(Occ(e)(t) \equiv ((e = Init \land t = 0) \lor \exists e',t' SCause(Occ(e')(t'),Occ(e)(t)))) \quad (14)$$

$$\forall \alpha,t(Change(\alpha)(t) \equiv$$
$$(\exists e SCause(Occ(e)(t),\alpha(t+1)) \lor \exists e SCause(Occ(e)(t),\neg\mathsf{T}\alpha(t+1)))) \quad (15)$$

4 The Theory of Causation

The formal definition of causation, which is given in Table 3, is expressed in the terms of the theory of events (event occurrences, preconditions, effects, success, failure, invocations, facts, inertia, change), the logical notions of consequence (semantic entailment in $\mathcal{CTC}$) and equivalence (semantic equivalence in $\mathcal{CTC}$), and the pragmatic notion of the context of occurrence. The definition assumes the setting of a finite event theory $\Theta = \Theta_E \cup \Theta_B$, where Θ_E is the theory of events and Θ_B is the background theory.

Axiom (9) states that any event e which succeeds at time t is a *prior sufficient cause* (a *PSCause*) of its (direct posterior) effects. Thus e is a *PSCause* of ϕ if e succeeds at t and any model of the background theory Θ_B is also a model of the instance $\mathit{Eff}(e)(t+1) \equiv \phi$ of the effects axiom for e

Axiom (10) states that the occurrence event e is a *contemporaneous sufficient cause* of the occurrence of event e' at t iff e invokes e' at t, and it is not true that there is an event e'' which (directly or indirectly) invokes e and which invokes e' at t. This requirement ensures that e' is causally dependent on e, and is illustrated by Example 1 below.

More abstractly, Axiom (11) states that e is a *sufficient cause* (an *SCause*) of effect ϕ at t if, at t, e is a prior sufficient cause of ϕ, or e is a contemporaneous sufficient cause of ϕ, or e is a sufficient cause of ψ which logically entails ϕ, or e is a sufficient cause of both ψ and χ and their conjunction is logically equivalent to ϕ.

The occurrence of e at t is a *direct cause* (a *Cause*) of effect ϕ iff e is the only sufficient cause of ϕ at t; Axiom (12).

Indirect causation results from causally linked chains of events, each of which may terminate in a fact. Accordingly, the indirect-causation relation *Causes* is defined to be the transitive closure of the direct-causation relation *Cause*, and is closed under conjunction of effects; Axiom (13).

The definition of sufficient causation makes it possible to give elegant statements of two new laws, which restrict changes and event occurrences to those which have sufficient causes. In order to do so, it is assumed that any initial conditions of the background theory are covered by a distinguished initial event *Init*, which occurs at time 0. The *law of occurrence*, Axiom (14), requires that any event occurrence other than *Init* must have a sufficient cause. The *law of change*, Axiom (15), requires that any change in the truth value of a fact must have a sufficient cause. The impact of these laws is discussed below.

Definition 11 *The* theory of causation, Θ_C, *consists of the axioms given in Table 3; thus* $\Theta_C = \{(9), \ldots, (15)\}$. *If* $\Theta_E \cup \Theta_B$ *is a finite event theory, then* $\Theta_C \cup \Theta_E \cup \Theta_B$ *is said to be a* causal theory.

Note that the definition of causation is reductive. In any given causal theory, all references to causation can be replaced by sentences containing only symbols from $\mathcal{TC}$ and the logical truth operator. As the interpretation of this operator is unaffected by the pragmatics for event theories, the same pragmatics can be used to interpret causal theories. The pragmatic interpretation of a causal theory Θ thus depends entirely on that of its constituent event theory and the (reductions of) the laws of occurrence and change. In particular, the *Causes* relation for Θ is determined by (is supervenient on) those of its event theory, formally echoing Hume's claim that the causal relation has no independent reality.

Clearly, if the occurrence of event e at time t is the *Cause* of effect ϕ, then e's occurrence is contextually sufficient for ϕ. In view of the laws of occurrence and change (axioms (14) and (15)), e's occurrence is also contextually necessary for ϕ. In order to see this, consider the case in which e is the *PSCause* of event occurrence or domain atom ϕ. Now, e can be removed from the context by removing all of its sufficient causes. This can be done without introducing new effects as the success of each of e's sufficient causes implies that none of them conflicts with any other simultaneous event. As e is the *Cause* of ϕ, it is its only *SCause* (Axiom (12)). So removing e from the context leaves ϕ without an *SCause*. So if ϕ is an event occurrence (a domain atom), then the law of occurrence (the law of change) requires that an additional event occurs at t as its *SCause*. Consequently, as the pragmatics minimizes event occurrences, removing e from the context also removes ϕ from it.

Proposition 1 *(Properties of the Causes relation) Let Θ be a pragmatically consistent causal theory. Then the sentences listed in Table 4 are true in all E-preferred models of Θ.*

In conclusion, two examples are given to illustrate the detailed workings of the theory.

Table 4. Properties of the *Causes* relation

Bivalence:	$\forall \phi, \psi (Causes(\phi, \psi) \vee \neg Causes(\phi, \psi))$
Transitivity:	$\forall \phi, \psi, \chi ((Causes(\phi, \psi) \wedge Causes(\psi, \chi)) \rightarrow Causes(\phi, \chi))$
Asymmetry:	$\forall \phi, \psi (Causes(\phi, \psi) \rightarrow \neg Causes(\psi, \phi))$
Actuality:	$\forall \phi, \psi (Causes(\phi, \psi) \rightarrow (\phi \wedge \psi))$
Consistency:	$\forall \phi, \psi (Causes(\phi, \psi) \rightarrow \neg Causes(\phi, \neg \mathsf{T}\psi))$
Conjunction:	$\forall \phi, \psi, \chi ((Causes(\phi, \psi) \wedge Causes(\phi, \chi)) \rightarrow Causes(\phi, \psi \wedge \chi))$
Consequence:	$\forall \phi, \psi, \chi ((Causes(\phi, \psi) \wedge \Box(\psi \rightarrow \chi)) \rightarrow Causes(\phi, \chi))$

Example 1. Consider the following simple blocks-world scenario. At time point 1, blocks $B1$ and $B2$ are at location $L1$, and $B2$ is on $B1$ (for simplicity, it is assumed that being above a location counts as being at it). Also at time 1, the event consisting of $B1$ moving to location $L2$ occurs. On the basis of this context of occurrence, it is expected that the move event will succeed, thereby causing $B1$ to move to $L2$. Moreover, as $B2$ is on $B1$, it is expected that the movement of $B1$ will cause $B2$ to move to $L2$. Finally, the movement of $B1$ should also cause $L1$ to become clear.

This scenario is represented by axioms (16)-(24) of Table 5. Let $\Theta_1 = \Theta_C \cup \Theta_E \cup \{(16), \ldots, (24)\}$, then

$\Theta_1 \models_E Cause(Occ(Move(B1, L1, L2))(1),$
$\qquad Occ(Move(B2, L1, L2))(1) \wedge Occ(Clear(L1))(1) \wedge At(B1, L2)(2))$
$\wedge\; Cause(Occ(Move(B2, L1, L2))(1), At(B2, L2)(2))$
$\wedge\; Cause(Occ(Clear(L1))(1), Clear(L1)(2))$

Proof. Let M be an E-preferred model of Θ_1. Then it follows by chronological minimization that $Pre(Init)(0)$ and $Occ(Init)(0)$ are the only domain, occurs or invocation atoms with temporal index $t \leq 0$ which are defined in M. As intended, it follows from axioms (1), (16) and (24) that $Succ(Init)(0)$ is true (in M). So it follows from axioms (9) and (11) that $Init$ is an $SCause$ of (16).

As $B2$ is on $B1$ at time 1 (Axiom (16)), the occurrence of the event $Move(B1, L1, L2)$ invokes the $Move(B2, L1, L2)$ event (Axiom (19)), which also occurs (Axiom (4)), and has the event $Move(B1, L1, L2)$ as sufficient cause (axioms (10) and (11)). Moreover, it follows from the minimization of occurs atoms at time 1 that $Move(B1, L1, L2)$ is the only sufficient cause, and hence is the *Cause* (Axiom (12)).

Both move events invoke the $Clear(L1)$ event at time 1. As the $Move(B1, L1, L2)$ event invokes the $Move(B2, L1, L2)$ event, the first of these move events is considered to be a $CSCause$ of the clear event, and the second is not (Axiom (10)). It follows that it is also an $SCause$ (Axiom (11)) and the *Cause* (minimization of occurrences, Axiom (12)).

The preconditions of the two *Move* events are true at time 1 (axioms (16), (17)), and it is consistent to assume that the $Move(B1, L1, L2)$ event succeeds, with the effect $At(B1, L2)(2)$ (axioms (1), (18)). It follows that the event is a $PSCause$ of this effect (Axiom (9)), an $SCause$ (Axiom (11)), and its *Cause* (Axiom (12), minimization of occurs atoms).

Table 5. Axioms for the examples

$$At(B1, L1)(1) \land At(B2, L1)(1) \land On(B2, B1)(1) \land Occ(Move(B1, L1, L2))(1) \qquad (16)$$
$$\forall x, l, l', t(Pre(Move(x, l, l'))(t) \equiv At(x, l)(t)) \qquad (17)$$
$$\forall x, l, l', t(\mathit{Eff}(Move(x, l, l'))(t) \equiv (At(x, l')(t) \land \neg At(x, l)(t))) \qquad (18)$$
$$\forall x, l, l', t((Occ(Move(x, l, l'))(t) \land On(y, x)(t)) \rightarrow$$
$$Inv(Move(x, l, l'), Move(y, l, l'))(t)) \qquad (19)$$
$$\forall l, t(Clear(l)(t) \equiv \neg \exists x At(x, l)(t)) \qquad (20)$$
$$\forall l, t(Pre(Clear(l))(t) \equiv \neg Clear(l)(t)) \qquad (21)$$
$$\forall l, t(\mathit{Eff}(Clear(l))(t) \equiv Clear(l)(t)) \qquad (22)$$
$$\forall x, l, l', t((Occ(Move(x, l, l'))(t) \land At(x, l)(t)) \rightarrow Inv(Move(x, l, l'), Clear(l))(t)) \quad (23)$$
$$Pre(Init)(0) \land Occ(Init)(0) \land (\mathit{Eff}(Init)(1) \equiv (16)) \qquad (24)$$
$$Pre(Init)(0) \land Occ(Init)(0) \land (\mathit{Eff}(Init)(1) \equiv ((16) \land At(B3, L1)(1))) \qquad (25)$$

It is also consistent to assume that the $Move(B2, L1, L2)$ event succeeds; in particular, Axiom (5) is satisfied by the success of the $Move(B1, L1, L2)$ event which invoked it. So it follows, as above, that the event succeeds and is the *Cause* of $At(B2, L2)(2)$.

Similarly the success of the $Move(B1, L1, L2)$ event at time 1 provides grounds for assuming the success of the $Clear(L1)$ event. It then follows, as above, that this is the *Cause* of $Clear(L1)(2)$. □

In this example it is assumed that locations may contain more than one object and that events may occur simultaneously. So it is typically the case that a location only becomes clear as a result of the combined effect of several events and the non-occurrence or failure of others. A location's becoming clear is thus typically a *global* ramification; an indirect effect of several events. But in the example the $Clear(L1)$ event is invoked *locally* by the movement of a block from a location; as only the occurrence of the move event and the location of the block are considered when invoking the clear event (Axiom (23)). This may appear reckless. Indeed, if the scenario is extended by adding another block, $B3$, at $L1$ but not *On* either of $L1$ or $L2$, then we expect the $Clear(L1)$ event to fail. However, as the pragmatics gives priority to change over inertia (by maximizing success assumptions before inertia assumptions), it seems that the $Clear(L1)$ event should succeed in all *E*-preferred models of the extended theory, with the mysterious side effect of $B3$ becoming locationless at time 2. However, this unintended outcome is prevented by the law of change (Axiom (15)). The law complements and completes the earlier representation of inertia. The inertia axiom (Axiom (7)) is still needed in order to represent the temporal projection of unchanged facts. By restricting changes to those which have a sufficient cause, the law of change curbs the effect of the success axiom (Axiom (1)) and strengthens the effect of the inertia axiom, thereby ensuring the proper balance between them. In particular, as illustrated by the following example, it forces the failure of events whose effects are not caused. Its presence thus means that events *can* be invoked locally and the consequences *can* be left to take care of themselves.

Example 2. Suppose that the background theory of Example 1 is extended by adding an axiom stating that block $B3$ is at location $L1$ initially. Then we expect that the movement of the other objects will not change the location of $B3$.

Let $\Theta_2 = \Theta_C \cup \Theta_E \cup \{(16), \ldots, (23), At(B3, L1)(1), (25)\}$, then:

$$\Theta_2 \models_E \neg Cause(Occ(Move(B1, L1, L2)(1), Clear(L1)(2)) \wedge At(B3, L1)(2).$$

Proof. Let M be an E-preferred model of Θ_2. Then, as in Example 1, that exactly three events occur at time 1 in M: $Move(B1, L1, L2)$, $Move(B2, L1, L2)$ and $Clear(L1)$. Moreover, it is consistent to assume that the two move events succeed (in M).

Suppose, for contradiction, that it is consistent to assume that the $Clear(L1)$ event also succeeds. Then $L1$ must be clear at time 2 (axioms (1), (22)). So it follows that $B3$ is no longer at $L1$ (Axiom (20)), and consequently it follows from the change and inertia axioms ((7) and (8)) that $Change(At(B3, L1))(1)$ is true. Consequently the law of change (Axiom (15)) requires that there is an $SCause$ for $\neg TAt(B3, L1)(2)$. Clearly none of the three events occurring at time 1 has this effect. So it follows from the law of change and the bivalence of the $Change$ relation that $\neg Change(Clear(L1))(1)$ is true. Consequently it follows from the change and inertia axioms that $\neg Clear(L1)(2)$ is true. But then it follows (axioms (1), (20), (22)) that $\neg Succ(Clear(L1))(1)$ is true.

On the other hand it is consistent to assume that $At(B, L3)$ is inert at time 1 and to conclude, by the inertia axiom, that $At(B3, L1)(2)$ is true. $\square$

As can be seen from this example, the law of change governs the success of events: while *caused* change is preferred to inertia, inertia is preferred to *uncaused* change. Its presence thus results in a Yin/Yang interplay between the opposite but complementary principles of change and inertia. The principle of change appears to be dominant; as change is preferred to inertia and events can be invoked whenever they might succeed. However, if the success of an event would give rise to changes which it did not cause, then the event fails and inertia prevails.

References

1. J. Bell (2001) Primary and Secondary Events. www.dcs.qmul.ac.uk/~jb.
2. R.E. Fikes and N.J. Nilsson (1971) STRIPS; a new approach to the application of theorem proving to problem solving, *Artificial Intelligence* 2, pp. 189–208.
3. D. Hume (1777) *Enquiry Concerning Human Understanding.* L.A. Selby-Bigge (ed.). Oxford University Press. Oxford.
4. D. Hume (1888) *A Treatise of Human Nature.* L.A. Selby-Bigge (ed.). Oxford University Press. Oxford.
5. S.C. Kleene (1952) *Introduction to Metamathematics.* North-Holland, Amsterdam.
6. D. Lewis (1973) Causation. *Journal of Philosophy* 70 (1973) pp. 556–567.
7. Y. Shoham (1988) *Reasoning About Change*, M.I.T. Press, Cambridge Mass.
8. G. White, J. Bell and W. Hodges (1998) Building Models of Prediction Theories. *Proc. KR'98*, Morgan Kaufmann, San Francisco, pp. 557–568.

How to Refer: Objective Context vs. Intentional Context

Claudia Bianchi

Philosophy Dept., University of Genoa, via Balbi 4, 16126 Genova - Italy
claudia@nous.unige.it
http://www.dif.unige.it/epi/hp/bianchi

Abstract. In "Demonstratives" Kaplan claims that the occurrence of a demonstrative must be supplemented by an act of *demonstration*, like a pointing (a feature of the objective context). Conversely in "Afterthoughts" Kaplan argues that the occurrence of a demonstrative must be supplemented by a *directing intention* (a feature of the intentional context). I present the two theories in competition and try to identify the constraints an intention must satisfy in order to have semantic relevance. My claim is that the analysis of demonstrative reference provides a reliable test for our intuitions on the relation between objective and intentional context. I argue that the speaker's intentions can play a semantic role only if they satisfy an *Availability Constraint:* an intention must be made available or communicated to the addressee, and for that purpose the speaker can exploit any feature of the objective context. This thesis implies the reconciliation between "Demonstratives" and "Afterthoughts".

1 Introduction

As it is well known, in "Demonstratives" David Kaplan claims that the occurrence of a demonstrative must be supplemented by an act of *demonstration*, like a pointing (a feature of the objective context). Conversely in "Afterthoughts" Kaplan argues that the occurrence of a demonstrative must be supplemented by a *directing intention*, the referential intention the speaker associate with the expression (a feature of the intentional context). In this paper, I will present the two theories in competition and try to identify the constraints an intention must satisfy in order to have semantic relevance. My claim is that the analysis of demonstrative reference provides a reliable test for our intuitions on communicative mechanisms, and more specifically on the relation between objective and intentional context. In particular, I will argue that the speaker's intentions can play a semantic role only if they satisfy an *Availability Constraint:* an intention must be made available or communicated to the addressee, and for that purpose the speaker can exploit any feature of the objective context (words, gestures, relevance or uniqueness of the referent in the context of utterance). This thesis implies the reconciliation between "Demonstratives" and "Afterthoughts".

The structure of the paper is the following:
In section 2. I present the distinction between indexicals and demonstratives.
In section 3. I analyse Kaplan's two theories of demonstratives.

In section 4. I offer a reconstruction of the objective perspective on context - according to which the reference of a demonstrative is determined by objective facts of the utterance context.

In section 5. I present a reconstruction of the intentional perspective on context - according to which the reference of a demonstrative is determined by adding certain features of the speaker's intention.

In section 6. I raise some objections against the intentional perspective on context.

In section 7. my analysis of demonstrative reference provides a test for our intuitions on communicative mechanisms, and more specifically on the relation between objective and intentional context.

In the conclusion, I argue that the speaker's intentions can play a role in semantics only if they satisfy an Availability Constraint, that is to say if they can be recognised by the addressee.

2 Indexicals and Demonstratives

Indexicals and demonstratives are referential expressions depending, for their semantic value, on the context of utterance: they have a reference only given a context of utterance. The conventional meaning of an indexical sentence like

(1) *I am drunk*,

independently of any context whatsoever, cannot determine the truth conditions of the sentence: to evaluate the sentence, the referent of *I* must be identified. The truth conditions of an indexical sentence are thus indirectly determined, as a function of the context of utterance of the sentence, and in particular as a function of the values of the indexicals. According to Kaplan and Perry, a function (or *character*) is assigned to each indexical expression as a type; given a context, the character determines the *content* of the occurrence – which is a function from circumstances of evaluation (possible world and time) to semantic values.

In "Demonstratives", Kaplan introduces the distinction between pure indexicals (expressions like *I, here, now*) and demonstratives (expressions like *this, that, she, he*). As I said, the language conventions associate with a pure indexical as a type a rule fixing the reference of the occurrences of the expression in context. The semantic value of an indexical (its content, its truth conditional import) is thus determined by a conventional rule and by a contextual parameter, which is a publicly available aspect of the utterance situation (the *objective context*). The character of an indexical encodes the specific contextual co-ordinate that is relevant for the determination of its semantic value: for *I* the relevant parameter will be the speaker of the utterance, for *here* the place of the utterance, for *now* the time of the utterance, and so on: the designation is then automatic, "given meaning and public contextual facts".[1]

Conversely, the meaning of a demonstrative, like *she* in the sentence

(2) *She is drunk*,

by itself doesn't constitute an automatic rule for identifying, given a context, the referent of the expression. The semantics of *she* cannot determine unambiguously its reference: if, for instance, in the context of utterance of (2) there is more than one woman, the expression *she* can identify any woman in the same way.

---

[1] [23], p. 595.

3 Demonstration vs. Intention

According to Kaplan in "Demonstratives", the occurrence of a demonstrative must be supplemented by a *demonstration*, an act of demonstration like a pointing: "typically, though not invariably, a (visual) presentation of a local object discriminated by a pointing".[2] The relevant semantic unit is then the demonstrative associated with a demonstration.[3] The act of demonstration is *semantically relevant* in order to complete the character of a demonstrative. The act of demonstration that could accompany a pure indexical is, in turn, either emphatic (as when one utters *I* pointing to oneself) or irrelevant (as when one utters *I* pointing to someone else: in this case, the referent of *I* remains the speaker): once the context of utterance is fixed, the linguistic rules governing the use of the indexicals determine completely, automatically and unambiguously their reference, no matter what the speaker's intentions are.

However, according to Kaplan, a demonstration does not always require an action on the speaker's part, as when we shout

(3) *Stop that man*

if there is only one man, or only one man rushing toward the door, or only one man running completely naked. Or there may be a convention identifying the *demonstratum* with any object appearing on a "demonstration platform"; or else the speaker may exploit a natural demonstration, as an explosion or a shooting star.[4] In this way, the speaker may exploit a gesture, or the uniqueness of the *demonstratum* in the context of utterance, or its saliency, or its relevance. Likewise, we can interpret in terms of uniqueness or relevance of the *demonstratum* the cases of non visual perceptual demonstratives, as in

That noise is driving me crazy[5];
This smell is delicious;
This flavour reminds me of something.

All the examples, in fact, are appropriate only if there is only one noise (or smell or flavour), or only one relevant noise in the context of utterance.

In "Afterthoughts", Kaplan modifies his own theory. He now acknowledges that even a gesture associated with an occurrence of a demonstrative, constituting the act of demonstration, may be insufficient to disambiguate the expression. Just imagine the sentence

(4) *I like that*

uttered by someone pointing clearly and unambiguously to a dog: the expression *that* could designate the dog, or his coat, or a button of the coat, or the colour of the coat or, for that matter, any spatial region or molecule between the speaker's finger and the dog. The gesture then does not have a semantic role anymore; for Kaplan the relevant factor is now "the speaker's directing intention". The demonstration has only the role of manifesting the intention, of externalising it – a role of pragmatic aid to communication: "I am now inclined... to regard the demonstration as a mere externalization of this inner intention. The externalization is an aid to communication, like speaking

---

[2] [16], p. 490.
[3] Cf. [16], p. 492: "The referent of a pure indexical depends on context, and the referent of a demonstrative depends on the associated demonstration".
[4] Cf. [16], p. 525f.
[5] Cf. [25], p. 200f.

more slowly and loudly, but is of no semantic significance".[6] Every occurrence of the same demonstrative as a type has to be associated not with an act of demonstration but with an intention.[7] In this sense, a demonstrative is different from an indexical: once the context of utterance is fixed, the linguistic rules governing the use of the indexicals determine completely, automatically and unambiguously their reference, no matter what the speaker's intentions are.[8]

Kaplan doesn't offer an explicit and fully satisfactory explanation of why he now favours IPC, and thinks demonstrations are not semantically significant. The arguments are made explicit by Marga Reimer and Kent Bach in a group of articles published at the beginning of the 90's in *Analysis* and *Philosophical Studies*. In what follows, I will reconstruct the two competing theories:
- the objective perspective on context (OPC): according to Kaplan 1977, the reference of a demonstrative is determined by objective facts of the context of utterance.
- the intentional perspective on context (IPC): according to Kaplan 1989, the reference of a demonstrative is determined by completing the character of the demonstrative with features of the speaker's intention.

We will see that, according to Bach, Reimer doesn't offer a fair reconstruction of IPC. In her reconstruction, the intentional perspective is reduced to a sort of Humpty Dumpty theory of language, according to which the speaker has a proposition in mind, and hopes that the addressee is a mind reader. I will first try to offer a better reconstruction of IPC and then try to identify the constraints an intention must satisfy in order to have semantic relevance.

4 Reimer and OPC

It is usual to distinguish between:
- the context in terms of intentional states of the participants, or shared assumptions[9] - what we can call the subjective context, or the cognitive context, or the *intentional* context;
- the context in terms of relevant states of affairs occurring in the world - the *objective* context.[10]

As I said, the reference of a demonstrative doesn't appear to be bound by semantic rules in the way the reference of an indexical seems to be: the semantic rule by itself doesn't determine the reference of the demonstrative expression in the light of the context of utterance. The question to be answered is: what do we have to add to semantic rules and context of utterance in order to have a complete proposition:

---

[6] [18], p. 582.
[7] [18], p. 588: "The directing intention is the element that differentiates the 'meaning' of one syntactic occurrence of a demonstrative from another, creating the *potential* for distinct referents, and creating the actuality of equivocation".
[8] Cf. [7]. For a different perspective on the pure indexicals/demonstratives distinction, see [9].
[9] Assumptions *actually* shared, as in [10], or only *supposedly* shared, as in [30].
[10] On the distinction between cognitive and objective context cf. [14], [21], [22], and [29].

- something like a demonstration – that is a feature of the *objective* context (OPC), or rather
- something like an intention – that is a feature of the *intentional* context (IPC)?

To answer this question, let's examine some of Reimer's examples. In all cases, the reference of the demonstrative seems to be individuated by the speaker's gesture, or else by an element of the context in the objective sense, by public contextual facts.

Case I. "Cases in which the demonstrated object is clearly not the object toward which the speaker has a 'directing intention'".[11] Suppose John grabs a bunch of keys on the desk, saying:

(5) *These are mine.*

He intends to refer to his own keys, but mistakenly grabs his officemate's keys. Intuitively, in this case, the reference is individuated by an objective aspect of the utterance situation, that is John's ostensive gesture. The keys on the desk belong to his officemate, hence (5) is false.

Case II. "Cases in which the demonstrated object is neither perceived by the speaker, nor the object the speaker 'has in mind'".[12] A classic example is provided by Kaplan in "Dthat". John points, without turning and looking, to the place on the wall which was occupied by a picture of Carnap and utters:

(6) *That is a picture of one of the greatest philosophers of the twentieth century.*

But, unbeknownst to him, the picture has been replaced by Spiro Agnew's portrait. Even if John intends to refer to Carnap's picture - or, as Kaplan writes, "has in mind" Carnap's picture[13] - he in fact refers to Agnew's picture: (6) cannot be taken as true.

Case III. "Cases in which there appears to be neither a demonstration nor a demonstratum, despite the presence of a 'directing intention'".[14] Suppose that John and Mary are in the park, observing several dogs (all equally salient) playing and running together. John intends to point and refer to his dog Fido, and utters

(7) *That dog is Fido*

but sudden paralysis prevents him from pointing or making any ostensive gesture, like nodding or glancing. According to Reimer, a supporter of IPC is committed to say that, if it is the speaker's intention that rules, then the reference of *that dog* is the dog John "has in mind". However, our intuitions are different. Since no dog was being demonstrated, no dog was referred to: like the description *the black dog* is empty if there is no black dog, the demonstrative description *that dog* is empty if no dog is demonstrated, and (7) doesn't express any proposition.

Case IV. If there is no demonstration, salience gets semantic significance in order to complete the character of the demonstrative. As in case III. John and Mary are in the park, observing several dogs playing and running together. John intends to point and refer to his dog Fido, and utters (7), but sudden paralysis prevents him from pointing or making any ostensive gesture. But suppose that Spot has made himself especially salient by his hysterical barking. In this case, intuitively, the reference of *that dog* seems individuated by salience. Mary is justified in taking John as referring to the most salient dog in the context of utterance, no matter what John's intentions are. The

---

[11] [25], p.189.
[12] [25], p.190.
[13] [17], p. 396.
[14] [25], p.190.

most salient dog in the context of utterance is Spot: (7) succeeds in expressing a proposition, but a false one.

Case V. However, the ostensive gesture generally overrides salience. As in case III. John and Mary are in the park, observing several dogs playing and running together. Suppose that Spot has made himself especially salient by his hysterical barking. John intends refer to his dog Fido, and, pointing directly to Fido, utters (7). Intuitively, in this case it is the gesture that has semantic significance and discriminates the referent from the other candidates: even if another dog, Spot, was more salient in the context of utterance, *that dog* refers to Fido and (7) is true.

Case VI. The ostensive gesture overrides the speaker's intentions. As in case III. John and Mary are in the park, observing several dogs playing and running together. John intends to point and refer to his dog Fido, and utters (7), but a nervous tic makes his arm move in the direction of another dog, Spot. Following the intentional perspective, one should say that if it is the speaker's intention that rules, then the reference is the dog John has in mind. But, intuitively, the reference seems individuated by John's gesture – even if unintentional – and his intentions seem irrelevant: (7) expresses a false proposition.

It seems, then, that in all the cases under examination, the speaker's intention doesn't play any essential role, that is any semantic role in determining the reference of the demonstrative – which is fixed (when it is fixed) by the objective context.

5 Bach and IPC

The main point of Bach's defence of IPC is to show that a communicative intention requires more than just 'having in mind'. According to Bach's theory of referential intentions "a referential intention is part of a communicative intention, an intention whose distinctive feature is that 'its fulfilment consists in its recognition'... A referential intention... involves intending one's audience to identify something as the referent by means of thinking of it in a certain identifiable way".[15]

Let's start with Kaplan's classic example (**Case II**). In Bach's reconstruction, two intentions must be attributed to the speaker:
  a. the intention to refer to Carnap's portrait;
  b. the intention to refer to the portrait on the wall behind him.
Although John intended to refer to Carnap's portrait, he didn't intend his addressee to recognise *that intention* (a.); the intention he intended the addressee to recognise was that referring to the portrait on the wall behind him (b.). The referential intention is this last one: "the one which you intend and expect your audience to recognize and rely on in order to identify a certain [picture] as the referent".[16]

The analysis of Kaplan's example is easily extended to **Case I** (John's keys). Although John intends to refer to his own keys, he doesn't intend Mary to recognise this intention; the intention he intends Mary to recognise is that referring to the keys he grabbed. The intention semantically relevant is this last one. Even if John intends to

---

[15] [3], p. 296. On referential intentions, see also [5] and [6]. As it is well known, Bach's theory is a development of Grice's, and of his intention-based and inferential view of communication.
[16] [4], p. 143.

refer to his own keys, he in fact refers to the keys he grabbed – which happen to belong to Mary. John's words express the proposition that the keys he grabbed are his: since they belong to Mary, (5) is false.

Let's now see **Case III** (the paralysis). Although John intends to refer to his own dog, he doesn't intend Mary to recognise this intention; the intention he intends Mary to recognise is that referring to the dog he is pointing at. But of course he has not done what it is necessary to enable Mary to recognise this very intention: so, Bach argues, the relevant intention is empty: "[IPC] does not say that such an intention can be fulfilled even if *no* act of demonstration is performed when, as in the example, the fulfilment of this intention requires such an act. After all, the intention in this case is to refer to what is being pointed at".[17]

Case IV (salience). Although John intends to refer to his own dog, he doesn't intend Mary to recognise this intention; the intention he intends Mary to recognise is that referring to *the relevant dog* in the context of utterance. The intention semantically relevant is this last one: there is no act of pointing, no explosion or falling star, in other words there is no further evidence – except relevance - permitting Mary to identify John's communicative intention. John's words express the proposition that the relevant dog in the context of utterance is his: since the relevant dog is the dog barking hysterically, and since Spot, and not Fido, is barking hysterically, (7) is false.

The same goes for **Case V** (the gesture overriding salience). Although John intends to refer to his own dog, he doesn't intend Mary to recognise this intention; the intention he intends Mary to recognise is that referring to the dog he is pointing at. IPC agrees here with OPC.

Case VI (John's tic). Although John intends to refer to his own dog, he doesn't intend Mary to recognise this intention; the intention he intends Mary to recognise is that referring to the dog he is pointing at. The intention semantically relevant is this last one, for the act of pointing (even if unintentional - but, and this is crucial, not recognised as such) is the only evidence permitting Mary to identify John's communicative intention. John's words express the proposition that the dog he is pointing at (Spot) is his: (7) is false.

Let's sum up. Suppose that the speaker utters the expression *that dog*: if the dog he intends to refer to is the only dog in the context of utterance, or the most salient dog, the demonstrative expression doesn't require any other action on the speaker's part. In all the other cases, if there are several dogs all equally salient, the speaker must complete the character of the demonstrative expression with an act of demonstration, like pointing, glancing, or nodding. The speaker has then the referential intention to refer to the dog he is pointing at: notice that pointing is only a way of making an object salient, and has no semantic significance, but only a pragmatic one - like speaking more slowly and loudly.

6 Some Objections against IPC

I agree with Bach analysis, and with his distinction between two kinds of intentions in a referential act: background intentions (as the intention of referring to Fido, or to Carnap's picture) and fundamental intentions (as the intention of referring to the dog

---

[17] [3], p. 298.

the speaker is pointing at, or to the portrait on the wall behind him). Yet, in my opinion, even if interpreted in this way, IPC may still raise some objections. Let's see some of them.

Case VII. Suppose that John and Mary are in the park, observing several dogs (all equally salient) playing and running together. John has the intention of showing Mary his dog Fido; to help her discriminate his dog among all the other dogs, he tells her that Fido has a bad limp. Then, pointing at Fido, he utters:

(7) *That dog is Fido.*

The reference of the expression *that dog* if Fido, hence (7) is true.

Case VIII. Like case VII, with the following exceptions: Fido clearly has no limp, but another dog, Spot, clearly has. Though Fido is in the most direct line with John's finger, John could possibly be taken as pointing, perhaps not too precisely, at Spot. Limping is the most relevant contextual information for discriminating the referent; in case VIII the reference of the expression *that dog* if Spot, hence (7) is false.

Case IX. Like case VII, with the following exceptions: John has been telling Mary many distinctive features Fido has: he has a bad limp, is huge, ferocious-looking, has a black leather collar with studs, and looks like a pit bull. All these things are true of Fido, except for the limp, and no other dog in the park is remotely like that, especially Spot, who has a bad limp, but is small, frail, with a red collar, and looks like a French poodle. In this scenario, Mary has enough independent contextual information to discriminate the reference of *that dog*: the reference is Fido, hence (7) is true.

It seems that the speaker's intentions are *neither necessary nor sufficient* to fix the reference of a demonstrative. In case VIII, the reference (Spot) is fixed *despite* John's intentions - which have Fido as object. In case IX, the reference (Fido) is fixed *independently* of John's intentions: even if John associates no intention with his use of the demonstrative, the reference would be discriminated by the information previously given. Not any intention, then, is a good candidate to fix the reference of a demonstrative. Let's examine one last case.

Case X. Like case VII, with the following exception: Spot has made himself especially salient by his hysterical barking. Suppose that John utters (7) with the intention of referring to Fido - a dog non-salient John is not pointing at. In this context, John's intention of referring to Fido, using no gesture, nodding, nor glancing, would be *bizarre*, i. e. unconnected with a context or a behaviour that would enable Mary to discriminate the intended dog.

7 Good Intentions

IPC, as I interpret it, requires communicative intentions to be non-arbitrary – that is connected with a behaviour that will enable the addressee to identify the referent.[18] In other words, an intention, to be semantically relevant, must satisfy what I propose to call an *Availability Constraint*, that is it must be communicated or made available to

---

[18] On this point, see [28], p. 198: Roberts speaks of "reasonable referential intentions", basing his argument on Donnellan's treatment of reasonable expectations and intentions: "On Donnellan's view... one's intentions are limited by reasonable expectations, which in turn are limited by established practices and particular stipulations" (p.196); cf. [11], pp. 212-214.

the addressee.[19] Mary can't recognise any intention John could have: she can't read John's mind. In case X, the only manifest basis for Mary to identify John's communicative intention is the presence, in the context of utterance, of a dog having made himself especially salient (for instance by his hysterical barking).

Let me state my point once again, in a slightly different way.[20] According to Reimer there are only two plausible accounts of the *proposition* John's words express in case X:
- a) Spot belongs to John;
- b) Fido belongs to John.

Following Bach's theory of communicative intentions, we should say that in case X the proposition John's words express is:
- c) the relevant dog belongs to John,

or
- c') the dog John succeeded in calling Mary's attention to belongs to John.

Since the relevant dog is the one barking hysterically, Spot, and since Spot doesn't belong to John, (7) is false.

Likewise in Kaplan's classic example

(6) *That is a picture of one of the greatest philosophers of the twentieth century,*

there are three accounts of the proposition expressed by John's words:
- a) the picture of Agnew is a picture of one of the greatest philosophers of the twentieth century;
- b) the picture of Carnap is a picture of one of the greatest philosophers of the twentieth century;
- c) the picture on the wall behind him is a picture of one of the greatest philosophers of the twentieth century.

c) is the proposition expressed by (6): since the picture on the wall behind John is Agnew's portrait, (6) cannot be taken as true. The proposition c) can account both for what John's words express and for what John wants to convey. b) is the proposition that John expects Mary to *infer* on the basis of the proposition c) – which is the proposition his words *express*: c) satisfies the Availability Constraint, but b) doesn't.

Not *any* intention satisfies the Availability Constraint, just the "good" ones. A "good" communicative intention is something an addressee, in normal circumstances, is able to work out using
1. external facts,
2. linguistic co-text,
3. background knowledge.

Of course, those three kinds of contextual information are nothing more than a way of spelling out relevance.[21]

---

[19] But, in my opinion, not to *any* competent speaker, as Garcia Carpintero proposes; cf. [12], p. 537: "I will take demonstrations to be sets of *deictical intentions* manifested in features of the context of utterance available as such to any competent user". On this point, see [8], chapter X; Marina Sbisà suggests to extend this availability constraint to all the "relevant participants" (personal communication).

[20] I am indebted to Chris Gauker for helping me reformulating my argument in the following way.

[21] I am well aware that relevance needs a definition far more accurate than the one given in this paper: for a more detailed analysis, see [8].

1. First, we have the information inferred from the extralinguistic or physical context - available to both speaker and addressee. As we said, the demonstrative expression *that dog* doesn't require any action on the speaker's part if the dog he intends to refer to is the only dog in the context of utterance, or the only dog among cats and birds, or the most salient dog (for "external" reasons, as, for example, his behaviour) in the context of utterance.

2. Second, we have the information inferred from the linguistic co-text. Suppose that, during the conversation in the park, John and Mary mention Spot; in this case a demonstrative (non anaphorical) use of

(8) *That dog costs a fortune*

will refer quite naturally to Spot. Notice that it is possible to build more sophisticated examples, referring not only to objects explicitly mentioned in the previous conversation, but only presupposed. In the same situation, if John utters

(9) *That collar costs a fortune*

the demonstrative expression *that collar* will refer to Spot's collar, even if no collar was already mentioned in the conversation.

3. Third, we have the information inferred from the knowledge shared by speaker and addressee, because they belong to the same community, or to the same subcommunity. Just think to the vertiginous amount of information two friends share, and may take as basis for the recognition of their interlocutor's communicative intentions. Suppose that John loves big, ferocious dogs, and Mary knows it. They are in a park observing several dogs all equally salient (for external reasons), and John utters

(10) *That dog is mine*:

Mary will easily determine the reference of *that dog* if there are dozens of French poodles but only one Rottweiler.

8 Conclusion

In my paper, I have presented two competing perspectives on the problem of the determination of the demonstrative reference - OPC and IPC - and I have tried to offer a fair reconstruction of IPC. According to Kaplan 1989, the addressee must take into account the speaker's intentions to identify the reference of the demonstratives. In my paper, the analysis of demonstrative reference has provided a reliable test for our intuitions on communicative mechanisms, and more specifically on the relation between objective and intentional context. Therefore, this analysis has been the starting point for a more general reflection on the notion of communicative intention. Examples have been provided to argue that the speaker's communicative intentions can play a semantic role only if they satisfy an Availability Constraint, that is to say if they are reasonable and not arbitrary, and can be recognised by the addressee: reference is determined by public behaviour, by intentional *acts* and not by intentions as mental objects.[22] In other words, to be semantically relevant, an intention must be made available or communicated to the addressee, and for that purpose the speaker can exploit any feature of the objective context - words, gestures, relevance or uniqueness of the referent in the context of utterance: elements of the intentional context can be identi-

---

[22] Cf. [28], p. 199.

fied only *through* the identification of elements of the objective context.[23] This thesis implies the reconciliation between "Demonstratives" – in which Kaplan claims that the occurrence of a demonstrative must be supplemented by a *demonstration*, like a pointing (a feature of the objective context) - and "Afterthoughts" – in which, conversely, Kaplan argues that the occurrence of a demonstrative must be supplemented by a *directing intention*, the referential intention the speaker associate with the expression (a feature of the intentional context).[24]

References

1. Almog, J., Perry, J. and Wettstein, H. (eds.): Themes from Kaplan. Oxford University Press, Oxford (1989)
2. Bach, K.: Communicative Intentions, Plan Recognition, and Pragmatics: Comments on Thomason and on Litman and Allen. In Cohen Ph. et al. (eds.): Intentions in Communication. M.I.T. Press, Cambridge (MA) (1990) 389–400
3. Bach, K.: Paving the road to reference. Philosophical Studies **67** (1992) 295–300
4. Bach, K.: Intentions and Demonstrations. Analysis **52** (1992) 140–146
5. Bach, K. and Harnish, R. M.: Linguistic Communications and Speech Acts. M.I.T. Press, Cambridge (MA) (1979)
6. Bach, K. and Harnish, R. M.: How Performatives Really Work: A Reply to Searle. Linguistics and Philosophy **15** (1992) 93–110
7. Bianchi, C.: Tree Forms of Contextual Dependence. In P. Bouquet et al. (eds.): Modeling and Using Context, Second International and Interdisciplinary Conference, Context'99, Proceedings. Springer-Verlag, Berlin Heidelberg New York (1999) 67–76
8. Bianchi, C.: La dipendenza contestuale. Per una teoria pragmatica del significato. Edizioni Scientifiche Italiane, Napoli (2001)
9. Bianchi, C.: Context of Utterance and Intended Context. In V. Akman et al. (eds): Modeling and Using Context. Third International and Interdisciplinary Conference, Context '01. Springer-Verlag, Berlin Heidelberg New York (2001) 73–86
10. Clark, H.: Arenas of Language Use. The University of Chicago Press & CSLI (1992)
11. Donnellan, K.: Putting Humpty Dumpty Together Again. The Philosophical Review **77** (1968) 203–215
12. Garcia-Carpintero, M.: Indexicals as Token-Reflexives. Mind **107** (1998) 529–563.
13. Gauker, Ch.: Domain of Discourse. Mind **106** (1997) 1–32
14. Gauker, Ch.: What is a context of utterance? Philosophical Studies **91** (1998) 149–172
15. Hale, B. and Wright, C. (eds.): A Companion to the Philosophy of Language. Blackwell, Oxford (1997)
16. Kaplan, D.: Demonstratives. An Essay on the Semantics, Logic, Metaphysics, and Epistemology of Demonstratives and Other Indexicals. In [1] 481–563
17. Kaplan, D.: Dthat. In P. French, T. Uehling and H. Wettstein (eds.): Contemporary Perspectives in the Philosophy of Language (1979) 383–400
18. Kaplan, D.: Afterthoughts. In [1] 565–614
19. Lewis, D.: General Semantics. Synthese **22** (1970) 18–67

---

[23] Of course one might object to my reconstruction - saying that, if only available intentions have semantic import, it is what makes them available and not their being real intentions in somebody's mind that counts. This would amount to say that the very notion of intention is problematic. I am indebted to an anonymous referee for this observation.

[24] I wish to thank Chris Gauker, Carlo Penco, Stefano Predelli, Marina Sbisà and Nicla Vassallo for extensive discussions on many points related to the topic of this paper.

20. Montague, R.: Pragmatics. In Montague, R.: Formal Semantics. Yale University Press, New Haven (1974) 95–118
21. Penco, C.: Objective and Cognitive Context. In P. Bouquet et al. (eds.): Modeling and Using Context, Second International and Interdisciplinary Conference, Context'99, Proceedings. Springer-Verlag, Berlin Heidelberg New York (1999) 270–283
22. Penco, C.: Context and Contract. In P. Bouquet and L. Serafini (eds.): Perspectives on Context. CSLI, Stanford (2003) forthcoming
23. Perry, J.: Indexicals and Demonstratives. In [15] 586–612
24. Perry, J.: Indexicals, Contexts and Unarticulated Constituents. Proceedings of the 1995 CSLI-Amsterdam Logic, Language and Computation Conference. CSLI Publications, Stanford (1998) 1–16
25. Reimer, M.: Demonstratives, Demonstrations, and Demonstrata. Philosophical Studies **67** (1991) 187–202
26. Reimer, M.: Do Demonstratives Have Semantic Significance? Analysis **51** (1991) 177–183
27. Reimer, M.: Demonstrating with Descriptions. Philosophy and Phenomenological Research **52** (1992) 877–893
28. Roberts, L. D.: How Demonstrations Connect with Referential Intentions. Australasian Journal of Philosophy **75** (1997) 190–200
29. Sbisà, M.: Speech acts in context. Language and Communication (2003) forthcoming
30. Stalnaker, R.: Context and Content. Oxford University Press, Oxford (1999)
31. Travis, Ch.: Saying and Understanding. Blackwell, Oxford (1975)
32. Travis, Ch.: The True and the False: the Domain of Pragmatics. Benjamins, Amsterdam (1981)

A SAT-Based Algorithm for Context Matching

Paolo Bouquet[1,2], Bernardo Magnini[2], Luciano Serafini[2], and Stefano Zanobini[1]

[1] Department of Information and Communication Technologies – University of Trento
Via Inama, 5 – 38100 Trento (Italy) {bouquet,zanobini}@dit.unitn.it

[2] ITC-IRST – Istituto per la Ricerca Scientifica e Tecnologica
Via Sommarive, 14 – 38050 Trento (Italy)
{magnini,serafini}@itc.it

Abstract. The development of more and more complex distributed applications over large networks of computers has raised the problem of *semantic interoperability* across applications based on local and autonomous semantic schemas (e.g., concept hierarchies, taxonomies, ontologies). In this paper we propose to view each semantic schema as a context (in the sense defined in [1]), and propose an algorithm for automatically discovering relations across contexts (where relations are defined in the sense of [7]). The main feature of the algorithm is that the problem of finding relationships between contexts is encoded as a problem of logical satisfiability, and so the discovered mappings have a well–defined semantic. The algorithm we describe has been implemented as part of a peer-to-peer system for Distributed Knowledge Management, and tested on significant cases.

1 Introduction

The development of more and more complex distributed applications over large networks of computers has created a whole new class of conceptual, technical, and organizational problems. Among them, one of the most challenging one is the problem of *semantic interoperability*, namely the problem of allowing the exchange of meaningful information/knowledge across applications which (i) use autonomously developed conceptualizations of their domain, and (ii) need to collaborate to achieve their users' goals.

Essentially, there are two main approaches for solving the problem of semantic interoperability. The first is based on the availability of shared semantic structures (e.g., ontologies, global schemas) onto which local representations can be totally or partially mapped. The second is based on the creation of a global representation which integrates local representations. Both approaches do not seem suitable in scenarios where: (i) local representations are updated and changed very frequently, (ii) each local representation is managed in full autonomy w.r.t. the other ones, (iii) local representations may appear and disappear at any time, (iv) the discovery of semantic relation across different representations can be driven by a user's query, and thus cannot be computed beforehand (runtime discovery) nor take advantage of human intervention (automatic discovery).

In this paper we propose an approach in which local schemas are viewed as contexts, namely as partial and approximate representations of the world from an individual's or a group's perspective [1] (two simple examples of schemas are the two directory structures from Google and Yahoo in Figure 1). This approach, which is motivated by the work on Distributed Knowledge Management (DKM) [4,3], is based on the assumption that a successful knowledge–based application should not "force" people to change their way of looking at things (encoded, for example, in a database schema or in the classification of a document management system), as the imposed schema would be perceived "either as oppressive or irrelevant" [13]. Thus, from our perspective, local schemas play the role of a lens through which people look at the world and make sense of it. In a word, a schema is the context in which facts are taken as true, decisions are made, objects are classified, relations among things are asserted and understood.

The problem of such a vision is that communication across different local schemas (contexts) becomes difficult. The algorithm we present in this paper is precisely a first solution to the problem of runtime and automatic discovery of semantic relations across autonomous contexts. More specifically, we start from a broad family of schemas (called concept hierarchies), and present a method for discovering the type of relation existing between two nodes (each representing a concept) belonging to different schemas. The main feature of the algorithm is that the problem of finding relations between concepts in different contexts is encoded as a problem of logical satisfiability of a set of formulae. This allows us to assign a precise semantic to each discovered mapping. In particular, we claim that the correct semantic for a mapping between concepts of different contexts is in terms of a compatibility relation (as defined in [7]), namely as a constraint on the local interpretations of the two contexts that are compatible with each others. In this sense, the algorithm we present is a first attempt to discover (rather than assume) relations over local models of two or more contexts (which, from a proof–theoretical point of view, corresponds to discover "bridge rules" [8] across contexts).

The paper goes as follows. First, we characterize the scenarios that motivate our approach, and explain why we use the theory of context as a theoretical background of the algorithm. Then, we describe the macro–blocks of the algorithm, namely semantic explicitation and context mapping via SAT. Finally, we describe the results of our preliminary tests and briefly compare our algorithm with some other proposals in the literature.

2 Motivating Scenarios

The work on the algorithm was originally motivated by a research on Distributed Knowledge Management [4], namely a distributed approach to managing corporate knowledge in which users (or groups of users, e.g. communities) are allowed to organize their knowledge using autonomously developed schemas (e.g., directories, taxonomies, corporate ontologies), and are then supported in finding relevant knowledge in other local schemas available in the corporate network.

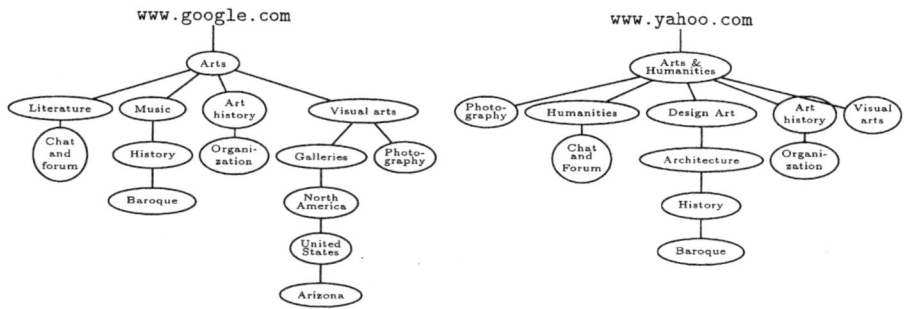

Fig. 1. Examples of concept hierarchies (source: Google and Yahoo)

In this scenario, the algorithm we present aims at solving the following problem. Let s (the *source schema*) and t (the *target schema*) be two autonomous schemas that different users (or groups) use to organize and access a local body of data. Given a concept k_s in s, and a concept k_t in t, what is the semantic relations between k_s and k_t? For example, are the two concepts equivalent? Or one is more (less) general than the other one? In addressing this problem, it is assumed that the basic elements of each schema are described using words and phrases from natural language (e.g., English, Italian); this reflects the intuition that schemas encode a lot of implicit knowledge, which can be made explicit only if one has access to the meaning of the words that people use to denote concepts in the schema.

Scenarios with similar features can be found in other important application domains, such as the semantic web (where each site can have a semantic description of its contents and services), marketplaces (where every participating company may have a different catalog, and every marketplace may adopt a different standard for cataloging products); search engines (some of them , e.g. Google and Yahoo, provide heterogeneous classifications of web pages in web directories); the file system on the PCs of different users (where each user stores documents in different directory structures). So the class of applications in which our algorithm can be applied is quite broad.

3 Local Schemas as Contexts

In many interesting applications, schemas are directed graphs, whose nodes and edges are labeled with terms or phrases from natural language. A typical example is depicted in Figure 1, whose structures are taken from the Google and Yahoo directories. In this section, we briefly argue why we interpret these schemas as contexts in the sense of [1] (see [7] for a formalization).

In schemas like the ones in the figure, the meaning of a label depends not only on its linguistic meaning (what a dictionary or thesaurus would say about that word or phrase), but also on the context in which it occurs: first, it depends on the position in the schema (e.g., the documents we as humans expect to find under the concept labeled *Baroque* in the two structures in Figure 1 are quite different, even if the label is the same, and is used in the same linguistic sense); second, it depends on background knowledge about the schema itself (e.g., that there are chat and forums about literature helps in understanding the implicit relation between these two concepts in the left hand side schema). These contextual aspects of meaning are distinct (though related) to purely linguistic meaning, and we want to take them into account in our algorithm.

For this purpose, the algorithm we present in this paper is applied to contexts rather than to schemas directly. In [1], a context is viewed as a box, whose content is an explicit (partial, approximate) representation of some domain, and whose boundaries are defined by a collection of assumptions which hold about the explicit representation. The notion of context we use in this paper is an special case of the notion above. A context is defined as a pair $c = \langle R_c, A_c \rangle$, where:

1. R_c is a graph, whose nodes and edges can be labeled with expressions from natural language;
2. A_c is a collection of explicit assumptions, namely attributes (parameter/value pairs) that provide meta-information about the content of the context.

In the current version of the algorithm, we restrict ourselves to the case in which R_c is a concept hierarchy (see Def. 1), and the explicit assumptions A_c are only three: the *id* of the natural language in which labels are expressed (e.g., English, Italian), the reference structure R_c of the explicit representation (the only accepted value, at the moment, is "concept hierarchy", but in general other values will be allowed, e.g., taxonomy, ontology, semantic network, frame), and the domain theory (see below for an explanation of this parameter). Their role will become apparent in the description of the algorithm.

A concept hierarchy is defined as follows:

Definition 1 (Concept hierarchy). *A concept hierarchy is a triple $H = \langle K, E, l \rangle$ where K is a finite set of nodes, E is a set of arcs on K, such that $\langle K, E \rangle$ is a rooted tree, and l is a function from $K \cup E$ to a set L of strings.*

Definition 2 (Hierarchical classification). *A hierarchical classification of a set of documents D in a concept hierarchy $H = \langle K, E, l \rangle$ is a function $\mu : K \to 2^D$.*

μ satisfies the following *specificity principle*: a user classifies a document d under a concept k, if d is about k (according to the user) and there isn't a more specific concept k' under which d could be classified[1].

---

[1] See Yahoo instruction for "Finding an appropriate Category" at http://docs.yahoo.com/info/suggest/appropriate.html.

Mappings between contexts are defined as follows:

Definition 3 (Mapping function). *A mapping function M from $H = \langle K, E, l \rangle$ to $H' = \langle K', E', l' \rangle$ is a function $M : K \times K' \to rel$, where rel is a set of symbols, called the* possible mappings.

The set rel of possible mappings we consider in this paper contains the following: $k_s \xrightarrow{\supseteq} k_t$, for k_s is more general than k_t; $k_s \xrightarrow{\subseteq} k_t$ for k_s is less general than k_t; $k_s \xrightarrow{*} k_t$ for k_s is compatible with k_t; $k_s \xrightarrow{\perp} k_t$ for k_s is disjoint from k_t; $k_s \xrightarrow{\equiv} k_t$ for k_s is equivalent to k_t. The formal semantics of these expressions is given in terms of compatibility between document classifications of H_s and H_t:

Definition 4. *A mapping function M from H_s to H_t is extensionally correct with respect to two hierarchical classifications μ_s and μ_t of the same set of documents D in H_s and H_t, respectively, if the following conditions hold for any $k_s \in K_s$ and $k_t \in K_t$:*

$$k_s \xrightarrow{\supseteq} k_t \Rightarrow \mu_s(k_s\downarrow) \supseteq \mu_t(k_t\downarrow)$$
$$k_s \xrightarrow{\subseteq} k_t \Rightarrow \mu_s(k_s\downarrow) \subseteq \mu_t(k_t\downarrow)$$
$$k_s \xrightarrow{\perp} k_t \Rightarrow \mu_s(k_s\downarrow) \cap \mu_t(k_t\downarrow) = \emptyset$$
$$k_s \xrightarrow{\equiv} k_t \Rightarrow \mu_s(k_s\downarrow) = \mu_t(k_t\downarrow)$$
$$k_s \xrightarrow{*} k_t \Rightarrow \mu_s(k_s\downarrow) \cap \mu_t(k_t\downarrow) \neq \emptyset$$

where $\mu(c\downarrow)$ is the union of $\mu(d)$ for any d in the subtree rooted at c.

The semantics introduced in Definition 4 can be viewed as an instance of the compatibility relation between contexts as defined in Local Models Semantics [7, 5]. Indeed, suppose we take a set of documents D as the domain of interpretation of the local models of two contexts c_1 and c_2, and each concept as a unary predicate. If we see the documents associated to a concept as the interpretation of a predicate in a local model, then the relation we discover between concepts of different contexts can be viewed as a compatibility constraint between the local models of the two concepts. For example, if the algorithm returns an equivalence between the concepts k_1 and k_2 in the contexts c_1 and c_2, then it can be interpreted as the following constraint: if a local model of c_1 associates a document d to k_1, then any compatible model of c_2 must associate d to k_2 (and vice versa); analogously for the other relations.

4 The Matching Algorithm

The algorithm has two main phases:

Semantic explicitation. In the schema level, a lot of information is implicit in the labels, and in the structure. The objective of this first phase is to make it as explicit as possible by associating to each node (and edge) k a logical formula $w(k)$ that encodes this information. Intuitively, $w(k)$ is an approximation of the human interpretation.

Semantic comparison. We encode the problem of finding mappings between two concepts k and k', whose explicit meaning is $w(k)$ and $w(k')$, into a problem of satisfiability, which is then solved by a SAT solver in a logic W (i.e., the logic in which $w(c)$ and $w(c')$ are expressed). Domain knowledge is also encoded as a set of formulas of W.

Since here we are mainly focussed on the second phase, we only provide a short description of semantic explicitation (details can be found in [10]), and then move to the SAT encoding.

4.1 Semantic Explicitation

The goal of the first phase is to make explicit all the semantic information which can be fruitfully used to define the SAT problem in a rich way. The main intuition is that any schema is interpreted (by its users) using two main sources of information: lexical information, which tells us that a word (or a phrase) can have multiple senses, synonyms, and so on; and a background theory, which provides extra-linguistic information about the concepts in the schema, and about their relations. For example, lexical information about the word "Arizona" tells us that it can mean "a state in southwestern United States" or a "glossy snake". The fact that snakes are animals (reptiles), that snakes are poisonous, and so can be very dangerous, and so on, are part of a background theory which one has in mind when using the word "Arizona" to mean a snake[2]. In the version of the algorithm we present here, we use WORDNET as a source both of lexical and background information about the labels in the schema. However, we'd like to stress the fact that the algorithm does not depend on the choice of any particular dictionary or theory (i.e., does not depend on WORDNET). Moreover, we do not assume that the same dictionary and background theory are used to explicit the semantic of the two contexts to be matched.

Semantic explicitation is made in two main steps: *linguistic interpretation* and *contextualization*.

Linguistic interpretation. Let $H = \langle K, E, l \rangle$ be a concept hierarchy and L_H the set of labels associated to the nodes and edges of a hierarchy H by the function l. In this phase we associate to each label $s \in L_H$ a logical formula representing the interpretation of that label w.r.t. the background theory we use.

---

[2] We are not saying here that there is only one background theory. On the contrary, theories tend to differ a lot from individual to individual, and this is part of the reason why communication can fail. What we are saying is that, to understand what "Arizona" means in a schema (such as the concept hierarchy in the left hand side of Figure 1), one must have a theory in mind.

Definition 5 (Label interpretation). *Given a logic W, a label interpretation in W is a function $I : L_H \to \mathit{wff}(W)$, where $\mathit{wff}(W)$ is the set of well formed formulas of W.*

The choice of W depends on the external assumptions of the context containing H. For concept hierarchies, we adopted a description logic W with $\sqcup$, $\sqcap$ and $\neg$, whose primitive concepts are the synsets of WORDNET that we associate to each label (with a suitable interpretation of conjunctions, disjunctions, multiwords, punctuation, and parenthesis). For example, WORDNET provides 2 senses for the label *Arizona* in Figure 1, denoted by #1 and #2; in this case, the output of the linguistic analysis is the following formula in W: `Arizona#1` $\sqcup$ `Arizona#2`

Contextualization. Linguistic analysis of labels is definitely not enough. The phase of contextualization aims at pruning or enriching the synsets associated to a label in the previous phase by using the context in which this label occurs. In particular, we introduce the concept of *focus* of a concept k, namely the smallest subset of H which we need to consider to determine the meaning of k. What is in the focus of a concept depends on the structure of the explicit representation. For concept hierarchies, we use the following definition:

Definition 6 (Focus). *The* focus *of a concept $k \in K$ in a concept hierarchy $H = \langle K, E, l \rangle$, is a finite concept hierarchy $f(k, H) = \langle K', E', l' \rangle$ such that: $K' \subseteq K$ contains k, its ancestors, and their direct descendants; $E' \subseteq E$ is the set of edges between the concepts of K'; l' is the restriction of l on K'.*

The *contextualization* of the interpretation of concept k of a context c is formula $w(k)$, called *contextualized interpretation* of k, which is computed by combining the linguistic interpretations associated to each concept h in the focus of k. The two main operations performed to compute $w(k)$ are sense filtering and sense composition.

Sense filtering uses NL techniques to discard synsets that are not likely to be correct for a label in a given focus. For example, the sense of *Arizona* as a snake can be discarded as it does not bear any explicit relation with the synsets of the other labels in the focus (e.g., with the synsets of *United States*), whereas it bears a part-of relation with `United States#1` (analogously, we can remove synsets of *United States*).

Sense composition enriches the meaning of a concept in a context by combining in linguistic interpretation with structural information and background theory. For concept hierarchies, we adopted the default rule that the contextual meaning of a concept k is formalized as the conjunction of the senses associated to all its ancestors. Furthermore, some interesting exceptions are handled. An example: in the Yahoo Directory, *Visual arts* and *Photography* are sibling nodes under *Arts & Humanities*; since in WORDNET photography is in a is–a relationship with visual art, the node *Visual arts* is re-interpreted as visual arts minus photography, and is then formalized in description logic as: `visual art#1`$\sqcup \neg$ `photography#1`

4.2 Computing Relations between Concepts via SAT

In the second phase of the algorithm, the problem of discovering the relationship between a concept k in a context c and a concept k' in a context c' is reduced to the problem of checking, via SAT, a set of logical relations between the formulas $w(k)$ and $w(k')$ associated to k and k'. The SAT problem is built in two steps. First, we select the portion T of the background theory relevant to the contextualized interpretation $w(k)$ and $w(k')$, then we compute the logical relation between $w(k)$ and $w(k')$ which are implied by T.

Definition 7. *Let $\phi = w(k)$ and $\psi = w(k')$ be the contextualized interpretation of two concepts k and k' of two contexts c and c', respectively. Let B be a theory (= logically closed set of axioms) in the logic where ϕ and ψ are expressed. The portion of B relevant to ϕ and ψ, is a subset T of B such that T contains all the axioms of B containing some concept occurring in ϕ or ψ.*

Clearly different contexts can be associated to different background theories, which encodes general and domain specific information. This information is stored in the context external assumptions under the field "domain". Furthermore, when we determine the mapping between two contexts c_s and c_t we can take the perspective (i.e., the background theory) of the source or that of the target. The two perspectives indeed might not coincide. This justifies the introduction of directionality in the mapping. I.e. $c_s \overset{\subseteq}{\longrightarrow} c_t$ means that c_s is more general than c_t according to the target perspective; while the relation $c_t \overset{\supseteq}{\longrightarrow} c_s$ represent the fact that c_s is more general that c_t according to the source perspective.

In the first version of our matching algorithm we consider a background theory B determined by transforming the WORDNET relations in a set of axioms in description logic, as shown in Table 1. In this table we introduce the notation $\equiv_w$, $\leq_w$, $\geq_w$, and $\perp_w$ to represent the following relation between senses stored in WORDNET.

1. s#k $\equiv_w$ t#h: s#k and t#h are synonyms (i.e., they are in the same synset);
2. s#k $\leq_w$ t#h: s#k is either a hyponym or a meronym of t#h;
3. s#k $\geq_w$ t#h: s#k is either a hypernym or a holonym of t#h;
4. s#k$\perp_w$t#h: s#k belongs to the set of opposite meanings of t#h (if s#k and t#h are adjectives) or, in case of nouns, that s#k and t#h are different hyponyms of the same synset.

In the extraction of the theory B from WORDNET we adopt a certain heuristic which turns out to perform satisfactory (see section on experimentation and evaluation). However, different sources as, specific domain ontologies, domain taxonomies, etc. and different heuristics can be used to build the theory B, from which T is extracted.

Going back to how we build the theory B, suppose, for example, that we want to discover the relation between *Chat and Forum* in the Google directory

Table 1. Encoding WORDNET relations in T-Box axioms

WORDNET relation	Domain axiom
t#k $=_w$ s#h	t#k $\equiv$ s#h
t#k $\leq_w$ s#h	t#k $\sqsubseteq$ s#h
t#k $\geq_w$ s#h	t#k $\sqsupseteq$ s#h
t#k $\perp_w$ s#h	$\neg$t#k $\sqsubseteq$ s#h

Table 2. Verifying relations as a SAT problem

relation	SAT Problem
$k_s \xrightarrow{\supseteq} k'_t$	$T_t \models w(k_t) \sqsubseteq w(k_s)$
$k_s \xrightarrow{\subseteq} k_t$	$T_t \models w(k_s) \sqsubseteq w(k_t)$
$k_s \xrightarrow{\perp} k_t$	$T_t \models w(k_s) \sqcap w(k_t) \sqsubseteq \perp$
$k_s \xrightarrow{\equiv} k_t$	$T_t \models w(k_t) \sqsubseteq w(k_s)$ and $T_t \models w(k_s) \sqsubseteq w(k_t)$
$k_s \xrightarrow{*} k_t$	$w(k_s) \sqcap w(k_t)$ is consistent in T_t

and *Chat and Forum* in the Yahoo directory in Figure 1. From WORDNET we can extract the following relevant axioms:

$$\text{art\#1} \sqsubseteq \text{humanities\#1}$$

(the sense 1 of 'art' is an hyponym of the sense 1 of 'humanities'), and

$$\text{humanities\#1} \sqsupseteq \text{literature\#2}$$

(the sense 1 of 'humanities' is an hyperonym of the sense 2 of 'literature').

The axioms extracted from WORDNET can now be used to check what mapping (if any) exists between k and k' looking at their contextualized interpretation. But which are the logical relations of $w(k)$ and $w(k')$ that encode a mapping function between k and k' as given in Definition 3? Again, the encoding of the mapping into a logical relation is a matter of heuristics. Here we propose the translation described in Table 2. In this table T_t is the portion of the background theory of c_t relevant to k_s and k_t. The idea under this translation is to see WORDNET senses (contained in $w(k)$ and $w(k')$) as sets of documents. For instance the concept art#i, corresponding to the first WORDNET sense of art, is though as the set of documents speaking about art in the first sense. Using the set theoretic interpretation of mapping given in definition 4, we have that mapping can be translated in terms of subsumption of $w(k)$ and $w(k')$. Indeed subsumption relation semantically corresponds to the subset relation.

So, the problem of checking whether *Chat and Forum* in Google is, say, less general than *Chat and Forum* in Yahoo amounts to a problem of satisfiability on the following formula:

$$\text{art\#1} \sqsubseteq \text{humanities\#1} \tag{1}$$

$$\text{humanities\#1} \sqsupseteq \text{literature\#2} \tag{2}$$

$$(\text{art\#1} \sqcap \text{literature\#2} \sqcap (\text{chat\#1} \sqcup \text{forum\#1})) \tag{3}$$

$$(\text{art\#1} \sqcup \text{humanities\#1}) \sqcap \text{humanities\#1} \sqcap (\text{chat\#1} \sqcup \text{forum\#1}) \tag{4}$$

It is easy to see that from the above axioms we can infer (3) $\sqsubseteq$ (4).

To each relation it is possible to associate also a quantitative measure. For instance the relation "c is compatible with d" can be associated with a degree, representing the percentage of models that satisfy $\phi \sqcap \psi$ on the models that satisfy $\phi \sqcup \psi$. Another example is the measure that can be associated to the relation "c is more general than d" which is the percentage of the models of that satisfy ϕ on the models that satisfy ψ. This measure give a first estimation on how much ψ is a generalization of ϕ, the lower percentage, the higher generalization.

5 Testing the Algorithm

In this section we briefly report from [11] the results of the first tests of the algorithm. We observe that the tests are performed on real schemas (i.e., pre-existing schemas that we found in real applications), and not on schemas created *ad hoc*.

5.1 Experiment 1: Generating Google's Links

The first test uses the Google web directory. It can be viewed as a concept hierarchy in which some paths in the hierarchical structure are linked to other paths (links are marked by the @-sign in the Google web page), a mechanism that allows "jumping" from a path to another in the hierarchy (a sort of symbolic link in a Unix file system). Our hypothesis is that these links can be viewed as human-defined relations between concepts, and thus can be used to validate the results of running our algorithm between concepts of the Google directory as if they were concepts of different contexts.

Since the Google directory is very large, the test was performed on the *News* sub–hierarchy, as it is relatively small and well covered by WORDNET. The result of computing 1740^2 (about 3,000,000) mappings are summarized and compared with Google's mappings in the following table:

Description	exact links found	%	wrong links found
Equivalence	7	5%	4
More + less general	3+81	56%	688
Total	91	61%	692

The table must be read as follows: the links provided by Google are on the whole 151, and the 61% has been found by the algorithm, while 60 links (39%) have been not found. Regarding the wrong links found, we can say that in the four cases the algorithm found an equivalence between concepts that were not linked in Google; we manually checked these cases, and concluded that the results of the algorithm were extremely plausible, and that the two concepts could be correctly linked in Google. For example, the algorithm found that the concept *News/Media/Media Producers/Television* is equivalent to *News/Media/Media Producers/Video*, based on the fact that one of the senses of television in WORDNET has video among its synonyms. The algorithm was not very accurate for the other two relations (precision = 11%), even though a manual verification of the "false positives" led us to conclude that in most cases they could be valuable suggestions for new Google links.

5.2 Experiment 2: Matching Google with Yahoo!

The aim of this experiment was to evaluate the CTXMATCH algorithm over pairs of overlapping structures from Google and Yahoo!. The test was performed on two pairs, those with root 'Architecture' and 'Medicine'. The results, expressed in terms of precision and recall, are reported in the following table:

Relations		Architecture Pre. Rec.	Medicine Pre. Rec.
equivalence	$\xrightarrow{\equiv}$	.71 .10	.78 .13
less general than	$\xrightarrow{\subseteq}$	.85 .49	.88 .46
more general than	$\xrightarrow{\supseteq}$	.51 .91	.60 .78

We observe that a content–based interpretation of contextual knowledge allows the discovery of non trivial mappings. For example, an inclusion mapping was computed between *Architecture/History/Periods_and_Styles/Gothic/Gargoyles* and *Architecture/History/ Medieval* as a consequence of the relation between *Medieval* and *Gothic* that can be found in WORDNET.

5.3 Experiment 3: Product Re-classification

The third test was in the domain of e–commerce. In the framework of a collaboration with a worldwide telecommunication company, the matching algorithm was applied to re-classify the catalog of the office equipment and accessories (used to classify company suppliers) into UNSPSC[3] (version 5.0.2). The validity of the relations found by the algorithm, shown in the following table, were double-checked manually.

---

[3] UNSPSC (Universal Standard Products and Services Classification) is an open global coding system that classifies products and services. UNSPSC is extensively used around the world for electronic catalogs, search engines, e–procurement applications and accounting systems.

	automatic classification[4]		after manual revision[5]	
Total items found	324	100%	324	100%
Rightly classified	197	60%	246	76%
Wrongly classified	67	21%	17	5%
Non classified	60	19%	61	19%

In particular, the automatic classification percentages are computed comparing the algorithm results with the pre-existent mappings. After manual review, the mappings automatically discovered by the algorithm improved the manual ones.

6 Related Work

Rahm and Bernstein [12] suggest that there are three general strategies for matching schemas: *instance based* (using similarity between the objects (e.g., documents) associated to the schema to infer the relationship between the concepts); *schema–based* (determining the relationships between concepts analyzing the structure of a hierarchy and the meanings of the labels); and *hybrid* (a combination of the two strategies above). Our algorithm falls in the second group. In this section, we briefly compare our method with some of the most promising schema–based methods recently proposed, namely MOMIS [2] a schema based semi automatic matcher, CUPID [9] a schema based automatic matcher and GLUE [6] an instance based automatic matcher.

The MOMIS (Mediator envirOnment for Multiple Information Sources) [2]) is a framework to perform information extraction and integration from both structured and semistructured data sources. It takes a global–as–view approach by defining a global integrated schema, starting from a set of sources schema. In one of the first phases of the integration, MOMIS supports the discovery of overlapping (relations) between the different source schema. This is done by exploiting the knowledge in a Common Thesaurus with a combination of clustering techniques and Description Logics. Another difference between the matching algorithm implemented in MOMIS and CTXMATCH is that MOMIS includes an interactive process as a step of the integration procedure, and thus does not support a fully automatic and run-time generation of mappings.

More similar to CTXMATCH is the algorithm proposed in [9], called CUPID. This is an algorithm for generic schema matching, based on a weighted combination of names, data types, constraints and structural matching. This algorithm uses a limited amount of linguistic knowledge, as it associates a thesaurus to each schema. However, unlike CTXMATCH, it does not exploit the whole power of a linguistic resource like WORDNET. Another difference between CUPID and CTXMATCH is that CUPID discovers relations between two schemas S and T only when S and the embedding of S in T are structurally isomorphic. As a

---
[4] Manually verified by ourselves.
[5] Manually verified by Alessandro Cederle Managing Director of Kompass Italia

consequence, CUPID cannot deal with concepts that are intuitively equivalent, but are represented as non isomorphic schemas.

A different approach to ontology matching has been proposed in [6]. Although the aim of the work (i.e. establishing mappings among concepts of overlapping ontologies) is in many respects similar to ours, the methodologies are significantly different. A major difference is that the GLUE system builds mappings taking advantage of information contained in instances, while the current version of the CTXMATCH algorithm completely ignores them. This makes CTXMATCH more appealing, since most ontologies currently available in the Semantic Web do not contain a significant collection of instances. A second difference concerns the use of domain-dependent constraints, which, in case of the GLUE system, need to be provided manually by domain experts, while in CTXMATCH they are automatically extracted from an already existing resource (i.e. WordNet). Finally, CTXMATCH provides a qualitative characterization of mappings in terms of the relation between two concepts, a feature which is not considered in GLUE. Even though a comparison with the results reported in [6] is rather difficult, the accuracy achieved by CTXMATCH can be roughly compared with the accuracy of the GLUE module which uses less information (i.e., the "name learner").

7 Conclusions

In the paper, we presented a first version of an algorithm for matching semantic schemas – viewed as contexts – via SAT.

We believe that this work can have a significant impact from a theoretical point of view. Indeed, the scientific challenge behind the algorithm is to determine what is the minimal common ground to enable communication between entities that do not share common meanings (at least, not in the sense of the approaches that assume the necessity of a shared ontology to enable communication). As a consequence, the relations discovered by the algorithm are always directional (from a concept in a context to concept in another context, but not vice versa), and this reflects the idea that what is a good mapping from the point of view encoded in a context might not be acceptable from the point of view encoded in the other context.

Of course, a lot of work remains to be done, and in particular: generalizing the types of structures we can match (beyond concept hierarchies); taking into account a larger collection of explicit assumptions; going beyond WORDNET as a source of linguistic and domain knowledge.

References

1. M. Benerecetti, P. Bouquet, and C. Ghidini. Contextual Reasoning Distilled. *Journal of Theoretical and Experimental Artificial Intelligence*, 12(3):279–305, July 2000.
2. Sonia Bergamaschi, Silvana Castano, and Maurizio Vincini. Semantic integration of semistructured and structured data sources. *SIGMOD Record*, 28(1):54–59, 1999.

3. M. Bonifacio, P. Bouquet, and R. Cuel. Knowledge Nodes: the Building Blocks of a Distributed Approach to Knowledge Management. In *To be appear in a special issue of the J.UCS journal and in the Proceedings of I-KNOW '02 - 2nd International Conference on Knowledge Management.*, Gratz - Austria, July 11–12, 2002. Springer Pub. & Co.
4. M. Bonifacio, P. Bouquet, and P. Traverso. Enabling distributed knowledge management. managerial and technological implications. *Novatica and Informatik/Informatique*, III(1), 2002.
5. A. Borgida and L. Serafini. Distributed description logics: Directed domain correspondences in federated information sources. In R. Meersman and Z. Tari, editors, *On The Move to Meaningful Internet Systems 2002: CoopIS, Doa, and ODBase*, volume 2519 of *LNCS*, pages 36–53. Springer Verlag, 2002.
6. A. Doan, J. Madhavan, P. Domingos, and A. Halevy. Learning to map between ontologies on the semantic web. In *Proceedings of WWW-2002, 11th International WWW Conference, Hawaii*, page ......, 2002.
7. C. Ghidini and F. Giunchiglia. Local models semantics, or contextual reasoning = locality + compatibility. *Artificial Intelligence*, 127(2):221–259, April 2001.
8. F. Giunchiglia and L. Serafini. Multilanguage hierarchical logics, or: how we can do without modal logics. *Artificial Intelligence*, 65(1):29–70, 1994. Also IRST-Technical Report 9110-07, IRST, Trento, Italy.
9. Jayant Madhavan, Philip A. Bernstein, and Erhard Rahm. Generic schema matching with cupid. In *The VLDB Journal*, pages 49–58, 2001.
10. B. Magnini, L. Serafini, and M. Speranza. Linguistic based matching of local ontologies. In P. Bouquet, editor, *Working Notes of the AAAI-02 workshop on Meaning Negotiation. Edmonton (Canada)*, Edmonton, Alberta, Canada, July 2002. AAAI, AAAI Press.
11. B. M. Magnini, L. Serafini, A. Doná, L. Gatti, C. Girardi, and M. Speranza. Large–scale evaluation of context matching. Technical Report 0301–07, ITC–IRST, January 2003.
12. Erhard Rahm and Philip A. Bernstein. A survey of approaches to automatic schema matching. *VLDB Journal: Very Large Data Bases*, 10(4):334–350, 2001.
13. S. L. Star. Working together: Symbolic interactionism, activity theory, and iformation systems. In *Communication and Cognition at Work*, pages 296–318. Cambridge University Press, 1997.

On the Difference between Bridge Rules and Lifting Axioms

Paolo Bouquet[1,2] and Luciano Serafini[2]

[1] Department of Information and Communication Technologies – University of Trento
Via Inama, 5 – 38100 Trento (Italy)
bouquet@cs.unitn.it

[2] ITC-IRST – Istituto per la Ricerca Scientifica e Tecnologica
Via Sommarive – 38050 Trento (Italy)
serafini@itc.it

Abstract. In a previous paper, we proposed a first formal and conceptual comparison between the two most important formalizations of context in AI: *Propositional Logic of Context* (PLC) and *Local Models Semantics/MultiContext Systems* (LMS/MCS). The result was that LMS/MCS is at least as general as PLC, as it can be embedded into a particular class of MCS, called MPLC. In this paper we go beyond that result, and prove that, under some important restrictions (including the hypothesis that each context has finite and homogeneous propositional languages), MCS can be embedded in PLC with generic axioms. To prove this theorem, we prove that MCS cannot be embedded in PLC using only lifting axioms to encode bridge rules. This is an important result for a general theory of context and contextual reasoning, as it proves that lifting axioms and entering context are not enough to capture all forms of contextual reasoning that can be captured via bridge rules in LMS/MCS.

1 Introduction

This paper continues the investigation of formal theories of context we started in [3]. In that paper, we compared two well-known formalizations of context, namely the *Propositional Logic of Context* (PLC) [5] and *Local Models Semantics* (LMS) [7], axiomatized via Multi Context Systems [9,8] (MCS)[1]. The main technical result was that LMS/MCS is at least as general as PLC, as it can be embedded into a particular class of MCS, called MPLC.

In this paper we go beyond that result, and analyze the claim that LMS/MCS is strictly more general than PLC. The main technical results are the following: (i) under some important restrictions (including the hypothesis that each context has finite and homogeneous propositional languages), LMS/MCS can be embedded in PLC with generic axioms; (ii) LMS/MCS cannot be embedded in PLC using only lifting axioms to encode bridge rules. These results are important for a general theory of context and contextual reasoning in two senses: first,

---

[1] Hereafter, we will refer to the general framework of LMS together with its axiomatization via MCS as LMS/MCS.

the restrictions needed to prove the first theorem have a significant impact on the fulfillment of the intuitive desiderata that were brought forward to motivate the formalization of context in AI (e.g., in [11]); second, they prove that lifting axioms and entering context are not enough to capture all forms of contextual reasoning that can be captured via bridge rules in LMS/MCS.

2 The Two Systems: PLC and LMS/MCS

In this section we quickly revise the two formalisms, and prepare the ground for the technical comparison between them[2].

2.1 Propositional Logic of Context

In this paper, we use the version of PLC presented in [4]. Given a set $\mathbb{K}$ of labels, intuitively denoting contexts, the language of PLC is a multi modal language on a set of atomic propositions $\mathbb{P}$ with the modality $ist(\kappa, \phi)$ for each context (label) $\kappa \in \mathbb{K}$. More formally, the set of well formed formulae $\mathbb{W}$ of PLC, based on $\mathbb{P}$, are

$$\mathbb{W} := \mathbb{P} \cup (\neg \mathbb{P}) \cup (\mathbb{P} \supset \mathbb{P}) \cup ist(\mathbb{K}, \mathbb{P})$$

The other propositional connectives are defined as usual. If κ is a context, then the formula $ist(\kappa, \phi)$ can be read as: ϕ is true in the context κ. PLC allows to describe how a context is viewed from another context. For this PLC introduces sequences of contexts (labels). Let $\mathbb{K}^*$ denote the set of finite contexts sequences and let $\overline{\kappa} = \kappa_1 \ldots \kappa_n$ denote any (possible empty) element of $\mathbb{K}^*$. The sequence of contexts $\kappa_1 \kappa_2$ represents how context κ_2 is viewed from context κ_1. Therefore, the intuitive meaning of the formula $ist(\kappa_2, \phi)$ in the context κ_1 is that ϕ holds in the context κ_2, from the point of view of κ_1. Similar interpretation can be given to formulae in sequences of contexts longer than 2. A model for PLC associates a set of partial truth assignments to a subset of context sequences and satisfiability is defined with respect to a context sequence.

Definition 1. *A model $\mathfrak{M}$ of PLC is a partial function which maps context sequences in $\mathbb{K}^*$ into a set of partial truth assignments for $\mathbb{P}$.*

$$\mathfrak{M} \in (\mathbb{K}^* \to_p \mathbf{P}(\mathbb{P} \to_p \{\text{true}, \text{false}\}))$$

where $A \to_p B$ denotes the set of partial *functions from A to B and $\mathbf{P}(A)$ denotes the powerset of A.*

The original intuition was that, partial truth assignments allow us to represent the fact that in different context sequences there are different sets of meaningful formulae. Indeed, a model $\mathfrak{M}$ defines a vocabulary, denoted by $\mathsf{Vocab}(\mathfrak{M})$, namely, a function that associates to each context sequence a set of meaningful

---
[2] An exhaustive presentation of the two formalisms is beyond the scope of this paper; interested readers can refer to the bibliography for more details.

formulae. Formally, a *vocabulary* is a relation $\mathsf{Vocab} \subseteq \mathbb{K}^* \times \mathbb{P}$ that associates a subset of primitive propositions with each context. $\mathsf{Vocab}(\mathfrak{M})$, i.e, the vocabulary defined by the model $\mathfrak{M}$, is the function that associates to each context sequence $\overline{\kappa}$ a subset of $\mathbb{P}$ for which all the assignments in $\mathfrak{M}(\overline{\kappa})$ are defined. That is, $\langle \overline{\kappa}, p \rangle \in \mathsf{Vocab}(\mathfrak{M})$ if and only if $\mathfrak{M}(\overline{\kappa})$ is defined and, for all $\nu \in \mathfrak{M}(\overline{\kappa})$, $\nu(p)$ is defined (where ν is a truth assignment to atomic propositions).

Satisfiability and validity of formulae are defined only for these models that provides enough vocabulary, i.e. the vocabulary which is necessary to evaluate a formula in a context sequence. Each formula ϕ in a context sequence $\overline{\kappa}$ implicitly defines its vocabulary, denoted by $\mathsf{Vocab}(\overline{\kappa}, \phi)$, which intuitively consists of the minimal vocabulary necessary to build the formula ϕ in the context sequence $\overline{\kappa}$. More formally, $\mathsf{Vocab}(\overline{\kappa}, \phi)$ is recursively defined as follows:

$$\mathsf{Vocab}(\overline{\kappa}, p) = \{\langle \overline{\kappa}, p \rangle\}$$
$$\mathsf{Vocab}(\overline{\kappa}, \neg\phi) = \mathsf{Vocab}(\overline{\kappa}, \phi)$$
$$\mathsf{Vocab}(\overline{\kappa}, \phi \supset \psi) = \mathsf{Vocab}(\overline{\kappa}, \phi) \cup \mathsf{Vocab}(\overline{\kappa}, \psi)$$
$$\mathsf{Vocab}(\overline{\kappa}, ist(\kappa, \phi)) = \mathsf{Vocab}(\overline{\kappa}\kappa, \phi)$$

Definition 2 (Satisfiability and Validity). *Let ϕ and $\mathfrak{M}$ be a formula and a model respectively. ϕ is satisfied in $\mathfrak{M}$ by an assignment $\nu \in \mathfrak{M}(\overline{\kappa})$ (notationally $\mathfrak{M}, \nu \models_{\overline{\kappa}} \phi$) according to the following clauses:*

1. $\mathfrak{M}, \nu \models_{\overline{\kappa}} p$ *iff* $\nu(p) = \mathsf{true}$;
2. $\mathfrak{M}, \nu \models_{\overline{\kappa}} \neg\phi$ *iff not* $\mathfrak{M}, \nu \models_{\overline{\kappa}} \phi$;
3. $\mathfrak{M}, \nu \models_{\overline{\kappa}} \phi \supset \psi$ *iff not* $\mathfrak{M}, \nu \models_{\overline{\kappa}} \phi$ *or* $\mathfrak{M}, \nu \models_{\overline{\kappa}} \psi$;
4. $\mathfrak{M}, \nu \models_{\overline{\kappa}} ist(\kappa, \phi)$ *iff for all* $\nu' \in \mathfrak{M}(\overline{\kappa}\kappa)$, $\mathfrak{M}, \nu' \models_{\overline{\kappa}\kappa} \phi$;
5. $\mathfrak{M} \models_{\overline{\kappa}} \phi$ *iff for all* $\nu \in \mathfrak{M}(\overline{\kappa})$; $\mathfrak{M}, \nu \models_{\overline{\kappa}} \phi$;
6. $\models_{\overline{\kappa}} \phi$ *iff for all* PLC-*model* $\mathfrak{M}$, *such that* $\mathsf{Vocab}(\overline{\kappa}, \phi) \subseteq \mathsf{Vocab}(\mathfrak{M})$, $\mathfrak{M} \models_{\overline{\kappa}} \phi$.

ϕ is valid *in a context sequence $\overline{\kappa}$ if* $\models_{\overline{\kappa}} \phi$; ϕ is satisfiable *in a context sequence $\overline{\kappa}$ if there is a* PLC-*model $\mathfrak{M}$ such that $\mathfrak{M} \models_{\overline{\kappa}} \phi$. A set of formulae T is satisfiable at a context sequence $\overline{\kappa}$ if there is a model $\mathfrak{M}$ such that $\mathfrak{M} \models_{\overline{\kappa}} \phi$ for all $\phi \in T$.*

According to the above definition, vocabularies affect truth in contexts making each formula outside the vocabulary false. This implies that a PLC-model $\mathfrak{M}$ presents a non classical semantics for all the formulas ϕ such that $\langle \overline{\kappa}, \phi \rangle \not\subseteq \mathsf{Vocab}(\mathfrak{M})$. For instance, if a proposition $\langle \overline{\kappa}, p \rangle \notin \mathsf{Vocab}(\mathfrak{M})$ then $\mathfrak{M} \not\models_{\overline{\kappa}} p \vee \neg p$. This "non classical" effect however disappear in the definition of validity. For validity of a formula ϕ is checked by considering only the models whose vocabularies contain ϕ. This means that validity and satisfiability can be formulated by considering only PLC-models with *complete vocabularies*, i.e. PLC-models $\mathfrak{M}$'s with $\langle \overline{\kappa}, p \rangle \in \mathsf{Vocab}(\mathfrak{M})$ for each $p \in \mathbb{P}$ and $\overline{\kappa} \in \mathbb{K}^*$.

Theorem 1 (Reduction to complete vocabulary). *A formula is valid in* PLC *if and only if it is satisfied by all the* PLC-*models with complete vocabulary. Similarly, a formula is satisfiable in* PLC *if and only if there is a* PLC-*model with complete vocabulary that satisfies it.*

(PL) $\vdash_{\overline{\kappa}} \phi$ If ϕ is an instance of a classical tautology
(K) $\vdash_{\overline{\kappa}} ist(\kappa, \phi \supset \psi) \supset ist(\kappa, \phi) \supset ist(\kappa, \psi)$
(Δ) $\vdash_{\overline{\kappa}} ist(\kappa_1, ist(\kappa_2, \phi) \vee \psi) \supset ist(\kappa_1, ist(\kappa_2, \phi)) \vee ist(\kappa_1, \psi)$
(MP) $\dfrac{\vdash_{\overline{\kappa}} \phi \quad \vdash_{\overline{\kappa}} \phi \supset \psi}{\vdash_{\overline{\kappa}} \psi}$
(CS) $\dfrac{\vdash_{\overline{\kappa}\kappa} \phi}{\vdash_{\overline{\kappa}} ist(\kappa, \phi)}$

Fig. 1. Axioms and inference rules for PLC

Ignoring vocabularies, PLC is a multi-modal K extended with the axiom (Δ), on the set of propositions $\mathbb{P}$. Indeed the Hilbert style axiomatization of validity proposed in [4]—presented in Figure 1—is the modal system K extended with the axiom (Δ).

2.2 Local Models Semantics and Multi-context Systems

The version of LMS we present here was presented in [7]. Let $\{L_i\}_{i \in I}$ be a family of languages defined over a set of indexes I (in the following we drop the index $i \in I$). Intuitively, each L_i is the (formal) language used to describe the facts in the context i. In this paper, we assume that I is (at most) countable. Let M_i be the class of all the models (interpretations) of L_i. We call $m \in M_i$ a *local model* (of L_i).

To distinguish the formula ϕ occurring in the context i from the occurrences of the "same" formula ϕ in the other contexts, we write $i : \phi$. We say that $i : \phi$ is a labelled wff, and that ϕ is an L_i-wff. For any set of labeled formulae Γ, $\Gamma_i = \{\phi \mid i : \phi \in \Gamma\}$.

Definition 3 (Compatibility chain[3]). *A compatibility chain* $\mathbf{c} = \{c_i \subseteq M_i\}_{i \in I}$ *is a family of set of models of L_i such that each c_i is either empty or a singleton. We call c_i the i-th element of $\mathbf{c}$. A compatibility chain is* nonempty *if one of its components is nonempty.*

A compatibility chain represents a set of "instantaneous snapshots of the world" each of which is taken from the point of view of the associated context. Due to the fact that contexts describe points of view of the *same world*, certain combinations of snapshots are possible while others can never happen. To distinguish between these two sets, LMS contains the notion of *compatibility relation*—defined in the following—represents the "admissible" combinations snapshots.

Definition 4 (Compatibility relation and LMS-model). *A compatibility relation is a set of compatibility chains. A LMS-model is a compatibility relation that contains a nonempty compatibility chain.*

---
[3] For the sake of this paper, we use a definiton of compatibility chain which is specialized and simpler than the one given in [7].

Definition 5 (Satisfiability and Entailment). *Let $\models$ be the propositional classical satisfiability relation. We extend the definition of $\models$ as follows:*

1. *for any $\phi \in L_i$, $c_i \models \phi$ if, for all $m \in c_i$, $m \models \phi$;*
2. $\mathbf{c} \models i : \phi$ *if* $c_i \models \phi$;
3. $\mathbf{C} \models i : \phi$ *if, for all $\mathbf{c} \in \mathbf{C}$, $\mathbf{c} \models i : \phi$;*
4. $\Gamma_i \models_{c_i} \phi$ *if, for all $m \in c_i$, if $m \models \Gamma_i$, then $m \models \phi$;*
5. $\Gamma \models_{\mathbf{c}} i : \phi$ *if, either there is a $j \neq i$, such that $c_j \not\models \Gamma_j$, or $\Gamma_i \models_{c_i} \phi$;*
6. $\Gamma \models_{\mathbf{C}} i : \phi$ *if, for all $\mathbf{c} \in \mathbf{C}$, $\Gamma \models_{\mathbf{c}} i : \phi$;*
7. *For any class of models $\mathfrak{C}$, $\Gamma \models_{\mathfrak{C}} i : \phi$ if, for all models $\mathbf{C} \in \mathfrak{C}$, $\Gamma \models_{\mathbf{C}} i : \phi$.*

We adopt the usual terminology of satisfiability and entailment for the statements about the relation $\models$. Thus we say that $\mathbf{c}$ satisfies ϕ at i, or equivalently, that ϕ is true in c_i, to refer to the fact that $c_i \models \phi$. We say that Γ entails $i : \phi$ in $\mathbf{c}$ to refer to the fact that $\Gamma \models_{\mathbf{c}} i : \phi$. Similar terminology is adopted for $\Gamma \models_{\mathbf{C}} i : \phi$ and $\Gamma \models_{\mathfrak{C}} i : \phi$.

MultiContext Systems (MCS) [9] are a class of proof systems for LMS[4]. The key notion of an MCS is that of bridge rule.

Definition 6 (Bridge Rule). *A bridge rule on a set of indices I is a rule of the form:*

$$\frac{i_1 : \phi_1 \quad \ldots \quad i_n : \phi_n}{i : \phi} \; br$$

where $i_1, \ldots, i_n, i \in I$, A bridge rule can be associated with a restriction, namely a criterion which states the conditions of its applicability.

Definition 7 (MultiContext System (MCS)). *A MultiContext System for a family of languages $\{L_i\}$, is a pair $\mathrm{MS} = \langle \{C_i = \langle L_i, \Omega_i, \Delta_i \rangle\}, \Delta_{br} \rangle$, where each $C_i = \langle L_i, \Omega_i, \Delta_i \rangle$ is a theory (on the language L_i, with axioms Ω_i and natural deduction inference rules Δ_i), and Δ_{br} is a set of bridge rules on I.*

MCSs are a generalization of Natural Deduction (ND) systems [12]. The generalization amounts to using formulae tagged with the language they belong to. This allows for the effective use of the multiple languages. The deduction machinery of an MCS is the composition of two kinds of inference rules: *local rules*, namely the inference rules in each Δ_i, and *bridge rules*. Local rules formalize reasoning within a context (i.e. are only applied to formulae with the same index), while bridge rules formalize reasoning across different contexts.

Deductions in a MCS are trees of formulae which are built starting from a finite set of assumptions and axioms, possibly belonging to distinct languages, and by a finite number of application of local rules and bridge rules.

---

[4] In this paper, we present a definition of MC system which is suitable for our purposes. For a fully general presentation, see [9].

2.3 Lifting Axioms and Bridge Rules

A crucial feature of a formal theory of context—contained both in LMS/MCS and PLC—is the possibility to specify relations between facts of different contexts. This is an essential feature of contextual reasoning, as contexts are not simply unrelated representations, but typically are different representations of the same world. For example, two contexts may describe the same piece of the world from the same perspective, but at different level of detail; or may describe the same piece of the world, only from different perspectives. PLC formalizes relations between contexts via *lifting axioms*, while LMS/MCS uses *bridge rules*. Lifting axioms are defined as

" ... axioms which relate the truth in one context to the truth in another context. Lifting is the process of inferring what is true in one context based on what is true in another context by the means of lifting axioms" [10]

The general form of lifting axioms is the following:

$$ist(\kappa_1, \phi_1) \wedge \ldots \wedge ist(\kappa_n, \phi_n) \supset ist(\kappa, \phi) \qquad (1)$$

As any formula in PLC, lifting axioms must be stated in a context. The lifting axiom above can be intuitively read as " ϕ is true in a context κ if the formulas $\phi_1, \ldots, \phi_n$ are true in the contexts $\kappa_1, \ldots, \kappa_n$ respectively".

Bridge rules, introduced in Definition 6, are inference rules whose premises and conclusion belong to different contexts. The general form of bridge rules is described in [9], and can be though as a generalization of a Natural Deduction inference rules [12] which involve more than one index. For the sake of this paper we consider only bridge rules of the following form.

$$\frac{\kappa_1 : \phi_1 \quad \ldots \quad \kappa_n : \phi_n}{\kappa : \phi} \, br \qquad (2)$$

The above bridge rules roughly formalize the same intuition as that formalized by lifting axiom (1).

The main difference between lifting axioms and bridge rules is that lifting axioms are stated in an external context, which must be expressive enough to represent facts of all the contexts involved (using *ist*-formulae), whereas bridge rules allow stating relations between contexts without the need of an external context. There are situations where having an external context may be an advantage (for example, when one needs to reason about lifting axioms themselves, e.g. to discover that a lifting axiom is redundant, or leads to inconsistent contexts). However, in general, specifying an external context can be very costly—especially when there are many interconnected contexts—as the external context essentially duplicates the information of each context. LMS/MCS allows both solutions. Indeed, instead of using bridge rules to lift a fact ϕ from κ_1 to κ_2, one can define a third context connected with κ_1 and κ_2 via bridge rules and explicitly add an axiom like (1) to this new context[5]. This very last observation

---
[5] This approach was used, for example, in the solution to the qualification problem presented in [2].

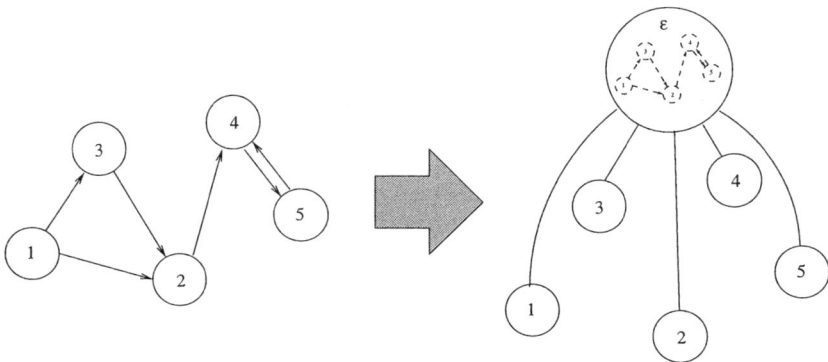

Fig. 2. Embedding LMS/MCS into PLC

constitutes the underlying idea of the proof of the fact that PLC can be embedded in LMS/MCS described in [3]. The converse question, i.e., if LMS/MCS can be reconstructed in PLC will be answered in the rest of this paper. As a consequence we will have a sharper intuition on the analogies and differences between bridge rules and lifting axioms.

3 Reconstructing LMS/MCS in PLC

Since a comparison of the two logical systems should be done on a common ground, we consider LMS/MCS with homogeneous languages, i.e., LMS/MCS whose contexts have all the same propositional language. Indeed, as it is shown by Treorem 1, PLC does not support contexts with different languages. Similarly we restrict the comparison to LMS/MCS in which all contexts have the same inference engine, which is contexts are all classical propositional theories.

The general intuition for encoding an MCS into PLC is shown in Figure 2. Given a MCS with I contexts, we define a PLC with I contexts (one for each context in MCS) and an additional (meta/external)-context ϵ. The content of each context in I and the compatibility relations (bridge rules) between contexts are described via *ist*–formulas in in ϵ. The representation of the content of the MCS contexts is quite straightforward: any formula $i : \phi$ in MCS is translated into a formula $\epsilon : ist(i, \phi)$ in PLC. For bridge rules, the translation is more tricky. Indeed, the intuition that a bridge rule like (2) is translated into the lifting axiom (1) does not work. Indeed, the following theorem proves a first important fact, namely that in general bridge rules cannot be modeled in PLC only as a set of lifting axioms. Let $\mathbb{BR}_I$ be the set of bridge rules between a set I of contexts with language $L_i = L_j$ (for $i, j \in I$).

Let $\mathbb{LA} \subset \mathbb{W}$ the set of lifting axioms among the contexts I expressed in a new context ϵ not in I. The notation $\Gamma \vdash_{br} i : \phi$ stands for: $i : \phi$ is derivable from Γ in the MCS with the set I of contexts, no axioms, and the set **br** of bridge rules.

Theorem 2. *There is no transformation* $la : \mathbb{BR} \to \mathbb{LA}$ *such that for any finite subset* $\boldsymbol{br} \subseteq \mathbb{BR}$ *of bridge rules:*

$$i_1 : \phi_1, \ldots, i_n : \phi_n \vdash_{\boldsymbol{br}} i : \phi$$
if and only if (3)
$$\vdash_\epsilon \bigwedge\nolimits_{br \in \boldsymbol{br}} la(br) \supset (ist(i_1, \phi_1) \wedge \ldots \wedge ist(i_n, \phi_n) \supset ist(i, \phi))$$

Proof. The theorem is proved by counterexample. Consider the following two bridge rules.

$$\frac{1:p}{2:q} \; br_{12} \qquad \frac{2:q}{1:r} \; br_{21} \qquad (4)$$

where p, q, and r are three distinct propositional letters. Let br_{12} and br_{21} be both unrestricted (i.e., always applicable). Considering br_{12} or br_{21} separately, they do not affect theoremhood in either context 1 and 2. Formally, for $i = 1, 2$, $\vdash_{br_{12}} i : \phi$ if and only if ϕ is a propositional tautology, and analogously $\vdash_{br_{21}} i : \phi$ if and only if ϕ is a tautology (see [6] for a proof of a similar fact). Instead, combining br_{12} and br_{21} in the same MCS, new theorems, which are not tautologies, can be proved. An example of such a theorem is $1 : p \supset r$, and its proof is the following:

$$\cfrac{\cfrac{\cfrac{1 : p^{(*)}}{2 : q} \; br_{12}}{1 : r} \; br_{21}}{1 : p \supset r} \supset I \text{(Discharging the assumption } {}^{(*)})$$

Let $la(br_{12})$ and $la(br_{21})$ be the following general conjunctions of lifting axioms:

$$la(br_{12}) = \bigwedge_{m=1}^{M} \left(\bigwedge_{k=1}^{K_m} ist(i_{mk}, \phi_{mk}) \supset ist(j_m, \psi_m) \right) \qquad (5)$$

$$la(br_{21}) = \bigwedge_{n=M+1}^{N} \left(\bigwedge_{k=1}^{K_n} ist(i_{nk}, \phi_{nk}) \supset ist(j_n, \psi_n) \right) \qquad (6)$$

where i_{mk}, i_{nk}, and j_n are either 1 or 2. Posing $\boldsymbol{br} = \{br_{12}, br_{21}\}$, we have that $\bigwedge_{br \in \boldsymbol{br}} la(br)$ is equivalent to the following formula:

$$\bigwedge_{n=1}^{N} \left(\bigwedge_{k=1}^{K_n} ist(i_{nk}, \phi_{nk}) \supset ist(j_n, \psi_n) \right)$$

Suppose, for contradiction, that equivalence (3) holds. Since $1 : p \supset r$ is derivable via br_{12} and br_{21}, we have that

$$\vdash_\epsilon \bigwedge_{br \in \boldsymbol{br}} la(br) \supset ist(i, p \supset r) \qquad (7)$$

Consider the PLC-model $\mathfrak{M}$ with $\mathfrak{M}(1)$ equal to all the assignments for L_1 and $\mathfrak{M}(2)$ equal to all the assignments for L_2. Since $p \supset r$ is not valid, there is an assignment ν such $\nu \not\models p \supset r$. By construction, $\mathfrak{M}(1)$ contains all the assignments to L_1. As a consequence $\mathfrak{M} \not\models_\epsilon ist(1, p \supset r)$. Soundness of PLC and (7) entail that $\mathfrak{M} \not\models_\epsilon \bigwedge_{br \in \boldsymbol{br}} la(br)$, and therefore, that there is an $n \leq N$ such that

$$\mathfrak{M} \models_\epsilon \bigwedge_{k=1}^{K_n} ist(i_{nk}, \phi_{nk}) \quad \text{and} \quad \mathfrak{M} \not\models_\epsilon ist(j_n, \psi_n) \tag{8}$$

The left part of (8) states that each ϕ_{nk} (with $1 \leq k \leq K_n$) is a tautology, as it must be true in all the assignments in $\mathfrak{M}(i_{nk})$. As a consequence we have that

$$\vdash_\epsilon \bigwedge_{k=1}^{K_n} ist(i_{nk}, \phi_{nk}) \tag{9}$$

The right part of (8) states that there is an assignment $\nu \in \mathfrak{M}(j_n)$ such that $\nu \not\models \psi_n$, i.e., ψ_n is not a tautology. Let us consider two cases $n \leq M$, and $n > M$. In the first case we have, due to the definiton of $la(br_{12})$, we have that

$$\vdash_\epsilon la(br_{12}) \supset \left(\bigwedge_{k=1}^{K_n} ist(i_{nk}, \phi_{nk}) \supset ist(j_n, \psi_n) \right) \tag{10}$$

while, in the second one we have:

$$\vdash_\epsilon la(br_{21}) \supset \left(\bigwedge_{k=1}^{K_n} ist(i_{nk}, \phi_{nk}) \supset ist(j_n, \psi_n) \right) \tag{11}$$

By applying Modus Ponens to (10) and (9), or to (11) and (9), we obtain one of the following two consequences:

$$\vdash_\epsilon la(br_{12}) \supset ist(j_n, \psi_n) \quad \text{or} \quad \vdash_\epsilon la(br_{21}) \supset ist(j_n, \psi_n)$$

If the equivalence holds we would have that, either $\vdash_{br_{12}} j_n : \psi_n$ or $\vdash_{br_{21}} j_n : \psi_n$, while ψ_n is not a tautology. But this is a contradiction.

Lifting axioms are not the only possible *ist*–formulas. There are *ist*–formulas, as for instance $\neg ist(i, \phi)$ or $ist(i, \phi) \supset ist(j, \psi) \vee ist(k, \theta)$, which are not lifting axioms but could be used to represent the compatibility relation formulated by bridge rules. So the question arises of whether bridge rules can be encoded by generic *ist*–formulas in some external context ϵ. In the following we show that this is the case for MCSs with a finite number of contexts and with finite languages.

Theorem 3. *There is a transformation $a(.)$ from finite sets $\boldsymbol{br} \in \mathbb{BR}_I$ of bridge rules to ist–axioms, and a context ϵ such that:*

$$\begin{array}{c} i_1 : \phi_1, \ldots, i_n : \phi_n \vdash_{\boldsymbol{br}} i : \phi \\ \textit{if and only if} \\ \vdash_\epsilon a(\boldsymbol{br}) \supset ist(i_1, \phi_1) \wedge \ldots \wedge ist(i_n, \phi_n) \supset ist(i, \phi) \end{array} \tag{12}$$

Proof. The proof is constructive, i.e., we define the transformation $a(.)$ for each set of bridge rules. The definition of $a(\boldsymbol{br})$ passes through a syntactic encoding of the LMS-models for $\boldsymbol{br}$.

Let C be a LMS-model (i.e. a set of chains), the set of PLC-models $\mathfrak{M}_C$ corresponding to C is defined as follows:

$$\mathfrak{M}_C = \left\{ \mathfrak{M}_{C'} \;\middle|\; \begin{array}{l} C' \text{ is a subset of } C \text{ such that for any } i \in I, \\ \mathfrak{M}(i) = \bigcup_{c \in C'} c_i \end{array} \right\} \quad (13)$$

Let $\mathbf{C}$ be the set of LMS-models for $\boldsymbol{br}$. The set $\mathfrak{M}_\mathbf{C}$ is defined as $\bigcup_{C \in \mathbf{C}} \mathfrak{M}_C$. Let us prove that the logical consequence defined by $\mathbf{C}$ can be represented by valid formulas in the set of models $\mathfrak{M}_\mathbf{C}$, i.e., that:

$$\begin{array}{c} i_1 : \phi_1, \ldots, i_n : \phi_n \models_\mathbf{C} i : \phi \\ \text{if and only if for all } \mathfrak{M} \in \mathfrak{M}_\mathbf{C} \\ \mathfrak{M} \models_\epsilon ist(i_1, \phi_1) \wedge \ldots \wedge ist(i_n, \phi_n) \supset ist(i, \phi) \end{array} \quad (14)$$

Suppose that $i_1 : \phi_1, \ldots, i_n : \phi_n \models_\mathbf{C} i : \phi$. Let $\mathfrak{M}_{C'} \in \mathfrak{M}_\mathbf{C}$, with $C' \subseteq C \in \mathbf{C}$. Suppose that $\mathfrak{M}_{C'} \models_\epsilon ist(i_k, \phi_k)$ for any $1 \leq k \leq n$. This implies that for all $c \in C'$, $c_{i_k} \models \phi_k$. From the hypothesis we have that $c_i \models \phi$, and therefore that $\mathfrak{M}_{C'} \models_\epsilon ist(i, \phi,)$.

Vice-versa, let us prove that $\mathfrak{M} \models_\epsilon ist(i_1, \phi_1) \wedge \ldots \wedge ist(i_n, \phi_n) \supset ist(i, \phi)$ for all $\mathfrak{M} \in \mathfrak{M}_\mathbf{C}$ implies that for any model C of $\boldsymbol{br}$ and for any chain $c \in C$, if $c_{i_k} \models \phi_k$ for $1 \leq k \leq n$, then $c_i \models \phi$. Notice that, for any $c \in C \in \mathbf{C}$ we have that $\mathfrak{M}_{\{c\}} \in \mathfrak{M}_\mathbf{C}$. By definition (see equation (13)), $\mathfrak{M}_{\{c\}}$ is such that $\mathfrak{M}(i) = c_i$. By hypothesis we have that $\mathfrak{M}_{\{c\}} \models ist(i_1, \phi_1) \wedge \ldots \wedge ist(i_n, \phi_n) \supset ist(i, \phi)$, which implies that if $c_{i_k} \models \phi_k$ for all $1 \leq k \leq n$, then $c_i \models \phi$.

To define $a(\boldsymbol{br})$ we proceed as follows: for any PLC model $\mathfrak{M} \in \mathfrak{M}_\mathbf{C}$ we find a formula $\phi_\mathfrak{M}$, that axiomatizes exactly $\mathfrak{M}$. Then the axiomatization of $\mathfrak{M}_\mathbf{C}$ can be obtained by the disjunction of all the axiomatization $\phi_\mathfrak{M}$ associated to each single PLC-model $\mathfrak{M}$ of $\mathfrak{M}_\mathbf{C}$ (this definition is possible because $\mathfrak{M}_\mathbf{C}$ is finite).

Let $\mathfrak{M} \in \mathfrak{M}_\mathbf{C}$, and let $\phi_\mathfrak{M}$ be the following formula

$$\bigwedge_{i \in I} \left(ist(i, \bigvee_{\nu \in \mathfrak{M}(i)} \phi_\nu) \wedge \bigwedge_{\nu \in \mathfrak{M}(i)} \neg ist(i, \neg \phi_\nu) \right) \quad (15)$$

where ϕ_ν is the conjunction of all the literals verified by the assignment ν. (15) is a finite formula, for the set I of context is finite and the set of literals in each context is finite too. By adding (15) as axioms in the context ϵ we obtain an PLC that is satisfied only by the model $\mathfrak{M}$. Let

$$a(\boldsymbol{br}) = \bigvee_{\mathfrak{M} \in \mathfrak{M}_\mathbf{C}} \phi_\mathfrak{M}$$

Let us now prove the equivalence (12). By soundness and completeness of $\boldsymbol{br}$, $i_1 : \phi_1, \ldots, i_n : \phi_n \vdash_{\boldsymbol{br}} i : \phi$ holds if and only if

$$i_1 : \phi_1, \ldots, i_n : \phi_n \models_\mathbf{C} i : \phi \quad (16)$$

By (14), we have that (16) holds if and only if for all $\mathfrak{M} \in \mathfrak{M}_C$,

$$\mathfrak{M} \models_\epsilon ist(i_1, \phi_1) \wedge \ldots \wedge ist(i_n, \phi_n) \supset ist(i, \phi) \tag{17}$$

By construction of $a(\boldsymbol{br})$, $\mathfrak{M} \models_\epsilon a(\boldsymbol{br})$, if and only if $\mathfrak{M} \in \mathfrak{M}_C$. This implies that (17) holds if and only if

$$\models_\epsilon a(\boldsymbol{br}) \supset ist(i_1, \phi_1) \wedge \ldots \wedge ist(i_n, \phi_n) \supset ist(i, \phi) \tag{18}$$

Finally, soundness and completeness of PLC implies that (18) holds if and only if $\vdash_\epsilon a(\boldsymbol{br}) \supset ist(i_1, \phi_1) \wedge \ldots \wedge ist(i_n, \phi_n) \supset ist(i, \phi)$, which concludes our proof.

Theorem 3 shows that the translation from bridge rules to generic *ist*-formulas is possible. However, it is still open the question if a *set of bridge rules* can be translated into *set of* ist-formulas which are lifting axioms. Here the answer is negative.

Theorem 4. *There does not exist a transformation $la(.)$ from finite sets $\boldsymbol{br} \in \mathbb{BR}_I$ of bridge rules to a conjunction of lifting axioms, and a context ϵ such that:*

$$\begin{array}{c} i_1 : \phi_1, \ldots, i_n : \phi_n \vdash_{\boldsymbol{br}} i : \phi \\ \text{if and only if} \\ \vdash_\epsilon la(\boldsymbol{br}) \supset ist(i_1, \phi_1) \wedge \ldots \wedge ist(i_n, \phi_n) \supset ist(i, \phi) \end{array} \tag{19}$$

Proof. The proof is by counterexample. Consider the following LMS/MCS composed of two languages L_1 and L_2 containing the single proposition p and q respectively. Consider the following set of bridge rules:

$$\frac{1 : \neg p}{2 : q} \; br_{12} \quad \frac{1 : p}{2 : \neg q} \; br_{12}^r \quad \frac{2 : \neg q}{1 : p} \; br_{21} \quad \frac{2 : q}{1 : \neg p} \; br_{21}^r \quad \frac{1 : \bot}{2 : \bot} \bot_{12} \quad \frac{2 : \bot}{1 : \bot} \bot_{21}$$

where all the rules but those indexed with r are non restricted. The chains that satisfies the un-restricted bridge rules are:

$$c = \langle p, \overline{q} \rangle, \quad d = \langle \overline{p}, q \rangle, \quad e = \langle p, q \rangle$$

where p denotes the model in which p is true and $\overline{p}$ the model in which p is false. Similarly for q and $\overline{q}$. The compatibility relations that satisfy the restricted bridge rules are:

$$\{c\}, \quad \{d\}, \quad \{e\}, \quad \{c, e\}, \quad \{d, e\}$$

Following the definitions given in the proof of Theorem 3 one can see that the *ist*-formulas associated to the set of LMS-models above is equivalent to the following:

$$\neg ist(1, \bot) \wedge \neg ist(2, \bot) \wedge (ist(1, p) \vee ist(2, q))$$

Notice that the above formula cannot be reduced in the form of a conjunction of lifting axioms.

4 Discussion

In the previous section we have given two somehow opposite results: namely Theorem 2 and Theorem 3. Intuitively the former states that bridge rules cannot be transformed into lifting axioms, so that this translation composes; the latter states that finite sets of bridge rules can be translated into a finite sets of *ist*-formulas. This two results constitutes two boundaries within which one can look for further correspondence results.

Theorem 2 states that a set of bridge rules cannot be translated into a set of lifting axiom simply by translating each single bridge rule into a lifting axiom. This is intuitively due to the fact that bridge rules allows for *inter-leaving of local reasonings*, while lifting axioms do not. By inter-leaving of local reasonings we mean the reasoning pattern composed by a sequence of chunks of local reasoning. This reasoning pattern allow for cyclic contextual reasoning. For instance, one starts in a context $\overline{\kappa}_1$ switches in a context $\overline{\kappa}_2$ then, switch back in the context $\overline{\kappa}_1$ and then again in the context $\overline{\kappa}_2$. Consider the bridge rules given in the counter-example of the proof of Theorem 2, plus the bridge rule:

$$\frac{1 : p \supset r}{2 : s} \ br'_{12}$$

An example of inter-leaving of local reasonings is the following proof of $2 : s$.

$$\frac{\dfrac{\dfrac{\dfrac{1 : p^{(*)}}{2 : q} \ br_{12}}{1 : r} \ br_{21}}{1 : p \supset r} \supset \text{I}_{(\text{Discharging the assumption }(*))}}{2 : s} \ br'_{12}$$

PLC does not support inter-leaving of local reasonings. The reasoning pattern implemented in PLC, instead, is "bottom up combination of local reasonings" in a tower of transcendent contexts. In this reasoning pattern one starts from the bottom of a tower of contexts, he locally reasons in a (set of) context(s), say in the context denoted by the sequence $\kappa_1 \ldots \kappa_n \kappa$, then he transcends to by (CS) to the context $\kappa_1 \ldots \kappa_n$ and he locally reasons there (e.g., by using the lifting axioms), then he transcends again to $\kappa_1 \ldots \kappa_{n-1}$. Eventually, he stops at some point of the tower. Theorem 2 shows that "inter-leaving of local reasonings" cannot be reduced to "bottom-up combination of local reasonings + lifting axioms".

Theorem 3, instead, provides a way to translate LMS/MCS into PLC. Furthermore, the counterexample provides in Theorem 4 show that the one proposed in Theorem 3 is the "simplest" translation, i.e., that any other translation cannot be reduced to a conjunction of lifting axioms. If one wants to rewrite bridge rules into lifting axioms he has to take into account the following two points:

1. in embedding LMS/MCS into PLC, bridge rules are not directly translated into implications, as one could expect. For instance the MCS containing the bridge rules (4) are not translated into the axioms of the form

$ist(1,p) \supset ist(2,q)$ and $ist(2,q) \supset ist(1,p)$ as shown by Theorem 2. Indeed, the PLC formalizing the bridge rules (4) is not computed by a direct (syntactic) translation of the bridge rules of MCS. The axioms (15) are determined by enumerating all the LMS-models of (4) and by axiomatizing them in a PLC-formula. This is not a problem of our translation, indeed any alternative translation which is equivalent to the axiom (15) with more than two contexts cannot be reduced to a set of lifting axioms.

2. the above translation is not compositional. This means that, if PLC_1 and PLC_2 are the representations of MCS_1 and MCS_2 respectively, then the translation of $MCS_1 \cup MCS_2$ (i.e., the MCS containing the axioms and the bridge rules of both MCS_1 and MCS_2) cannot be defined as the union of the axioms of PLC_1 and PLC_2.

5 Conclusions

This paper concludes the technical and conceptual comparison between LMS/MCS and PLC we started in [3]. The results presented in this paper will help clarify the technical and conceptual differences between the two approaches, by showing how bridge rules can be represented in lifting axioms or in ist-formulas. In particular we have shown that:

1. Bridge rules cannot be translated into lifting axioms;
2. sets of bridge rules can be translated into set of ist-formulas which cannot be reduced to a conjunction of lifting axioms.

We stress the fact that the two formalisms do not provide equivalent solutions, even if they share some of the intuitive motivations for having a formal theory of context in AI. The technical results we provide in the previous paper [3] and in this paper allow us to justify the conclusion that LMS/MCS is more general than PLC, and that it captures some patterns of contextual reasoning in a more intuitive and straightforward way. Moreover, in our opinion, the restrictions needed to reconstruct LMS/MCS in PLC have a significant impact on the appropriateness of PLC to capture the intuitive desiderata of a logic of context in AI.

Acknowledgements. We thank Massimo Benerecetti, Chiara Ghidini and Fausto Giunchiglia for the useful discussions on the paper. We also want to thank the anonymous reviewers for the useful comments, and in particular the reviewer who suggested the re-formulation of Theorem 2 and Theorem 3 in terms of relative interpretations [13]. For lack of time, we could not take the advice in this paper, but we plan to explore this idea in future versions of the paper.

References

1. V. Akman, P. Bouquet, R. Thomason, and R.A. Young, editors. *Modeling and Using Context*, volume 2116 of *Lecture Notes in Artificial Intelligence*. Springer Verlag, 2001. Proceedings of CONTEXT'2001 – Third International and Interdisciplinary Conference on Modeling and Using Context (27–30 July 2001, Dundee, Scotland).
2. P. Bouquet and F. Giunchiglia. Reasoning about theory adequacy: A new solution to the qualification problem. *Fundamenta Informaticae*, 23(2–4):247–262, June,July,August 1995. Also IRST-Technical Report 9406-13, IRST, Trento, Italy.
3. P. Bouquet and L. Serafini. Two formalizations of context: a comparison. In Akman et al. [1], pages 87–101. Proceedings of CONTEXT'2001 – Third International and Interdisciplinary Conference on Modeling and Using Context (27–30 July 2001, Dundee, Scotland).
4. S. Buvač, V. Buvač, and I.A. Mason. Metamathematics of Contexts. *Fundamentae Informaticae*, 23(3), 1995.
5. S. Buvač and Ian A. Mason. Propositional logic of context. In R. Fikes and W. Lehnert, editors, *Proc. of the 11th National Conference on Artificial Intelligence*, pages 412–419, Menlo Park, California, 1993. American Association for Artificial Intelligence, AAAI Press.
6. G. Criscuolo, F. Giunchiglia, and L. Serafini. A Foundation for Metareasoning, Part I: The proof theory. Technical Report 0003-38, IRST, Trento, Italy, 2000. To be published in the Journal of Logic and Computation.
7. C. Ghidini and F. Giunchiglia. Local Models Semantics, or Contextual Reasoning = Locality + Compatibility. *Artificial Intelligence*, 127(2):221–259, April 2001.
8. F. Giunchiglia. Contextual reasoning. *Epistemologia, special issue on I Linguaggi e le Macchine*, XVI:345–364, 1993. Short version in Proceedings IJCAI'93 Workshop on Using Knowledge in its Context, Chambery, France, 1993, pp. 39–49. Also IRST-Technical Report 9211-20, IRST, Trento, Italy.
9. F. Giunchiglia and L. Serafini. Multilanguage hierarchical logics or: how we can do without modal logics. *Artificial Intelligence*, 65(1):29–70, 1994. Also IRST-Technical Report 9110-07, IRST, Trento, Italy.
10. R.V. Guha. Contexts: a Formalization and some Applications. Technical Report ACT-CYC-423-91, MCC, Austin, Texas, 1991.
11. J. McCarthy. Notes on Formalizing Context. In *Proc. of the 13th International Joint Conference on Artificial Intelligence*, pages 555–560, Chambery, France, 1993.
12. D. Prawitz. *Natural Deduction - A proof theoretical study*. Almquist and Wiksell, Stockholm, 1965.
13. A. Tarski, A. Mostowski, and R.M. Robinson. *Undecidable theories*. North-Holland, 1968.

Context Dynamic and Explanation in Contextual Graphs

Patrick Brézillon

LIP6, Case 169, University Paris 6, 8 rue du Capitaine Scott, 75015 Paris, France
Patrick.Brezillon@lip6.fr

Abstract. This paper discusses the dynamic of context through the use of a context-based formalism called contextual graphs that has been initially developed in the SART application for the development of a support system in incident solving on a subway line. First, we present the formalism of contextual graphs through its new implementation. Second, we discuss the dynamic of context in contextual graphs. Third, we present two characteristics of contextual graphs as they relate to the dynamic of context, the incremental knowledge acquisition and the explanation generation. We conclude by a discussion of the key properties and the potential of contextual graphs for other applications.

Keywords: Contextual graphs, explanation, visual explanations, context dynamic, applications

1 Introduction

Brezillon [1, 2] defined context as a collection of relevant conditions and surrounding influences that make a situation unique and comprehensible. Based on this initial work, Pomerol and Brezillon [19] showed strong relationships between context and knowledge. Pasquier et al. [16] gave an example of the application of these ideas in the SART application in the monitoring of a subway line. A large volume of knowledge (about trains, electricity, people reaction, and so on) contributes to make each situation unique, while some more particular conditions about the time, the day, the weather and so on, influence many decisions. Brezillon and Pomerol [7] proposed three types of context called external knowledge, contextual knowledge and proceduralized context.

At a given step of a decision process or of the accomplishment of a task, we distinguish between the part of the context which is relevant at this step, and the part which is irrelevant. The latter part is called external knowledge. The former part is called contextual knowledge, and obviously depends on the individual agent and on the decision at hand. Moreover, there is a part of the contextual knowledge that is proceduralized at this step, which we refer to as the proceduralized context. The proceduralized context is invoked, structured and situated according to a given focus.

An important issue is the passage from contextual knowledge to proceduralized context. This proceduralization process [17, 18] is task-oriented and provides a consistent explanatory framework to anticipate the results of a decision or an action.

This point is particularly salient when a company establishes procedures and its employees contextualize these procedures to develop efficient practices.

Companies establish procedures that are collections of safety action sequences permitting to solve a given problem in a wide set of circumstances. These procedures are supposed to cover large classes of problems whatever the conditions in which problems must be solved. This is a kind of uniformization in problem solving but it often results in sub-optimal solutions for problem solving. Conversely, each operator develops their own practice, tailoring the procedure in order to take into account the current context, which is particular and specific.

The modeling of operators' reasoning (practices) is a difficult task because operators use a number of contextual elements, and because procedures for solving complex problems have some degree of freedom. Thus, it would be better to store advantages and disadvantages rather than the complete decision.

This discussion points out that if it is relatively easy to model procedures, the modeling of the corresponding practices is not an easy task because they are as many practices as contexts of occurrence. Moreover, it is not possible to establish a global procedure for complex problem solving, but only a set of sub-procedures for solving different parts of the complex problems.

Based on the design of the contextual graphs for the SART application (e.g. see [19]), we present in this paper a new development of our context-based formalism that (1) goes beyond the SART application, (2) is relevant for problems dealing with procedures, practices and context, and (3) presents new functionality in terms of incremental acquisition of practices and explanation generation. Hereafter, the paper is organized in the following way. First, we present the formalism of contextual graphs through its current implementation. This version of the contextual graphs differs of the version presented previously [15] because we suppress assumptions concerning storage and update of data that darken the expressiveness of the formalism about the dynamic of context, the incremental acquisition of practice and the capacity of explanation generation. Second, we discuss the dynamic of context as represented in contextual graphs as a movement of elements between the contextual knowledge and the proceduralized context, with introduction of new elements from the external knowledge when a new practice has to be acquired. Third, we introduce the types of explanation on practices and problem solving that can be generated from contextual graphs. We conclude with a discussion of the properties and potentialities of contextual graphs.

2 Contextual Graphs

2.1 Introduction

The contextual-graph formalism has been developed initially for an application for incident solving on a subway line ([8], http://www.lip6.fr/SART/). The general observation is that the company establishes procedures for incident solving, and the operator in charge of a subway line adapts the procedure for solving an incident to the context in which each incident occurs. This contextualized procedure is called a practice, and thus there are as many practices as different contexts encountered by the operator.

In our tests operators appreciated the easy understanding of the system's behavior through the use of contextual graphs in which knowledge and reasoning are used in a manner very close to the manner in which they solve incidents [15]. An extension of this project could be for the training of the future operators, thanks to the expressiveness of the contextual graphs, their manipulation (aggregation and expansion of parts of the contextual graph, etc.) and the possibility to replay some incident solving to study potential variants. However, the use of contextual graphs is not limited to the SART application, but is relevant in all domains where operators' reasoning deals with the need to contextualize "official" procedures in order to develop efficient practices by accounting for the context in which the practice is elaborated.

A contextual graph (also noted hereafter CxG) allows a context-based representation of a given problem solving for operational processes by taking into account the working environment [3]. The initial structure of a CxG (its skeleton) is defined by the procedure that is established by the company. The CxG is then progressively enriched by the practices used by operators by applying the procedure in different contexts.

A path in a contextual graph represents a practice in which operator's actions are intertwined with the contextual elements considered explicitly by the operator. A practice differs generally from another one by few actions that are discriminated by a contextual element that has different instantiations for the two practices. Once the divergence between the two practices disappears, the two practices are recombined in a unique path.

2.2 Elements of a Contextual Graph

A contextual graph is an acyclic directed graph with a unique input, a unique output, and a serial-parallel organization of nodes connected by oriented arcs. A node can be an action, a contextual node, a recombination node, or a sub-graph (an activity).

2.2.1 Actions and Activities

An action is an executable method. An activity is a complex action assembling different elements such as a contextual graph with a unique input and a unique output. Mechanisms of aggregation and expansion, as in conceptual graphs, allow users to have different views on a contextual graph and transform an activity into action.

An activity is identified as such by operators as a recurring structure observed in different contextual graphs. The identification of an activity is interesting because a change in an activity appears automatically in all the contextual graphs where the activity has been identified. Activities are organized in a directed hierarchy, an activity possibly calling sub-activities, to maintain the status of acyclic directed graph to the structure.

2.2.2 Contextual Nodes and Recombination Nodes

A contextual element is represented by two types of node, namely a contextual node and a recombination node. A contextual node corresponds to the explicit instantiation of the contextual element. For example, a contextual element could corresponds to be in a hurry with the instantiations "yes" and "no." A contextual node is represented by C(1, n) where n is the number of exclusive branches corresponding to known practices. The associated recombination node R(n, 1) corresponds to the abandon of the instantiation of the contextual element once the action on the branch is accomplished. Then, there is a convergence of the different alternatives towards the same action sequence to execute after.

Thus, at the contextual node, a piece of contextual knowledge becomes instantiated and enters the proceduralized context. At a recombination node, that last piece entered in the proceduralized context goes back to the contextual knowledge. Thus, a change in the context correspond to the movement of a piece of contextual knowledge into the proceduralized context, or conversely from the proceduralized context to the contextual knowledge.

Contextual and recombination nodes give to contextual graphs a general structure of spindle or series of spindles, with a divergence of branches at contextual nodes initiated by a diagnosis, and a convergence at recombination nodes, thanks to actions or activities realized.

2.2.3 Sub-graphs
A sub-graph represents a local reasoning (a diagnosis/action structure) corresponding to intermediate goals. A sub-graph can be an action, a sequence of actions, or a pair of contextual and recombination nodes. A sub-graph is itself a contextual graph, directed, acyclic, with one input and one output. If a sub-graph contains on a branch a contextual node, it contains necessarily its recombination node on the same branch. Conversely, if a subgraph is on a branch, it contains at most all the items on the branch.

2.2.4 Parallel Action Grouping
A parallel action grouping represents a set of m steps in a problem solving that can be realized in parallel or in any order but all must be accomplished before to continue. For example, a coffee preparation requires to take coffee, filter and the reservoir, these actions can be executed in any order but must be accomplished before to switch on the machine, the order in which these three actions must be executed does not matter. The activity is judged globally with respect to a high-level goal. For example, the type of coffee machine generally does not appear explicitly in the example of the coffee preparation, when it would aloow to order the previous actions (e.g. if the place where to put the filter is fix on the coffee machine). The ordering of the actions to execute in a parallel action grouping depends on contextual elements that does not appear in the contextual graph because they are not at the same level of description and constitutes a dense net of contextual nodes leading to few solutions (see [4] for a discussion on this point). This is a way to deal with the incompleteness or complexity of the local information.

2.2.5 An Example
Figure 1 gives an example of contextual graph. An action is represented by a square box. A contextual node is represented by a large circle and Cj.k is the instance k of the contextual node Cj (1, n). A recombination node Rj is represented by a small

black circle. (Subgraph and parallel action grouping are not represented and discussed in this example.)

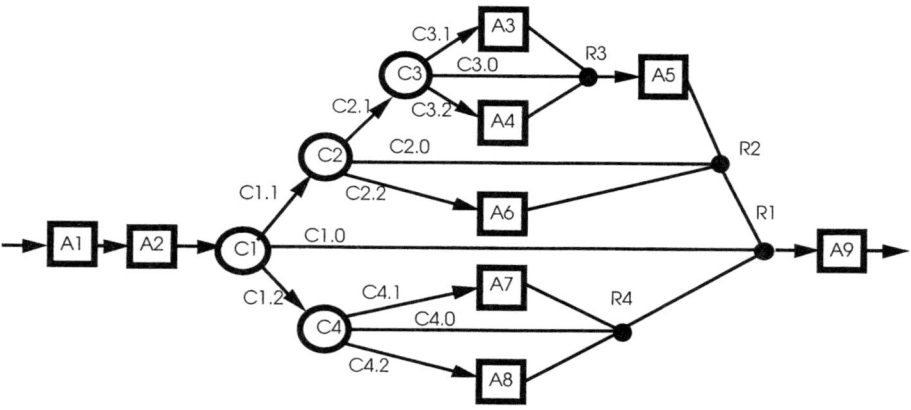

Fig. 1. An example of contextual graph

The operator provides a practice as a sequence of actions such as {A1, A2, A3, A5, A9}. The corresponding path is given by the sequence of actions intertwined with contextual and recombination nodes as {A1, A2, C1.1, C2.1, C3.1, A3, R3, A5, R2, R1, A9} on the upper path in Figure 1. A sub-graph can be an action (e.g. A3), a sequence of actions (e.g. A1-A2), a pair of contextual and recombination nodes and all the items between them (e.g. C3-A3/A4-R3), all the branches between a contextual node and recombination node (e.g. the upper branch of C2 for the value C2.1 with C3-A3/A4-R3-A5).

2.3 Practical Aspects: Implementation

We developed a software for exploiting the formalism of contextual graph. This implementation is realized actually as a prototype written in Java, with a storage of all the data in a database. It presents usual functionality as: switching between different language (French and English at any moment of a session), identification of the user (two types of users, namely the "super-user" who can create a new graph and the "user" who can only enrich the graph with new practices), enrichment and correction of all texts (immediately visible), an online help, different types of visualization (graph resizable according to the dimensions of the window, aggregation and expansion of parts of the graph), comparison of graphs, coloring sub-graph (e.g. for identifying an activity found in different contextual graphs), visualization of the growth of the graph (an addition after the previous one, or the series of additions), comparison of action sequences, explanation on all the items (history, contextual information, etc.), identification of the context of each action, etc.

2.4 Related Works

The contextual graph approach presents some common points with the Ripple Down Rule (RDR) technique [9, 10]. The RDR technique is a hybrid case-based and rule-based approach in which context is an important aspect of RDR, captured in associated cases and the exception structure. In the RDR technique, a case can fall under only one rule, and an error may be corrected by adding one exception rule taking into account only cases that previously fell under the rule to be corrected. There is a two-way dependency relation between rules (with if-true and if-false) such that rule activation is investigated only in the context of other rule activation. Ripple down rules form a binary decision tree that differs from standard decision trees in that compound clauses are used to determine branching, and these clauses need not exhaustively cover all cases so that it is possible for a decision to be reached at an interior node. The RDR technique relies on the fact that people cope with the acquisition and maintenance of complex knowledge structures by making incremental changes to them within a well-defined context such as the effect of changes is locally contained in a well-defined manner [11]. Thus the knowledge that is introduced is highly contextualized. The recommendation given by an expert depends on the context in which it is given and does not consist of a description of the expert's thought processes but is a justification of why this recommendation was made.

Gonzalez and Ahlers [12] describe a knowledge representation paradigm to model the intelligent behavior of simulated agents in a simulator-based tactical trainer. Their hypothesis is that tactical knowledge is highly dependent upon the context (i.e. the situation being faced) and proposed a system called context-based reasoning (CxBR). CxBR encapsulates knowledge about appropriate actions and/or procedures, as well as possible new situations, into contexts. This paradigm has been tested in an application for submarine tactical officers on a patrol mission. Tactical knowledge is required in order to endow autonomous intelligent agents with the ability to act, not only intelligently, but also realistically, in light of a trainee's action. Gonzalez and Ahlers' work is based on the idea that by associating the possible situations and corresponding actions to specific contexts, the identification of a situation is simplified because only a subset of all possible situations are applicable under the active context.

Turner [22] developed a system--an adaptive reasoner--to make context explicit for autonomous underwater vehicles to tackle unanticipated events in complex environments. Contextual knowledge is represented as a set of contextual schemas (c-schemas), then retrieving the most appropriate of those and using them to help the reasoner behave appropriately for its current context. Turner describes context-mediated behavior (CMB) that is based on the idea that an agent have explicit knowledge about contexts in which it may find itself, then use that knowledge when in those contexts. CMB is implemented in the Orca program, an intelligent controller for autonomous underwater vehicles.

3 Movement between Contextual Knowledge and the Proceduralized Context

3.1 The Three Types of Context in a Contextual Graph

Brézillon and Pomerol [18] proposed three types of context, namely external knowledge, contextual knowledge and the proceduralized context. This distinction is expressed in the formalism of contextual graphs in the following way. External knowledge is the knowledge that does not intervene in the contextual graph (i.e. belong to another contextual graph or does not exist in the database). This is a source of contextual knowledge through the incremental acquisition of new practices, as discussed below. Contextual knowledge exists in the CxG (the context of the CxG is composed of all the contextual elements in the graph) but is not considered through an instantiation. At the level of a practice (a given path in a contextual graph), all the contextual elements that are not on the path represent contextual knowledge. At the level of an action, contextual knowledge corresponds to contextual nodes out of the path where is the action, when the contextual elements belonging to the path are ordered in a sequence and considered through their instantiations, and thus constitute the proceduralized context. The proceduralized context is an ordered sequence of instantiated contextual elements on the path.

3.2 Context at a Step of a Practice Execution

The context of the contextual graph in Figure 1 is given by the elements {C1, C2, C3, C4}. The context of the action A3 is composed of two parts: the contextual elements used on the path from the input to the action and the other elements. The later elements are contextual knowledge (e.g. C4). The former contextual elements are instantiated, C1 with the value C1.1, C2 with the value C2.1 and C3 with the value C3.1. Thus, the context of the action 3 is defined by:
- The proceduralized context: {C1 with the value C1.2, C2 with the value C2.1, C3 with the value C3.1}, supposing that the actions A1 and A2 are realized.
- The contextual knowledge: {C4}

The context of the action A3 is described in a fixed and static way.

Consider now the context of the path where is the action A3 Once the action A3 is executed, the value C3.1 of C3 does not matter anymore (i.e. at the recombination node R3). The contextual element C3 leaves the proceduralized context at the recombination node R3 and goes back to contextual knowledge. Thus, the context of the action A5, which follows the recombination node R3, is described by:
- The proceduralized context: {C1 with the value C1.2, and C2 with the value C2.1}, and
- The contextual knowledge: {C3, C4}.

The context of the action A5 is also described in a fixed and static way. It differs from the context of action A3 by the contextual element C3 that moved from the proceduralized context to the contextual knowledge at the level of the practice. Thus, during the progress of the practice execution from action 3 to action 5, the context of the practice evolves when the contexts of A3 and A5 are static.

3.3 Context Evolution During the Progress of a Practice Execution

The dynamics of the context appears at the practice level when the focus of attention moves. The contextual knowledge and the proceduralized context evolve during the progress of a practice execution (along a path). For example, consider the upper path in Figure 1: {A1, A2, A3, A5, A9}. Its context presents the following dynamic along the practice execution (each line of the Table represents a step in the application of the practice, a step corresponding to a change in the context):

Table 1. Dynamics of the context along the upper path in Figure 1

L	Context from	Contextual Knowledge	Proceduralized context
1	0	{C1, C2, C3, C4}	{Ø}
2	C1	{C2, C3, C4}	{C1.1}
3	C2	{C3, C4}	{C1.1, C2.1}
4	C3	{C4}	{C1.1, C2.1, C3.1}
5	R3	{C3, C4}	{C1.1, C2.1}
6	R2	{C2, C3, C4}	{C1.1}
7	R1	{C1, C2, C3, C4}	{Ø}

The movement between the contextual knowledge and the proceduralized context follows the rule "last in, first out." The contextual elements are instantiated in the order C1, C2, and C3 (in the proceduralized context) and return to the contextual knowledge as C3, C2, and C1. The progress of the practice execution until an item itself is an element of the context. Thus, two contexts having the same contextual knowledge and proceduralized context (as at lines 3 and 5 of the Table 1) are different by their history in the practice.

3.4 Incremental Knowledge Acquisition in Contextual Graphs

An important part of our system is the identification of a sequence of actions used by the operator for a problem solving. This is realized by interaction between the operator and the system through a graphical representation of the current state of the contextual graph. Once a problem is solved, the operator reports the problem solving by providing the system with the action sequence used for the problem solving. Then, the operator tells the system which known practice is the closest of the entered sequence of actions. The entered sequence can be a known sequence or not. This is determined by the system that matches actions of the sequences in an ordered way. Once a discrepancy is detected (an action is different, new or missing between the two sequences), the system ask the operator the reason of the difference. The operator provides the system with the missing contextual element (definition), its location (position of the contextual and recombination nodes on the path), its instantiations for the known practice and the entered practice. The contextual element that is added, generally comes from the external knowledge. The reason is that the instantiation of this contextual element was not relevant before, but is instantiated in the new practice in a specific way. Thus, the movement from the external knowledge to the contextual knowledge of a contextual graph goes through its use in a proceduralized context.

This is a way to tackle the infinite dimension of the context because the external knowledge is considered only when needed [14].

Thus, contextual graphs have the capacity of evolving by accommodation and assimilation of practices. A new practice differs of a known practice in the CxG by few elements (generally an action). As a consequence, a contextual graph will possess more and more of practices as a kind of corporate memory. The acquisition of a new practice corresponds to the addition in a contextual graph of the minimum number of elements (generally one pair contextual node – recombination node and an action). Never there is to copy a large part of the contextual graph as in a decision tree [15]. (The complete algorithm is under study now.)

4 Explanation Generation in Contextual Graphs

Explanation generation was based on the domain knowledge, i.e. the task at hand, the actions (definition, input, output) in our case. Since about ten years it is known that such explanations bring few to the user and nothing to operators because of the lack of consideration for the context [6]. In contextual graphs, context is represented explicitly, the knowledge is acquired in its context of use, and thus an explanation can be generated from all the items in a contextual graph (contextual elements, actions, activities, their ordering, etc.)

The explanation of a practice is mainly the presentation of the different contextual elements intervening along the path, the order in which they intervene, their temporary instantiations and the temporal chronology in which they have been incorporated in the contextual graph. The explanation of an action in a practice relies mainly on the proceduralized context of this action. The system can thus explain the reasoning hold in the practice (until this action) by presenting (1) the contextual elements explicitly used in the practice until the action, (2) the instantiations of these contextual elements, (3) the order in which are instantiated the different contextual elements, and, the most important, (4) the order in which (and the reasons why) the contextual elements have been introduced in the contextual graph. These two last points are a way to take into account in the explanation the context dynamics that leads to the action to explain. This leads to view explanation as a process in progress along the reasoning progress, rather than deriving it from known and static factors. Our position is close from Leake's position [13] about explanation in case-based reasoning, but the explanation generated in a contextual graph is at different levels of detail (the proceduralized context or the order and the reasons of the introduction of each contextual element).

As the system and the user interact on the same contextual graphs, each one can provide the other with relevant explanation. In the case of explanations provide by the user, the system enters a phase of incremental acquisition of practices that will improve later its reasoning. In this way, the task at hand, the incremental acquisition and the generation of explanations must be intertwined, an important issue in cooperation [5]. Conversely, explanations enable the contextual knowledge to be proceduralized at the right place by supporting the process of incremental acquisition of practices.

Moreover, the mechanisms of aggregation and expansion, as in conceptual graphs [20, 21], allow the user to focus on one part of the contextual graph or another according to his focus of attention. For example, it is possible to study parts of a reasoning and all the variants (the practices). This is particularly interesting to understand the differences between two practices, the role plays by a contextual element in the choice of an action instead of another, etc.

In conclusion, the context-based formalism called contextual graphs gives a uniform representation of actions series and contextual elements. Thus explanations are easier to produce because knowledge on which explanation relies is explicit in the representation. Contextual elements explains the reasons of the choice of an action on another action. With the history of changes in a contextual graph, it is possible to produce different types of explanation.

At the level of a practice, explanation is a way to present the progress of the application of a practice, the movement between the contextual knowledge and the proceduralized context, the changes between the practice and the previous one, the variants added after. It is possible to generate explanations at another level to present contextual information as the creation date of the practice, the author, the problem solving requiring this change for the first time, etc.

With the graphical interface for representing contextual graphs, the system can generate visual explanation on the path from the source to a given element, the ways in which a practice has been progressively specialized, the growth of a contextual graph (with the incremental addition of practice, the practices introduced by a given operator, etc. This aspect, thanks to the incremental practice acquisition, is a new way to generate explanations.

This shows that the task at hand, the incremental acquisition of practices and explanation generation are three aspects of the same thing (the task at hand in the large).

5 Discussion

Context-based formalisms allow a representation of knowledge and reasoning in a way that is directly comprehensible by users. The structures in a contextual graph put at the same level actions and activities (complex action structures). Thus, two people having to interpret the same activity at different levels can understand each other. For example, "Empty the train of travelers" is interpreted as a simple action by the operator who is responsible of the subway line and a complex activity by the driver (stop at the next station, announcement to travelers to leave the train, go and check that nobody is still in the train, close the doors and leave the station).

At the action level, making explicit contextual elements allows to explain the reasons for the choice of an action on another one. Thus, information in contextual graphs is useful and useable for operators.

After using a contextual graph for a while, most of the possible practices would be recorded. By analyzing the whole contextual graph (the initial procedures and all the practices), this would allow the company to improve its procedures, and thus reinforce the value of operators' practices on too general procedures. Another consequences of the expressiveness of the practice representation and explanation capability in contextual graphs is for training purpose (1) of future operators by discussing subtlety of the task accomplishment, and (2) of operators by exchanging and discussing experiences.

In contextual graphs, incremental acquisition of practices and explanation generation are naturally intertwined with the task at hand. Moreover, the progressive evolution of the contextual graph and its graphical representation interact continuously and thus can not be separated according to the rules of the software engineering (separation of the interface from the program).

There are however some limits in this context-based representation at the level of the parallel action grouping. In the example of the coffee preparation (see [4]), we observe a dense net of contextual nodes for the selection of one of the three actions "Take reservoir", "Take filter" and "Take coffee" that can be executed in a sequence that depends on factors such as the places where are the items (e.g. filters are in the cupboard and coffee in the refrigerator) and the user's preferences (if I take coffee box in second, I have just to keep in my hand the coffee box to put it in the filter that will be made ready just before). The type of the machine also may intervene. For example, in some machines the filter is put on the reservoir, when on other machines it is fixed on the body of the machine. In the former case, it is necessary to first pour water in the reservoir, install the filter on the reservoir, and then put coffee in the filter. In the same spirit, it could be important to make explicit the number of persons interested by the coffee preparation (the choice of the machine depends on it), the place where are things and their relative location (place: home or at work; filter: in the cupboard or near the coffee machine; coffee: near the coffee machine or in the refrigerator; reservoir: near the coffee machine or in the "kitchen or wash room"), and the relationships between things. For example, is the filter on the receptacle or not? Is the cup a part of the receptacle itself? Must we take all the coffee machine to fill the reservoir with water? The availability of the resources in the operating environment (water, coffee, electricity source), etc. also intervene implicitly at the level of the parallel action grouping. All these factors constitute a set of choices that can be different from one day to another one (optimization of movement, of the duration of the operation, of the number of operations).

Another weakness of the context-based representation concerns the representation of time. Now, time is represented by the fact that contextual graphs are directed, and that there is an ordering of the actions to execute. However, it is not possible to represent the fact that an action must be accomplished, say, ten minutes before starting the execution of another one. For example, one needs to stick successively two objects, the second object once the first one is definitely stuck.

However, even now contextual graphs present some potentialities to exploit. It is clear that the more a system based on contextual graphs is used, the more it will preserve corporate memory. As a side-effect, it is possible to revise the procedures according to all the variants developed by operators. The new procedures would be more robust. As a contextual graph could describe all the ways in which something can be used (say, as the access to a server), it could be possible then to determine secure and sensible paths of access and forbid sensitive ones after to identify what a user is doing. We are currently studying such lines of use of contextual graphs.

Acknowledgments. We thank Laurent Pasquier, then Ph.D. student with which we have developed contextual graphs for the SART application, and Juliette Brézillon that has improved the software under several aspects. We also thank the French Foreign Ministry that provides grants for the SART application.

References

1. Brézillon, P.: Context in problem solving: A survey. The Knowledge Engineering Review, **14(1)** (1999) 1–34
2. Brézillon, P.: Modeling and using context: Past, present and future. Rapport de Recherche du LIP6 2002/010, Université Paris 6, France.
http://www.lip6.fr/reports/lip6.2002.010.html (2002)
3. Brézillon, P.: Using context for Supporting Users Efficiently. Proceedings of the 36th Hawaii International Conference on Systems Sciences, HICSS-36, Track "Emerging Technologies", R.H. Sprague (Ed.), Los Alamitos: IEEE, CD-Rom (2003a)
4. Brézillon, P.: Contextual graphs: A context-based formalism for knowledge and reasoning in representation. To appear in Research Report, LIP6, University Paris 6, France (2003b)
5. Brézillon, P., Cases, E.: Cooperating for assisting intelligently operators. Design of Cooperative Systems (COOP-95). INRIA (Publisher) (1995) 370–384
6. Brézillon, P., Pomerol, J.-Ch.,: User acceptance of interactive systems: Lessons from Knowledge-Based and Decision Support Systems. International Journal on Failures & Lessons Learned in Information Technology Management **1(1)**(1997) 67–75.
7. Brézillon, P., Pomerol, J.-Ch.: Contextual knowledge sharing and cooperation in intelligent assistant systems. Le Travail Humain, **62(3)** (1999) 223–246
8. Brézillon P., Pasquier L., Pomerol J.-Ch.: Reasoning with contextual graphs. European Journal of Operational Research, **136(2)** (2002) 290–298
9. Clancey W.J. Model construction operators. Artificial Intelligence, **53** (1992) 1–115
10. Compton, P., Jansen, R.: Knowledge in context: a strategy for expert system maintenance. Barter, C.J. and Brooks, M.J., Ed. AI'88: 2nd Australian Joint Artificial Intelligence Conference, Adelaide Australia, November 1988, Proceedings. Berlin, Springer (1990a) 292–306
11. Compton, P., Jansen, R.: A philosophical basis for knowledge acquisition. Knowledge Acquisition **2(3)** (1990b) 241–258
12. Gaines, B.R., Compton, P.: Induction of Ripple-Down rules applied to modeling large databases. Journal of Intelligent Information Systems, **5(3)** (1995) 211–228
13. Gonzalez, A.J., Ahlers, R, Context-based representation of intelligent behavior in training simulations. *Transactions* **15**(4) (1997) 153–166
14. Leake, D B,: Case-based reasoning: Experiences, lessons, and future directions. Chapter I. *CBR in context: The present and future* Menlo Park: AAAI Press/MIT Press (1996)
15. McCarthy, J,: Notes on formalizing context *Proceedings of the 13th IJCAI* **1** (1993) 555–560
16. Pasquier L.: Modélisation de raisonnement tenus en contexte. Application à la gestion d'incidents sur une ligne de métro. Thèse de l'Université Paris 6, juillet (2002)
16. Pasquier, L., Brézillon, P., Pomerol J.-Ch.: Context and decision graphs in incident management on a subway line. Modeling and Using Context (CONTEXT-99). In: Lecture Notes in Artificial Intelligence, N° 1688, Springer Verlag (1999) 499–502
17. Pomerol J-Ch.,: Decision Making Biases and Context, Brussels DSS Conference, *Journal of Decision Systems* (2003, to appear)
18. Pomerol J.-Ch.., Brézillon P. Dynamics between contextual knowledge and proceduralized context, in *Modeling and Using Context* (CONTEXT-99), Lecture Note in Artificial Intelligence n° 1688, Springer Verlag, (1999) 284–295
19. Pomerol J.-Ch., Brézillon P. About some relationships between knowledge and context. In: P. Bouquet, L. Serafini, P. Brézillon, M. Benerecetti, F. Castellani (Eds.): Modeling and Using Context (CONTEXT-01). Lecture Notes in Computer Science, Springer Verlag, N° 1688, (2001) 461–464. (Full paper at
http://www-poleia.lip6.fr/~brezil/Pages2/Publications/CXT01/index.html)
20. Sowa, J.F.: "Conceptual Structures: Information Processing in Mind and Machine", Addison Wesley Publishing Company (1984)

21. Sowa, J.F.: Knowledge Representation: Logical, Philosophical, and Computational Foundations, Brooks Cole Publishing Co., Pacific Grove, CA, ©2000. Actual publication date, 16 August 1999. 594 + xiv pages; ISBN 0-534-94965-7
22. Turner, R M,: "Context-mediated behavior for intelligent agents" *International Journal of Human-Computer Studies* Special Issue on Using Context in Applications **48**(3): (1998) 307–330
23. Walker, R.J.,d Murphy, G.C.: Implicit context: easing software evolution and reuse. In: D.S. Rosenblum (ed.) Proc. of the ACM SIGSOFT Eighth Int. Symp. on the Foundations of Software Engineering (FSE-8). ACM Press, (2000) 69–78

A Deduction Theorem for Normal Modal Propositional Logic

Saša Buvač

buvac@post.harvard.edu

Abstract. We develop a Hilbert style calculus for modal propositional logic which allows for the deduction theorem. Labels are used to keep track of both modalities that have been entered and assumptions that have been made. The main technical result of this paper is the equivalence of the labelled deductive calculus to the normal calculus for modal propositional logic.

We assume a standard modal propositional language: a standard propositional language with a countable number of modalities designated by placing an integer in square brackets in front of a formula. By convention ϕ, ψ, and χ range over formulae, T ranges over sets of formulae, and n, m, and i range over integers. A normal calculus, a staple of modal logics since [1], is now defined in the usual way:

Definition(derivation): a formula, ϕ, is derivable, written

$$\vdash \phi$$

iff ϕ is an element of the least set which contains all the instances of the following axiom schemata:

(PL) $\quad \phi \quad$ provided ϕ is a propositional tautology
(K) $\quad [n](\phi \supset \psi) \supset [n]\phi \supset [n]\psi$

and is closed under the following rules of inference:

(MP) $\quad$ from ϕ and $\phi \supset \psi$ infer ψ
(RN) $\quad$ from ϕ infer $[n]\phi$.

Although quite elegant, a normal calculus does not allow for the deduction theorem.[1] We define a new calculus, for which we write $T \vdash: \phi$, say that ϕ is deducible from T, and prove a deduction theorem:

---

[1] We could, of course, extend derivability of a normal calculus to allow for assumptions: $T \vdash \phi$ if ϕ is an element of the least superset of T which contains **PL** and **K** and is closed under **MP** and **RN**. Then, the deduction theorem, $T, \phi \vdash \psi \Rightarrow T \vdash \phi \supset \psi$, implies the derivability of a typically unacceptable schema

$$\phi \supset [n]\phi$$

(the converse of schema **T**): start with $\phi \vdash \phi$, apply **RN** to get $\phi \vdash [n]\phi$, and then the deduction theorem.

Theorem(deduction): $T, \phi \vdash: \psi \Leftrightarrow T \vdash: \phi \supset \psi$.

To ensure that our deductive calculus indeed calculates modal propositional logic we show its equivalence to the normal calculus:

Theorem(normalization): $\vdash \phi \Leftrightarrow \vdash: \phi$.

The basic idea is to introduce labels to keep track of both modalities that have been entered and assumptions that have been made.

Definition(label): a label is any finite sequence of formulae and modalities. By convention, the letters b and c range over labels.

With this notion of label in place we now define our deductive calculus.

Definition(deduction): a formula, ϕ, is deducible with a label, c, from assumptions, T, written
$$T \vdash c : \phi$$
iff the tuple (c, ϕ) is an element of the least superset of $\epsilon \times T$ which contains all the instances of the following axiom schemata:

(pl) $c : \phi$ provided ϕ is a propositional tautology
(k) $c : [n](\phi \supset \psi) \supset [n]\phi \supset [n]\psi$

and is closed under the following rules of inference:

(mp)   from $c : \phi$ and $c : \phi \supset \psi$ infer $c : \psi$
(exit)   from $c, [n] : \phi$ infer $c : [n]\phi$
(enter)   from $c : [n]\phi$ infer $c, [n] : \phi$
(assume)   from $c : \psi \supset \phi$ infer $c, \psi : \phi$
(discharge) from $c, \psi : \phi$ infer $c : \psi \supset \phi$.

We write $T \vdash: \phi$ for $T \vdash \epsilon : \phi$, where ϵ is the empty sequence, and say that ϕ is deducible from T.

1 Proofs of Theorems

We first introduce and investigate the notion of labelling which is needed in the proofs of both theorems.

Definition(labelling): a labelling of a formula, ϕ, with a label, c, written
$$c \sqsupset \phi$$
is the formula defined inductively to be

ϕ if $c = \epsilon$,
$b \sqsupset [n]\phi$ if $c = b, [n]$, and
$b \sqsupset (\psi \supset \phi)$ if $c = b, \psi$.

Lemma(labelling):

1. $\vdash \phi \Rightarrow \vdash c \sqsupset \phi$
2. $T \vdash b : c \sqsupset \phi \Leftrightarrow T \vdash b, c : \phi$.

Furthermore, the following is a derived rule of the normal calculus:

(MPc)   from $c \sqsupset \phi$ and $c \sqsupset (\phi \supset \psi)$ infer $c \sqsupset \psi$.

It is derived by induction on the structure of c. The base case, for $c = \epsilon$, is just **MP**. We assume the rule holds for c and show it for $c, [n]$ first, and for c, χ later.

Case(from $c, [n] \sqsupset \phi$ and $c, [n] \sqsupset (\phi \supset \psi)$ infer $c, [n] \sqsupset \psi$): We begin by assuming
$$c, [n] \sqsupset (\phi \supset \psi).$$
Therefore, by definition of labelling,
$$c \sqsupset [n](\phi \supset \psi).$$
By inductive hypothesis and **labelling** lemma 1 applied to **K** we now get
$$c \sqsupset ([n]\phi \supset [n]\psi).$$
Together with $c \sqsupset [n]\phi$ (which follows from $c, [n] \sqsupset \phi$ by definition of labelling) by inductive hypothesis we get
$$c \sqsupset [n]\psi.$$
Therefore, by definition of labelling,
$$c, [n] \sqsupset \psi.$$

Case(from $c, \chi \sqsupset \phi$ and $c, \chi \sqsupset (\phi \supset \psi)$ infer $c, \chi \sqsupset \psi$): Begin by assuming
$$c, \chi \sqsupset (\phi \supset \psi).$$
Therefore, by definition of labelling,
$$c \sqsupset \chi \supset (\phi \supset \psi).$$
By inductive hypothesis and propositional logic we now get
$$c \sqsupset (\chi \supset \phi) \supset (\chi \supset \psi).$$

Together with $c \sqsupset (\chi \supset \phi)$, (which follows from $c, \chi \sqsupset \phi$ by definition of labelling) by inductive hypothesis we get

$$c \sqsupset (\chi \supset \psi).$$

Therefore, by definition of labelling

$$c, \chi \sqsupset \psi.$$

Proof(labelling): We first articulate and prove, in turn, two equalities which are needed in the proof of the lemma. The first equality,

$$[n]c \sqsupset \phi = [n], c \sqsupset \phi,$$

is proved by induction on the structure of c. The base case, for $c = \epsilon$, is trivial. For the inductive hypothesis, assume the lemma for c. We now begin with $[n]c, [m] \sqsupset \phi$. By definition of labelling this is equal to $[n]c \sqsupset [m]\phi$. By inductive hypothesis, this is $[n], c \sqsupset [m]\phi$, which, again by definition of labelling, is equal to $[n], c, [m] \sqsupset \phi$. Next, we begin with $[n]c, \chi \sqsupset \phi$. By definition of labelling this is equal to $[n]c \sqsupset \chi \supset \phi$. By inductive hypothesis, this is $[n], c \sqsupset \chi \supset \phi$, which, again by definition of labelling, is equal to $[n], c, \chi \sqsupset \phi$. The second equality,

$$\psi \supset c \sqsupset \phi = \psi, c \sqsupset \phi,$$

is proved in a similar way, by induction on the structure of c. The base case, for $c = \epsilon$, is trivial. For the inductive hypothesis, assume the lemma for c. We now begin with $\psi \supset c, [m] \sqsupset \phi$. By definition of labelling this is equal to $\psi \supset c \sqsupset [m]\phi$. By inductive hypothesis, this is $\psi, c \sqsupset [m]\phi$, which, again by definition of labelling, is equal to $\psi, c, [m] \sqsupset \phi$. Next, we begin with $\psi \supset c, \chi \sqsupset \phi$. By definition of labelling this is equal to $\psi \supset c \sqsupset \chi \supset \phi$. By inductive hypothesis, this is $\psi, c \sqsupset \chi \supset \phi$, which, again by definition of labelling, is equal to $\psi, c, \chi \sqsupset \phi$.

We now turn to the first part of the lemma.

Case(1): Proof is by induction on the structure of c. The base case where $c = \epsilon$ is trivial. For inductive hypothesis, we assume $\vdash c \sqsupset \phi$. We consider two inductive cases. To prove $\vdash [n], c \sqsupset \phi$ we use **RN** on the inductive assumption $\vdash c \sqsupset \phi$ to get $\vdash [n]c \sqsupset \phi$, which in turn by the first of the above equalities gives $\vdash [n], c \sqsupset \phi$. The other inductive case concerns $\vdash \psi, c \sqsupset \phi$. Assume again $\vdash c \sqsupset \phi$. Therefore by propositional logic $\vdash \psi \supset c \sqsupset \phi$. Now the second of the above equalities gives $\vdash \psi, c \sqsupset \phi$.

Case(2): This proof is by induction on the structure of b. The base case where $b = \epsilon$ is again trivial. For the inductive hypothesis we assume the lemma holds for b, and we consider two inductive cases. In the first case we begin with

$$T \vdash b, [n] : c \sqsupset \phi.$$

By **exit** and **enter** rules this is equivalent to

$$T \vdash b : [n]c \sqsupset \phi.$$

Now, the first equality above gives

$$T \vdash b : [n], c \sqsupset \phi,$$

and by the inductive hypothesis, we get equivalence to

$$T \vdash b, [n], c : \phi.$$

In the second case we begin with

$$T \vdash b, \psi : c \sqsupset \phi.$$

By **assume** and **discharge** rules this is equivalent to

$$T \vdash b : \psi \supset c \sqsupset \phi.$$

Now, the second equality above gives

$$T \vdash b : \psi, c \sqsupset \phi.$$

and again by the inductive hypothesis, we get equivalence to

$$T \vdash b, \psi, c : \phi.$$

[labelling]

We now turn to showing the deduction theorem.

Proof(deduction): We prove each direction in turn, starting with the shorter proof.

Case(if): This direction is just **mp** observing the following two structural rules:

$$T, \phi \vdash : \phi$$

and

$$T \vdash : \chi \;\Rightarrow\; T, \phi \vdash : \chi.$$

We let χ in the latter be $\phi \supset \psi$ to get

$$T \vdash : \phi \supset \psi \;\Rightarrow\; T, \phi \vdash : \phi \supset \psi.$$

The left hand side is the premise; the right hand side together with former structural rule $T, \phi \vdash : \phi$ via **mp** gives

$$T, \phi \vdash : \psi.$$

Case(only if): We prove the equivalent form

$$T, \phi \vdash c : \psi \;\Rightarrow\; T \vdash \phi, c : \psi.$$

(The **only if** direction of the deduction theorem follows from the above with $c = \epsilon$ via one application of **discharge**.) The proof is by induction on the structure of the deduction of $c : \psi$.

Case(base with $\phi \neq c \sqsupset \psi$): $c : \psi$ must be an instance of an axiom, in which case $\phi, c : \psi$ is too.

Case(base with $\phi = c \sqsupset \psi$): We begin with an instance of the propositional tautology
$$T \vdash : \phi \supset \phi.$$
By **assume** we get
$$T \vdash \phi : \phi.$$
Since $\phi = c \sqsupset \psi$ this is
$$T \vdash \phi : c \sqsupset \psi.$$
Therefore, by the **labelling** lemma 2,
$$T \vdash \phi, c : \psi.$$

Case(mp): Assume
$$T, \phi \vdash c : \psi \;\Rightarrow\; T \vdash \phi, c : \psi$$
$$T, \phi \vdash c : \psi \supset \chi \;\Rightarrow\; T \vdash \phi, c : \psi \supset \chi$$
and that $T, \phi \vdash c : \chi$ had been deduced from the left hand sides of the two inductive hypotheses above via **mp**. Then $T \vdash \phi, c : \chi$ can be deduced from the right hand sides of the hypotheses also via **mp**.

Case(exit): Assume
$$T, \phi \vdash c, [n] : \psi \;\Rightarrow\; T \vdash \phi, c, [n] : \psi$$
and that $T, \phi \vdash c : [n]\psi$ had been deduced from the left hand side of the inductive hypothesis above via **exit**. Then $T \vdash \phi, c : [n]\psi$ can be deduced from the right hand sides of the inductive hypothesis also via **exit**.

The proof is similar for other rules of inference. [deduction]

To prove normalization we need to manipulate derivations and deductions as objects; to this end we introduce the notion of a formula tree: a tree whose every node contains a formula. By convention A, B, and C range over formula trees. We establish the following notation for formula trees:

A_ϕ means that A is a formula tree with root ϕ
A^ϕ means that A is a formula tree with a single node, and that node is ϕ
A^B_ϕ means that A is a formula tree with root ϕ and exactly one branch: the formula tree B
$A^{B\ C}_\phi$ means that A is a formula tree with root ϕ and exactly two branches: the formula trees B and C.

To illustrate our notation we observe that A_ϕ implies one of the following: A^ϕ, or that there exists a B such that A_ϕ^B, or that there exist B and C such that $A_\phi^{B\ C}$.

We extend the above meanings of superscripts and subscripts on formula trees, to trees whose every node contains a (c, ϕ) pair. only in such cases we use $\alpha, \beta,$ and γ instead of $A, B,$ and C;

Proof(normalization): We show that every derivation can be transformed into a deduction and vice versa.

When transforming a derivation into a deduction we need to keep track of all the applications of **RN** starting from the leaf nodes.

Definition(@):

$$A^\phi @ c = \alpha^{c:\phi}$$

$$A^B_{[n]\phi} @ c = \alpha^{B@c,[n]}_{c:[n]\phi}$$

$$A^{B\ C}_\phi @ c = \alpha^{B@c\ C@c}_{c:\phi}.$$

Lemma(deduction construction): if A_ϕ is a derivation then $\alpha_{c:\phi} = A@c$ is a deduction, for any c.

Intuitively, to transform a deduction into a derivation we simply replace every : with a $\sqsupset$.

Definition([[·]]):

$$[[\alpha^{c:\phi}]] = A^{c\sqsupset\phi}$$

$$[[\alpha^\beta]] = [[\beta]]$$

$$[[\alpha^{\beta\ \gamma}_{c:\phi}]] = A^{[[\beta]]\ [[\gamma]]}_{c\sqsupset\phi}.$$

Lemma(derivation construction): if $\alpha_{c:\phi}$ is a deduction then $A_{c\sqsupset\phi} = [[\alpha]]$ is a derivation. [normalization]

All that remains to be proved are the two construction lemmas.

Proof(deduction construction): Construction is by induction on the structure of the derivation. The base case is trivial, as every axiom of the normal calculus is an axiom with any label of the deductive calculus.

Case(MP): Assume that

1. if B_ψ is a derivation then $\beta_{c:\psi} = B@c$ is a deduction
2. if $C_{\psi\supset\phi}$ is a derivation then $\gamma_{c:\psi\supset\phi} = C@c$ is a deduction
3. $A^{B_\psi\ C_{\psi\supset\phi}}_\phi$ is a derivation.

Following the definition of @ we let

$$\alpha_{c:\phi}^{B@c\ C@c} = A_{\phi}^{B\ C}@c.$$

Since $A_{\phi}^{B_\psi\ C_{\psi\supset\phi}}$ is a derivation, then so are both B_ψ and $C_{\psi\supset\phi}$; therefore, by inductive hypothesis we get that $\beta_{c:\psi} = B@c$ and $\gamma_{c:\psi\supset\phi} = C@c$. Thus

$$\alpha_{c:\phi}^{\beta_{c:\psi}\ \gamma_{c:\psi\supset\phi}} = A_{\phi}^{B\ C}@c.$$

Furthermore, by inductive hypothesis we get that $\beta_{c:\psi}$ and $\gamma_{c:\psi\supset\phi}$ are both deductions; therefore by **mp** $\alpha_{c:\phi}^{\beta_{c:\psi}\ \gamma_{c:\psi\supset\phi}}$ must also be a deduction, which completes this case as the choice of c was arbitrary.

Case(RN): Assume that

1. if B_ψ is a derivation then $\beta_{c,[n]:\psi} = B@c,[n]$ is a deduction
2. $A_{[n]\phi}^{B_\psi}$ is a derivation.

Following the definition of @ we let

$$\alpha_{c:[n]\phi}^{B@c,[n]} = A_{[n]\phi}^{B_\psi}@c.$$

Since $A_{[n]\phi}^{B_\psi}$ is a derivation, then so is B_ψ; therefore, by inductive hypothesis we get that $\beta_{c,[n]:\psi} = B@c,[n]$. Thus

$$\alpha_{c:[n]\phi}^{\beta_{c,[n]:\psi}} = A_{[n]\phi}^{B_\psi}@c.$$

Furthermore, by inductive hypothesis we get that $\beta_{c,[n]:\psi}$ is a deduction; therefore by **exit** $\alpha_{c:[n]\phi}^{\beta_{c,[n]:\psi}}$ must also be a deduction, which completes this case as the choice of c was arbitrary. **[deduction construction]**

Proof(derivation construction): Construction is by induction on the structure of the deduction.

Case(base): Assume that $\alpha^{c:\phi}$ is a deduction. Then $c:\phi$ is an instance of **pl** or **k**. Therefore, ϕ is an instance of **PL** or **K**, and thus we get $\vdash \phi$. Now **labelling lemma 1** gives $\vdash c \sqsupset \phi$.

Case(mp): Assume that

1. if $\beta_{c:\phi}$ is a deduction then $B_{c\sqsupset\phi} = [[\beta]]$ is a derivation
2. if $\gamma_{c:\phi\supset\psi}$ is a deduction then $C_{c\sqsupset\phi\supset\psi} = [[\gamma]]$ is a derivation
3. $\alpha_{c:\psi}^{\beta\ \gamma}$ is a deduction.

Following the definition of $[[\cdot]]$ we let

$$A^{[[\beta]]\ [[\gamma]]}_{c\sqsupset\psi} = [[\alpha^{\beta\ \gamma}_{c:\psi}]].$$

The last assumption implies that both β and γ are deductions and the first two assumptions thus yield that $B_{c\sqsupset\phi} = [[\beta]]$ and $C_{c\sqsupset\phi\supset\psi} = [[\gamma]]$. Therefore

$$A^{B_{c\sqsupset\phi}\ C_{c\sqsupset\phi\supset\psi}}_{c\sqsupset\psi} = [[\alpha^{\beta\ \gamma}_{c:\psi}]].$$

From the same combination of assumptions we also conclude that $B_{c\sqsupset\phi}$ and $C_{c\sqsupset\phi\supset\psi}$ are both derivations, and therefore by **MPc** so is $A^{B_{c\sqsupset\phi}\ C_{c\sqsupset\phi\supset\psi}}_{c\sqsupset\psi}$.

Case(exit): Assume that

1. if $\beta_{c,[n]:\phi}$ is a deduction then $B_{c,[n]\sqsupset\phi} = [[\beta]]$ is a derivation
2. $\alpha^{\beta}_{c:[n]\phi}$ is a deduction.

The latter assumption implies that β is a deduction, and the former assumption thus yields that $[[\beta]] = B_{c,[n]\sqsupset\phi}$ is a derivation. We let $A = [[\alpha]]$ and then, by the definition of $[[\cdot]]$, $A = B_{c,[n]\sqsupset\phi}$ is a derivation. Therefore, by definition of labelling, so is $A_{c\sqsupset[n]\phi}$.

The other 3 cases are similar. [derivation construction]

2 Related Works

Combining the actions of entering and exiting with natural deduction style inference (assuming and discharging) was proposed, defined, and utilized for AI examples in [2]. However, the latter paper assumed but never proved the deduction theorem. Our deductive calculus is a quantifier free version of [2] aimed at showing the deduction theorem. We have done this by distinguishing between modalities and labels, both of which had been grouped under the single category of *context* in [2].

Adding labels to deductive systems has been studied by many authors in the past (see [3]); however, comparisons to all the resulting logics are beyond the scope of this paper.

References

1. Gödel, K.: Eine Interpretation des intuitionistischen Aussagenkalküls. In: Ergebnisse eines mathematischen Kolloquiums. Volume 4. Verlag Franz Deuticke, Vienna (1933) 39–40 English translation, An Interpretation of the Intuitionistic sentential logic, in *The Philosophy of Mathematics*, edited by Jaakko Hintikka, Oxford University Press, pages 128–129, 1969.
2. McCarthy, J., Buvač, S.: Formalizing Context (Expanded Notes). In Aliseda, A., Glabbeek, R.v., Westerståhl, D., eds.: Computing Natural Language. Volume 81 of CSLI Lecture Notes. Center for the Study of Language and Information, Stanford University, Stanford, California (1998) 13–50
3. Gabbay, D.M.: Labelled Deductive Systems, Part 1. Volume 33 of Oxford Logic Guides. Oxford University Press (1996)

Natural Deduction and Context as (Constructive) Modality

Valeria de Paiva

PARC, 3333 Coyote Hill Road, Palo Alto CA 94304, USA

Abstract. This note describes three formalized logics of context and their mathematical inter-relationships. It also proposes a Natural Deduction formulation for a *constructive* logic of contexts, which is what the described logics have in common.

Keywords: Logical formalisms for context, Representing context and contextual knowledge, Context in knowledge representation.

1 Introduction

The word "context" has too many different meanings, so we should start by explaining that we are interested in logics of context designed to help automated reasoning in Artificial Intelligence (AI), more specifically, in knowledge representation. Thus we are interested in mathematically understanding and clarifying work that, starting with McCarthy's seminal papers[McC96,McC93,McCB97], aims at giving the (informal) notion of context the role of a first-class object in a logical system.

Our goal is a mathematically well-behaved logical system that models reasoning that happens when we say, for example, that in the context of Sherlock Holmes stories it is true that Sherlock Holmes lives in Baker Street, London. For a traditional mathematical logician, this informal notion of context is modeled by considering different logical theories and the burden of deciding how these logical theories interact is shifted to the metalogic and the human reasoner. In this paper we take for granted that the reader has been convinced by McCarthy's, Giunchiglia's (and others') arguments that context should be a first-class object in a logical system and that the question to be solved is *which* logical system should one use. Narrowing our focus, we concentrate not in deciding which logical system to use, but on the much smaller question of comparing, in terms of their mathematical properties, the systems[1] in the literature where context is modeled via a modality operator, usually written as $ist(\kappa, A)$. Here the basic intuition is that formulas, such as A, are true not in absolute terms, but in certain contexts, in particular, in the context named by the constant κ. There

---

[1] A referee has rightly complained that we do not discuss how well these systems match the intuitions they are trying to model. While this task seems very important, this author does not have the right intuitions to carry it out. Moreover, the project [CC+02] that our theoretical investigation underpins has moved to a new direction.

are many reasons why this is a good idea for AI and these, as well as examples of the applications of these ideas, are discussed in the literature. But even narrowing down the problem to choosing between systems and considering only the systems based on some kind of modality, the task is daunting. The literature on notions of context and on formalizations thereoff, i.e. logics of context is really vast[AS96]. This paper discusses three propositional[2] systems: Buvac and Mason's propositional logic of contexts, henceforth PLC [BBM95], Nayak's system (here called $\mathcal{N}$) a logic of contexts for multiple domain theories [Nay94]and Massacci's system $\mathcal{T}$ [Mas95], described as a tableaux version of PLC. The Trento group framework for logics of context, called LMS/MCS, for Local Model Semantics/MultiContext systems [BS00,SG00] was also originally considered, but that comparison is now in a companion paper[deP]. This is because, strictly speaking, MCS/LMS has no explicit modality. However, it is well-known that the *bridge rules* of their main system MR correspond, technically, to a K necessity operator.

In the next section we discuss why worry about Natural Deduction, why constructivity is important for us, what constitutes a Natural Deduction formulation of a logic and why obtaining a Natural Deduction formulation for logics of context is problematic and worthwhile. Then, in the following sections, we give succint descriptions of the systems of contexts we consider. After that we compare and evaluate those systems. The upshot is that we can produce a very stringent Natural Deduction formulation for what these systems have in common. The natural deduction formulation for this core constructive language is spelled out in detail in the following section.

2 Natural Deduction: Why?

McCarthy, when first discussing the idea of contexts in AI, suggested that a "strong form of Natural Deduction" should hold for an intuitively appealing logic of contexts. His suggestion of a logic of contexts is based on the notion of a modality $\text{ist}(\kappa, A)$. The intuition of using a modality operator to deal with logics of context is common to all the systems we discuss (and many others we do not). But the systems differ along three different dimensions. First they differ on which *properties* the modalities are supposed to have, then they differ on *how* they are described mathematically, e.g. whether one uses axioms or tableaux systems or Natural Deduction rules and finally they differ on which properties do they prove of the system they consider, whether they have soundness and completeness and with respect to what kind of model.

We advocate the view that a logic should be independent of its different presentations, that is, that one should be able to give different presentations (using axioms, sequents, rules) for any *decent* logic, as we can do for e.g. classical or constructive first-order logic. Moreover, since these formalizations are only different presentations of the same logic, we believe that one must be able to prove them all equivalent, using syntactic translations between the systems. Thus our first aim is to prove that there is a decent logic of contexts, that is, there is

---

[2] There are first-order systems in the literature, but we restrict our attention to propositional systems.

a formal logic of contexts which can be given in several different presentations, all proved equivalent.

McCarthy's and Guha's intuitions were formalized by S. Buvac, V. Buvac and I. Mason in[BBM95]. Their formalization was done in a Hilbert-style system, usually the easiest kind of formalism as far as modal logics are concerned. That paper leaves open how to formalize their propositional logic of contexts in a Natural Deduction setting. Actually, when discussing future work they say:

> We also plan to define non-Hilbert style formal systems for context. Probably the most relevant is a natural deduction system, which would be in line with McCarthy's original proposal of treating contextual reasoning as a strong version of natural deduction. In such a system, entering a context would correspond to making an assumption in natural deduction, while exiting a context corresponds to discharging an assumption.

But this future work has not, as yet, come to fruition, which is not surprising, considering the amount of controversy surrounding Natural Deduction for Modal Logics in general. For some of this controversy (and a detailed explanation) the reader is directed to [BdPR01].

A formal description of what constitutes a Natural Deduction formalism will not be attempted here, but we take as paradigmatic the work of Prawitz[Pra65], which is sometimes described as Gentzen-style Natural Deduction, by contrast to Fitch-style Natural Deduction. Gentzen-style Natural Deduction derivations are tree-shaped, usually with one introduction and one elimination rule for each logical connective. More importantly, the introduction and elimination rules give rise to a notion of normalization (elimination of the 'detour' in the proof, that consists of one introduction rule followed immediately by the elimination of the same connective). For intuitionistic logic this paradigm works very well, both for first-order and for higher-order calculi. For other logics, especially modal logics, the formalism does not work so well. Prawitz, for example, only deals with the systems called S4 and S5 in his treatise and even that treatment is not optimal[BdP00]. In a nutshell, the problem is that it is hard[3] to provide introduction and elimination rules for a K-style necessity ($\Box$) operator that satisfies only the sequent calculus (Scott's) rule:

$$\frac{\Gamma \vdash B}{\Box \Gamma \vdash \Box B}$$

For a start, this rule is clearly both an introduction and an elimination rule. But the crux of the problem is how to write, using a tree-like derivation that, after the use of the necessitation rule, all the premises become boxed. Proof-theoretic trees only grow downwards, not upwards. If instead of usual Prawitz-style trees, one tries to use Natural Deduction in sequent-style, as advocated by Martin-Loëf (which corresponds to writing the rule as above) the problem persists. One essential component of Natural Deduction is its ability to put proofs together. If you have proofs $\pi: A_1, \ldots, A_K \vdash B$ and $\sigma: C \vdash A_1$, you must be able to compose

---
[3] So hard that Bull and Segerberg in [BS84] discuss whether modal logic is not *natural* enough to have a Natural Deduction formulation.

them in ND, obtaining $\sigma;\pi$ a proof of $C, A_2, \ldots A_k \vdash B$. But if you apply the Box rule to π, obtaining $\Box A_1, \ldots, \Box A_k \vdash \Box B$, then you cannot compose it with σ anymore. This is an unfortunate situation and there are several very different solutions to this problem in the literature. Most of the solutions build-in some of the semantics into the syntax of modal logic: this is the case for Gabbay's labelled deductive systems, Simpson's framework and Basin et al's framework. The solution we prefer is merely syntactic, see section 6, but there are tradeoffs, discussed later.

The (proof theoretic) received wisdom about logical formalisms is that:

- Axiomatic systems are the easiest ones to devise and also the ones where it is easier to prove theorems *about* the system;
- Sequent calculi are the systems that are easy to mechanize and
- Natural Deduction systems are the ones most similar to the way humans construct proofs.

It is also the case that given a Gentzen-style Natural Deduction system one can, automatically derive both sequent calculus and axiomatic systems from it, but the converses are not always true. Hence Natural Deduction systems are the most informative formalism. But exactly what constitutes a Natural Deduction system and, given that modal logics must depart somehow from the traditional setting, what are the most important properties to preserve is subject to personal taste and warrants discussion.

Given that sequent calculi (and tableaux systems) are, arguably, better formalisms for automatic *proof search*, whereas Natural Deduction comes into its own when dealing with proof normalization, one may wonder why we worry about a Natural Deduction version of a constructive logic of contexts. In the one hand, we are interested in *deep* understanding of the logic in question and a Natural Deduction formalization gives the ability to change formalisms as explained above. Since the different formalisms do not constitute different systems, but are simply different presentations of a given system, a Natural Deduction presentation, together with its translations, affords logical respectability. On the other hand, our emphasis on *constructivity* of the logic explains an ulterior (and eventual) goal: we would like to use the Curry-Howard correspondence to provide a functional programming language for dealing with proofs of statements in context.

But even discounting the motivation of a Curry-Howard system for contexts, it is true that the exercise of comparing logics tends to clarify our understanding. This explains the emphasis on the comparison of the systems in this paper. Both Buvac, Buvac and Mason's PLC and Nayak's $\mathcal{N}$ are given as axiomatic systems, while Massacci's calculus is given as a tableaux system – a close cousin of a sequent calculus. Thus we start by describing PLC and $\mathcal{N}$ and then we discuss Massacci's system. After that we introduce our own Natural Deduction system.

3 The Propositional Logic of Contexts PLC

Buvac, Buvac and Mason's paper "Metamathematics of Contexts"[BBM95] is the most developed formalization of McCarthy's ideas [McC93] about a propositional logic of contexts. Their propositional logic of contexts extends *classical* propositional logic in (at least) two ways: first, it adds a new modality $\texttt{ist}(\kappa,\phi)$, used to express that the sentence ϕ holds or is true in context κ. Second, they postulate that each context has its own vocabulary, ie a set of propositional atoms that is meaningful in that context. They describe a basic logic of contexts, describe a semantics for this basic logic similar to the traditional semantics for first-order logic, discuss various extensions of this basic logic and give a correspondence theory, relating axioms to extensions of the basic semantics.

We start with two given, distinct, countably infinite sets $\mathcal{K}$, a set of labels (intuitively denoting some basic contexts) and $\mathcal{P}$ the set of all propositional atoms. Then well-formed formulas $\mathcal{F}$ are built from the sets $\mathcal{K}$ and $\mathcal{P}$ by negation and implication, together with the $\texttt{ist}(\kappa,\phi)$ operator.

$$\mathcal{F} := \mathcal{P} \cup (\neg \mathcal{F}) \cup (\mathcal{F} \to \mathcal{F}) \cup \texttt{ist}(\mathcal{K}, \mathcal{P})$$

Instead of using simply the set $\mathcal{K}$ of basic labels PLC uses the set of finite sequences over $\mathcal{K}$, $\mathcal{K}^*$. A context, denoted $\overline{\kappa}$, consists of a finite sequence $(\kappa_1...\kappa_n)$ of elements of $\mathcal{K}$ (or in the degenerate case ϵ, the empty sequence). But when one writes $\texttt{ist}(\overline{\kappa}, A)$ this actually means $\texttt{ist}(\kappa_1,(\texttt{ist}(\kappa_2,\ldots \texttt{ist}(\kappa_n, A)\ldots)))$. This use of sequences of basic contexts corresponds to PLC's intuition that what holds in a context depends on how you arrived at this context, so that $\kappa_1\kappa_2$ represents how context κ_1 is seen from context κ_2.

We also need to explain the role of vocabularies. The intuitive idea is that a vocabulary (the set of meaningful propositional atoms) is defined for each context. Thus we have a relation $\textsf{Vocab}$ between $\mathcal{K}^*$ and $\mathcal{P}$. The notion of derivability ($\vdash_{\overline{\kappa}} A$) that defines PLC is also dependent on the vocabulary used, so it should be written as $\vdash_{\overline{\kappa}}^{\textsf{Vocab}}$, but PLC makes the simplifying assumption that given any formula A and context $\overline{\kappa}$ we can calculate the vocabulary of the formula A in context $\overline{\kappa}$ using a function $Vocab(\overline{\kappa}, A)$. Moreover, PLC's Definedness Condition asserts that whenever we state $\vdash_{\overline{\kappa}} A$, we implicitly assume that the $Vocab(\overline{\kappa}, A)$ is contained in (the previously given and forever fixed) $\textsf{Vocab}$.

Buvac, Buvac and Mason assume the following axioms:

(**taut**) $\vdash_{\overline{\kappa}} A$ for all classical tautologies A

(**K**) $\quad \vdash_{\overline{\kappa}} \texttt{ist}(\kappa, A \to B) \to (\texttt{ist}(\kappa, A) \to \texttt{ist}(\kappa, B))$

(Δ) $\quad \vdash_{\overline{\kappa}} \texttt{ist}(\kappa, \texttt{ist}(\kappa_1, A) \vee B) \to \texttt{ist}(\kappa, \texttt{ist}(\kappa_1, A)) \vee \texttt{ist}(\kappa, B)$

together with the proof rules of context switching (CS) and Modus Ponens (MP) below.

$$(CS) \frac{\vdash_{\overline{\kappa}\kappa_1} A}{\vdash_{\overline{\kappa}} \texttt{ist}(\kappa_1, A)} \qquad (MP) \frac{\vdash_{\overline{\kappa}} A \to B \quad \vdash_{\overline{\kappa}} A}{\vdash_{\overline{\kappa}} B}$$

The axioms[4] and rules above constitute the Hilbert-style system for PLC. Note that derivations are always in context, i.e. the turnstile is always decorated with the context sequence where the derivation occurs. We say A is provable in

---

[4] The axiom (**taut**), valid for all systems considered in this note, is disputed by a referee, who suggests that truth in a context should be constrained by relevance to a context. But relevance is a much harder problem than localization of truth, which is the simplified aim of these logics of context.

context $\overline{\kappa}$ iff $\vdash_{\overline{\kappa}} A$ is an instance of an axiom schema or follows from provable formulae by one of the inference rules.

The axiom schemas (**taut**) and (**K**) are traditional, in that logics with modalities usually satisfy all tautologies of the basic (in their case classical) logic and the axiom **K** is generally considered the bare minimum to require of a modality. The Modus Ponens rule (MP) is also traditional, but adapted to hold in each and every context $\overline{\kappa}$.

The context switching rule (CS) and the axiom (Δ) deserve some discussion. It is easy to see that the context switching rule is more general than the usual modal necessitation rule. If one erases contexts from the derivability relation the context switching rule becomes the necessitation rule. But it is not immediately clear that whenever one uses the context switching rule in a PLC proof, the modal necessitation rule could have been used instead.

Let us call *localized multimodal* **K**, the system consisting of two axiom schemas:

(**taut**) $\vdash_{\overline{\kappa}} A$ for all classical tautologies A

(**K**) $\quad \vdash_{\overline{\kappa}} \mathtt{ist}(\kappa, A \to B) \to (\mathtt{ist}(\kappa, A) \to \mathtt{ist}(\kappa, B))$

together with rules

$$(Nec^*) \frac{\vdash_{\overline{\kappa}} A}{\vdash_{\overline{\kappa}} \mathtt{ist}(\kappa_1, A)} \quad (MP) \frac{\vdash_{\overline{\kappa}} A \to B \quad \vdash_{\overline{\kappa}} A}{\vdash_{\overline{\kappa}} B}$$

Proposition 1 (Serafini) *Assume that all contexts have the same vocabulary. Given a proof π of A in PLC, there exists a proof π' of A in the system localized multimodal* **K** *plus Δ.*

Proof: Consider the first appearance of the context switching rule in π. Assume it uses $\vdash_{\overline{\kappa}\kappa_1} A$ to give $\vdash_{\overline{\kappa}} \mathtt{ist}(\kappa_1, A)$. The proof till this use of context switching (CS) was all done in the context $\overline{\kappa}\kappa_1$. Since all axioms in the context $\overline{\kappa}\kappa_1$ are also axioms in $\overline{\kappa}$ and whatever uses of (MP) in $\overline{\kappa}\kappa_1$ are also uses in $\overline{\kappa}$ we can remove κ_1 from the whole proof and after this transformation the proof looks like

$$\frac{\vdots}{\frac{\vdash_{\overline{\kappa}} A}{\vdash_{\overline{\kappa}} \mathtt{ist}(\kappa_1, A)}}$$
$$\vdots$$

Applying this transformation to all occurrences of the context switching rule, we obtain a proof that only uses the localized necessitation rule.$\square$.

The reader will have noticed the assumption of all contexts having the same vocabulary. Recent work[BS00] of Bouquet and Serafini's shows semantically that the vocabularies of PLC play no essential logical role. They say that their "Reduction to Complete Vocabularies" theorem allows them to conclude that PLC really is the normal multimodal logic K extended with the extra axiom Δ.

The axiom (Δ) is problematic from the proof-theoretic perspective. Buvac, Buvac and Mason say that axiom Δ corresponds to the validity reading of the modality ist(κ, A). They justify their adoption of axiom (Δ) (for their most generic logic of contexts) by saying that if we disregard vocabulary restrictions then Δ can be written as

$$(\Delta') \qquad \text{ist}(\kappa_1, \text{ist}(\kappa_2, A)) \vee \text{ist}(\kappa_1, \neg\text{ist}(\kappa_2, A))$$

which they read as saying that "it is true in knowledge base κ_1 that A is valid in the knowledge base κ_2, or it is true in knowledge base κ_1 that A is not valid in the κ_2 knowledge base". Thus each knowledge base behaves as if it can *see* into another knowledge base and decide for any formula A whether or not it is valid in the second knowledge base. But it is not clear that this kind of property is *essential* (or even sensible) for a basic logic of contexts. Actually [CP98] states: "This axiom [Δ] does not seem justified, even for the applications they consider. There is no reason why a database should have complete information about the contents of other databases."

3.1 Evaluating PLC

Buvac and Mason say that

> Modelling truth or validity in a context by a Kripke model, ie by a relation between worlds would not be intuitive, because we want contexts to be reified as first class objects in the semantics. This will allow us (in the predicate case) to state relations between contexts, define operations on contexts and specify how sentences from one context can be lifted into another contexts.

But PLC is a propositional logic and its extension to the 1st order case is far from trivial. Also in the context of PLC no relations, nor operations between contexts are specified. Thus the only reason given by Buvac and Mason for not considering a Kripke-style semantics, that "it is not intuitive to model validity in a context by a relation between worlds" seems too vague. A matter of taste, like saying that you should always use first-order logic, if you can.

It is satisfying to have a sound and complete (first-order-like) semantics for PLC, and for some of its reasonable extensions, but it is not clear how much the semantics presented forces one to accept axiom Δ^5. It is also not clear to me, why such a first-order-like semantics is or would be better than a possible-worlds semantics. Thirdly the role of vocabularies and whether one should have contexts modelled as sequences of basic contexts (or not) is still unclear.

Finally note that to consider a constructive version of PLC we need to take as basis any axiomatization of constructive logic and if we decide that the axiom (Δ) is not required, we just keep $(CS), (MP)$ and **K**, nothing more needs to be done.

---

[5] This is actually an usual problem with any axiomatic system, it is always the case that other axioms might be better, less redundant or more informative. This is another reason for considering other formalisms for a "minimal" logic of contexts.

4 The Logic of Contexts for Multiple Theories $\mathcal{N}$

Nayak [Nay94] takes a different view of the problem of devising a useful logic of contexts: he suggests that, for the purposes of representing and reasoning with multiple domain theories, rather than developing new syntax and new semantics for a logic, we can simply stick with a natural (multimodal) extension of a traditional modal logic. Nayak suggests to write a necessity modal operator for each context (contexts are simply labeled by natural numbers) and to allow different contexts to have different vocabularies.

Nayak presents two main reasons for treating contexts as modal operators, instead of extended terms, as in PLC. First, he says, in the propositional case the context operators and terms are effectively equivalent. Second, the advantage of contexts as terms is that it allows reasoning about contexts *within* the logic, but, he contends, most of the reasoning he wants to do about contexts and about relations between contexts can be done in a meta-theory. Hence it should be worthwhile investigating the properties of a simpler logic of contexts.

The syntax of Nayak's logic of contexts has a set of propositions $\mathcal{P}$, as before, as contexts $\mathcal{K}$ it has natural numbers $\{1, 2, 3, \ldots, n, \ldots\}$, and instead of $\texttt{ist}(i, A)$ for A in $\mathcal{P}$, Nayak denotes that formula A is valid in a context i, by an indexed necessity operator $C_i(A)$. To faciliate the comparison we will use PLC's notation instead. Well-formed formulae are given by

$$\mathcal{F} := \mathcal{P} \cup (\neg \mathcal{F}) \cup (\mathcal{F} \to \mathcal{F}) \cup \texttt{ist}(i, \mathcal{F}), i \in \mathcal{K}$$

Because Nayak's logic wants to pay attention to different vocabularies for different contexts, it defines a function vocabulary, which maps contexts to the collection of propositions defined for that context, $voc: \mathcal{K} \to 2^{\mathcal{P}}$. Since some propositions are not part of the vocabulary of some contexts, we say that a well-formed formula A is *meaningful* with respect to voc if for any propositional letter p occurring in A, if p is immediately within a context $\texttt{ist}(i,)$ then p must be in the vocabulary of that context.

Nayak assumes the following axioms:

(**A1**) $\vdash A$ for all (classical) *meaningful* tautologies A

(**A2**) $\vdash \texttt{ist}(i, (A \to B)) \to (\texttt{ist}(i, A) \to \texttt{ist}(i, B))$, for $1 \leq i \leq n$

(where all formulae in axiom **A2** are assumed meaningful) together with the proof rules of Necessitation and Modus Ponens below, where $\texttt{ist}(i, A)$ is assumed meaningful.

$$(Nec) \frac{\vdash A}{\vdash \texttt{ist}(i, A)} \qquad (MP) \frac{\vdash A \to B \quad \vdash A}{\vdash B}$$

In a nutshell Nayak proposes using a normal multimodal system **K** as the basic logic, but goes on to say that this axiomatization does not restrict enough the properties of contexts or their inter-relationships. For the purpose of modelling these extra properties, he introduces three new axioms:

(**A3**) $\texttt{ist}(\kappa, A) \to \texttt{ist}(\kappa_1, \texttt{ist}(\kappa, A))$
(**A4**) $\neg \texttt{ist}(\kappa, A) \to \texttt{ist}(\kappa_1, \neg \texttt{ist}(\kappa, A))$
(**A5**) $\texttt{ist}(\kappa, A) \to \neg \texttt{ist}(\kappa, \neg A)$

The system consisting of multimodal **K** together with axioms **A3, A4, A5** (called $\mathcal{F}_n$ in Nayak's work) is called here $\mathcal{N}$, for Nayak. Note that axiom **A5** is a generalization of modal axiom **D**, ie it is **D** for every context operator, discussed in the extensions of PLC. Axioms **A3** and **A4** are the generalizations of positive introspection and negative introspection that appear in converse form in other extensions of PLC. Nayak's logic $\mathcal{N}$ is greatly simplified, it does not need to deal with sequences of contexts and these generalizations "ensure that every context knows about what every other context does and does not know, i.e. the facts true in a context are context independent".

Making Nayak's system *constructive* is a matter of making the propositional basis constructive and the basic modal operators constructive. Thus it is clear that it depends on deciding which shape of constructive modal logic one prefers.

4.1 Evaluating $\mathcal{N}$

There is much to recommend the use of 'off the shelf' logical systems. But it must be pointed out that the draconian simplifications brought about, especially by axioms **A4** and **A5** make Nayak's theory applicable only to situations where the contexts are almost not related at all, as in his example of Saturn's moon Titan and tropical forests.

The simplifications brought about by the extra axioms seem too strong for a minimal logic of contexts. Having said that, it would be good to have Nayak's system at one end of the spectrum of useful context logics. One problem is providing a natural deduction formulation for axioms **A4** and **A5**.

5 Massacci's Tableaux System

Massacci's papers [Mas95,Mas96] deal with a tableaux version of a logic of contexts. Massacci seems to be referring to PLC, as defined by Buvac, Buvac and Mason, but as we will discuss his logic proves more theorems than basic PLC. To describe the system Massacci calls $\mathcal{T}$ (for tableaux) we start with two distinct countably infinite sets $\mathcal{K}$ and $\mathcal{P}$, the set of all basic contexts and the set of all propositional atoms. Then well-formed formulas $\mathcal{F}$ are built from the sets $\mathcal{K}$ and $\mathcal{P}$ by negation and implication, together with the $\text{ist}(\kappa, A)$ operator. As in PLC, contexts are actually sequence of basic contexts and contexts determine the *vocabulary* of an application or theory. The vocabulary is as before described by a function $\text{vocab}: \mathcal{K}^* \to 2^{\mathcal{P}}$ assigning to each context sequence $\overline{\kappa}$ a subset of the basic propositions that are supposed to be meaningful in that context.

But instead of axioms, Massacci introduces tableaux rules, together with a semantics in terms of "superficial valuations". Massacci's tableaux system uses formulae with labels and labelled deduction rules. The labels on the formulae have a double role: given a contextualized formula $\langle \overline{\kappa}[n]: A \rangle$, $\overline{\kappa}$ is a sequence of basic contexts, n is an integer and A is a well-formed formula as above. Intuitively the prefix $\overline{\kappa}[n]$ 'names' the n-th superficial valuation, where A holds.

The first three rules correspond to the propositional *classical* basis and are standard for tableaux systems, except that they carry annotations telling you in which context/world you are working.

$$(\&)\frac{\overline{\kappa}[n]\colon A\&B}{\overline{\kappa}[n]\colon A,\overline{\kappa}[n]\colon B} \quad (\neg/\vee)\frac{\overline{\kappa}[n]\colon \neg(A\vee B)}{\overline{\kappa}[n]\colon \neg A \mid \overline{\kappa}[n]\colon \neg B} \quad (\neg\neg)\frac{\overline{\kappa}[n]\colon \neg\neg A}{\overline{\kappa}[n]\colon A}$$

The next two rules, called "databases rules" require some explanation. The local contextual database LB is a set of formulae holding in the initial context κ_0. The global contextual database GB contains the formulae holding in every context sequence $\overline{\kappa}_0$ extending the initial sequence κ_0. These rules are necessary to deal with logical consequence in modal logics, but are not related to the essence of contexts.

$$(\text{Loc})\ \frac{\vdots}{\kappa_0[1]\colon A}\ \text{If } A \text{ is in } LB, \text{ where } \kappa_0 \text{ is the initial context}$$

$$(\text{Glob})\ \frac{\vdots}{\overline{\kappa}[n]\colon A}\ \text{If } A \text{ is in } GB, \overline{\kappa} \text{ is present and extends the initial context}$$

The last two rules, *positive and negative lifting* deal with the essence of contexts. They somehow reproduce the effects of the modal axiom **K** and of the necessitation rule, plus the effect of the extra axiom Δ.

$$(\text{P-lift})\ \frac{\overline{\kappa}[n]\colon \mathtt{ist}(\kappa,A)}{\overline{\kappa}\kappa[m]\colon A}\ \text{If } \overline{\kappa}\kappa[m] \text{ is present in the branch}$$

$$(\text{N-lift})\ \frac{\overline{\kappa}[n]\colon \neg\mathtt{ist}(\kappa,A)}{\overline{\kappa}\kappa[m]\colon \neg A}\ \text{If } \overline{\kappa}\kappa[m] \text{ is new for the branch}$$

Massacci shows that the axiom (Δ) is derivable in his tableaux system, but does not prove syntactic equivalence between the systems PLC and $\mathcal{T}$: ideally we should like a theorem like $\vdash_{\mathsf{PLC}} A$ iff $\vdash_{\mathcal{T}} A$. To obtain such a theorem we need to show how to derive the rules of positive and negative lifting, using the axioms and rules of PLC.

5.1 Evaluating Massacci's Systems

Massacci claims two main advantages for his system: Firstly that the rules reflect "epistemic properties (lifting, use of assumptions, etc)". This seems too subjective. But secondly he proves computational properties: the system allows for local and incremental computation, satisfies strong confluence and can be adapted to different search heuristics. These advantages are clear, usually tableaux calculi are better for proof search than axiomatic systems. Also his kind of tableaux were devised for efficient automated theorem proving, which is always useful. Hence it would seem a good idea to constructivize $\mathcal{T}$ and to try to prove the conjecture above that $\vdash_{PLC} A$ iff $\vdash_{\mathcal{T}} A$ But I do not see how to mimick the positive and negative lifting rules of $\mathcal{T}$ using PLC's axioms and rules, and I guess the

6 Our Natural Deduction Formulation

The Natural Deduction system of contexts we have developed works only for the normal multimodal **K** fragment of PLC, $\mathcal{N}$ and $\mathcal{T}$, that is, for a system we could call $\mathbf{K}_n$. Of course one could always add up the axiom Δ to this system, but adding any axiom to a natural deduction system seems a bad idea. Our natural deduction system comprises the usual natural deduction rules for the propositional connectives, plus the following schema of rules, one for each modality $\mathtt{ist}(\kappa, \_)$.

$$\frac{\Gamma \vdash \mathtt{ist}(\kappa, A_i) \quad A_1, A_2, \ldots, A_k \vdash B}{\Gamma \vdash \mathtt{ist}(\kappa, B)} \Box_\kappa$$

where by $\Gamma \vdash \mathtt{ist}(\kappa, A_i)$ we mean a sequence of derivations $\Gamma \vdash \mathtt{ist}(\kappa, A_1)$, $\Gamma \vdash \mathtt{ist}(\kappa, A_2), \ldots \Gamma \vdash \mathtt{ist}(\kappa, A_k)$. This is an old formulation of normal modal **K**, dating back at least to the mid-eighties [Bel85]. People familiar with the formulation of the necessitation rule for system **K** in sequent calculus, need to note that the new rule $\Box_\kappa$ "builds-in" substitutions.

The monomodal system using rule $\Box_\kappa$ (for a single modality $\Box$) over a constructive basis is discussed in detail in [BdPR01]. On that paper, several possible formulations of a natural deduction formulation for a basic notion of necessity are discussed and compared. In particular a discussion of Fitch-style Natural Deduction[F52], and its formulation as a framework for constructive modal logics versus Prawitz-style Natural Deduction and why we prefer the latter is sketched[6]. The reason is simply that it is not obvious how to provide categorical semantics for Fitch-style Natural Deduction formalisms, whereas it is so for Prawitz-style natural deduction. It is also briefly mentioned in [BdPR01] that we do not discuss approaches to constructive modal logics that use the semantics of modal logics, in terms of Kripke models and accessibility relations, as part of the syntactic information used to characterize these systems. Using the intended semantics to define your syntax may not be cheating, but feels somehow underhand, especially when proving soundness of your system. Approaches along these lines include Gabbay's labelled deductive systems and Simpson's framework. Clearly our system is not a framework: we can only do a few modal systems (**K** and **S4**, possibly **KT, KD, K4**) and indeed rules change according to the system that we are considering. Our only advantage at the moment, when compared to the frameworks mentioned before, is to produce *semantics of proofs* for the systems we can deal with. This was the goal from the beginning.

---

[6] One preliminary answer to how would the Curry-Howard isomorphism help context logics is that a type theory with context modalities could be easily implemented in an interactive theorem prover such as Isabelle or PVS and this would facilitate the creation/interconnection of large repositories of theories.

The multimodal extension of the system does not appear to present any problems. Localized derivation (ie. derivation in context) can also de done, by labelling the turnstyle with a given context $\overline{\kappa}$.

$$\frac{\Gamma \vdash_{\overline{\kappa}} \texttt{ist}(\kappa, A_i) \quad A_1, A_2, \ldots, A_k \vdash_{\overline{\kappa}} B}{\Gamma \vdash_{\overline{\kappa}} \texttt{ist}(\kappa, B)} \, loc\Box_\kappa$$

We call the system without localization $\mathcal{B}$, because of Bellin's 1984 paper and we can prove that $\mathcal{B}$ satisfies strong normalization/cut-elimination, subformula property and also enjoys a simple categorical semantics, in terms of a cartesian closed category together with a finite collection of endofunctors, one for each modality. We expect similar properties to hold for the localized system, but have not had time to verify it.

7 Comparing Systems

Both PLC and Nayak's system $\mathcal{N}$ are given as axiomatic systems and comparing them first seems natural. Nayak's system $\mathcal{N}$ is clearly too simplified to compare with PLC, but given the system $\mathcal{N}$ without axioms **A3**, **A4**, **A5** and PLC, without Δ, do we have the same system? The question hinges on the effect of sequences of contexts, versus individual modalities, decorating the derivability relation. As we have seen the context switching rule of PLC can be substituted by the necessitation rule of localized multimodal **K**, if differences of vocabulary are disconsidered. But is this as general as usual multimodal **K** ?

For instance, if we have two unrelated contexts κ_1 and κ_2, which can only be concatenated to form $\kappa_1 * \kappa_2$ and A is a theorem, the following derivations are perfectly fine in $\mathbf{K}_n$:

$$\frac{\dfrac{\vdash A}{\vdash C_{\kappa_1} A}}{\vdash C_{\kappa_2} C_{\kappa_1} A} \qquad \frac{\dfrac{\vdash A}{\vdash C_{\kappa_2} A}}{\vdash C_{\kappa_1} C_{\kappa_1} A}$$

But presumably in PLC, only one of them, would be valid, as if $\kappa_1 * \kappa_2$ is a valid context sequence whereas $\kappa_2 * \kappa_1$ is not, then the context switching rule can only be applied to $\kappa_1 * \kappa_2$. At least for PLC this seems to be the case, as if the sequence of contexts doesn't matter, they describe it a *flat* model. Only if all contexts sequences formed from a given set $\mathcal{K}$ are valid and if only the distinct elements of any context sequence matter, ie if the derivability relation denoted by the context sequence $\vdash_{\kappa_1 * \kappa_2 \ldots * \kappa_n}$ is equivalent to the derivability relation denoted by any permutation of the sequence $\kappa_1 * \ldots * \kappa_n$ then Bouquet and Serafini's claim that "PLC is just the normal multimodal logic **K** extended with the Δ axiom" is justified. If "the new theorems proved in PLC with respect to normal multimodal **K** are only due to Δ" then PLC is indeed a sublogic of $\mathcal{N}$ and of Massacci's $\mathcal{T}$ and in terms of provability exactly equivalent to Bouquet and Serafini's MPLC.

Comparing Massacci's system $\mathcal{T}$ to PLC, we see that since the axiom **D** is not provable in PLC, but is directly derived from negative lifting rule in $\mathcal{T}$, $\mathcal{T}$ seems to prove more theorems than PLC. Also Massacci has proved that $\mathcal{T}$ proves all theorems that PLC proves. But it is not clear whether $\mathcal{T}$ proves only these.

8 Conclusions

Comparing the four systems in this note, it seems that a designer/user of a logic of contexts has plenty of choice between systems. He may choose not to have the axiom (Δ) at all, in which case our (localized) system, the restricted version of PLC and some variation of the tableaux system $\mathcal{T}$ should all be proved equivalent. This corresponds to a decent minimal logic of contexts. A context logic can also have the axiom Δ explicitly, as in PLC, or have its effect via multimodal **KD**, as Massacci's system seems to indicate. If the effect of Δ is desired, the second route may be best, as one has at least the axiomatic and the tableaux versions already in place. Finally our context logic user may opt for a simplified logic of contexts along the lines of Nayak's system. In that case, I don't know how to provide a sequent calculus or a Gentzen-style Natural Deduction formulation. Proof-theoretic tricks, as taking the formulas of the system considered only up to the equivalence relation that identifies $\Box_i A$ with $\Box_j \Box_i A$ can be used, but the effect is not elegant. Lastly the comparison between our system and the MCS/LMS work deserves more discussion that could be given here [deP]. Briefly the MCS/LMS systems seem to be able to embed all traditional modal logics, constructively or not, very easily, in what is a generalization of Natural Deduction. But it is not clear to me how to decorate MCS/LMS proofs with terms in a Curry-Howard isomorphic way. This, as well as proof semantics for those systems is subjecto for further work.

Acknowledgments. I would like to thank T. Altenkirch, C. Condoravdi, D. Crouch, R. Guha, L. Serafini and, especially, N. Alechina and T. Braüner for discussions.

References

[AS96] V. Akman and M. Surav. Steps Toward Formalizing Context.*AI Magazine.*

[AMPR98] N. Alechina, M. Mendler, V. de Paiva and E. Ritter. Categorical and Kripke Semantics for Constructive Modal Logics. In *Computer Science Logic (CSL'01)*, Paris, September 2001.

[BMV98] D. Basin, S. Matthews and L Vigano. Natural Deduction for Non-Classical Logics. *Studia Logica*, 60(1): 119–160, 1998.

[Bel85] G. Bellin A system of natural deduction for GL. *Theoria*, 2: 89–114, 1985.

[BdPR01] G. Bellin, V. de Paiva and E. Ritter. Extended Curry-Howard Correspondence for a Basic Constructive Modal Logic. In Proceedings of Methods for Modalities II, November 2001.

[BdP00]   G. M. Bierman and V. de Paiva. On an Intuitionistic Modal Logic. *Studia Logica*, 65: 383–416, 2000.
[BS00]   P. Bouquet and L. Serafini. Comparing Formal Theories of Context in AI. Submitted to Artificial Intelligence.
[BS84]   R. Bull and K. Sergerberg. Basic Modal Logic In *Handbook of Philophical Logic*, eds Gabbay and Guenthner, vol II, 1984.
[BBM95]   S. Buvac, V. Buvac, and I. Mason. Metamathematics of Contexts. *Fundamenta Informaticae*, 23(3), 1995.
[CP98]   T. Costello and A. Paterson Quantifiers over Contexts. CommonSense'98.
[CC+02]   D. Crouch, C. Condoravdi, R. Stolle, T. King, V. de Paiva, J. O.Everett and D. Bobrow Scalability of redundancy detection in focused document collections. Proceedings First International Workshop on Scalable Natural Language Understanding (ScaNaLU-2002), Heidelberg, Germany, May 2002.
[F52]   F. Fitch Symbolic Logic.
[F83]   M. Fitting Proof Methods for Modal and Intuitionistic Logic. 1983.
[G95]   R. Guha. Contexts: A Formalization and Some applications. Ph.D Thesis, Stanford University, 1995.
[McC93]   J. McCarthy. *Notes on Formalizing Context.* In *Proc. of the 13th Joint Conference on Artificial Intelligence (IJCAI-93)*, 555–560, 1993.
[McCB97]   J. McCarthy and S. Buvac. *Formalizing Context (Expanded Notes).* Available from http://www-formal.stanford.edu/jmc/
[McC96]   J. McCarthy. *A Logical AI Approach to Context.* Available from http://www-formal.stanford.edu/jmc/
[Mas94]   F. Massacci. *Strongly Analytic Tableaux for Normal Modal Logics.* In *Proceedings 12th CADE*, 1994, vol 814, LNAI 723-737.
[Mas95]   F. Massacci. *Superficial Tableaux for Contextual Reasoning.* In *Proceedings of the AAAI-95 Fall symposium on "Formalizing context"*, 60–66, 1995.
[Mas96]   F. Massacci. *Contextual Reasoning is NP-complete.* In Proceedings 13th AAAI 1996, pages 621–626. AAAI Press.
[Nay94]   P. Nayak. *Representing Multiple Theories.* In Proceedings of the Eleventh National Conference on Artificial Intelligence (AAAI'94), 1994.
[deP]   V. de Paiva Natural Deduction Systems for Contexts, manuscript, 2003.
[Pra65]   D. Prawitz. *Natural Deduction: A Proof-Theoretic Study.* Almqvist and Wiksell, 1965.
[SG00]   L. Serafini and F. Giunchiglia. ML systems: A Proof Theory for Contexts. To appear in the Journal of Logic Language and Information, July 2001. ITC-IRST Technical Report 0006-01.
[Ser02]   L. Serafini. Personal communication, 2002.

Communicative Contributions and Communicative Genres: Language Production and Language Understanding in Context

Anita Fetzer

University of Stuttgart
Institute of English Linguistics
D-70174 Stuttgart
Germany
anita.fetzer@po.uni-stuttgart.de

Abstract. This contribution examines the connectedness between the production and understanding of language, and between language use and context. It is firmly anchored to a relational conception of context and adapts Clark's [4] conception of language use as both cognitive and social. The introduction spells out the basic premises for assigning language production and understanding the statuses of social actions. The second part discusses the connectedness between language production, language understanding and communicative contribution by examining the premises for the differentiation between linguistic competence and communicative performance. The third part extends the micro frame of investigation by accommodating a further layer of context and contextual constraints: communicative genre [18]. Contrary to a communicative contribution, communicative genre is a collectively oriented macro category based on collaboration, cooperation [13], We-intentionality [23] and social intelligence [12]. It functions as a filter by constraining the production and understanding of possible micro communicative contributions in accordance with a particular macro goal.

1 Introduction

Language is neither generally produced at random, nor is it understood at random. Rather, with the intention to communicate a speaker produces one or more utterances for a particular hearer[1]. Utterances are not only produced and directed at a coparticipant in a particular situation, but they are also produced to achieve one or more communicative goals. Following Clark [4], language use is conceived of as both cognitive and social: coparticipants perform cognitive and social actions by producing language, viz. they perform one or more communicative acts, and they understand language by performing cognitive and social actions through which they infer their fellow coparticipants' communicative intentions. Against this background, language

---

[1] Speakers and hearers are assigned a dual function in communication: a speaker also acts as a hearer, and vice versa. Following Schegloff [21], I refer to them and their dual function by the term of coparticipant which is intended to denote the fuzziness of the two categories.

use is an intentional and rational endeavour, which represents a social action par excellence: it is anchored to minimally two coparticipants, minimally one utterance and to context. As regards the production and understanding of language, both are anchored to cognitive contexts; as regards the retrieval of particular presupposed information in order to produce or calculate communicative meaning, the relevant propositions are anchored to (1) linguistic context, which frames the actual language output, (2) cognitive context, which frames the actual language input, and (3) social context[2], which frames the linguistic context in which language output is produced. These different types of context are conceived of in a relational manner and are, for this reason, interconnected: social-context information interacts with linguistic-context information, linguistic-context information interacts with cognitive-context information, and cognitive-context information interacts with linguistic-context and social-context information. As a consequence of this relational perspective, context denotes a dynamic concept which is constantly updated in communication, and, if necessary, revised and recontextualized [6].

The goal of this contribution is to examine the connectedness between the production and understanding of language on the one hand, and between language use and context on the other hand. It is firmly anchored to an ethnomethodological perspective, according to which social context is constructed in and through the process of communication [7]. The following section discusses the connectedness between language production, language understanding and communicative contribution by examining the premises on which the differentiation between linguistic competence and communicative performance is based. It shows that an interactive conception of these types of competence allows for the accommodation of cognitive, linguistic and social contexts which are required for assigning the minimal unit of language use, the utterance, the status of a communicative contribution. The third section extends the micro frame of investigation in order to accommodate macro-oriented layers of context and their contextual constraints: communicative genres [18] and networks [23; 19]. The production and understanding of a micro contribution must be in accordance with the corresponding communicative-genre goals, and this is the reason why a communicative genre is assigned the function of a filter. Contrary to the micro notion of communicative contribution, a communicative genre is a collectively oriented macro category based on collaboration, cooperation [13], We-intentionality [23] and social intelligence [12]. In conclusion, a relational and interactive conception of language use in context requires both micro and macro categories, such as individual and social intelligence, I-intentionality and We-intentionality, communicative contribution and communicative genre, and coordination and cooperation.

---

[2] This contribution does not explicitly distinguish between social context and sociocultural context. Instead, it employs the superordinate term of social context. This is because sociocultural context is interpreted as representing a particular subset of social context specified by culture-specific instantiations of, for example, basic deictic categories of space and time, or ideological conceptual meanings, such as democracy, freedom, autonomy or individual.

2 Language Production, Language Understanding and Communicative Contributions

One of the fundamental premises in natural-language research is the differentiation between linguistic competence and communicative performance, and their underlying pillars of well-formedness and grammaticality on the one hand, and acceptability and appropriateness on the other hand. Contrary to the distinction of language as two independent systems[3], namely internal competence and external performance, this contribution adopts a relational and dynamic outlook on language and argues for interactive internal and external systems which are interdependent on cognitive, linguistic and social contexts. But of what nature is the connectedness between linguistic competence and communicative performance? Following Allen and Seidenberg [1:117], there is no clear-cut answer to this question: "The mapping between competence grammar and performance is at best complex, as we have noted; it is also largely unknown. A problem arises because the primary data on which the standard approach [generative linguistics] relies - grammaticality judgements - are themselves performance properties." Against this background, the premise of two autonomous systems, viz. external E-language which exists outside the mind and is independent of it, and internal I-language which exists inside the mind and is connected with it, does no longer seem reasonable. Instead, a functional-grammar outlook on language is adopted, according to which internal sentences are related to external utterances. But where do internal sentences and external utterances meet, and where do they depart? Following Givón [10:1], internal and external grammars serve different purposes:

> Perhaps the best way of saying what grammar is from a functional perspective is to say first what grammar is not. Grammar is not a rigid set of rules that must be followed in order to produce **grammatical sentences**. Rather, grammar is a set of strategies that one employs in order to produce **coherent communication**.
> Nothing in this formulation should be taken as denial of the existence of rules of grammar. Rather, it simply suggests that rules of grammar – taken as a whole - are not arbitrary; they are not just for the heck of it. The production of rule-governed grammatical sentences is the *means* by which one produces coherent communication.

Thus, the rules of an internal grammar play a decisive role in the production and interpretation of utterances in context, to which they are expected to be connected in a coherent manner. But is their connectedness to context a sufficient condition for the definition of rule-governed grammatical sentences? To answer this question, we have to be more precise about the notions of coherence, context and rule-governed grammatical sentence.

---

[3] Some frameworks differentiate between an internal I-language, which is not affected by potential performance-related inadequacies resulting from interfacing logical form (LF) and phonetic form (PF) modules, and an external language, which is the realization of the interfacing modules of I-language in external linguistic and social contexts.

Contrary to the mutually exclusive definitions of internal language and external language, functional grammar assigns a dynamic status to language and assigns it the status of a constitutive part of the cognitive system [10]. Here, a rule-governed grammatical sentence is one means amongst others to produce coherent discourse. This functional outlook on rule-governed grammatical sentences requires the explicit accommodation of (1) coherence, (2) social, linguistic and cognitive contexts, to which the sentences and propositions are expected to be connected in a coherent manner, and (3) communicative intention, which is a basic requirement for the production of coherent or dovetailed conversation. Against this background, the production and interpretation of sentences in context is directly related to the production and interpretation of communicative contributions, which are produced and interpreted in accordance with the Cooperative Principle (CP), namely "such as is required, at the stage at which it occurs, by the accepted purpose of direction of the talk exchange" [13: 45]. In the framework of the CP, coherence is spelled out by the concepts of "such as is required" and "dovetailed" [13:48], and in the framework of functional grammar, it is explicated by the surface phenomena of reference, ellipsis and substitution, conjunction, cohesive links and lexical cohesion, and by the macro phenomenon of discourse topic. As regards their connectedness to context, both coherence and the Gricean CP are anchored to some type of sequence and thus to cognitive, linguistic and social contexts. As a consequence of this, rule-governed sentences and their instantiations in context, viz. communicative contributions, must be examined with regard to their connectedness to cognitive, linguistic coherence and social contexts.

Assigning rule-governed sentences and their instantiations in context, i.e. language production and language understanding, the status of a communicative contribution connects it with the fundamental pragmatic premises of rationality, intentionality and appropriateness. The former manifest themselves in the coparticipant's production and interpretation of communicative contributions set in a game of giving and asking for reasons, which is explicitated by Brandon [2:xxi] as follows: "In a weak sense, any being that engages in linguistic practices, and hence applies concepts, is a *rational* being; in the strong sense, rational beings are not only linguistic beings but, at least potentially, also *logical* beings". Brandon is even more explicit in his conception of rationality [2:117]: "Rationality consists in mastery of those practices [the game of giving and asking for reasons, as Sellars calls it]. It is not to be understood as a logical capacity. Rather, specifically logical capacities presuppose and are built upon underlying rational capacities." Against this background, appropriateness is necessarily anchored to coparticipants, communicative contributions and contexts, and is calculated with regard to the connectedness between the linguistic representation of a communicative intention and its social and linguistic contexts. For this reason, appropriateness feeds on both external and internal languages: it draws from internal language for the formulation and interpretation of coparticipant-intended meaning, and it draws from external language with regard to the connectedness between a communicative contribution and its linguistic and social contexts. The anthropologist Muriel Saville-Troike [20:53-54] specifies the concept of appropriateness as follows: "The choice of appropriate language forms is not only dependent on static categories, but on what precedes and follows in the communicative sequence, and on information which emerges within the event which may alter the relationship of participants." Put differently, a necessary condition for the definition of appropriate language use is the notion of choice, which manifests itself in the coparticipant's ability to choose between a more appropriate utterance and a less appropriate utterance. This is also reflected in the principle of sociolinguistic variation [3], according to which language is used differently in different situations[4]. Implicit in this principle is the premise that a particular communicative intention, for instance a request, can be expressed by an almost infinite number of utterances and thus in a more and less explicit, and in a more and less polite manner. As a consequence of this connectedness between the language system (linguistic expressions and grammatical constructions), language use (utterances and communicative contributions), social practice (what is considered as appropriate or inappropriate by a speech community) and context, Brown [3:169] points out that

> [o]ne cannot mechanistically apply the Brown and Levinson model of politeness strategies to discourse data; particular linguistic realizations are not ever intrinsically positively or negatively polite, regardless of context. Politeness inheres not in forms, but in the attribution of polite intentions, and linguistic forms are only part of the evidence interlocutors use to assess utterances and infer polite intentions.

She then elaborates on the prerequisites of a linguistically and socioculturally competent coparticipant. In order to coparticipant-intend politeness, coparticipants have to be able to monitor their fellow coparticipants' and their own actions, and possible perlocutionary effects in the framework of AIP (anticipatory interactive planning):

> To operate according to the model, speakers have to be able to modify the expression of their communicative intentions so as to take account of what they see as their interlocutor's views of what they might be taken to be wanting to communicate, including what impositions to face might be on the table, as well as his or her assessments of the speaker's and hearer's relative power and social distance. [3:154]

The foundations, on which AIP is based, are (1) the calculation of the coparticipants' social actions, (2) their possible perlocutionary effects, and (3) the degree of politeness communicated. According to the view taken here, this is only possible if coparticipants are intentional and rational agents who act in accordance

---

[4] What is of interest here is the fact that the principle of sociolinguistic variation is not restricted to the domains of language use and social practice. It is also inherent in Brandon's [2:425] philosophy-of-language approach and his premise, that

> [a] language cannot refer to an object in one way unless it can refer to it in two different ways. This constraint will seem paradoxical if referring to an object by using a singular term is thoughtlessly assimilated to such activities as using a car to reach the airport or using an arrow to shoot a deer: even if only one car or one arrow is available and impossible to reuse, what one is doing can still genuinely be driving to the airport or shooting the deer.

with the principle of sociolinguistic variation by selecting the most appropriate form for their goals. That is, they choose particular lexical expressions amongst other possible expressions, particular grammatical constructions amongst a number of other possible constructions, and particular intonational contours amongst a number of other possible contours.

To conclude, it does not seem reasonable to see communicative performance as an activity independent of the mind. Instead, it must be re-evaluated and recontextualized as expressing a speech community's conception of what is appropriate and what is not. Appropriateness manifests itself in social practice and its underlying rules and regularities and therefore denotes a far more complex phenomenon than had been anticipated: it is anchored to (1) the linguistic units of sentence, proposition and utterance, (2) the principle of sociolinguistic variation, which is a constitutive part of intentional and rational communication and manifests itself in the coparticipant's anticipatory interactive planning, (3) the exchange of communicative contributions, (4) the coparticipants seen from *I - we* and *I - thou* perspectives [2:508], and (5) their micro and macro linguistic and social contexts. As a consequence of this change of perspective, linguistic competence is a subset of communicative competence and denotes the coparticipant's ability to differentiate between grammatical and ungrammatical sentences, appropriate and inappropriate communicative contributions, and coherent and incoherent dialogue, as well as to her /his ability to produce coherent dialogue and interpret dialogue in a coherent manner, which is examined in the following section.

3 Language Production, Language Understanding, and Communicative Genre

Coherence is a macro concept which is anchored to the macro-oriented domain of communicative genre. From a relational viewpoint, it is also reflected in its communicative contributions. Thomas Luckmann [18:177] explicates the functions and forms of a communicative genre as follows:

> Communicative genres operate on a level *between* the socially constructed and transmitted codes of 'natural' languages and the reciprocal adjustment of perspectives, which is a presupposition for human communicative interaction. They are a universal formative element of human communication.
> (...)
> Human communicative acts are *pre*defined and thereby to a certain extent *pre*determined by an existing social code of communication. This holds for both the 'inner' core of that code, the phonological, morphological, semantic and syntactic structure of the language, as well as for its 'external' stratification in styles, registers, sociolects and dialects. In addition, communicative acts are predefined and predetermined by explicit and implicit rules and regulations of the *use* of language, e.g. by forms of communicative etiquette.

Luckmann's references to "*pre*defined" and "*pre*determined" are also implicit in John Heritage's [15:242] notion of *doubly contextual* which designates the dual function of linguistic context in conversational interaction: on the one hand, a communicative contribution invokes linguistic context by constructing it itself, on the other hand, its sole production and interpretation provide the context for the subsequent talk. That is, a communicative contribution relies upon existing context for its production and interpretation, and it is in its own right an event which shapes a new context for the action that will follow. Thus, the act of speaking and interpreting constructs contexts and at the same time constrains the construction of contexts. In the following Luckmann's outlook on the macro category of a communicative genre and its constitutive pillar of coherence are refined by the explicit accommodation of (1) the intentionality of social action, (2) I-thou sociality [2] and we-intentionality [23], (3) intersubjectivity, and (4) practical reasoning, AIP and social intelligence.

Natural-language communication is frequently defined by interlocutors, their communicative intentions and by the performance of unilateral speech acts in context. But context, viz. cognitive, linguistic and social contexts, is not a unilateral, but rather a relational notion which can not be reduced to a micro context only. Rather, it is represented by interdependent layers, to employ the onion metaphor [25], or by interdependent frames [11]. For this reason, the retrieval of a communicative contribution's micro contextual references is necessarily connected with the macro category of coherence and thus with a communicative-genre frame of reference and its constitutive discourse topic(s). This extension of frame and the change of perspective from a micro-contextual, bottom-up approach to a macro-contextual, top-down frame of reference has the necessary consequence that the coparticipants' production and interpretation of communicative meaning can no longer be restricted to their individual intentions only. Instead, the production and interpretation of intersubjective meaning must be based on the Searlean conceptions of collective intentionality and We-intention [22, 23, 24], on Dascal's [5] conception of collective we-intention and on Brandon's [2] conception of I-thou sociality. Against this background, communicative meaning is calculated in accordance with the particular macro constraints of a communicative genre, which filter the production and interpretation of intersubjective meaning accordingly. But is a communicative genre performed intentionally?

The intentionality of action is not only a core concept in the research paradigm of natural-language communication but also in the field of artificial intelligence with respect to the question of how longer stretches of talk are processed. What is of relevance for this investigation about the connectedness between intentions and context is Litman and Allen's [17:376] process-oriented differentiation between discourse intentions or plans of a speaker and plans generated by these plans: "*Discourse intentions* are purposes of the speaker, expressed in terms of both the task plans of the speaker (the domain plans) and the plans recursively generated by the plans (the discourse plans)". In other words, a speaker may have a particular discourse intention, for instance to conduct an interview about renting an apartment, but s/he can not plan every single action or task by him/herself because a plan is a dynamic construct which may generate different subplans, if the immediate context requires a change of action. Moreover, if Litman and Allen's conception of a discourse intention is adapted to a dialogue setting, discourse intentions are not only postulated, but also have to be ratified in interaction by accepting or rejecting them. Furthermore, like any other pragmatic presuppositions, discourse intentions can be represented both explicitly and implicitly. Analogously to the explicit or implicit linguistic representation of intentions in speech act theory, Dascal [5] differentiates between overt and covert collective we-intentions.

The dynamic nature of discourse is also reflected in the processing mode employed. Contrary to the bottom-up processing of single actions, discourse processing, as is argued by Litman and Allen [17:380], requires a top-down approach: "Once a set of discourse and domain plans is recognized, each is expanded top down by adding the definitions of all steps and substeps (based on the plan libraries), until there are no unique expansions for any of the remaining substeps." If the artificial-intelligence setting is adapted to the constraints and requirements of natural-language communication, the recognition of 'a set of discourse and domain plans' can be compared and contrasted with the hearer's calculation and recognition of the speaker's communicative intention and the corresponding inference processes involved. But are these two tasks really equivalent? As regards non-complex plans, such as a request to pass the vinegar, the two domains can be equated. As regards complex plans, however, for instance the performance of the communicative genre of an interview, we have to differentiate between a discourse or a macro intention anchored to the macro category of a communicative genre as a whole, and a communicative or micro intention anchored to the performance of an individual communicative contribution as a part (of a whole). Yet what kind of relationship is there between a micro or an individual I-intention and a macro or a collective we-intention?

In his investigation about the connectedness between intentionality and conversation Searle [22:400] explicitly stresses the fact that:

> [c]ollective intentional behavior is a primitive phenomenon that cannot be analyzed as just the summation of individual intentional behavior; and collective intentions expressed in the form "we intend to do such-and-such" or "we are doing such-and-such" are also primitive phenomena and cannot be analyzed in terms of individual intentions expressed in the form "I intend to do such-and-such" or "I am doing such-and-such"

Against this background, we-intentions of collective intentionality can not be reduced to the summation of individual I-intentions. Instead, they are intrinsically linked to the macro category of communicative genre, or to employ Searle's [22:406] own words: "The reason that we-intentions cannot be reduced to I-intentions, even I-intentions supplemented with beliefs and beliefs about mutual beliefs, can be stated quite generally. The notion of a we-intention of collective intentionality, implies the notion of *cooperation*." For this reason, we-intentions are necessarily anchored to dialogue, viz. to a frame of reference that goes beyond an individual communicative contribution, which, if conceived as a part in a whole, "is derivative from the collective intentionality 'we are doing act A'" [22:403]. Thus, the concept of we-intentionality is a context-dependent notion par excellence. Yet it is not only anchored to cognitive contexts but also, as explicated and specified in Searle [23], to social contexts and social reality, which are further refined in Searle [24:109] into background presuppositions anchored to sociocultural context that vary from culture to culture. Social reality and social context are constructed through collective representation in accordance with constitutive rules and collective acceptance. Moreover, we-intentionality is anchored to the dialogue principle of cooperation: "Collective intentionality presupposes a Background sense of the other as a candidate for cooperative agency; that is, it presupposes a sense of the others as more than just conscious agents, indeed as actual or potential members of a cooperative activity." [22:414].

The premise of cooperation, on which Searle's definition of we-intentionality is based, is further refined by Brandon [2:508] with regard to the more basic category of I-thou sociality:

> The social distinction between the fundamental deontic attitudes of undertaking and attributing is essential to the institution of deontic statuses and the conferral of propositional contents. This is, (...) an *I-thou* sociality rather than an *I-we* sociality. Its basic building block is the relation between an audience that is attributing commitments and thereby keeping score and a speaker who is undertaking commitments, on whom score is being kept. The notion of a discursive *community* - a we - is to be built up out of these *communicating* components.

This intersubjective stance is further refined by Grosz and Sidner [14:427], who underline the cooperative and collaborative nature of dialogue as follows: "To account for extended sequences of utterances, it is necessary to realize that two agents may develop a plan together rather than merely execute the existing plan of one of them. That is, language use is more accurately characterized as a collaborative behavior of multiple active participants."

Collaborative behaviour is also a key concept in the language-as-social action paradigm, where it applies to the joint production of a communicative genre, the joint production of communicative contributions and the joint production of utterances. Following Goody [12:26], the macro category of a communicative genre is conceived of as "socially constructed models for the solution of specific types communicative problems". That is, a speech community provides particular 'plans', or particular we-intended macro propositions, in and through which specific types of communicative actions are performed. For instance, the task of seeking and providing information is generally performed in and through the communicative genre of an interview, the task of influencing and persuading people is generally performed in and through the communicative genres of a speech or a debate.

The inferencing processes involved in natural-language communication are the standard ones, namely deduction, by which one infers specific instances from a general rule, induction, by which one presumably discovers the general rule from a representative sample of specific instances, and abduction, by which one reasons by hypothesis from instances or general rules to their wider context. Givón [8:14] stresses the decisive difference between abduction and the deductive procedure of inferring specific instances from a general rule, "[t]his mode of **hypothesis** [abduction, A.F.] often involves **analogical reasoning**, and thus the pragmatic, context-dependent notions of *similarity* and *relevance*". The importance of abductive reasoning is also stressed by Levinson's [16:230] reference to Aristotle: "As Aristotle argued, the logic of action is a distinct species of non-monotonic (defeasible) reasoning, a *practical reasoning* (PR) as it has been dubbed by philosophers." Thus, practical actions do not exist as such but are performed by social actors who act in a rational manner and are, for this reason, in a position to account for their social actions. Garfinkel [7:3] explicates the process and product of the accountability of social action as follows:

> (1) Whenever a member is required to demonstrate that an account analyses an actual situation, he invariably makes use of the practices of "et cetera", "unless" and "let it pass" to demonstrate the rationality of his achievement. (2) The definite and sensible character of the matter that is being reported is settled by an assignment that reporter and auditor make to each other that each will have furnished whatever unstated understandings are required. Much therefore of what is actually reported is not mentioned. (3) Over the time for their delivery accounts are apt to require that "auditors" be willing to wait for what will have been said in order that the present significance of what has been said will become clear. (4) Like conversations, reputations, and careers, the particulars of accounts are built up step by step over the actual uses of and references to them. (5) An account's materials are apt to depend heavily for sense upon their serial placement, upon their relevance of the auditor's projects, or upon the developing course of the organizational occasions of their use.

As regards the chronology of accounting for one's actions and performing actions, social actors can only account for their social actions once they have processed and contextualized them by filling the gaps and finding the grounds to argue for the appropriateness of a social action. Through the process and product of accounting for social actions, which is anchored to a retrospective-prospective outlook on communication, social actors demonstrate substantive rationality and daily life rationalities. So far, reasoning has been primarily seen from a hearer viewpoint who calculates the meaning of an utterance with regard to the question of what the speaker intends to communicate. But what happens if a dialogue stance is adopted?

Dialogue is based on collaboration and cooperation and therefore requires social intelligence, which manifests itself in interactive thinking [3] and AIP [12] as well as in the social-interaction notion of a learned program. The dyadic and dialogic conception of AIP is explicated by Goody [12:12] as follows:

> Both through inner speech, which is the sort of dialogue with ourselves (between me and I), and through our close attention to conversational partners, spoken language seems to have constructed a dialogue template for social cognition. In inner speech and in conversation, dialogue and the dyad are built into human cognition.
> (...)
> This highlights the complexity of AIP, which must both model contingent responses and model strategies for securing actions from others which are favourable to ego's goals. AIP moves constantly back and forth between modelling and strategic action.

To conclude, a social conception of language based on the (partly) joint production of communicative contributions and the joint production of communicative genres can no longer be based on an autonomous outlook on language. Instead, it requires a relational conception of language anchored to the collective category of communicative genre based on collaboration, cooperation [13], We-intentionality [23] and social intelligence [12]. Against this background, it seems more plausible to adopt a network perspective which employs parallel distributed processing [24, 25].

4 Conclusions

A relational conception of language use in context requires both micro and macro categories, such as individual and social intelligence, I-intentionality and We-intentionality, communicative contribution and communicative genre, and coordination and cooperation. The dual status of communication as both cognitive and social and its consequences for the actual production and interpretation of utterances in context is succinctly formulated by Levinson [16:238]:

> Linguistic communication is fundamentally parasitic on the kind of reasoning about others' intentions that Schelling and Grice have drawn attention to: no-one says what they mean, and indeed they couldn't - the specificity and detail of ordinary communicated contents lies beyond the capabilities of the linguistic channel: speech is a much too slow and semantically undifferentiated medium to fill that role alone. But the study of linguistic pragmatics reveals that there are detailed ways in which such specific content can be suggested - by relying on some simple heuristics about the 'normal way of putting things' on the one hand, and the feedback potential and sequential constraints of conversational exchange on the other.

References

1. Allen, J., Seidenberg, M.: The emergence of grammaticality in connectionist networks. In: MacWhinney, B. (ed.): The Emergence of Language, Lawrence Erlbaum, Mahwah (1999) 115–152
2. Bar-Hillel, Y.: Indexical expressions. In: Kasher, A. (ed.): Pragmatics: critical concepts. Routledge, London (1998) 23–40
3. Brandon, R.: Making it Explicit: Reasoning, Representing, and Discursive Commitment. Harvard University Press, Cambridge (1994)
4. Brown, P.: Politeness strategies and the attribution of intentions: the case of Tzeltal irony. In: Goody, E. (ed): Social Intelligence and Interaction, Cambridge University Press, Cambridge (1995) 153–174
5. Clark, H.: Using Language, Cambridge University Press, Cambridge (1996)
6. Dascal, M.: On the pragmatic structure of conversation. In: Searle, J. et al. (eds): (On) Searle on Conversation, Benjamins, Amsterdam (1991) 35–56
7. Fetzer, A.: Non-acceptances: re- or un-creating context. In P. Bouquet, P. Brezillon, L. Serafini (eds.): 2nd international and interdisciplinary conference on modeling and using context (Context'99). Springer, Heidelberg (1999) 133–144

8. Garfinkel, H.: Studies in Ethnomethodology, Polity Press, Cambridge (1994)
9. Givón, T.: Mind, Code and Context, Laurence Erlbaum, Hillsdale (1989)
10. Givón, T.: English Grammar: a Function-Based Introduction, Benjamins, Amsterdam (1993)
11. Givón, T.: Generativity and variation: the notion 'rule of grammar' revisited. In: MacWhinney, B. (ed): The Emergence of Language, Lawrence Erlbaum, Mahwah (1999) 81–114
12. Goffman, E.: Frame Analysis, North Eastern University Press, Boston (1974)
13. Goody, E.: Introduction: some implications of a social origin of intelligence. In: Goody, E. (ed): Social Intelligence and Interaction: Expressions and Implications of the Social Bias in Human Intelligence, Cambridge University Press, Cambridge (1995) 1–36
14. Grice, H.: Logic and conversation. In: M. Cole, Morgan, J. (eds): Syntax and Semantics. Vol. III, Academic Press, New York (1975) 41–58
15. Grosz, B., Sidner, C.: Plans for discourse. In: Cohen, P., Morgan, J.,Pollack, M. (eds): Intentions in Communication, MIT Press, Cambridge (1992) 417–444
16. Heritage, J.: Garfinkel and Ethnomethodology, Polity Press, Cambridge (1984)
17. Levinson, S.: Interactional bias in human thinking. In: Goody, E. (ed.): Social Intelligence and Interaction, Cambridge University Press, Cambridge (1995) 221–260
18. Litman, D., Allen, J.: Discourse processing and commonsense. In Cohen, P., Morgan, J.,Pollack, M. (eds): Intentions in Communication, MIT Press, Cambridge (1992) 365–388
19. Luckmann, T.: Interaction planning and intersubjective adjustment of perspectives by communicative genres. In: Goody, E. (ed.): Social Intelligence and Interaction, Cambridge University Press, Cambridge (1995) 175–188
20. Perry, J.: Reference and Reflexibility, CSLI Publications, Stanford (2001)
21. Saville-Troike, M.: The Ethnography of Speaking, Blackwell, Oxford (1989)
22. Schegloff, E.: To Searle on conversation: a note in return.In: Searle, J. et al. (eds): (On) Searle on Conversation, Benjamins, Amsterdam (1991) 133–128
23. Searle, J.: Collective Intentions and actions. In: Searle, J. et al. (eds): (On) Searle on Conversation, Benjamins, Amsterdam (1991) 401–415
24. Searle, J.: The Construction of Social Reality, The Free Press, New York (1995)
25. Searle, J.: Mind, Language and Society. Doing Philosophy in the Real World, Weidenfeld & Nicolson, London (1999)
26. Sperber, D., Wilson, D.: Relevance: Communication and Cognition, Blackwell, Oxford (1996)

Explanation as Contextual Categorization

Leslie Ganet[1], Patrick Brézillon[2], and Charles Tijus[1]

[1] Laboratoire Cognition & Usages, CNRS FRE 2627
2, rue de la Liberté 93523 Saint-Denis Cedex
ganet@netcourrier.com, tijus@univ-paris8.fr
[2] LIP6, Case 169, University Paris 6, 8 rue du Capitaine Scott, 75015 Paris, France
Patrick.Brezillon@lip6.fr

Abstract. Our concern is the explanation generation in a representation based on contextual categorization. We point out that the explicit consideration of the context is necessary for the generation of relevant explanations. We present how the model captures the context necessary for explanation and we report some results compatible with the hypothesis that explanation is based on categorical networks according to a model based on the Galois lattice.

1 Introduction

In the past 30 years, Artificial Intelligence research aimed at developing automated reasoning systems and programs for solving problems. In 80's, the question was to develop systems that were able to explain their line of reasoning. Although at that time Artificial Intelligence was the science that explored the most deeply the explanation process, an implicit assumption was to consider explanations as a vehicle for transferring information and knowledge from the machine to the user, supposing that the machine was the oracle and the user the novice [1]. Feedback were used by the system to tailor its explanation to user's needs. However, such feedback were limited to acceptance signs, and users rarely may intervene in the generation of explanations. An opposite position was taken in the SEPT application by letting the user build alone his explanation [2]. This was not a better solution because users had to tackle complex commands as an additional task with their work and temporal constraints. An actual perspective is that the user and the system must cooperate to solve jointly the problem and to co-construct the explanation of the solution [1].

Because a good explanation is a contextualized explanation, the lack of contextual information about the task at hand has been recognized as a weakness of rule-based systems[4]: there is a lack of consideration for context, a recurrent problem in knowledge engineering. Our position is that explanation is based on context, and that context can be processed through a contextual categorization mechanism.

Consider, for instance the sentence "*I heard a lion in my office this morning*" [4]. This is an ambiguous sentence since either the lion (interpretation a) or the person (interpretation b) can be in the office and the other in the near outside. Additional context such as "*This is what the man says when he enters the Police station*" will favor the former while "*This is what the man says after he explains he was watching on a TV program on animals in his office*" will favor the latter. Our proposal is to demonstrate that and how an explanation rest on categories of actual objects using actual relational properties that the human cognitive system builds up. Starting with sentence 1, object-person and object-lion are put in the same category because they have the property of *"being close from each other"* and to be both located in the office place (the building that comprises the office). Subcategories are "the person" and "the lion" (Figure 1).

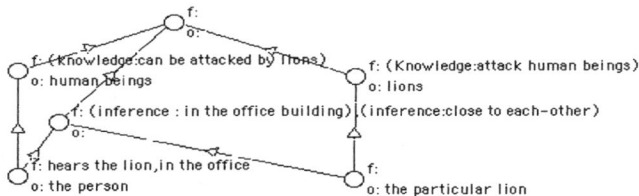

Fig. 1. According to the contextual categorization theory, when listening "I heard a lion in my office this morning", 1. actual objects "the person who is talking" and "the particular lion" are depicted as objects (o), 2. they are instantiated as subordinate categories with their specific features (f), 3. the categories /person/ and /particular lion/ are linked to their known categories and properties /human beings – can be attacked by lions/ and /lions-attacks human beings/ and are linked to their contextual superordinate category of being close to each-other (by inference because the person hears the lion) and located in the same building (by inference because they are close to each-other in a place that comprises the office). Notice that alternative interpretation locating the actual lion in the office instead of the person would exhibit the same network of contextual categories

From the network of contextual categories of Figure 1, one can tell the story, as "This is a story about human beings and lions. In an office's building there were a person and a lion and the person hears the lion". First, note that this description matches both interpretation a and b. This is one of the features of contextual categorization, which is the capability of extracting commonalties while it compute contextual diversity. Other important feature is that description/explanation can be generated by top-down parsing the network of categories.

Another context of the "hearing a lion" situation might be provided as follows: "I work in an university near a zoo that I can see from the window of my office. There are lions in that zoo. I often hear lions roar. It just was the case this morning." Then, the network of categories relating to the situation is rectified: the places of the lion (the zoo) and of the person (the office at the university) are differentiated, and the university will be near the zoo, and the superodinate known category for the actual lion will be "lions in zoo" (that do not attack human-beings").

The question addressed in this paper concerns the way in which relevant explanation can be built from such a network of categories. One general purpose is to define the content and the format of an explanation according to the context.

The paper is organized as follows. Section 2 introduces explanation. We propose an inventory of current work about explanation, mainly about how explanation is generated and we detail the parameters and factors that influence explanation. Section 3 presents how context plays a key role in explanation generation by identifying different contexts. Then we describe the contextual categorization framework for building explanation (Section 4) and we report data that support our proposal. Finally, we conclude (section 5) by discussing how our framework might be completed in order to be a general model of the construction of contextually based explanation.

2   What Is Explanation?

Derived from the Latin word *explicatio(nem)*, and from *explicare* which means literally "to unfold", from *plicare*, "to fold" (used until the XVIIe century). This word appeared in the French language in 1322, and was borrowed from the French into the English language in 1528.

According to different dictionaries as sources (Larousse, Hachette) but also to dictionaries of psychology, explanation is both the action of explaining as well as a development intended to make something comprehensible. It can also mean an account of something or a clarification concerning a series of actions taken. In French, it can also mean the supervision of someone, a discussion, quarrels concerning the supervision of someone. In cognitive psychology, an "explanation" occurs when a subject adopts a system of meaning, coherence, a presumption of a structure in a text or of a phenomenon. In a general sense, explanation has the aim of solving a problem of comprehension. Let us retain, in the extended meaning, that explanation corresponds to any operation implied in the constitution of the understanding of a phenomenon.

According to the goal of explaining, we can distinguish different kinds of explanation: account, alibi, annotation, apology, clarification, comment, commentary, definition, demonstration, description, elucidation, excuse, gloss, exemplification, explication, exposition, illumination, illustration, interpretation, justification, plea, reason, solution, indication. There are other kinds of explanation in French such as analogy, answer, argument, causes, controversy, debate, development, discussion, dispute, exegesis, explanation, exposure, hermeneutics, information, key, motive, note, notices, paraphrase, precision, reason, study, talk, translation.

All the different kinds of explanation do have a content. They are also constructed according to a format and they are communicated through a medium. Explanation also depends on other factors such as the purpose of the explanation, the explainer and the explainee, feedback and task constraints. In our approach, we consider explanation between two actors, two people or a person and a machine. Our research is restricted to the content of explanation and its format.

Content. The content of any explanation comprises the phenomenon to be explained either explicitly or implicitly, and additional information. Additional information (1) has to be related to what is to be explained either as knowledge or as inference, and (2) has to be drawn from the same domain or, if drawn from a different domain, should show similar relations between elements (analogy). Finally, the content has to integrate contextual information in order to highlight the phenomenon to be explained.

Format. The content has a certain structure, a format that sequentially structures the content. For example, "*A whale is said to be mammal because whales breast-feed, their children*" is one kind a format. "*Whales breast-feed their children. A whale is said to be a mammal*" is another kind of format.

3 The Context, a Key Factor for Explanation Generation

3.1 Preliminary

Mackie [5] has already stressed the context-dependency of explanation as a process of making a distinction between some current situation and another class of situations. Thus, context – involving both explainer beliefs and goals – is crucial in deciding how good an explanation is, and a theory of contextual influences can be used to determine which explanations are appropriate.

Leake [6] considers the relationships between explanations and context in the framework of case-based reasoning. An explanation is required when there is a conflict between an event and a model that we have of the place where the event occurs. Leake argues that such a conflict is a property of the interaction between events and context: Any particular fact can be anomalous or non-anomalous, depending on the situation and on the processing we are doing. For example, 25°C may be considered hot weather for Paris, France, but cold for Rio de Janeiro, Brazil. To be relevant to an anomaly, explanations must resolve a belief conflict underlying the anomaly. To resolve an anomaly, an explanation must account for why prior reasoning led to false expectations or beliefs. Any anomaly would allow the retrieval of explanation for identical anomalies, provided that the same anomaly was always described the same way and that distinct anomalies always received distinct characterization. Finally, Leake lists ten major explanation purposes triggered by anomalies that rely on several elements of context (expected/believed conditions, previously unexpected conditions, possible repair points, actor's motivations, etc.).

Thus, explanation and context are strongly intertwined. Explanations make context explicit in order to clarify a step in the reasoning process. They are a means to point out the links between the problem at hand and shared knowledge in its current state.

The way in which an explanation must be chosen and generated depends essentially on the context in which the two actors find themselves. An explanation always takes place relative to a space of alternatives that require different explanations according to the current context. Comparing two explanations leads seeing how their contextual spaces differ. Thus, taking into account context is necessary to study explanation [7].

3.2 The Different Contexts

From an engineering point of view, the context is a collection of relevant conditions and surrounding influences that make a situation unique and comprehensible [8]. However, there are other points of view on context. In the accomplishment of a task, a person identifies which knowledge is relevant to the job based on previous experience. What Brézillon and Pomerol [3] call "contextual knowledge" are pieces of knowledge judged relevant and which can be mobilized at a specific step in the decision making process. A subset of the contextual knowledge at that step is invoked, structured and situated according to the focus corresponding to the step in the decision making process. This subset is called the proceduralized context.

An important issue is the transition from contextual knowledge to the proceduralized context. Proceduralization depends on the focus on a task, even for a political conversation which can be based on how a political program should work. Thus, proceduralization, like "know how", is task-oriented and is often triggered by an event or primed by the recognition of a pattern. Another aspect of proceduralization is that people transform contextual knowledge into functional knowledge or causal and consequential reasoning in order to anticipate the result of their own actions. Proceduralization requires a consistent explicative framework in order to anticipate the results of a decision or an action. This consistency is obtained by reasoning about causes and consequences in a given situation. We can thus separate the reasoning between diagnosing the real context and anticipating the follow up. The second step requires conscious reasoning about causes and consequences.

A second aspect of proceduralization concerns a kind of instantiation. This means that the contextual knowledge or background context needs further specification to fit the task at hand. Precision and specification brought to bear on the contextual knowledge are also a part of the proceduralization process. For instance, it has been shown [9] that there are different levels of context, from the more general to the more specific and heterogeneous. A context at one level (e.g. the group context) contains rules that are instantiated at the level below (e.g. individual contexts). For example, when the rule is a speed limit of 50 km/h in a city (group context), a driver will control the speed of his vehicle using the accelerator and brake pedals (individual context).

4 Contextual Categorization

Contextual categorization is a component of diverse theories or models about human cognition that comprises perceptual categorization [10], categorization for text understanding [11], or task oriented categorization [12]. We describe how contextual categorization Theory works and how it can help modeling the generation of explanation and how it can be used for modeling the transition from contextual knowledge to proceduralized knowledge. Contextual categorization is founded on a basic and simple mechanism that is applied to process environmental inputs of any kind. It is of interest because it has two main results: (i) it shows the organization of both present objects and of present properties, or, more precisely, how properties are distributed to form categories of objects and (ii) it reflects the organization of the world. The contextual categorization model operates on Galois Lattices to create a single hierarchy of categories with transitivity, asymmetry and irreflexivity, when given the $O_n \times P_m$ boolean table which indicates for each of the n objects, O, whether it does or doesn't have each of the m properties, P. The maximum number of categories is either 2^n-1, or m if $m < 2^n-1$, in a lattice whose complexity depends on the way properties are distributed across objects (table 1). For instance, having three objects (a, b c), the maximum number of categories is seven: the categories that factorizes respectively abc, ab, ac and cd shared properties and the categories that comprehend respectively a, b, and c unshared properties.

The Galois lattice corresponding to the binary description in table 1 is shown in Figure 1 as a hierarchy of Categories of objects defined by properties. The link between categories is a "KIND-OF" link: Y is a kind of X. Due to the inheritance principle; category Y includes properties of category X. This can be seen in the Boolean table in table 1. The Galois Lattice can be used to build a hierarchy of categories that merges when factorizing the properties [13]. The categories are contextual because they are a function of what objects are in the current situations and rather than simply a function of pre-existing categories in long-term memory. This is an alternative to case-based theories that need to encode each of the contexts in which an object could be met. The hierarchy of categories provides a circumstantial and contextual structure of the objects present. What is fundamental to contextual categorization, is that contextual categorization computes each unique object in the context of all the other objects that form its unique context.

Table 1. A binary description of "I heard a lion in my office this morning" that corresponds to the Galois Lattice in Figure 1

a	human beings	the person	lions	the particular lion
hears the lion		1		
(Knowledge :attack human beings)			1	1
(inference : in the office building)			1	1
(inference :close to each-other)		1		1
(knowledge :can be attacked by lions)	1	1		
in the office		1		

Explanation is based on description. We propose, first, that contextual categorization is the mechanism that is used to describe the phenomenon to be explained and, secondly, that the description is obtained by constructing the Galois lattice of the situation including the context. Third, that explanation is constructed syntactically by parsing the Galois lattice.

4.1 Building a Description for Explanation

Consider the material shown in Figure 2. It is an example of a set of characters we use in our experiments. We use such material for simplicity's sake. However, the objects in other situations might be, for instance, the cars involved in an accident, or different results obtained from experiments, etc. Each display contains a number of objects. One object differs from all the others (i.e., an intrusive object) and participants are asked to detect the intrusive object [14] and to explain the way in which it is different. The intrusive object is the only specimen in its category. For example, in the sample presented Figure 2, the intrusive object is *i* because the set of objects are letters; both vowels and consonants; however, *i* is the only vowel and thus it is the intrusive object.

The goal of this particular experiment was to demonstrate that both detection and description are based on contextual categorization.

From a cognitive point of view, seeking an intrusive object is seeking an object that has fewer characteristics in common with the other objects in the set. To complete the task, the participant has to use a categorization process considering the relevant properties of the situation (contextual properties). Once the contextual network of categories has been built, subjects easily evaluate the number of properties shared by each object in the set. Thus, they are able to indicate the object that has the fewest properties in common with the others. We call this object the intrusive object.

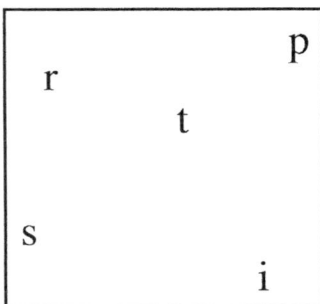

Fig. 2. An example of a set of objects to be described

Participants can be asked to explain their choices and to justify why they chose an object as being the intruder. Thus, we can observe whether they exploit the network of contextual categories that can be built from the display. In other words, we are interested in the process of construction of explanation with the hypothesis that the structure of explanation is related to the structure of the information considered, which is to say the relation between contextual knowledge and proceduralization.

Consider the following set of 9 objects: "3,2", "a small rectangle", "E", "a beetle", "π", "51", "a dog", "H", "a large square". They can be put in categories that both group and differentiate them. For instance, "3,2" and "51" might be considered as "numbers" while "rectangle" and "square" can be grouped as "geometric shapes" and "H" and "E" grouped as "roman letters." But in addition, "3,2", "51", "H", "E" and "π" can be seen as "characters".

Figure 3 provides the network of categories in which the objects can be simultaneously grouped and differentiated. Our assumption is that the participant having to name the category to which the intrusive object belongs has to differentiate it from its context and will use the most specific level that differentiates the intrusive object from its context. For instance, "E" surrounded by squares and rectangles can be called a letter and the justification for choosing "E" as intrusive can be "because it is the only letter." In contrast, "E" surrounded by "H" and "π" will be considered as "the only vowel".

First, this task of recognition makes it possible to control the context of the intruding object. Indeed, placing "E" among consonants or large rectangles does not lead to the same effects of context. Second, we can vary the context surrounding the intruder with objects belonging to categories more or less distant from it. In the example referred to above, if "E" is surrounded by consonants, then it is placed in a context that is semantically close: there are only two arrows joining the vowel category to the consonant category. In contrast, if we place "E" in a set of "large rectangles", seven arcs are necessary to connect these two categories. The context of the intrusive object "E" is more distant.

We tested the material in a pre-experiment. 84,05 % of the responses we obtain from 41 participants were fitting the predictions. These results are compatible with Treisman & Gelade's pop-out theory [15], although we explain the effect in terms of categorization and context rather than in terms of filter theory. In addition we found that more the context goes away from the intrusive object, more the explanation was enriched. In this case, the participants don't systematically observe that quoting the distinctive category of the intrusive object provides a sufficient explanation. Moreover, we observe another kind of effect of context. It seems more obvious to detect a consonant among vowels than a vowel among consonants. In consequence, there are contexts that simplify the task. We suppose that this effect comes from the number of objects included in the contextual category. Indeed, in the vowel category, we count six instances: *a, e, i, o, u* and *y* in opposite to the consonant category that comprises 20 of them.

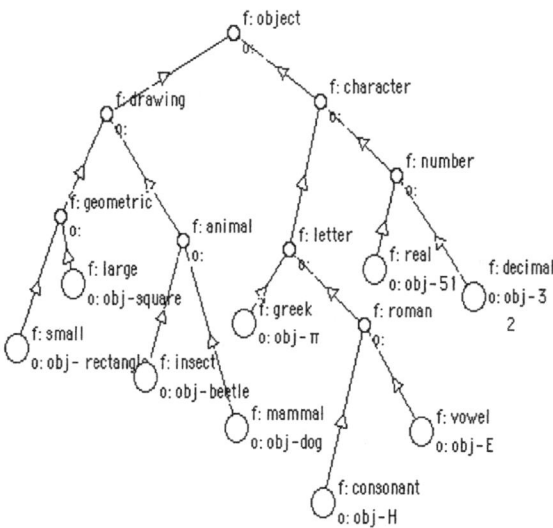

Fig. 3. The hierarchy of the set of categories that structures the set of 9 objects: "3,2", "a small rectangle", "E", "a beetle", "π", "51", "a dog", "H", "a large square".

4.2 Building a Structure for Explanation

Explanation is the process of providing information to someone who already does have some prerequisite knowledge (which should be first evaluated). For instance, in order to explain what is a duck, "a duck" can be defined as being an "animal", like a "chicken", but going in the water. This example is based on a specific explanation: the *description* where causal bonds between events and objets, so the time, are not considered. The description just rest on our categorized knowledge to provided names of objects (categories) and properties.

Such explanation-based verbalization was studied by Ganet [16], from data collected by Faure [17] with participants having to judge similarity between two sounds (sound 1 and sound 2) and to explain their judgment. To do so, twelve sounds were presented in couples providing 12 X 11 verbalizations by each of 20 listeners. We reasoned as if explanation-based verbalization were computed as predicted by contextual categorization, then participants should provide properties in their description in a strict order, revealing how they proceduralize their contextual knowledge.

We labeled "A" the properties common to both sounds (as in the sentence "the two sounds are soft"), "B" the properties describing sound 1 (as in the sentence "the first sound is rich and hot"), "C" the properties describing sound 2 (as in the sentence "the second sound is dry"), "D" the relational properties used to compare sound 2 to sound 1 (as in the sentence "but the second sound is more brilliant than the first") and "E" the relational properties used to compare sound 1 to sound 2 (as in the sentence "the first sound was longer than the second"). An explanation such as "I found the two sounds dissimilar because if they are both quite hot, the second is brilliant and longer than the first" was then coded an "ACD" explanation while an explanation such as "the second sound was dry but shorter than the first which was soft and the two are low" was then coded as a "CDBA" explanation. contextual categorization and listening order from first to second sound predict that participants should build their explanation in a "A then B then C then D then E" order, which means than the following structured explanation based verbalization "ABCDE". Among 325 possible formats, 31 (for example as "ADE", "AC", "A", "DE", "D") are compatible with predictions while the remaining 294 formats ("BA", "CDBA", "ACBDE", "ED") are not predicted by contextual categorization.

Table 2 show the percentage of each type of feature (A, B, C, D and E) given as a first, second, third, four and fifth feature given by the participants. The first feature the participants enounced the most was of type A (67 %). The second was of type B, and so on. Although only 31 of 325 types (9 %) of possible verbalizations were compatible with contextual categorization, 53 % of the verbalizations corresponded to our strict predictions. More over, the more frequent verbalization was of the ABC form (10%), followed by ABCD (6%), by A (5.2 %), AD (5%) and BC (4.2. %). Each of the predicted form was six times more frequent than the non-predicted form. In addition the 13 most frequent forms of explanation were 54% of the total amount of verbalization. Among them, ten were predicted (unpredicted were AED, BCA and ABDA) and corresponded to 78 % of them. In summary, contextual categorization appears to be a good approximation of modeling such human descriptive explanation. More important, these results permit us to continue our research and exploring more deeply the explanative process with its other type of explanation.

Table 2. Percentage of feature of type A, B, C, D and E as a function of the order in the verbal explanation. Type A features were given 63 % as the first feature. Type B features were given 37 % as the second feature and so on

	t1	t2	t3	t4	t5
A	63,12	4,38	10,98	9,86	9,35
B	20,50	37,37	11,53	9,13	5,61
C	4,12	23,71	40,07	21,88	21,50
D	3,95	17,61	27,88	36,78	34,58
E	8,31	16,92	9,55	22,36	28,97

5   Conclusion

Earlier expert systems, like MYCIN, align their explanation facility directly with the reasoning paths that define movement across contexts of the diagnostic system. Thus, all traces of reasoning that represent the traversed contexts are kept and their contents provided to the user for explanation. Here, the definition of context is restricted to knowledge and to steps of inferences but MYCIN is not selective in its construction of explanations.

In other systems, such as that described by Wick [18], an explanation facility is aligned only periodically with the reasoning of the system. In Wick's system, only some parts of the contexts that the system reasons with, are explained to the user. In this approach, additional explanatory knowledge (the domain knowledge and the expertise that are not directly necessary for the task at hand) may be used to generate enhanced explanations. This implies that the explanation path separates from the path of reasoning to produce effective explanations. Context is here an extended version of the previous one because it also contains domain and task knowledge not directly considered in the reasoning of the problem solving, and eventually some information on users through a model. One problem with such an approach is that it may be unsuitable for critical applications whose results may affect the safety of processes and people.

Another approach for explanation is to accept that the reasoning of the system is often different from that of the user. Thus, the user and the system may have different interpretations on the current state of the problem solving. The differing interpretations will be compatible if the user and the system make proposals, explain their viewpoints and spontaneously produce information [2]. In order to align the system's reasoning with that of the user and vice versa, the user and the system must co-construct the explanation in the current context of the problem solving. People who are trying to understand something often may offer an explanation that embodies their current understanding, expecting to have it corrected [19]. Thus, explanations become an intrinsic part of the problem solving and, as a consequence, the line of reasoning of the system may be modified by explanation. This leads to cooperative problem solving. Again, context here is an extended version of the context in the previous approach because it also integrates direct information from users, mainly on the basis of their actions on the system and on the real-world process.

If it seems acceptable that explanations intervene in the evolving context of interaction, it is difficult to say more about this for two reasons. Firstly, the co-building of explanations is an accepted idea but rather very few studies consider it. Secondly, context being not a mature domain of research, its dependency upon explanations is not really considered. For example, Lester and Porter [7] propose a model of explanation generation that includes simple methods for representing and updating context. However, their model makes assumptions about the representation of the context, not about how it is processed.

An alternative can be found in the line of Bever & Rosembaum work [20], we can see the semantic analysis as the form of a semantic hierarchy of features characterized by a principle of cognitive economy of which the first effect is to bring back the infinite diversity of the environment to a finished number of category. A formalism which seems adapted for that is the lattice of Galois [21] [13]. In the line of the theory on the categorization known as "based on the properties", we use the Galois lattice for description and the formation of the categories, a category being defined by a set of surface properties (visible properties), structural properties (parts and fitting of the parts of the object), functional properties (for what the object is used) and procedural properties (how we use it). These categories are used to group the objects which are gathered because they share common properties. The properties are used to form categories but the categories are also used to allot properties [22] [36]. Our general assumption is that the lattice of Galois is also a suitable formalism to explain the cognitive construction of the verbal production of the comparisons which are made to build up explanation.

Our approach might be useful for Automatic Generation Of Explanation if we could diagnose the user description of the data. As shown with our current experiments, the labels the participants use to describe the objects might reflect the level of differentiation, which is to say the way they conceptualize the data. The context appears very useful and deserve to be considered.

References

1. Karsenty, L. and Brézillon, P.: Coopération Homme-Machine et Explication. Le Travail Humain, 58. (1995) 291–312
2. Brézillon P, Bau D-Y, Fauquembergue P, Hertz A & Maizener A: Elaboration of the SEPT expert system as a coupling of a simulator and a diagnostician. Third Industrial and Engineering Applications of Artificial Intelligence and Expert Systems. Charleston, South Carolina, USA, 1 (1990) 54–60.
3. Brézillon, P., Pomerol, J.-C.: Misuse and Nonuse of knowledge-bases systems: the past experiences revisited. In: Implementing systems for supporting management decisions. P. Humphrey, L. Bannon, A. McCosh, P. Migliarese, J.-C; Pomerol (eds).Chapman and Hall, London (1996) 44–60
4. Brézillon, P.: Context needs in cooperative building explanations. First europan Conference on Cognitive Science in Industry. Luxembourg (1994) 343–350
5. Mackie, J.: Causes and Conditions. American Philosophical Quaterly. 2 (1965) 245–264
6. Leake, D. B.: Evaluating Explanations. Lawrence Erlbaum Associates Inc. (1992)
7. Lester, J.-C.,Porter, B. W.: Generating context sensitive explanations in interactive knowledge-based system. Proceeding of the AAAI'91 Workshop on comparative Analysis of Explanation Planning Architectures. (1991) 27–41
8. Anderson, J. R.: Rules of the Mind. Hillsdale N. J.: Lawrence Erlbaum. (1995)
9. Brézillon, P.: Using context for Supporting Users Efficiently. Proceedings of the 36th Hawaii International Conference on Systems Sciences, HICSS-36, Track "Emerging Technologies" (2003, to appear)
10. Schyns, P., Goldstone, R.L., & Thibaut, J.-P.: The development of features in Sjoberg, L. (1972). A cognitive theory of similarity. Goteborg: Psychological Reports, 2 (10) (1998)
11. Tijus, C.A., Moulin, F.: L assignation de signification à partir de textes d histoires drôles. L'Année Psychologique, 97 (1997) 33–75.
12. Barsalou, L.W.,: Deriving Categories to Achieve Goals. In A. RAM & D.B. LAKE (eds.), Goal-Driven Learning. Cambridge: The M.I.T. Press. 1995
13. Poitrenaud, S.: The Procope Semantic Network: an alternative to action grammars. International Journal of Human-Computer Studies, 42 (1995) 31–69.
14. Tijus, C. A.: Contextual Categorization and Cognitive Phenomena. In: P. B. V. Akman, R. Thomason, R.A. Young (Eds.) (ed):Modeling and Using Context. Third International and Interdisciplinary Conference, CONTEXT, 2001. Dundee, UK (2001)

15. Treisman, A.M., & Gelade, G. A feature-integration theory of attention. Cognitive Psychology **12** (1980) 97–136.
16. Ganet, L., Faure, A. and Tijus, C. A.: Étude de la Comparaison d'Objets et de Sons à Partir de Jugements de Dissemblance: Prédictions à Partir du Treillis de Galois. Laboratoire Cognition et Activité Finalisées, Saint Denis., (1999)
17. Faure, A., McAdams, S.: Des sons aux mots: comment parle-t-on du timbre musical ?. In: Langage et pensée: 10ème Séminaire Jean-Louis Signoret, Edited by: Viader, F., Eustache, F., Brussels: De Boeck. (2001)
18. Paris, C. L., Wick, M. R., Thompson, W. B.: The line of reasoning versus the line of explanation. Proceedings of AAAI Workshop on explanation. (1988) 4–7
19. Mark, B.: Explanation and interactive knowledge acquisition. Proceeding of AAAI'88 Workshop on Explanation. (1988)
20. Bever, T. G., Rosembaum, P. S.: (eds) R. A. Jacobs and P. S. Rosembaum. Reading in English Transformational Grammar. Some lexical structures and their empirical validity. Blaisdell (1970) 586–599
21. Guénoche, A., Van Mechelen, I.: Galois Approach to the Induction of Concepts. In:, R. Michalski et al. (Eds.). Categories and concepts: Theorical views and inductive data analysis, Academic Press (1993) 287–308
22. Tijus, C., Poitrenaud, S. & Richard, J. F.: Propriétés, objets, procédures: les réseaux sémantiques d'actions appliqués à la représentation des dispositifs techniques. Le Travail Humain, **59** (1996) 209–229.

Effects of Context on the Description of Olfactory Properties

Agnès Giboreau[1], Isabel Urdapilleta[2], and Jean-François Richard[2]

[1]ADRIANT
54 rue Lamartine 75009 Paris, France
a.giboreau@adriant.com
[2]University Paris VIII, CNRS FRE 2308,
Dpt Cognition et Activités Mentales Finalisées, 2 rue de la Liberté
93526 St Denis cedex 02, France
{isabel.urda, richard}@univ-paris8.fr

Abstract. This experiment examines the effects of context on olfactory descriptions. Odors are difficult to describe, and their verbalization results in strong individual variation. Sixty subjects were asked to describe 12 floral perfumes in two environmental contexts: an isolated context in which the odors were presented one by one, and a comparative context in which they were presented in groups of three. The results show contextual effects on the verbalization of olfactory properties. When the odors were presented in groups of three, 1) the subjects generated a larger number of olfactory descriptors, 2) there were fewer unique properties, i.e. generated by only one subject, and 3) subjects were more likely to verbalize general properties than specific properties. We discuss these results in light of categorization theories and the role of perceived properties in the assignment of objects to a specific category on the basis of context.

1 Introduction

We are interested in the role of context in the identification and description of sensory properties related to olfaction. In particular, we focus on the olfactory description by naïve subjects of fragrances in various environmental conditions: isolated or simultaneous presentation of samples.

The study of olfaction presents special difficulties. The concept of sensory properties tends to be more complex than in the case of vision or hearing, particularly because olfactory properties are often associated with taste. It is often difficult to describe olfactory properties because of the many processes involved in the cognitive integration of sensory information, including physical stimulation, sensory verbalization and perception. Traditionally, there are said to be five sensory modalities: hearing, touch, vision, taste and smell (this classification originated with Aristotle). Sensory properties are associated with a specific sensory modality; they are classified as a function of the modality to which they belong. For example, transparency and translucence are visual properties of objects and are unequivocally associated with vision. Visual information lends itself to the classification of properties by modality: the appearance of objects is easy to describe (brightness, color, shape, etc.) and psychophysics effectively describes the relationship between physical stimulation and perception [1]. The same holds true for auditory information. There is even greater ambiguity over the classification of properties by sensory modality with respect to the chemical senses – taste and olfaction. When subjects are asked to give examples of olfactory perceptions, the only ones that come readily to mind are flowers and fruits, descriptions related to other perceptions as mild, unripe and sweet, and hedonic descriptors [2], which, moreover, reflect great inter-individual variability [3].

What is the reason for this difficulty in describing olfactory stimulation?

Olfactory perception thresholds are highly variable from one individual to another [4, 5], and there are many different types of anosmia: many people are completely insensitive to certain molecules. But researchers control for these perceptual believes that the encoding of olfactory information results from a personally significant event. Many authors [6, 7] support the idea of a special olfactory memory, different from other sensory modalities because of its very strong association with personal experience, which endows it with an individual character. However, the experiments of Chu and Downes [8] show that deeply rooted, odor-related personal memories mainly result from childhood experiences during the ages of six to 10. Whatever the case may be, odors are difficult to describe. Berglund *et al.* [2] conducted similarity tests between odor samples and demonstrated that odors could be classified along a hedonic dimension (*bad* versus *pleasant*), but the samples could not be characterized by any other descriptive dimension. The same odor sample presented to several subjects generated a wide variety of descriptors. These observations tend to refute the existence of a "true label," which designates a specific term associated with a specific odor. Elmers [9] proposes the existence of an "inner nose" that allows one to have internal olfactory images without necessarily possessing a specific vocabulary. In this way, he explains the deficiency of our olfactory language.

Cognitive theories of categorization that integrate the effects of context could shine new light on the problem of olfactory description. But what do we mean by the "effects of context"?

1.1 Effects of Context

One can point to many different contextual effects:

One such effect results from the interactions among sensory modalities, including the halo effect, which involves the influence of one sensory perception on another even though they are physiologically different. Confusions of this type between color and odor and between taste and odor are well known, For example, a green drink will be described as having a mint smell, vanilla milk will be described as sweet [10] and a sucrose solution will be judged sweeter if a pineapple, strawberry or caramel aroma is added. Degel and Koster's experiment [11] illustrates this effect. Subjects were asked to match an odor (presented in a bottle) with a visual context. The pictures contained an image associated with odor (for example, a cup of coffee on a table and leather jackets in a department store). The results demonstrate the interaction between odor and vision, an interaction that involves implicit memory.

A second effect concerns the effects of mental context that clearly emerge when the test instructions are changed and the same test is conducted over different sessions (task repetition). Indeed, certain tasks seem to produce large intra-individual variations. Barsalou [12] shows that during a task in which subjects were asked to verbalize olfactory properties, only one-third of descriptors were common to the two participants and the descriptors given by the same subject varied according to the point of view the subject was asked to take (his/her own or the point of view thought to be held by a third person). These results also show that, for the same subject, just over half the descriptors overlapped between the two sessions. This strong variability has been confirmed by other authors [13, 14].

The mental context can also lead the subjects to favor either specific or general responses. For example, in response to the instruction, "describe the properties of this object," the participants can answer, "it's a fruit, it's an apple, it's a Golden Delicious, it has a skin, etc." This inter-individual difference in describing an object's properties can be explained by pragmatic factors [15], e.g. the subjects use conversational rules that allow them to be as informative as possible. They express themselves at a specific level so that properties can be used to distinguish one category from another. That is why subjects favor specific answers in response to a general question about an object's characteristics, while the general characteristics are considered much less informative, based on the rationale of providing information that can clearly differentiate objects.

A third effect of context concerns the effects resulting from the range of stimuli presented to the subject. Among these effects are simple contrast (objects that seem strong when compared to other objects with low intensity and vice versa); range mapping (subjects match the range of objects to the available notation scale); frequency bias (subjects tend to use all response categories possible the same number of times); contraction bias (subjects tend to match the center of the scale to the mid-range stimulus); centering bias (the responses tend to gravitate around the center of the scale); and, finally, transfer effects (experiences from the previous sessions affect the responses of the session in progress). These effects mainly involve the use of scales of intensity. With respect to odors, Lawless and Heymann [10] demonstrate, for example, that variations in sensory quality are caused by simple effects of contrast: the same odor (dihydromyrcenol) is judged to be more woody in a lemony context than in a woody context. The contexts were thus modified by the presence of other molecules (more woody or more lemony).

We decided to concentrate on the last of these three contextual effects, e.g. the influence of objects present at the same time. We will examine the effect of this context, which we call "environmental context," on the verbalization of specific or general properties and not on their classification.

1.2 Context and Verbalization of Properties

The description of sensory properties can be studied in the context of cognitive theories of categorization. Categorization can be defined as the perception of similarities and differences among objects in the perceptual scene, using categories stored in memory that define the properties on which perceived similarities and differences are based. Categorization is one of the essential characteristics of human cognition. It allows individuals to order the environment according to classifications that help them deal with non-identical stimuli by using the most relevant properties in the specific context. Categorization therefore seems to be a very important mechanism that plays a role in all our daily activities, from the perception of a stimulus to its behavioral response. It is essential for explaining the functioning of semantic memory, cognitive development and, in general, the major cognitive activities, such as planning, memorization and perception [16, 17].

The various theories of categorization allow us to underscore the importance of object properties in cognitive processes. Some researchers assign a dominant role to objects in the construction and use of categories [16]. In this case, properties are associated with categories. The features that are characteristic of objects belonging to a basic category, *e.g.* "*table,*" are easy to describe [18] and convey more information than the features characteristic of superordinate categories, *e.g.* "*furniture*" or subordinate categories, e.g. "*coffee table*" [19]. Other authors suggest that objects are organized according to the properties they share [20]. In such a semantic network, the properties that describe objects are organized in such a way that certain properties are more generic while others are more specific. However, we believe that their description varies according to the context. Consider the following example: let's say that the letter 'a' is the object studied. Several properties can be used to categorize it: a character, a letter, a vowel, the first letter of the alphabet, a lower-case letter, etc. If we place this object in different contexts and subjects are asked to underline the object in question, the instructions will have to be precise to avoid any ambiguity. In the following examples, it will be necessary to ask subjects to underline, respectively, the letter [385a]; the lower-case letter [AaAA]; and the vowel [cat]. Thus the object is the same but the distinctive property varies depending on the context.

In tasks involving the generation of object characteristics, effects of context related to the level of specificity of properties can be expected. When the environmental context involves two or more categories, we assume that a larger number of properties will be generated, including not only distinctive properties but also shared properties. Vrignaud [14] compared the number of common properties expressed when categories were presented alone or in pairs. During the presentation of pairs, an average of 3.35 properties was obtained, compared to 0.78 properties when objects were presented in isolation. The author suggests that this effect is associated with the structural alignment effect described by Markman and Gentner [21]; common properties that share characteristics at a deep level can be aligned during a comparative task while more superficial characteristics are taken into account during a simple evaluative task.

The purpose of our experiment is to study the effects of environmental context on the olfactory description of 12 perfumes. The effect of environmental context is observed by comparing olfactory descriptors generated in the conditions, "isolated object (S = solo)" and "set of objects (T = trio)." The dependent variables are the number of properties, their level of specificity and the number of unique properties.

We have formulated the following three hypotheses:
1. Specific properties are dominant when an isolated stimulus is presented and general properties are dominant when several objects are presented. The subject tends to express the general properties of objects during a comparative presentation, while the presentation of a single object favors the expression of more specific properties.

2. When specific properties are dominant (when an isolated stimulus is presented), individual differences are greater: the mental context of the subject plays a major role since the environmental context presents few constraints.

3. A larger number of properties is generated when the objects are presented simultaneously. We assume that the simultaneous presentation of several objects encourages the subject to verbalize the general properties shared by the different objects and the specific properties of each object.

2 Materials and Methodology

2.1 Participants

Sixty participants (male and female, university students) volunteered for the experiment.

2.2 Materials

Twelve samples were chosen to represent a floral fragrance universe. The fragrances were flower essences: patchouli, a thousand flowers, jasmine, geranium, ylang ylang, chamomile, lavender, lilac, rose and three flower blends. The concentration of the ethanol solutions was determined in order to assure equivalent perceived intensities. The solutions were applied to strips of paper and once the alcohol had evaporated, presented to the subjects.

2.3 Protocol

In half the cases, the subjects received three samples at the same time (condition T). In the other half of the cases, the subjects received the samples one after the other (condition S). In condition T, the subject described the samples by comparing them to each other. In condition S, the subject described the samples one after the other. Verbalizations were noted in full.

2.4 Data Analysis

For each perfume and each subject, we noted the total number of properties generated per condition (in order to study the effect of context on the generation of properties) as well as the number of unique descriptors per condition (in order to study the variability among subjects according to the context). Finally, a level of specificity was assigned to each descriptor. To do so, we relied on a classification developed during previous studies [22]. This olfactory classification is based on a categorization scheme applied to the organization of olfactory descriptors [23, 24]. The classification is in the form of a tree diagram: general properties are the properties found at the top of the tree, such as *"food"* (level 1). Specific properties are the properties found on the farthest branches of the tree, like *"lemon"* (level 5). Thus, for each property communicated by subjects during the experimental task, a level of specificity-generality is assigned. Level 1 corresponds to the most general categories, while level 5 corresponds to the most specific. Table 1 provides examples of properties corresponding to the five levels.

Table 1. Examples of properties for each level of specificity

1	2	3	4	5
Civilization	House	Homecare product	Detergent	Washing dish liquid
Nature	Vegetal	Forest	Wood	Pine
Food and drinks	Fruits / vegetable	Fruit	Citrus	Lemon
Perception	Touch	Thermal	Hot	Burning
Feeling	Unpleasant	Sickening	Emetic	Extremely emetic
Value judgment	I like	I prefer	One of the best	The best

The percentage of properties in each of the five levels of specificity was calculated for each of the experimental conditions (conditions S and T) in order to study the effect of context on the specificity of descriptors produced in either an isolated situation or a comparative situation.

3 Results

We will preset, respectively, the results concerning the effect of context on the level of specificity of properties, the number of properties and the number of properties verbalized by a single subject (unique properties).

3.1 Effect of Environmental Context on the Level of Specificity

The percentages of properties by category (from 1 = general to 5 = specific) are presented in Table 2 according to the method of presentation of objects (S = solo, T = trio).

Table 2. Percentage of properties from each level of specificity according to context condition (S: single sample presentation, T: three samples presentation). The sum of proportions for levels 2 + 3 and 4 + 5 are indicated in brackets (Σ).

Level		General		Specific		Not coded	Total
	1	2	3	4	5		
Single sample (S)	0.40	6.40 (Σ=22.40)	16.00	33.20 (Σ=72.00)	38.80	5.20	100
Three samples (T)	1.44	16.91 (Σ=37.41)	20.5	29.86 (Σ=56.48)	26.62	4.68	100

The results are as follows:
- Few properties belong to the most general level (level 1). The most general categories of the classification are: civilization, nature, food, physical state, perception, feeling, value judgments. There are few olfactory descriptors that correspond to such a high level of generality. A chi-square test conducted on the two distributions based on the five levels resulted in a significant difference ($p = 6.63 \cdot 10^{-3}$) between the percentages by level of specificity of the two conditions.
- There is a larger proportion of properties in levels 2 and 3 in the comparative condition: 37% of general properties when the odors are presented in groups of three and 22% when the odors are presented alone. Conversely, a larger proportion of specific properties is generated when the odors are presented in an isolated manner (72%) than in a comparative context (56%).

These results confirm our main hypothesis about the effect of context, e.g. that more general properties are generated when several objects are presented together. Conversely, more specific properties are generated when the objects are presented alone. These are the object's distinctive properties at the level of the basic category to which it belongs.

3.2 Effect of Environmental Context on the Total Number of Properties

The 60 subjects generated a total of 652 descriptors for the 12 aromas in the isolated context and the comparative context.

The descriptors can be broken down in the following manner based on the two contexts (see Table 3):

Table 3. Total number of generated properties according to context conditions (S: single sample presentation, T: three samples presentation).

Conditions	S	T
Number of subjects (N)	39	25
Total number of properties (P)	374	278
Average number of properties by subject (P/N)	9.59	11.12

In the isolated condition, 39 subjects completed the experiment and generated a total of 374 descriptors, which comes to an average number of 9.59 descriptors per subject. This number rose to 11.12 descriptors per subject when the samples were presented in groups of three. A Student's t test conducted on the two conditions shows that the probability of erroneously rejecting the null hypothesis is 0.1028.

In line with our hypothesis, the results indicate that a larger number of properties were generated in the comparative context.

3.3 Effect of Environmental Context on the Number of Unique Properties

Out of 379 properties generated in the isolated condition, 227 were unique, e.g. they were used only once, by one subject and for one odor. That number represents 60.7% of all descriptors. In the simultaneous presentation condition, this percentage dropped to 42.1% (Table 4).

Table 4. Number of unique properties according to context conditions (S: single sample presentation, T: three samples presentation).

Conditions	S	T
Number of subjects (N)	39	25
Number of unique properties (UP)	227	117
Percentage of unique properties (UP/P)	60.7%	42.1%

A Student t test comparing the two conditions indicates that the probability of incorrectly rejecting the null hypothesis is 0.0382. Therefore, there were significantly more unique properties generated in the isolated context. These results confirm that the variability between subjects is greater when the stimuli are presented alone (condition S) than when they are presented together (condition T). Inter-individual variability is therefore larger when the property is described at a more specific level.

Moreover, there is less variability in the condition involving the simultaneous presentation of odors. In the comparative condition, the effect of context is more pronounced: one might conclude that the objects presented generate a complex level of properties that include the specific properties of the objects as well as properties common to all or part of the objects. Conversely, in the isolated condition, the network of properties generated is simpler, conditioned by the single object. The subject's mental context therefore carries more weight and is probably the source of the greater variability in the isolated context.

These results therefore confirm the hypothesis of a greater contextual variability when the object is presented by itself.

4 Conclusions

This study of the number of properties, the level of specificity of properties and the number of unique properties generated in conditions of isolated context (single object) and comparative context (three objects) confirmed our hypotheses concerning the effects of context on the olfactory terms used to describe 12 perfumes.

The results demonstrate the effects of environmental context on the verbalization of properties generated by the objects presented:

- The subjects generate more olfactory properties when several odors are presented together: the distinctive properties of each object within each category, but also the distinctive properties of the objects in comparison with each other.
- The number of unique properties is larger when an odor is presented by itself: one observes greater variability due to the expression of each subject's mental context, which is less influenced by a restrictive environmental context as opposed to the situation in which several objects are presented at the same time.
- The subjects tend to generate more specific properties than general properties in the isolated condition. Conversely, when several odors are presented simultaneously, the subjects tend to verbalize more general than specific properties.

We can therefore conclude from this series of results that the study of context can contribute to a better understanding of the difficulty in describing odors. However, categorization, verbalization of property and context are linked. Like Poitrenaud [25] and Richard and Tijus [26] we believe that the variability of categorization results from the fact that the properties perceived in the objects differ according to the context, which is comprised of the other objects presented simultaneously in the immediate environment. For example, in the context of a meal, the word "tomato" evokes the category, "vegetable": one perceives that this is something eaten as an appetizer, entrée or side dish. In an agricultural or horticultural context, however, tomatoes are perceived as fruits and the properties attributed to them are that they grow on plants, they come from flowers and they're picked during a specific season. These are not the same properties that are perceived in a stimulus situation in which other stimuli are presented at the same time. The reason is that there is a reciprocal generation of properties among the objects comprising the context, so that certain properties stand out while others remain hidden. More specifically, the properties selected are 1) properties shared by all the objects, which allows subjects to assign them to a category and 2) the properties that distinguish these objects and which allow subjects to determine sub-categories.

We have examined the effects of context with respect to the environmental context comprised of one or three objects. Other contextual situations should be studied in order to develop a broader understanding of the influence of environmental context. For example, the influence of the number of objects present on the generation of sensory properties associated with other sensory modalities should be investigated. It is also necessary to examine other types of context to better understand the effects of context on sensory descriptors – particularly the mental context and the type of task, which could be varied by, for example, changing the types of objects present while maintaining an equivalent olfactory context (odor on a strip of cloth or in bottles, natural objects, etc.). For a broader point of view, experiments could be completed with sensory professionals such as trained panels, oenophile, perfume creators, in order to relate natural categorisation with expert categorisation of odors.

References

1. Bagot, J.D. (ed.): Information, sensation et perception. Armand Colin, Paris (1996)
2. Berglund, B., Berglund, U., Engen, T., Ekman, G.: Multidimensional scaling analysis oftwenty-one odors. Scandivian Journal of Psychology, 14 (1973) 131–137
3. Holley, A.: Eloge de l'odorat. Odile Jacob, Paris (1999)
4. Stevens, D.A., O'Connell, R.J.O.: Individual thresholds and quality reports of human subjects to various odors. Chemical Senses, 16 (1991) 57–67
5. Baird, J.C., Berglund, B., Olsson, M.J.: Magnitude estimation of perceived odor intensity: empirical and theoritical properties. Journal of Experimental Psychology – Human Perception and Performance, 22, 1 (1996) 244–255
6. Aggleton, J.P., Waskett, L.: The ability of odours to serve as state-dependent cues for real-world memories: can Viking smells aid the recall of Viking experiences? British Journal of Psychology 90 (1999) 1–7
7. Wrzesniewski, A., McCauley, C., Rozin, P.: Odor and affect: individual differences in the impact of odor on liking for places, things and people. Chemical Senses. 24 (1999) 713–721
8. Chu, S., Downes, J.J.: Long live Proust: the odour-cued autobiographical memory bump. Cognition. 75 (2000) 41–50
9. Elmers, D.G.: Is there an inner nose? Chemical Senses 23 (1998) 443–445

10. Lawless, H.T., Heymann H.: Sensory evaluation of food: principles and practices. Chapman & Hall, Aspen, Maryland. (1999) 301–340
11. Degel, J., Koster, E.P.: Implicit memory for odors: a possible method for observation. Perceptual and Motor Skills, 86 (1998) 943–952
12. Barsalou, L.W.: Intraconcept similarity and its implications for interconcept similarity. In: Vosniadou, S. & Ortony, A. (eds.): Similarity and Analogical Reasoning. Cambridge University Press, New York. (1989) 76–121
13. Belleza, F.S.: Reliability of retrieval from semantic memory: common categories. Bulletin of the Psychonomic Society. 22 (1984) 324–326.
14. Vrignaud, P.: Approche différentielle de la typicalité. Unpublished doctoral dissertation, Université de Paris V (1999)
15. Sperber, D., Wilson, D.: La pertinence. Communication et cognition. Editions de Minuit, Paris (1986)
16. Rosch, E., Mervis, C.B.: Family resemblances: studies in the internal structure of categories. Cognitive Psychology. 7 (1975) 573–605
17. Urdapilleta, I., Nicklaus, S., Tijus, C.: Sensory evaluation based on verbal judgments. Journal of Sensory Studies, 14 (1999) 79–95
18. Rosch, E. (ed.): Principles of categorization in cognition and categorization. Laurence Erlaum Associated Publishers Hillsdale NJ (1978)
19. Zacks, J., Tversky, B.: Event structure in perception and conception. Psychological Bulletin, 127 (2001) 3–21
20. Collins, A.M., Quillian, M.R.: Retrieval time from semantic memory. Journal of Verbal Learning and Verbal Behaviour, 8 (1969) 240–247.
21. Markman, A.B., Gentner, D.: Splitting differences: a structural alignment view of similarity. Journal of Memory and Language, 32 (1993) 517–535
22. Giboreau, A., Urdapilleta, I., Richard, J.F.: Naming olfactory properties: designing an odor description space. (submitted)
23. Schleidt, M., Neumann, P., Morishita, H.: Pleasure and disgust, memories and associations of pleasant and unpleasant odors in Germany and Japan. Chemical Senses, 13 2 (1988) 279–283
24. Dubois, D.: Categories as acts of meaning: the case of categories in olfaction and audition. Cognitive Science Quaterly, 1 (2000) 35–68
25. Poitrenaud, S.: La représentation des procédures chez l'opérateur: description et mise en œuvre des savoir-faire. Unpublished doctoral dissertation, University Paris VIII, Paris (1998)
26. Richard, J.F., Tijus, C.A.: Modelling the affordances of objects in problem solving. In Quelhas C. & Perera F (eds.): Cognition and Context. Special Issue of Analyse Psychologica, (1998) 293–315

Varieties of Contexts

R. Guha[1] and John McCarthy[2]

[1] IBM Research, San Jose, USA
[2] Stanford University, Stanford, USA

Abstract. We believe that a deeper understanding of the uses of contexts, in terms of its impact on knowledge representation structures, as reflected by a corpus of examples, is vital to the programme of formalizing contexts in Artificial Intelligence. In this paper, we examine a number of examples from the literature from the perspective of identifying general usage patterns. We identify four important varieties of contexts — Projection Contexts, Approximation Contexts, Ambiguity Contexts and Mental State Contexts. We define each type, describe sub-types, list benchmark examples of each sub-type, discuss their practical uses and the requirements they make of the underlying logic. We pay particular attention to the problem of lifting, i.e., of using information obtained from one context in another and describe how these different varieties of contexts tend to require different kinds of lifting rules.

1 Introduction

Mathematics has developed and used logic to express the "eternal truths" [19], such as Peano's axioms, from which the rest of mathematics follows. For that programme, it is essential that the meaning of these axioms not depend on *anything* other than the logic in which they are stated. In particular, the circumstances of their statement should not impact their meaning. Consequently, traditional logic has expressly avoided contextuality. Only recently have logicians and philosophers started developing logics [2] which explicitly account for situations.

In contrast, human communication exploits the situation or context of the communication, often to an extreme degree, leaving much implicit. Processing on these communications also exploits context, i.e., we don't *completely* decontextualize what we hear into a global frame of reference before we reason with it. In fact we argue that a complete decontextualization is not just undesirable, but impossible. We do however have a deep understanding of the role of the situation on the meaning of an utterance. Factoring this in plays an important role when we use information obtained from one situation in a different situation.

We believe that logical formulae and other knowledge representations used by AI programs are more akin to human communication than to Peano's axioms. Therefore, understanding and coping with the effects of context on representation structures is is important for AI programs.

Since their first introduction into the logical AI programme in [14], a substantial amount of work has gone into formalizing contexts. A number of different logics ([9], [6], [17], [8]) have been proposed to deal with some (but not all!) of the motivating examples from [14], [9], [15][1].

Unfortunately, the different proposed extensions to traditional logic (both propositional and quantified) have very different forms. [9] and [4] interpret contexts as theories with the *ist* predicate corresponding to validity. [17] and [6] provide a modal interpretation for contexts. [8] provides a different kind of semantics based on the concepts of locality and compatibility. Consequently, it is very difficult to compare and evaluate these different approaches in terms of their appropriateness for representing knowledge. A complicating factor is that while most of the examples require a quantified logic of contexts, most of the proposed new logics[2] only deal with the propositional case.

We believe that a deeper understanding of the phenomenon, in terms of knowledge representation structures, as reflected by a corpus of examples, is vital to making progress. In this paper[3], we examine a number of examples from the literature from the perspective of identifying general usage patterns. We identify four important varieties of contexts. Three of the varieties of context we present — Projection Contexts, Approximation Contexts and Mental State Contexts are loosely correlated with the three kinds of contexts identified by Beneceretti, Bouquet and Ghidini [3]. In this paper, we go one step further and give precise definitions of each of these types, describe sub-types, list benchmark examples[4] of each sub-type, discuss their practical uses and the requirements they make of the underlying logic. We pay particular attention to the problem of lifting, i.e., of using information obtained from one context in another. We describe how these different varieties of contexts tend to require different kinds of lifting rules. We also identify an important fourth variety of contexts, namely, Ambiguity Contexts.

2 Lifting: A Framework for Categorizing Contexts

A computer system, especially an AI system, will work with several, perhaps many, contexts. The relations between sentences in different contexts are specified by *lifting relations*, and so are the relations between the values of terms in different contexts.

For example, we may have two contexts $A3$ and $A5$ specializing time to 3pm and 5pm respectively. Some sentences will be true in $A5$ if and only if they

---

[1] Some researchers [11] have argued that the phenomena of contexts is not one, but several different unrelated phenomenon. We believe this is a reaction to the lack of a good technical characterization of the problems that are being addressed by the introduction of contexts.

[2] It is not clear that we need an extension to traditional first order logic to handle contexts. In fact, it be very desirable if we did not require an extension.

[3] [21], an earlier paper with the same title explores varieties of *logics* of contexts. In contrast, our focus is on varieties of *uses* of contexts.

[4] Due to space constraints, we are only able to give brief descriptions of the examples in this paper. An extended version of this paper, available on the web, will contain the details of the examples.

are true in $A3$. Other sentences have more complicated relations. Thus if both contexts involve driving from San Francisco to Los Angeles it may be that the distance to Los Angeles is 140 miles less in $A5$ than in $A3$.

In traditional symbolic AI systems (ranging from deployed expert systems to systems such as Advice Taker) there is a single uniform database containing the programs knowledge. It is uniform in the sense that all sentences have the same contextualization. This enables the program to freely combine and use its information, i.e., if it has ϕ and $\phi \Rightarrow \beta$, it can combine the two to conclude β.

As we argued earlier, representations used by AI programs are similar to human communications in that they have context dependence. An AI program that has knowledge about a broad range of topics or takes inputs from a variety of sources has to cope with different subsets of its knowledge having different contextual dependencies. The situation/context in which a formula (or other representation) is given to the program makes assumptions and simplifications which in turn affects the statement of the formula. We expect the program to receive chunks of knowledge of varying sizes from different sources. Some will be from interactions akin to discourses where the contextual dependencies pertain to the situation and topic of the discourse. Some will be in the form of task-specific knowledge bases such expert systems where the dependencies will pertain both to the fragment of the world pertinent to the system and task being performed. As illustrated by these two examples, context dependencies come in a wide range of styles and shapes. The different kinds of assumptions and simplifications and the ways in which they affect formulae give us the different kinds of contexts.

The program will contain a number of databases, each corresponding to formulae with a different set of contextual dependencies, each pertaining to a different fragment of the world. It uses a set of terms denoting contexts to keep track of and deal with these dependencies. Each context (c) corresponds to a database (Δ_c) that pertains to a fragment (D_c) of the overall universe of discourse (D) that the program has knowledge about. Properties of this fragment may allow the program to make assumptions and simplifications that are not warranted by the larger D. These assumptions and simplifications are reflected in Δ_c.

If the statements ϕ and $\phi \Rightarrow \beta$ have different contextual dependencies, the program can't always combine them to conclude β. Before combining two sentences with different contextual dependencies, the program need to reconcile relative contextual dependencies. This relative decontexualization is done using a set of axioms we call *Lifting Formulae*.[5]

We believe that the main problem of contexts for an advice-taking AI program is that of coping with contextuality of the advice given to it. To cope with the contextuality, it needs to be able to factor out the relative contextuality so that it can use knowledge gathered in one context in another. To put it another way, the central issue in modeling contexts is that of lifting. We therefore look at the problem of determining varieties of contexts from the perspective of

---

[5] The name "lifting formula" came by analogy with topology. When there is a many-one map f from one space A to another B, some facts about B can be lifted to A. The analogy has not paid off so far.

lifting. We determine the varieties of contexts based on their assumptions and simplifications and the effects of these on lifting formulae into and out of them.

A brute-force approach of writing a separate lifting rule for every pair of contexts, for every relation will require a number of axioms that makes the Frame Problem look minor in comparison. We need a combination of good defaults, and very general lifting rules to solve this problem.

Just as the Frame Default [13] captures the intuition that most events don't affect most fluents, we need a similar default that most contextual factors don't affect the representation of most facts. This is the default that allows us to assume that most of what we know applies even in completely new and strange situations. An early attempt at capturing this intuition is described in [9], but clearly much more work is required.

For writing general lifting rules, we need a good understanding of what each (type of) lifting rule is doing. This is our real goal in this paper. Our goal in categorizing contexts is not just to organize them, but to help develop patterns of general and widely applicable lifting rules. In the next four sections we present the four varieties of contexts.

3 Projection Contexts

Sometimes, we can make assumptions about certain objects that occur often in the sentences in the database Δ_c corresponding to the context c. These assumptions allow us to simplify Δ_c by dropping portions of sentences pertaining to these assumptions. In the extreme, if the assumptions are strong enough, we can start dropping parameters to functions/predicates corresponding to these objects. The resulting simplified Δ_c can be seen as a *projection* of a more general Δ which makes these assumptions explicit. The actual projection operator depends on the structural form of the assumption. We now examine some of the popular examples in the literature that correspond to Projection Contexts.

Ex. 1: Normalcy/Kindness Assumptions

The use of contexts for making *Kindness Assumptions* [16] in planning and for making *Normalcy Assumptions* [9] in Cyc's Microtheories fall into this category. In such uses, we assume "normal" conditions, e.g., people are acting rationally, the physical location of the objects is on/near earth's surface, etc. We then leave these assumptions unstated in the database fragment Δ_c. These conditions might be made explicit only when the system tries to lift statements out of Δ_c. These assumptions might allow us to simplify the vocabulary of Δ_c to a point where the assumptions can no longer be expressed in Δ_c. [16] shows how a fully qualified axiom such as (using the syntax of [15])

$$C_0 : (\forall x) haveTicket(x) \wedge atAirport(x) \wedge clothed(x) \wedge conscious(x)... \Rightarrow canFly(x) \quad (1)$$

can be simplified in an appropriate *TravelContext* to

$$TravelContext : (\forall x) haveTicket(x) \wedge atAirport(x) \Rightarrow canFly(x) \quad (2)$$

In another example, Cyc [9] has the following axiom, which pushes qualifications such as the person acting rationally, the workplace not being a beach, there not being a fire, etc. onto the WorkplaceContext.

$$WorkplaceContext : (\forall x) Person(x) \land atWork(x) \Rightarrow clothed(x) \qquad (3)$$

Lifting into such contexts requires verifying that the normalcy assumptions are satisfied. Similarly, lifting out of such contexts requires qualifying the lifted axioms with normalcy assumptions. If the context c_1 makes the assumptions γ_1, γ_2, ..., and c_2 does not, and c_2 makes the assumptions β_1, β_2, ... which c_1 does not, the these lifting axioms have the form:

$$C_0 : (\forall \overline{x}) ist(c_1, \phi(\overline{x})) \land \gamma_1(\overline{x}) \land \gamma_2(\overline{x}) \land ... \Rightarrow ist(c_2, \phi(\overline{x})) \qquad (4)$$

$$C_0 : (\forall \overline{x}) ist(c_2, \phi(\overline{x})) \land \beta_1(\overline{x}) \land \beta_2(\overline{x}) \land ... \Rightarrow ist(c_1, \phi(\overline{x})) \qquad (5)$$

Arguably, this use of contexts can often be replaced by the use of nonmonotonic defaults. However, this use of contexts is not just for "normalcy" conditions. They are useful for capturing open-ended bundles of assumptions pertaining to classes of situations, including non-normal ones. For example, we might have a context corresponding to "War Time Conditions", analogous to the above *WorkplaceContext*. Since there is nothing normal about war time, it is probably not reasonable to always assume the assumptions made by such a context.

This example also illustrates how contexts can be very useful when combining knowledge built for different purposes. A knowledge base built for battlefield management might very reasonably assume that there is a war going on. On the other hand, a medical diagnosis system would probably not want to make that assumption. Contexts provide a useful mechanism encapsulating axioms based on their origin.

This example uses contexts as a solution to the qualification problem. It pushes the qualification onto the context parameter which acts as a hook for stating the qualifications in an incremental fashion, without rewriting all the axioms in the context.

Contexts are also required, for this kind of use, when the assumptions and simplifications cause a change in the language of Δ_c. Amarel's [1] example of reformulating the missionaries and cannibals problem, in which the assumption that the identity of particular missionaries and cannibals is not relevant allows for a reformulation that significantly reduces the computational complexity, is an example of this.

Ex. 2: Parameter Suppression

When the assumption is strong enough, the objects (about which assumption is being made) may become irrelevant and we may be able to drop references to them from our predicates and functions. *AboveTheory* ([15], [16]) is an example of this where the situation parameter gets suppressed. If we can assume that the world is static, i.e., if the situation argument is the same for all statements of the form $holds(\phi, <situation>)$, we can simply drop the reference to the situation and write just ϕ. Contexts such as $specSit(s_1)$ [16] obtained by focussing on a particular situation s_1 also fall into this category. Here are some examples of axioms in the *AboveTheory*:

$$AboveTheory : (\forall xy)(on(x,y) \Rightarrow above(x,y)) \qquad (6)$$

$$AboveTheory : (\forall xyz)(above(x,y) \land above(y,z) \Rightarrow above(xz)) \qquad (7)$$

Lifting involving contexts such as $specSit(s_1)$ is quite straightforward. We only lift into the context those axioms which have the appropriate value for that parameter. Similarly, when lifting out, we reinstate the appropriate value for the parameter.

$$blocks : (\forall xys)(holds(on(x,y),s) \Leftrightarrow ist(specSit(s), on(x,y))) \qquad (8)$$

$$blocks : (\forall xys)(holds(above(x,y),s) \Leftrightarrow ist(specSit(s), above(x,y))) \qquad (9)$$

If the parameter is not fixed to a particular value but is only assumed to be the same (whatever it is) across all axioms, as in the case of *AboveTheory*, when lifting out, reinstate all occurrences of the parameter with a universally quantified variable.

$$C_0 : ist(AboveTheory, (\forall (\overline{x})\phi(\overline{x}))) \Leftrightarrow ist(blocks, (\forall \overline{x}, s)holds(\phi(\overline{x}), s)) \qquad (10)$$

Lateral lifting, in which axioms are lifted without any modification from contexts such as *AboveTheory* into contexts such as $specSit(s_1)$ are accomplished with axioms like the following:

$$C_0 : (\forall s)(ist(AboveTheory, (\forall (\overline{x})\phi(\overline{x}))) \Leftrightarrow ist(specSit(s), (\forall (\overline{x})\phi(\overline{x})))) \qquad (11)$$

As we mentioned in section 2, writing lifting axioms specific to particular contexts such as *AboveTheory* will not scale. We need more general axioms such as the following, which will work across all static theories.

$$C_0 : (\forall sc)StaticTheory(c) \land (ist(c, (\forall (\overline{x})\phi(\overline{x}))) \Leftrightarrow ist(specSit(s), (\forall (\overline{x})\phi(\overline{x})))) \qquad (12)$$

Contexts such as *AboveTheory* and $specSit(s_1)$ suppress the situation (i.e., temporal) parameter and correspond to static models of the world. Similarly, parameters corresponding to location can be suppressed to create spatially local models of the world.

Ex. 3: Database Partitioning/Segmentation

Nayak [17] describes SIGMA [20], a knowledge-base of scientific domain knowledge that supports building executable domain models. SIGMA contains knowledge describing two very different application domains: modeling the atmosphere of one of Saturn's moons (Titan) and modeling a forest ecosystem. By separating the axioms of these two very different domains into different contexts, reasoning can be focused on just the axioms in the domain.

Cyc makes similar use of its *Microtheory* mechanism. The Naive Physics Microtheory, which contains a simple model of the physical world, has little in common with the US Legal microtheory which contains a simple model of the US legal process. By separating the axioms into different Microtheories, both knowledge entry and subsequent reasoning can be simplified.

In addition to long lived contexts such as the Naive Physics Microtheory, Cyc also uses shorter lived, *Problem Solving Contexts* [9] for focussing on a particular problem that it is trying to solve. When Cyc is given the description of a scenario (about which it will be asked questions), Cyc creates and uses a specialized Problem Solving Context (PSC) for that scenario. The scenario description, typically a set of ground facts, is entered into the PSC and general axioms are lifted into this PSC from Microtheories such as the Naive Physics Microtheory. Based on differences in normalcy assumptions, approximations, etc. between the PSC and Microtheory, the lifting might involve changing the axioms. The PSC serves to focus the inference engine on only the relevant objects.

In this role, contexts act like a "package" mechanism, not unlike the package mechanism found in programming languages such as Java and Lisp, with the caveat that lifting into Problem Solving Contexts is substantially more sophisticated and complex than importing lisp or java objects between packages.

3.1 Discussion

Projection Contexts are probably the most widely used type of contexts in implemented KR systems. In fact, systems such as KRL had a form of contexts for this purpose long before the introduction of contexts into logical AI.

The primary demand that Projection Contexts make of the underlying logic comes when the simplifying assumptions are strong enough to warrant a change in the vocabulary. In such a case, we need the underlying logic to be much less restrictive than traditional first order logic (FOL) about well-formedness, etc. For example, in traditional FOL, $holds(s_1, on(a,b)) \wedge on(a,b)$ is typically not a well formed formula. But with the kind of vocabulary simplifications introduced by Projection Contexts, we might very well require it in our database. Non-traditional variants of FOL such as those described by Hayes and Menzel [10] seem capable of providing this functionality.

4 Approximation Contexts

Often, the task for which a database is used permits us to use approximate models for representation and reasoning. The most well known approximate model is Newtonian mechanics. Subfields of AI such as qualitative physics have extensively used approximations which make it feasible to model complex phenomenon. Approximations are widely used in the the common sense world too, e.g., we often approximate the price of object by ignoring tax, shipping, etc.

Ex. 4: Attribute Approximation

In the simple case, the value of a particular attribute (f) of an object is approximated. These approximations have the general form

$$f(\overline{x}) = f_1(\overline{x}) \approx f_1(\overline{x}) + f_2(\overline{x}) + f_2(\overline{x})... \qquad (13)$$

To obtain the approximation context, we substitute occurrences of the right hand side with the left hand side. Similarly, when lifting out of such approximation contexts, we add back the terms that were dropped out.

The database example given in [16], in which the Navy, Airforce and GE all have databases of prices, each a different approximation of the total price paid by the tax payer, is an example of Attribute Approximation. The price of an object in the GE database is approximated to not include spare parts, unlike the Navy database which includes spare parts, and the Airforce database includes both spare parts and inventory costs. Presumably, somewhere else is the true price of the object including discounts, shipping costs, etc.

When lifting formulae from Approximation Contexts to contexts which don't make the assumption, we need to factor in the terms that were approximated out. When combining data in the GE database with data in the Navy database, we have to be careful about the approximations made by each. This is done with lifting formulae like the following:

$$C_0 : (\forall xy) \qquad ist(c_{navy}, price(x) = y) \Leftrightarrow$$
$$ist(c_{GE}, y = price(x) + price(spares(x))) \Leftrightarrow$$
$$ist(c_{airforce}, y = price(x) + price(spares(x)) + cost(inventory(x)))$$

Ex. 5: Structural Approximation

Attribute Approximations only affect a particular attribute (typically numeric) of some class of objects. More complex are *structural approximations* such as approximating a car as a cuboid, a somewhat curved road as a line and processes as being instantaneous. These map one set of objects (such as a car, a road or processes) to a corresponding, more easily modeled set of objects (such as a cuboid, line or instantaneous event). Structural approximations are very common when modeling the physical properties of objects. Objects get approximated into regular shapes that are characterizable by simple geometric formula, into shapes with lower dimensions, etc. Having done this, the Approximation Context may altogether dispense with the original object, often using the *same* symbol to denote the approximate model.

When lifting conclusions from contexts which make structural approximations, it is important to not lift axioms pertaining to the approximation itself. So, for example, if for the purpose of calculating the distance to a certain galaxy we approximate earth to a point, it would be inappropriate to lift a conclusion that earth's volume is zero. In fact, most structural approximations are targeted at computing a particular attribute(s) of the approximated object (e.g., distance to the galaxy) and these are the only attributes of the object can be lifted out.

4.1 Discussion

Approximate models are probably the most widely used case of people using different explicit models of a phenomenon. This is especially the case in engineering and science.

Projection Contexts and Approximation Contexts are similar in that they exploit assumptions to formulate simpler theories. However, there is an important distinction between these two classes of contexts. The assumptions made by Projection Contexts are typically consistent with the database Δ that the context Δ_c is projected from. In contrast, the assumptions made by Approximation Contexts are usually logically inconsistent with the context they are derived from. So, in addition to the requirement made by Projection Contexts, Approximation Contexts also impose the requirement that the system tolerate the inconsistency between the more accurate model and its approximation. In particular, Approximation Contexts need some form of referential opacity so that formulas such as the following are not invalid.

$$ist(c_1, volume(Earth) = 0) \wedge (volume(Earth) = 1,097 \times 10^{18} m^3) \qquad (14)$$

5 Ambiguity Contexts

Sometimes, the reference of a symbol might be unambiguous in a narrow scope or situation in which certain constraints may be assumed, but ambiguous in a larger scope without the aid of these additional, often implicit constraints. The goal of Ambiguity Contexts is to capture this scope so that statements containing the ambiguous references can be given to the program without full disambiguation. The narrowness of the scope can also be used to advantage to perform more efficient reasoning. The scope could be defined by the situation, by a discourse or by the problem solving goals of the program.

Ex. 6: Indexicals

Indexicals (such as *he, she, it, now* and *here*) are the best examples of the use of Ambiguity Contexts. [9], [16] and many others have shown how a logical formulas such as $hungry(He, Now)$, which contains the unresolved indexicals He and Now can be added to a database in a limited *Discourse Context*. The advantage of doing this is that disambiguation can be postponed, while the reasoning engine can profitably derive conclusions from the statement.

The following examples, adapted from [9], illustrate the use of linguistic terms and predicates in a formal language. It is assumed that a natural language front end, using a lexicon and linguistic knowledge rephrases the natural language utterance as a formula. This formula might be heavily context dependent in a manner illustrated below.

Pronouns and Indexicals such as He, She, It, Now, I, etc. are terms in the language. The sentence "he is hungry" translates to $hungry(He)$, "it is now 4pm" to $(Now = 4pm)$, and so on.

The language includes the functions The and A to handle definite and indefinite references. The function A is similar to the article A. The sentence "the lady owns a bag" would be translated into $owns(The(Lady), A(Bag))$.

Constraints such as the following, together with an appropriate minimization of the predicate $present$ ([9], [15] which specifies whether a context includes a certain object in its domain) enable a program to use a wide range of knowledge and deduction techniques to determine the denotation of indexicals.

$$(\forall c_i y) ist(c_i(y = It)) \Rightarrow ist(c_i, \neg Person(y)) \wedge present(c_i, y) \qquad (15)$$

Ex. 7: Homonymy

Buvac [5] describes the use of contexts to capture a different kind of ambiguity than that exhibited by indexicals. Consider the statement "He went to the bank", where it is not clear whether the word 'bank' denotes a financial bank or river bank. In typical natural language systems, this disambiguation would have to be done before the parse of the statement can be added to the database. Buvac shows how with contexts, the natural language front end can add the ambiguity preserving translation into the database. Buvac considers the statement "Vanja is at the bank". The denotation of "bank" is ambiguous. The statement can be added to an appropriate $Discourse\ Context\ C_{d0}$ as:

$$C_{d0} : at(Vanja, The(Bank)) \qquad (16)$$

Next we are told that he got money from the bank

$$C_{d0} : gotMoney(Vanja, The(Bank)) \qquad (17)$$

Based on common sense axioms such as the following, Buvac shows how the system can infer that the bank must be a financial bank and not a river bank.

$$C_0 : (\forall cxy) ist(c, gotMoney(x, y)) \Rightarrow FinancialBank(y)) \qquad (18)$$

[9] shows how the same approach can be used to treat prepositions such as at, to, for, etc. The introduction of "predicates" such as for allows us to translate the sentence "Fred bought the rose for Jane" as,

$$(\exists e)(Buying(e) \wedge object(e, The(Rose)) \wedge for(e, Jane)) \qquad (19)$$

Similarly a variadic function Etc can be used to represent ellipsis. The sentence "Fred likes ice cream, softees, etc." would be translated as

$$likes(Fred, Etc(IceCream, Softee)) \qquad (20)$$

Ex. 8: Metonomy & Polysemy

With homonymy, the different denotations of the word denote very different and unrelated things. With metonomy and polysemy, the different denotations are closely related. Consider the two sentences "flight UAL201 landed in San Francisco at 1.35pm" and "San Francisco elected John Smith to ...". In the first sentence, 'San Francisco' denotes San Francisco's airport (which isn't even in the city of San Francisco), and in the second, it denotes the electoral community of the city of San Francisco. There are many other such denotations of the term 'San Francisco' (the actual land mass, all the people, the executive branch of the city, the city's economy, ...). Common natural language usage typically does not use different terms for these different concepts. The usage of the term is usually adequate to distinguish between them.

As with indexicals and homonymy, we can use contexts to preserve this kind of ambiguity as well. However, unlike indexicals and homonymy, in many cases, statements with metonomy/homonymy ambiguities can be lifted without resolving the ambiguity. Indexicals and homonymy are purely linguistic phenomenon. Metonomy and polysemy are not just linguistic, but also epistemic phenomenon. Consider the following example. In a simple theory about wars, attacks, etc., we might not distinguish between a country, its government and its armed forces. So, we might have axioms such as the following, which says that before a country attacks another, the head of state of that country has to approve it.

$$occurs(s_i, attacks(x, y)) \Rightarrow occurs(prior(s_i), approve(headOfState(x), attacks(x, y))) \tag{21}$$

This model of the world, where we don't distinguish between the different branches of the government etc. is adequate for a great many tasks. Now, consider a context describing a coup or mutiny in which one arm of the state fights another. Clearly, our simple representation breaks down. In particular, axioms like the one given above are clearly wrong. At this point, we would like to switch to the finer grained representation.

5.1 Discussion

Ambiguity Contexts enable the database to contain logical statements which still have indexicals and homonymous references in them. This provides a great deal of flexibility in when and how these references are disambiguated. In particular, it becomes easier to use domain knowledge and the logical inferencing apparatus for disambiguation.

That said, this use of contexts can be replicated without resorting to contexts by the introduction of new terms. For example every new reference to an indexical (such as he) could be mapped into a new term (such as he-3994), with the appropriate constraints added to he-3994. Alternately, one could introduce a term such as $he(utterance_i)$ which refers to the denotation of the word 'he' in $utterance_i$. Homonymy can be similarly treated.

The use of contexts for metonomy and polysemy on the other hand is much more significant and powerful. As experience with Cyc [12] shows, broad, large scale knowledge bases, which cover many different aspects of a set of objects have to make many subtle distinctions. For example, Cyc distinguishes between the land mass associated with a city, its populace, different branches of its government, its head, and so on. While the ability to make these subtle distinctions is useful, it makes the task of knowledge entry substantially more difficult. Further, most of these distinctions are non-essential in most circumstances. Being forced to make them all the time complicates both knowledge entry and subsequent reasoning. Contexts provide a mechanism by which we can use the simplest formalism, i.e., the one that makes the fewest distinctions, most of the time, transcending to the more expressive representation only when we need to.

The main requirement that Ambiguity Contexts impose on the underlying logic is that of referential opacity. In other words, the formula $ist(c_1, (He = John)) \land ist(c_2, (He = Jane)) \land (Jane \neq John))$ should not be invalid.

6 Mental State Contexts

Mental State Contexts correspond to the use of contexts to capture propositional attitudes and knowledge of other kinds of "alternate" states of affairs such as fiction.

Unlike the previous four kinds of contexts, these contexts are not characterized by what they contain, but in terms of their provenance. Consequently, there is little that we can say in general about lifting into/from them.

Ex. 9: Fictional Contexts

In [15] McCarthy gives the example of using contexts to make statements that are true in the fictional context corresponding to Sherlock Holmes stories. Such a context could include statements like:

$$SherlockHolmesContext : Detective(Holmes) \qquad (22)$$

$$SherlockHolmesContext : partner(Holmes, Watson) \qquad (23)$$

We rarely, if ever, lift axioms out of fictional contexts. However, we may lift axioms from non-fictional contexts into with fictional contexts. So for example, even though the SherlockHolmesContext does not explicitly state that Rome is in Italy, we can lift this from a non-fictional context into the SherlockHolmesContext.

Ex. 10: Perspectives, Counterfactuals, and Propositional Attitudes

Contexts may be used to represent the world from the perspective of an agent. These non-fictional contexts are closely related to different kinds of propositional attitudes which have been widely studied in philosophy and AI. There is a rich body of examples from those fields. Contexts have been proposed in [15], [8] and elsewhere as a mechanism for handling perspectives and propositional attitudes in general. Ghidini and Giunchiglia [8] provide an example of a "Magic Box", which contains 6 sectors, each possibly containing a ball. Two agents, Mr. 1 and Mr. 2 each have different views of the box, based on their physical locations. They also consider the case of each agent having a partial view of the box. It is further possible to consider each agents view of the other agents view and so on.

Costello and McCarthy [7] treat counterfactuals using contexts. For example, consider the sentence "If another car had come over the hill when you passed there would have been a head-on collision." "If another car had come over the hill when you passed" defines a counterfactual context. Note that the context is highly incomplete - it doesn't say exactly when or what kind of car.

6.1 Discussion

Of the different varieties of contexts, Mental State Contexts are probably the most demanding of the underlying logic. In addition to the requirements imposed by the earlier categories, they also bring in the requirements imposed by propositional attitudes [18]. It is indeed possible that this variety of contexts is a different phenomenon, best dealt with different machinery.

7 Conclusion

In this paper we looked at a number of examples of contexts and distilled them into four important categories, each of which has distinct properties in terms of lifting and each of which imposes different requirements on the underlying logical machinery. We hope that these categories and examples will be useful in comparing and evaluating different approaches to dealing with contexts.

Acknowledgements. We would like to thank Valeria dePaiva for comments on a draft of this paper.

References

1. Saul Amarel. On representation of problems of reasoning about action. In Donald Michie, editor, *Machine Intel. 3*, pages 131–171. Edinburgh Univ. Press, 1971.
2. J. Barwise and J. Perry. *Situations and Attitudes*. MIT Press, 1983.
3. Massimo Beneceretti, Paolo Bouquet, and Chiara Ghidini. On the dimensions of context dependence: Partiality, approximation, and perspective. In *Modeling and Using Context*, pages 59–72. Springer-Verlag, Berlin, 2001.
4. Sasa Buvač. Quantificational logic of context. In Howard Shrobe and Ted Senator, editors, *AAAI 1996*, pages 600–606, Menlo Park, California, 1996. AAAI Press.
5. Saša Buvač. Ambiguity via formal theory of context. Available from http://www-formal.stanford.edu/buvac/, 1995.
6. Saša Buvač and Ian Mason. Propositional logic of context. In *AAAI*, pages 412–419, 1993.
7. Tom Costello and John McCarthy. Useful counterfactuals. *Linköping Electronic Articles in Computer and Information Science*, 4(12), 1999.

8. Chiara Ghidini and Fausto Giunchiglia. Local models semantics, or contextual reasoning = locality + compatibility. *Artificial Intelligence*, 127(2):221–259, 2000.
9. Ramanathan V. Guha. Contexts: a formalization and some applications. Technical Report STAN-CS-91-1399, Stanford CS Dept., Stanford, CA, 1991.
10. P. Hayes and C. Menzel. A semantics for the knowledge interchange format.
11. Patrick J. Hayes. Contexts in context. In Sasa Buvač and Lucia Iwańska, editors, *AAAI Fall Symposium on Context*, pages 71–81, Menlo Park, CA, 1997. AAAI.
12. Douglas B. Lenat and R. V. Guha. *Building Large Knowledge-based Systems: Representation and Inference in the Cyc Project*. Addison-Wesley, 1990.
13. John McCarthy. Applications of circumscription to formalizing common sense knowledge. In Vladimir Lifschitz, editor, *Formalizing Common Sense: Papers by John McCarthy*, pages 198–225. Ablex Publishing Corporation, Norwood, New Jersey, 1990.
14. John McCarthy. Generality in artificial intelligence. In Vladimir Lifschitz, editor, *Formalizing Common Sense: Papers by John McCarthy*, pages 226–236. Ablex Publishing Corporation, Norwood, New Jersey, 1990.
15. John McCarthy. Notes on formalizing contexts. In Ruzena Bajcsy, editor, *Proceedings of the Thirteenth International Joint Conference on Artificial Intelligence*, pages 555–560, San Mateo, California, 1993. Morgan Kaufmann.
16. John McCarthy and Saša Buvač. Formalizing context (expanded notes). Available from http://www-formal.stanford.edu/buvac., 1995.
17. P. Pandurang Nayak. Representing multiple theories. In B. Hayes-Roth and R. Korf, editors, *AAAI-94*, pages 1154–1160, Menlo Park, CA, 1994.
18. N.J.Nilsson and M. Genesereth. *Logical Foundations of Artificial Intelligence*. Addison-Wesley, 1987.
19. W.V.O. Quine. *Propositional Objects*. Columbia Univ. Press, 1969.
20. R.M.Keller, M. Rimon, and A. Das. A knowledge based prototyping environment for construction of scientific modeling software. *Automated Software Engineering*, 1994.
21. Yoav Shoham. Varieties of context. In Vladimir Lifschitz, editor, *Artificial Intelligence and Mathematical Theory of Computation*, pages 393–407. Academic Press, San Diego, 1991.

Ubi-UCAM: A Unified Context-Aware Application Model*

Seiie Jang and Woontack Woo

KJIST U-VR Lab.
Gwangju 500-712, S. Korea
{jangsei,wwoo}@kjist.ac.kr

Abstract. Context-aware application plays an important role in the ubiquitous computing (ubiComp) environment by providing the user with comprehensive services even without any explicitly triggered command. In this paper, we propose a unified context-aware application model which is an essential part to develop various applications in the ubiquitous computing environment. The proposed model affirms the independence between sensor and application by using a unified context in the form of Who (user identity), What (object identity), Where (location), When (time), Why (user intention/emotion) and How (user gesture), called 5W1H. It also ensures that the application exploits a relatively accurate context to trigger personalized services. To show usefulness of the proposed model, we apply it to the sensors and applications in the ubiHome, a test bed for ubiComp-enabled home applications. According to the experimental results, without loss of generality, we believe it can be extended to various context-aware applications in daily life.

1 Introduction

Ubiquitous computing (ubiComp) allows users to get comprehensive services with ubiquitous computing resources in daily life [1][2]. The sensors and applications in ubiComp-enabled environment will be more intelligent with the development of related technologies, such as embedded networking, pervasive sensing, and intelligent processing. Such a smart environment potentially provides the personalized intelligent services without any explicit user's commands in the near future. In order to achieve such intelligent services, the environment needs to obtain user-centered context information without distracting the users.

Over the last few years, various research activities on context-aware applications have been reported. For example, ACE (Adaptive Control of Home Environment) is a system to control temperature and lighting conditions at home by training the daily life patterns of the residents using Neural Net [3]. Both EasyLiving [4] and AwareHome [5] have showed how context information can be used in the home environment. Meanwhile, MIM (Multimedia Interface Manager) showed how to recognize the user's context through various modalities (i.e. seeing, hearing, touching) through camera, microphone, and haptic glove [6]. Note, however, that contexts used in those applications have different meanings and formats according to the chosen applications.

---

* This work was supported by University Research Program by MIC in Korea.

In this paper, we propose a unified context-aware application model that can be used in the ubiComp-enabled home environment [7]. The proposed model consists of two main blocks, i.e. ubiSensor and ubiService. The ubiSensor creates a preliminary context in the form of Who (user identity), What (objects identity), Where (location), When (time), Why (user intention/emotion), and How (user gesture), called 5W1H, by monitoring the user in the environment. The ubiService determines an integrated context by merging preliminary contexts from various ubiSensors and generates the final context that triggers a user-centered service.

The proposed model has various advantages over conventional context models. For example, like a Context Toolkit [8][9], it does not use any mediation for context. However, it maintains independence between sensors and applications by separating the role of Context Toolkit into ubiSensor and ubiService. Then, the ubiSensor generates a preliminary context instead of directly passing the sensed raw data to the Context Toolkit. The resulting context can be shared by all ubiServices and, thus, by all applications. As a result, the context reusability also can be guaranteed. The ubiService also ensures that the application exploits a relatively accurate context to trigger personalized services by feeding back the integrated context to ubiSensors.

This paper is organized as follows: In Section 2, we explain basic terminologies used in this paper. In Section 3, we describe the proposed unified context-aware application model in the ubiComp-enabled home environment. The implementation and experimental results are explained in Section 4 and 5, respectively. Finally, the conclusion and future works are discussed in Section 6.

2 Context for the UbiComp-Enabled Home

Smart Home plays an important role as an application in UbiComp-enabled services. However, the present state of Smart Home focuses on home automation to control doors, lights, elevators, etc. automatically by device-controlling technology such as LONWORKS [10][11], or home networking to connect various information appliances together. UbiComp-enabled Home shall support not only home automation and home networking but also personalized services based on context. To implement the UbiComp-enable Home, we have to overcome the restrictions from which existing context-aware application model suffers, especially dependence between sensor and application, and chaos of context definitions.

Nowadays a sensor of UbiComp-enabled Home depends on its own services. Because of the dependence, developers of context-aware application suffer from adding/replacing/deleting a sensor(s) and from modifying many source codes. Also it is hard to reuse a sensor(s) in other applications.

This dependence can be reduced by using smart sensor in UbiComp which has capability in sensing, processing, and networking. The sensor is indirectly connected to application through the networking and generates unified information for several applications through the processing. It is easy for a sensor to be added, deleted, or replaced by another and reused by other applications. This paper shows that smart sensor converts signals into high level context and transmits this context to application. Specifically, it changes sensed signals to context in forms of 5W1H by its own processing and transmits the context to various applications through its own networking. Therefore, it may guarantee the independence of sensor as well as the reusability for application.

UbiComp-enabled Home requires well-defined context. However, in most of the applications reported, the context is not well defined. Previous context-aware applications mainly use ad-hoc definitions according to the selected applications. For example, Schilt et al. defines context as information about the user and object such as identity and location [12]. Dey et al. defines context as sensed information by the application such as identity, location, activity and state of people, groups and objects [13]. Note however that those definitions may be inconsistent, i.e. changing depending on the selected applications, since such definitions are only suitable for the specific applications.

To solve the problem, we define 5W1H as a unified context so that it can be applicable to all applications in ubiComp environment [1][2]. In general, many context-aware applications retrieve information or trigger a service according to a part of 5W1H such as user identity, location, and time. One theory suggests a unified context, in the form of 5W1H, provides information enough to be used by several applications. Therefore, the unified context model exploiting 5W1H may work in most context-aware applications without loss of generality.

It is necessary that applications of UbiComp-enable Home analyze context to support the user-centered service. To get precise context, we define different levels of context, i.e. preliminary, integrated and final context. The preliminary context generated from a sensor is not enough to trigger a proper service. In general, the extracted context from a sensor may not be accurate or even incomplete since a sensor may not generate all 5W1H. Thus, we introduce integrated context and, thus, final context. The integrated context is completely filled with 5W1H by merging preliminary contexts from a set of sensors. The final context is refined to trigger a user-centered service, which is a set of parameters to be used by a service function. As a result, an application developer may easily design context-aware applications by specifying the condition of the service trigger as a 5W1H.

3 Ubi-UCAM: Unified Context-Aware Application Model

The proposed ubi-UCAM, a unified context-aware application model in ubiquitous computing environment, consists of ubiSensor and ubiService, as shown in Fig. 1. The ubiSensors generate a set of preliminary context. Then the ubiSensors and the ubiServices exchange context through embedded networking modules. The ubiService yields the integrated context by merging the preliminary contexts from a set of ubiSensors and generates the final context by refining the integrated context with the current state of ubiService. Besides, ubiService multicasts the integrated context to ubiSensors, currently connected to ubiService, to help ubiSensor update the preliminary context. The final context is used to trigger the user-centered service.

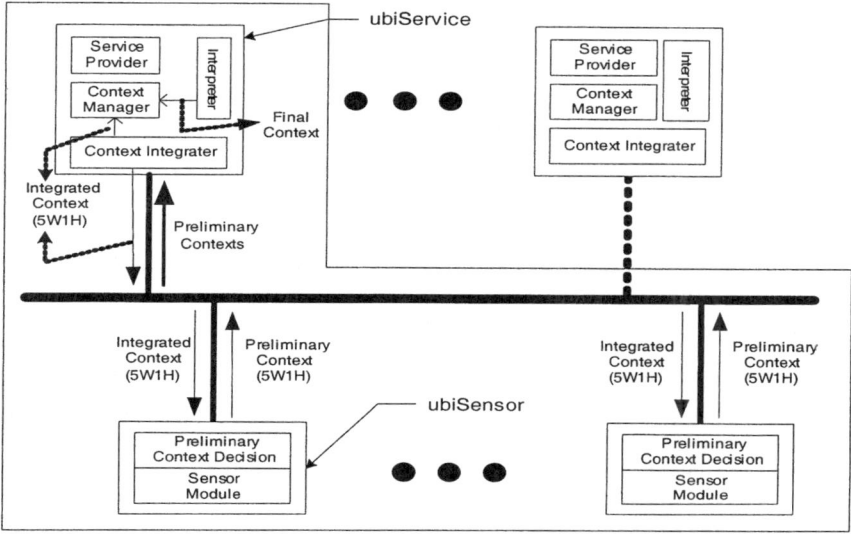

Fig. 1. Ubi-UCAM Architecture

3.1 UbiSensor

The ubiSensor, as shown in Fig. 1, consists of both sensor and preliminary context decision modules. The sensor module monitors the activities of the user in the environment. Then, the context decision module creates the preliminary context in the form of 5W1H by analyzing the sensed signals. As shown in Table 1 the preliminary context is decided using the predefined 'context library' for a specific application. Note that both 'how' and 'why' components among 5W1H, corresponding to the gesture/action and intention/emotion of user, may require more complicated processing. However, to make the problem simple, all 5W1H is determined by the predetermined context library. Accordingly, the ubiSensor referring the same context library generates the same preliminary context.

The resulting preliminary context can be represented in the message format, as shown in Fig. 2. It is more flexible to express preliminary context by using tab character to separate each element of 5W1H. The '-' character also presents empty element, which results from the fact that a sensor module cannot determine the whole 5W1H at a time.

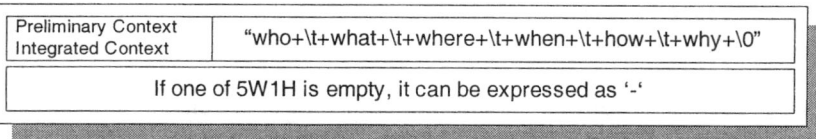

Fig. 2. Context Message Format

Table 1. Context Library

Preliminay Context	Definition
Who	User identity (User Name)
What	Real or virtual object related to user intention
Where	Location of the user or the object
When	Time (YearDateHourMinute)
How	User gestures or action
Why	User intention/emotion to control some services

3.2 Ubiservice

The ubiService, as shown in Fig. 1, consists of four main modules; context integrator (CI), context manager (CM), interpreter (INT), and service provider (SP). CI collects preliminary contexts for a given time (ΔT) from a set of ubiSensors connected to the ubiService, and decides the integrated context. As shown in Fig. 3, the preliminary contexts are aligned and elements of 5W1H in the same column are merged into the integrated context by voting. In case of 'why', we use simple linear mapping, which can be improved by adapting Neural Net. The resulting integrated context has a complete user-centered 5W1H and is forwarded to CM. Simultaneously, the integrated context is multicasted to all ubiSensors.

CM compares the integrated context with all context conditions in a hash table to trigger SP, as shown in Fig. 4. If a context condition is matched, CM calls a function of SP that is associated with the context condition. Otherwise, CM discards the integrated context. The hash table manages context conditions as a key and information of function as a value. The table supports both 1:1 and N:1 relations between a key and a value and also guarantees fast search of integrated context in context conditions. After delivering the selected information and corresponding function to INT, CM runs the service function with the final context from INT.

INT provides CM with the final context, e.g. function name and parameters to trigger specific SP. The final context is generated by mapping the context condition to the parameters based on the current state of ubiService. SP is a set of functions to be triggered as service of ubiService. Each function is associated with a context condition in the Hash table and requires parameters to work. Fig. 5 shows context flow among CI, CM, INT, and SP.

3.3 Networking

The ubiSensor is connected to a network that provides a lookup service maintaining attributes of ubiSensors such as state of connection with ubiService, a sort of preliminary context, etc. The ubiService requests ubiSensors to the lookup service with the needed attributes, and the lookup service returns information of ubiSensor that can provide a preliminary context satisfying the attributes. After receiving information, ubiService directly connects to ubiSensor based on the information. The connection between ubiSensor and ubiService is implemented with middleware such as JINI [14]. Each ubiSensor notifies its own state of connection to the lookup service whenever a change occurs.

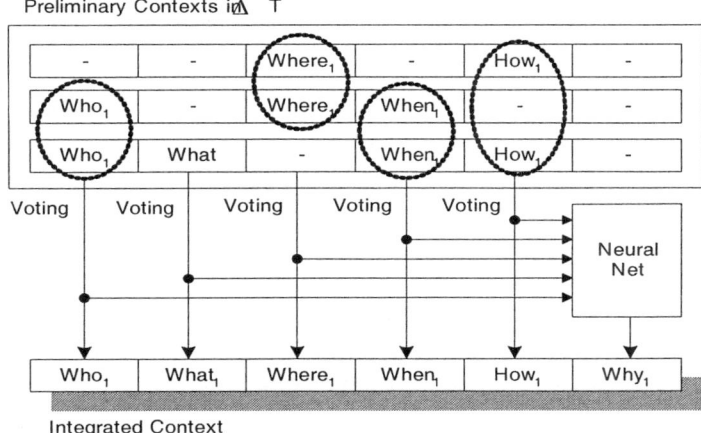

Fig. 3. Integrated Context Processing

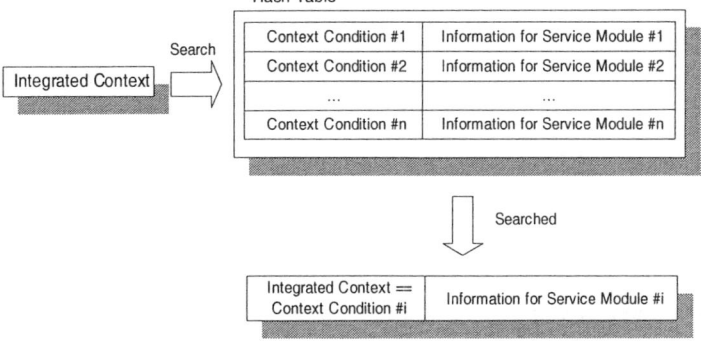

Fig. 4. Searching Context Condition in Hash Table

4 Application: Context-Aware Movie Player

We applied the proposed ubi-UCAM to 'ubiHome', a testbed for ubiComp-enabled home applications. In ubiHome, several ubiSensors (e.g. portable memory, IR sensor, on/off sensor, 3D camera, etc.) provide the preliminary contexts in the form of 5W1H corresponding to user/object identity, location, gesture, time etc. To show the usefulness of the proposed model, we developed a ubiService, which is called c-MP (Context-aware Movie Player).

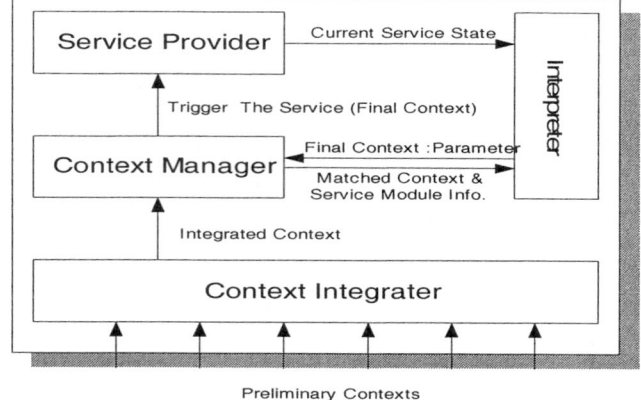

Fig. 5. Context Flow in ubiService Components

The c-MP provides residents in ubiHome with context-aware services. The c-MP provides user-centered services based on the context such as user's identity (Who), user's location (Where), time (When), gesture (How), object for movie player (What) and user's intention to control movie player (Why). For example, after a resident enters a living room with ubiKey, he/she sits down on a sofa in front of the TV. Then, a ubiService menu automatically appears on the monitor. If the resident selects movie player from the menu, the c-MP displays a list of movie titles with user-wise history, as shown in Fig. 6. When the resident rises from his/her sofa, c-MP automatically pauses the movie. If he/she comes back and sits down within 30 seconds from the kitchen for snacks or beverages, c-MP resumes the movie. While, he/she does not come back in 30 seconds or goes out of ubiHome, c-MP saves the paused status and time and automatically stops. The resident can control the movie player by his/her gestures as well as by remote controller [15] . For example, he/she can increase volumes by raising a right hand up and decrease volumes by putting it down. He/she can enlarge screen size by raising a left hand up and lessen screen size by putting it down.

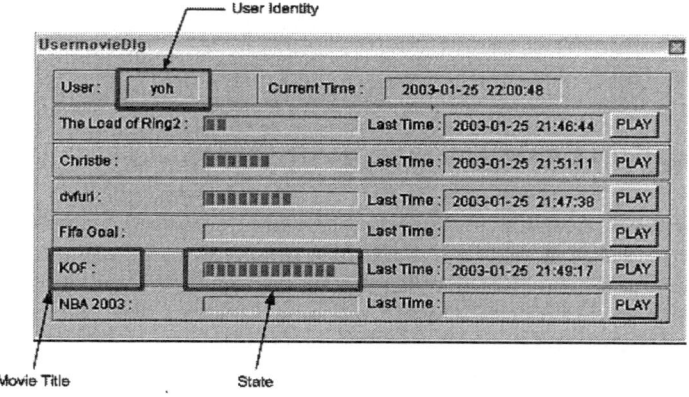

Fig. 6. Example of Context-aware Service

The c-MP gets the preliminary contexts from several ubiSensors such as ubiKey [16], ubiFloor [17], CoachSensor [16] and SpaceSensor[15], as shown in Fig. 7. The ubiKey using portable memory as sensor module generates the identity of the user (Who), location (Where), entering/exiting information (How) and entering/exiting time (When). The ubiFloor, where an on/off sensor is attached per 2cm * 5cm space, yields relative position to TV (How) and time (When). The CoachSensor, where three on/off sensors are embedded in coach, determines the pose of the body, standing up/sitting down (How), and time (When). The SpaceSensor using a 3D camera analyzes hand/body gestures (How) and time (When).

As shown in Table 2, each ubiSensor generates the preliminary context based on context library of ubiHome. For example, when the user, S.Jang (a resident of ubiHome) enters the living room, the ubiKey makes a context message such as "sJang\t-\t LivingRoom\t200301271940\tEnter\t-". When he sits down on a sofa in front of the TV, the CoachSensor generates a context message such as "-\t-\tCoach\t200301271942\t SitDown\t-". If he stands up on the ubiFloor and moves toward the TV, ubiFloor generates a context message such as "sJang\tTV\t-\t200301271944\Comming\t-". If he raises his right hand, the SpaceSensor generates a context message such as "-\t-\t200301271800\t RightHandUp\t-". Finally, all context messages are delivered to the c-MP.

Table 2. Example of Context Library for ubiHome

Preliminay Context	Definition
Who	Name of resident in ubiHome
	i.g. wWoo, sJang, yOh, sLee, dHong, sKim, yLee, ySuh, sOh, mLee, sjOh, wLee, shLee, kKim, smJung
What	Service Object in ubiHome
	- real object : Light, TV, MoviePlayer, AV Player, Movie Title
	- virtual object: Volume, Speed, Size, Luminosity
Where	Location information of ubiHome
	- LivingRoom, Kitchen, BedRoom
When	Time (YearDateHourMinute)
	- 200301271900
How	User gestures which are emuerated in a predefined form for ubiHome
	- Enter, Exit, SitDown, StandUp, Coming, Going, G(Select), G(Play), G(Stop), G(Pause), G(FastFoward), G(VolumeUp), G(VolumeDown), G(SizeUp), G(SizeDown), G(TurnOn), G(TurnOff), G(Bright), G(Dark)
Why	User intenstion and emotion
	- Intention: to-Play, Select, Stop, Pause, Increase, Decrease, Select, TurnOn, TurnOff,
	- Emotion: Happy, Angry, Sleepy, Active

The c-MP consists of CI, CM, INT and SP. The CI, as shown in Fig. 8, gathers preliminary contexts every 0.5 seconds. Then it fills an integrated context with 4W1H determined by voting and an empty 'Why'. The remaining element 'Why' can be determined by lookup table or Neural Networks. The CM searches a context condition in the Hash table to find a matched integrated context. If matching occurs, CM triggers a service function of c-MP with the final context. The INT translates the resulting integrated context into the final context in the form of parameters considering the current state of ubiService.

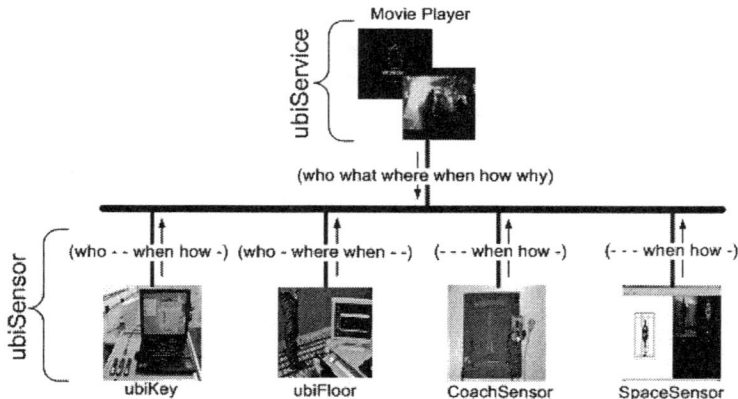

Fig. 7. Example of ubiSensor and ubiService

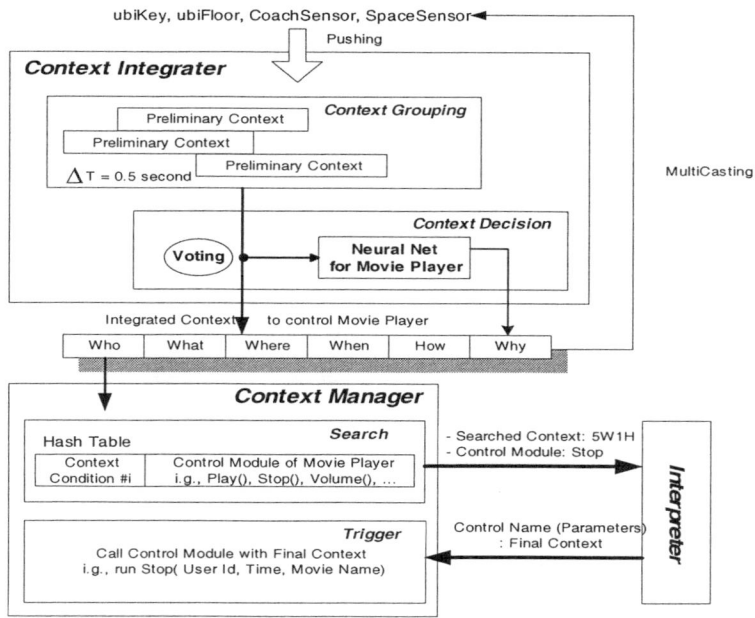

Fig. 8. Context Flow of c-MP

Fig. 9 shows the relationship between the states of c-MP and the integrated contexts. The SP supports control functions such as Play(), Stop(), Select(), Size(), Volume(), Pause(), and FF() according to the context condition.

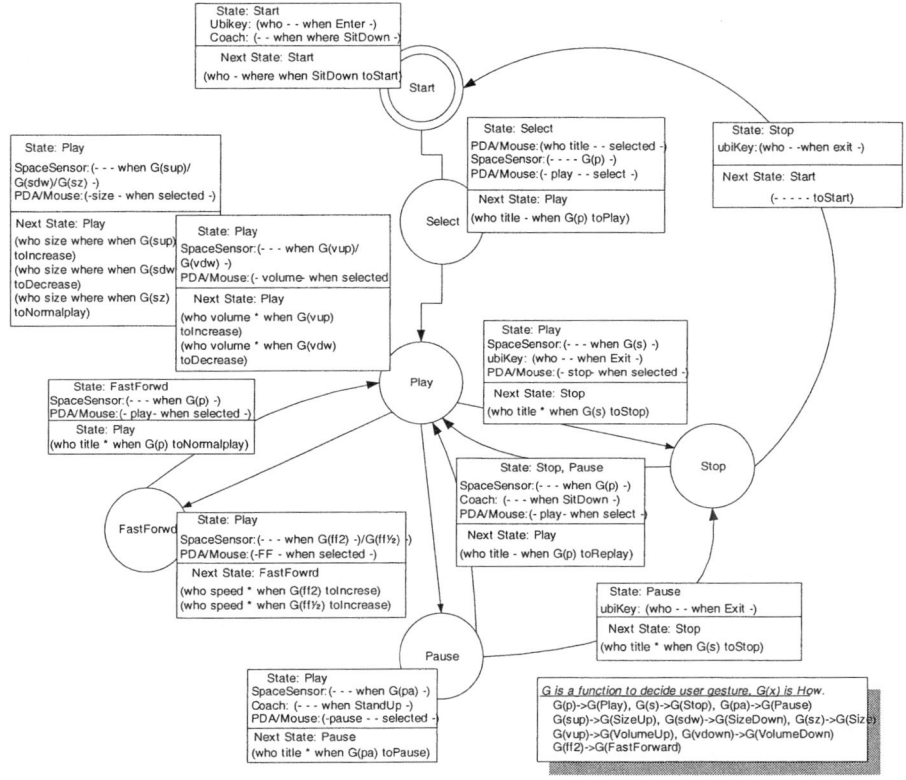

Fig. 9. States & Integrated Contexts of c-MP

5 Experiments

To show the usefulness of the proposed ubi-UCAM, we applied it to a context-aware application, c-MP. And we compared it with a noncontext-aware application, WinAmp, a normal movie player with new skin (java-juke) [18]. Fourteen volunteers (the fifties: 2 persons, the forties: 1 person, the thirties: 3 persons, the twenties: 6 persons, the teens: 2 persons) tested both applications and reported the convenience and satisfaction.

With an assumption that a user was in ubiHome to watch a movie, we measured the time and the number of explicit commands required to start a movie on TV, waiting time per explicit command, and CPU usage of the computer (CPU: PentiumIII

800MHz, Memory: 256MB, OS:WindowsXP). The results are shown in Table 3. As shown in Table 3, when a user watches a movie with Winamp, he/she must click serveral times to selcet a movie. As a result, it requires much attention and long time duration. While with c-MP, he/she needs only two or three explicit commands to select it, since the c-MP automatically provides a user-centered list of movies according to his/her preferences and previous activities. Therefore, c-MP requires relatively less attention and shorter time duration than those of Winamp. The main tradeoffs are waiting time and CPU usage because the c-MP requires processing to get a proper context.

Table 3. Quantatitive Factors

	WinAmp	c-MP
Time duration	20~35 sec	8~12sec
# of Explicit Command	5~12	2~3
Waiting Time per Explicit Command	100~350ms	500~1200ms
CPU Usage	10~15%	30~40%

We have also analyzed the degree of complexity in learning and usage of each player, and the results are summarized in Table 4. As shown in Table 4, most of the participants had no problem in learning how to contol WinAmp with new skin because they were familiar with WinAmp, while they spent some time to learn the instruction of the c-MP, gesture-based commands. Note, however, that after getting familar with the c-MP, they quickly adapted to the new interfaces. Additionally, most of them were satisfied with the personalized movie-playing list that showed the status of the movie (to be watched, to be paused, not to be watched) with time information. Especially, the fifties were positive about controlling movie player by their gesture because they could give attention to a movie without an annoying remote controller. The teens were interested in the auto-play/pause/stop functions because they often movied around ubiHome.

Table 4. Qualitative Factors

	WinAmp	c-MP
Learning Complexity	Easy	Easy
Learning Time	1 minute	2~3 minutes
Usage Complexity	Normal	Easy
Satisfaction	60%	80%

6 Discussion

In this paper, we proposed the ubi-UCAM, a unified context-aware application model in the ubiquitous computing environment and applied it to an ubiComp-enabled home application. The proposed model introduces a unified context in forms of 5W1H that can be shared by various applications and specifies the role of context (i.e., preliminary context, integrated context, final context). According to the experimental results, the proposed model affirms the independence between sensor and application by using a unified context in the form of 5W1H, and exploits relatively accurate context to trigger personalized servieces. However, we must expand on the expression of context because it is difficult to represent the complex context for all application through only context in the form of 5W1H.

References

1. Jang, S., Lee, S., Woo, W.: Research Activities on Smart Environment. IEEK, Magazine. Vol. 28. (2001) 85–97.
2. Yoon, J., Lee, S., Suh, Y., Ryu, J., Woo, W.: Information Integration System for User Recognition and Location Awareness in Smart Environment. KHCI 2002.
3. Mozer, M.: The Neural Network house: An environment that adapts to its inhabitants. In M. Coen (Ed.), Proceedings of the American Association for Artificial Intelligence Spring Symposium on Intelligent Environments. Menlo, Park, CA: AAAI Press. (1998) 110–114
4. Shafer, S., Brumitt, B., Meyers, B.: The EasyLiving Intelligent Environment System. CHI Workshop on Research Directions in Situated Computing (2000)
5. Dey, A., Salber, D., Abowd, G.: A Context-aware Infrastructure for Smart Environments. In Proceedings of the 1st International Workshop on Managing Interactions in Smart Environments (MANSE '99). Dublin, Ireland (1999)
6. Marsic, Medl, I. A., Flanagan, J.: Natural communication with Information Systems. Proceedings of the IEEE. Vol.88, No. 8. (2000) 1354–1366
7. Jang, S., Woo, W.: Architecture of Context based Application in Ubiquitous Computing Environment. KHCI (2003) 346–351
8. Dey, A. K., Abowd, G.: The Context Toolkit: Aiding the Development of Context-Aware Applications. Proceedings of the Workshop on Software Engineering for Wearable and Pervasive Computing. Limerick, Ireland (2000)
9. Salber, D., Dey, A. K., Abowd, G. D.: The Context Toolkit: Aiding the Development of Context-Enabled Applications. In the Proceedings of the 1999 Conference on Human Factors in Computing Systems. Pittsburgh, PA. (1999) 434–441
10. LONWORKS, http://www.ktechon.co.kr/spbook/sp_lonworks.html
11. LONWORKS, http://echelon.com
12. Schilit, B., Theimer: M. Disseminating Active Map Information to Mobile Hosts. IEEE Networks. Vol.8. No.5. (1994) 22–32
13. Dey, A. K.: Understanding and Using Context. Personal and Ubiquitous Computing, Special issue on Situated Interaction and Ubiquitous Computing. Vol.5. No.1 (2001)
14. JINI, http://www.sun.com/software/jini
15. Hong, D., Woo, W.: A Vision-based 3D Space Sensor for Controlling ubiHome Environment. In Proceeding of HCI (2003)
16. Oh, Y., Jang, S., Woo, W.: User Authentication and Environment Control using Smart Key. KSPC 2002. Vol. 15. No. 1. (2002) 264
17. Lee, S., Song, H., Ryu, J., Woo, W.: Music Player with the Smart Floor. KSPC 2002. Vol. 15. No. 1. (2002) 158
18. Java-duke, http://www.winamp.com/skins/

Contextual Effects on Word Order: Information Structure and Information Theory

Nobo Komagata

Department of Computer Science, The College of New Jersey
PO Box 7718, Ewing, NJ 08628, USA
komagata@tcnj.edu

Abstract. To account for a type of contextual effect on word order, some researchers propose *theme-first* (old things first) principles. However, their universality has been questioned due to the existence of counterexamples and the possibility of arguably *rheme-first* (new things first) languages. Capturing the contextual effects on theme-rheme ordering (information structure) in terms of information theory, this paper argues that word order is affected by the distribution of informativeness, an idea also consistent with counterexamples and rheme-first languages.

1 Introduction

Various contextual effects on word order have been a topic of active research since at least the eighteenth century [1]. Many have noted that *old* information comes before *new* information [2–4]. The old and new components in an utterance are often called *theme* and *rheme*, respectively, and the theme-rheme organization is called *information structure*.[1] Accordingly, the idea of "old thing first" is also called the *theme-first principle*.

The theme-first principle seems to be able to account for certain word order phenomena, especially in *free-order* languages such as Czech [7]. Nevertheless, the proposal cannot be maintained in the stated form, because there are a number of counterexamples, such as the following. Note that bold face represents phonological prominence.

(1)   *a.*   Who knows the secret?

       b.   [**Peter**]$_{Rheme}$ [knows it]$_{Theme}$.

In the response in this example, the sentence-initial position is the rheme with new information corresponding to the *wh*-word in the question.

Furthermore, Lambrecht points out that a greater problem for the theme-first principle is the existence of arguably rheme-first languages [1]. For example, Mithun reports data from the Siouan, Caddoan, and Iroquoian languages and

---

[1] The contrast between theme and rheme is also referred to as the contrast between *topic* and *focus*, respectively. This paper uses the terms *theme* and *rheme*, focusing on the essence found in the contrast observed in many studies. Note that we assume that information structure is a binary partition at the utterance level [5, 6].

argues that these languages have a rheme-first tendency [8]. Similar data in other languages have also been reported [9–11]. Although it is not obvious that these are indeed rheme-first languages, the data still show a consistent pattern rather different from more theme-first languages.

Now that we cannot maintain the theme-first principle, at least as stated earlier, we must question whether something general can still be said about the contextual effects on word order in connection to information structure. Counterexamples in languages like English do not seem to be abundant. In addition, the rheme-first languages seem to be limited to a small number of languages. If different word order principles apply to different languages in an ad hoc way, it would pose a challenge to developing a universal account of language as a human cognitive process. Since information structure has been associated with word order in various forms, e.g., the Prague school [7] and strict theme-rheme ordering of Halliday [4], the above observation may undermine the role of information structure.

This paper develops an idea in Vallduví [12], who cites Dretske [13], regarding the notion of *information* (in terms of *entropy*) and analyzes word order from that point of view. In this connection, we also discuss the definition of information structure based on information theory.

The main hypothesis discussed here is that information structure is a means to even out the information load carried by the theme and the rheme of an utterance (referred to as *information balance*). Then, we can show that the ordering of a low-entropy theme followed by a high-entropy rheme is more desirable than the other ordering, which is considered the universal principle behind the theme-first tendency. However, if the theme is totally predictable (i.e., zero entropy), the ordering does not affect the information balance. This situation appears to correspond to apparent exceptions to the theme-first principle.

Word order is a complex phenomenon involving lexical, syntactic, and pragmatic constraints [14]. This paper inevitably leaves out certain important aspects, such as word order within a phrase, where morpho-syntax tends to fix word order quite rigidly.

The rest of this paper is organized as follows. Section 2 introduces an analysis of the theme-first principle based on information theory. Section 3 discusses various rheme-first cases and analyzes whether they are accountable within the current approach. Section 4 presents an information-theoretic definition of information structure.

2 Information-Theoretic Analysis of Word Order

In this section, we discuss the idea of applying information theory to the analysis of the theme-first principle using the following short discourse, where the second utterance is partitioned into a theme and a rheme.

(2) *i.* John has a house.

ii. [The **door**]$_{Theme}$ [is **purple**]$_{Rheme}$.

Compared to the above example, the following alternative appears less natural.

(3) i. John has a house.
    ii. [**Purple**]$_{Rheme}$, [the **door** is]$_{Theme}$.

The difference will be analyzed later in this section.

2.1 Basic Entropy Computation

This subsection discusses a way to compute the entropies of a theme and a rheme as independent events, using example (2). Immediately after the first utterance in the example, the speaker might want to talk about either the roof or the door, something related to the house, or even a completely different subject. For each of these subjects, there may be a variety of possible predicates, e.g., large, wooden, flat, expensive, purple, and so on. Although it is possible to demonstrate the computation of entropy for an arbitrarily complicated case, we use the following simplified scenario for presentation purposes: two choices for the theme between the door and the roof, and five choices for the rheme among yellow, red, orange, pink, and purple.

Roughly speaking, with more choices, the likelihood of choosing a particular option is smaller. In other words, the informativeness of a single choice among many would be higher than the one from fewer choices. This idea can be formally represented using the notion of *entropy* (good introductions include [15, 16]). Informally, high entropy is associated with high informativeness, low predictability, high uncertainty, more surprise, etc. The use of entropy has been discussed even in linguistics and philosophy [17–19]. For example, while Cherry suggests usefulness [19], Bar-Hillel is more cautious, saying that *information* is different from *meaning* [17]. Naturally, the focus of this paper is not on meaning, but on word order.

Under a very special case where all the events are equally likely (uniform distribution), the entropy of an event is directly related to the number of choices. In terms of probability, the chance of hitting a particular choice out of n choices is $1/n$. Entropy is a measure related to this probability, but it is also adjusted logarithmically so that it is *additive*, in accordance to human sense. For n equally-likely outcomes, $x_1, ..., x_n$, the entropy is defined as a function $H_{uniform}$ on real numbers:[2]

$$H_{uniform}(p) = \log_2 n = -\log_2(1/n) = -\log_2 p.$$

For example, under the current scenario for example (2), the entropy of the theme with two choices is $\log_2 2 = 1.0$, and the entropy of the rheme with five choices is $\log_2 5 \simeq 2.322$.

Entropy is a general function that can also be applied to an event X with n outcomes $[x_1, ..., x_n]$ and the corresponding probability distribution $[p_1, p_2, ..., p_n]$. Here, p_i is the probability of x_i, i.e., the shorthand for $P(X = x_i)$ or $P(x_i)$. Naturally, we must have $\sum_{i=1}^{n} p_i = 1$. For a particular outcome x_i, the (pointwise)

---
[2] The use of base 2 is convenient as it enables us to measure entropy in terms of *bit*.

entropy is $-\log_2 p_i$. We now compute the weighted average of the information for all the outcomes. That is, we multiply the ith entropy with its own probability, p_i, and then add them all (averaging makes sense due to the logarithmic conversion). Let us denote the probability distribution in question as **p** (bold face represents a vector, a *list* of values). Then, the entropy function H is defined as follows:

(4) $$H(\mathbf{p}) = -\sum_{i=1}^{n} p_i \log_2 p_i$$

For example, if the five choices of the rheme in example (2) have a probability distribution $\mathbf{r} = [0.275, 0.15, 0.15, 0.15, 0.275]$, the entropy $H(\mathbf{r})$ is $-(2 \times 0.275 \log_2 0.275 + 3 \times 0.15 \log_2 0.15) \simeq 2.256$.

2.2 Dependency between Two Events

Although the entropies for a theme and a rheme were assumed independent in the previous subsection, the choice of the latter component would naturally depend on the choice of the former. For instance, in example (2), the predicates for the roof and those for the door are likely to have different probability distributions. In order to analyze the dependency between theme and rheme, this subsection introduces some basic ideas about entropies of two events.

We now consider two events X and Y. Suppose that event X has two possibilities, x_1 and x_2, and event Y, two possibilities y_1 and y_2. Then, the *joint probability* for each combination of x_i and y_i can be summarized as follows:

(5)

	y_1	y_2
x_1	$p_{1,1}$	$p_{1,2}$
x_2	$p_{2,1}$	$p_{2,2}$

Naturally, the sum of all the probabilities must satisfy: $\sum_{j=1}^{n} \sum_{i=1}^{m} p_{i,j} = 1$.

At this point, we consider extending the definition of entropy (4) to a two-event situation, summing over both of the events. For events X and Y with m and n possibilities, respectively, we have joint probability $p_{i,j}$ for x_i and y_j. Then, the *joint entropy* of the two events is defined as follows:

$$H(X, Y) = -\sum_{j=1}^{n} \sum_{i=1}^{m} p_{i,j} \log_2 p_{i,j}$$

As an example, let us consider the following joint probability distribution for X and Y:

(6)

	y_1	y_2
x_1	0.1	0.2
x_2	0.3	0.4

Then, the joint entropy can be computed as follows:

$$H(X, Y) = -(0.1 \log_2 0.1 + 0.2 \log_2 0.2 + 0.3 \log_2 0.3 + 0.4 \log_2 0.4) \simeq 1.846 .$$

Since the joint probability already contains the complete information about the two events, knowing X and Y separately would generally lead to some redundancy. For example, since $H(X) + H(Y) \simeq 0.881 + 0.971 = 1.852$, we see that $H(X,Y) < H(X) + H(Y)$.

We now consider the information measure that corresponds to $H(X,Y) - H(X)$. Since Y is conditional to X, it is called the *conditional entropy*, represented as $H(Y|X)$. Analogously, we can also consider $H(X|Y)$. Then, the following equation relates the information measures discussed so far.

$$H(X,Y) = H(X) + H(Y|X) = H(Y) + H(X|Y)$$

Returning to example (6), we have $H(X,Y) = H(X) + H(Y|X) \simeq 0.881 + H(Y|X)$. Thus, we know that $H(Y|X)$ is 0.965, which is less than $H(Y) \simeq 0.971$. Since conditional information never increases the uncertainty, we have the following inequality: $H(X|Y) \leq H(X)$.

Another measure is used to indicate the degree of dependence between two events, called *mutual information*, which is defined by the following equation: $I(X;Y) = H(X) + H(Y) - H(X,Y)$.

2.3 Information Balance

We now apply the ideas introduced in the previous subsections to our analysis of word order. We use example (2) with the following probability distribution for the theme and the rheme (t_1 and t_2 refer to the two theme choices and r_i refers to one of the five rheme choices).

(7)

	r_1	r_2	r_3	r_4	r_5	$\sum t_i$
t_1	0.25	0.125	0.075	0.025	0.025	0.5
t_2	0.025	0.025	0.075	0.125	0.25	0.5
$\sum r_i$	0.275	0.15	0.15	0.15	0.275	

How we can actually come up with such a probability distribution is a difficult question. Since some possibilities can be related to the context through inference (linguistic and extra-linguistic), it naturally involves the kind of difficulty faced in many pragmatic studies. Next, there is a question of whether the probability distribution under discussion should be understood only from the speaker's point of view. In addition, the notion of joint entropy involves the connection between two events, which also requires analysis. For the present discussion, we assume that the probability distributions for the theme and the rheme are available, and we will build arguments based on this assumption.

The entropies for the theme, the rheme, and the entire utterance (independently) are $H(T)$, $H(R)$, and $H(T,R)$, respectively. If the rheme is delivered after the theme, we consider the conditional entropy of the rheme after excluding the effect of the theme, i.e., $H(R|T)$. Then, $H(T,R) = H(T) + H(R|T)$. On the other hand, if the utterance is made in the rheme-theme order, we have $H(T,R) = H(R) + H(T|R)$. In the following, as the entropy of the latter component, be it the rheme or the theme, we always use the conditional entropy. The basic information measures for example (7) are computed as follows:

$$H(T) = -(0.5\log_2 0.5 + 0.5\log_2 0.5) = 1.000$$
$$H(R) = -(2 \times 0.275\log_2 0.275 + 3 \times 0.15\log_2 0.15) \simeq 2.256$$
$$H(T,R) = -(\,2 \times 0.25\log_2 0.25 + 2 \times 0.125\log_2 0.125$$
$$+2 \times 0.75\log_2 0.75 + 4 \times 0.025\log_2 0.025) \simeq 2.843$$
$$H(R|T) = H(T,R) - H(T) \simeq 1.843$$
$$H(T|R) = H(T,R) - H(R) \simeq 0.587$$
$$I(T;R) = H(T) + H(R) - H(T,R) \simeq 0.413\,.$$

In order to compare the evenness of the information distribution between theme and rheme, we introduce a measure, *information balance*, defined as follows:

Definition 1. *Information balance: The standard deviation of the entropies of the theme and the rheme (of an utterance) for a particular ordering.*

Note again that the entropy of the latter component is a conditional entropy. With this definition, the main proposition of this paper can be described as follows:

Proposition 1. *The information structure with a lower information balance is preferred.*

Next, let us compute the information balance of the theme-rheme (rheme-theme) ordering, denoted as σ_{TR} (σ_{RT}). To do so, we first compute the average of the entropies for the theme and the rheme (identical for both orders): $E_{TR} = E_{RT} = H(T,R)/2 \simeq 1.421$.

$$\sigma_{TR} = \sqrt{\left(|H(T) - E_{TR}|^2 + |H(R|T) - E_{TR}|^2\right)/2} \simeq 0.421$$

$$\sigma_{RT} = \sqrt{\left(|H(R) - E_{RT}|^2 + |H(T|R) - E_{RT}|^2\right)/2} \simeq 0.835$$

Thus, we have $\sigma_{TR} < \sigma_{RT}$.

For both of the word orders, the relevant entropy measures and information balances are summarized below.

(8) a. Theme Rheme   Information Balance
       $H(T)$ $H(R|T)$ σ_{TR}
       1.000   1.843   0.421
   b. Rheme Theme   Information Balance
       $H(R)$ $H(T|R)$ σ_{RT}
       2.256   0.587   0.835

This shows that the theme-rheme order has a more even distribution of entropies than the rheme-theme order. That is, it would be easier for the listener to process the information in the theme-rheme order.

Now, we can formulate the principle underlying the theme-first tendency as follows:

Theorem 1. *(Informally) If the entropy of the theme is lower than that of the rheme, the theme-rheme ordering is never worse than the other ordering with respect to information balance. (Formally) If* $H(T) \leq H(R)$, $\sigma_{TR} \leq \sigma_{RT}$.

The above theorem is interesting on the following two points: (i) it predicts that the theme-rheme ordering is preferred, and (ii) it can also specify under what condition there is no difference between the two orders. Here is a proof.

Proof. First, the information balance for the two events X and Y in that ordering is computed as follows:

$$\sigma_{XY} = \sqrt{\left(|H(X) - E_{TR}|^2 + |H(X,Y) - H(X) - E_{TR}|^2\right)/2}$$
$$= \sqrt{\left(|H(X) - E_{TR}|^2 + |-H(X) + E_{TR}|^2\right)/2}$$
$$= |H(X) - E_{TR}|.$$

Let us consider the (independent) entropies for T and R as $H(T)$ and $H(R)$, respectively. Since $H(T) \leq H(R)$, we have $\sigma_{TR} = |H(T) - E_{TR}|$ and $\sigma_{RT} = |H(R) - E_{TR}|$. Then, applying $H(X,Y) = H(Y) + H(X|Y)$ and $H(X|Y) \leq H(X)$, we have the following.

$$\sigma_{TR} - \sigma_{RT} = [E_{TR} - H(T)] - [H(R) - E_{TR}] = H(T,R) - H(R) - H(T)$$
$$= H(T|R) - H(T) \leq 0$$

Therefore, $\sigma_{TR} \leq \sigma_{RT}$. □

2.4 Special Cases

As suggested in the previous subsection, information balance can be the same for both the theme-rheme and rheme-theme orders in certain cases.

First, the theme and the rheme could have exactly the same information (or are completely dependent), i.e., $H(T,R) = H(T) = H(R)$. However, this case is unlikely in reality.

Second, if the theme and the rheme are completely independent, i.e., $I(X;Y) = 0$, the joint entropy is the sum of $H(T)$ and $H(R)$, i.e., $H(T,R) = H(T) + H(R|T) = H(T) + H(R)$. Thus, the information balance would not depend on the theme-rheme ordering. As we noted earlier, it is more likely that the theme and the rheme have some informational dependency, and thus this case would be atypical. However, there is an important special subcase. If the theme is completely predictable, i.e., $H(T) = 0$, the entire information solely depends on $H(R) = H(T,R)$, i.e., $\sigma_{TR} = \sigma_{RT}$. The information balance is now between zero and $H(R)$ regardless of the word order. The situation corresponds to Lambrecht's statement: if theme (his topic) is established, there is no need for it to appear sentence-initially [1]. The symmetrical case where $H(R) = 0$ is unlikely because we can assume that the rheme always has some information.

In summary, assuming that the theme has a lower entropy than the rheme, the theme-rheme ordering is never worse than the other with respect to information balance. Exceptions to the theme-first principle occur when the theme is completely predictable, i.e., $H(T) = 0$.

3 Analysis of Rheme-First Cases

In this section, we examine various rheme-first cases. The first subsection deals with exceptions in English, a language that is not considered rheme-first. The second subsection deals with examples in an arguably rheme-first language.

3.1 Exceptions in English

In example (1), the theme is completely predicable. Thus, its entropy is zero. As a result, it falls into the special case discussed in the previous section, where the position of the theme does not affect the information balance. Exceptions to strict theme-principles like this are still consistent with the present hypothesis.

There is another point regarding the status of contrastive theme, as in the following example.[3]

(9) Q: Well, what about the **beans**? Who ate **them**?

A: [**Fred**]$_{Rheme}$ [ate the **beans**]$_{Theme}$.

Here, the word "beans" is stressed because of the potential contrast between beans and, say, potatoes. One might question whether the entropy of such a theme is zero. But as long as the theme is completely predictable as in the above example, its entropy is still zero. Thus, the above example is consistent with our analysis. The existence of contrastive elements does not necessarily increase the entropy. In this respect, entropy computation is different from analyzing the set of alternatives as discussed in Steedman [22].

Lambrecht argues that contrastive themes (his topic) must appear sentence-initially because they must announce a new topic or mark a topic shift [1]. But example (9) is a counterexample to this analysis. Unlike Lambrecht, the present hypothesis predicts and accepts the existence of a contrastive theme after the rheme as long as it has zero entropy.

In written texts in English, it is generally more difficult to find a rheme-first pattern. Here is an attempt to create a text comparable to example (9).

(10) *i.* Once upon a time, the villagers planted beans and potatoes. One day, they noticed that someone ate the beans. Someone must have ate them.

ii. Fred ate the beans.

iii. Fred was a monk who ...

Although utterance (10*i*) provides basically the same information as question (9Q), utterance (10*ii*), which is the same as (9A), sounds less natural in this

---

[3] Predicates like "eat" imply the existence of a (possibly deleted) event argument [20], which may affect the information-theoretic analysis [21]. This situation can be avoided by using another type of verb, such as "know."

text. An alternative, "the one who ate beans was Fred," sounds more natural. This suggests that the entropy of "ate the beans" is not zero. Unlike the context generated by a question, utterances in a written text tend to leave a variety of options after them. This seems to explain why the theme-first tendency is observed more commonly in the written form of English.

The present hypothesis predicts the following: it is preferable for an unpredictable theme to precede a rheme. However, it is always possible to violate such a preference. As an example, consider the following abstract taken from a medical journal (utterances are numbered for reference purposes).

Title: [0]Overuse Injuries in Children and Adolescents

[1]The benefits of regular exercise are not limited to adults. [2]Youth athletic programs provide opportunities to improve self-esteem, acquire leadership skills and self-discipline, and develop general fitness and motor skills. [3]Peer socialization is another important, though sometimes overlooked, benefit. [4]Participation, however, is not without injury risk. [5]While acute trauma and rare catastrophic injuries draw much attention, overuse injuries are increasingly common.

In utterance 3, between the phrases (A) "peer socialization" and (B) "another important, though sometimes overlooked, benefit," phrase (B) seems to connect to the context more strongly due to the word "benefit," which already appeared in utterance 1. While the choice of "benefit" is among other contextually linked alternatives, the choice of "peer socialization" is among more diverse possibilities. Then, the entropy of phrase (B) must be lower than that of phrase (A). If the phrases are reversed as in "Another important, though sometimes overlooked, benefit is peer socialization," the information balance of this utterance would be lower and more appropriate than the original utterance 3 in this context.

3.2 Rheme-First Languages

Although some have claimed certain languages to be rheme-first, we need to be careful about identifying rheme-first patterns. First, depending on the way it is defined, typological classification of verb-initial language may simply mean that the pattern occurs more frequently than others. Second, being verb-initial does not automatically mean that the language is full of rheme-first patterns [23].

The discussion below focuses on Iroquoian data taken from Mithun [8], which seems to represent the most prominently rheme-first case (*newsworthiness*-first, to use her term). The utterances are taken from Tuscarora stories. The background is as follows: the speaker first describes a long journey on the ice, discovery of land, and preparation for a sacrifice (some phonetic symbols have been replaced for font availability reasons: "a" for right-hooked schwa and "$?$" for glottal stop).

(11) i. $[ha?\ uh\acute{a}?na?\ ru?n\acute{a}?ah]_{Rheme},\ wahr\acute{a}hra?,\ ...$
 the head man he said
 "the headman said, ..."

 ⋮ (after the sacrifice is made)

ii. ą̀:waeh tihruyą́hw*ąh   haení:ką: uhą́*ną* ru*ną́*ąh*
    where   he has learned from that   head   man
    "Where had he learned it, that headman?"

    ⋮ (the speaker begins his recipe for cornbread)
iii. Tyahraetšíhą ką:θ   [uhsaéharœh]$_{Rheme}$ ... wa*kkúhae*.
     first   customarily ash   I went after
     "First, I usually would go after ashes."

    ⋮ (after a kettle is prepared and is boiling)
iv. U:ną ką:θ   [yahwa*kką*naé:ti*]$_{Rheme}$ hä̞jthu ha*uhsaéharaeh.
     then customarily there I poured   there   the ash
     "Then I would pour the ashes in there."

We exclude utterance (*ii*) from discussion because analysis of the information structure of a question is beyond the scope of this paper. First, (*iii*) and (*iv*) include an adverbial at the beginning of the utterance. Thus, they do not have rheme-first patterns in a strict sense. On the other hand, the last constituent is a part of the theme in each utterance. Thus, we see some type of rheme-theme pattern consistently, which is strikingly different from *more* theme-first languages. The constituents after the rhemes are either a pronoun, a definite expression, or a fairly light verb. That is, these constituents are highly predictable and their entropies are very low, if not zero.

Let us examine other utterances from the same story. The following is an introductory sentence to begin a war story.

(12) U:nąha* kyaení:ką: tikahà:wi* kyaení:ką: [kayą*rì:yus
     long ago this   so it carries this   they fight
     kyaení:ką: wahstąhá:ka:*, tisną* kuráhku:]$_{Rheme}$.
     this   Bostonians   and   British
     "One time long ago the Americans and the British were at war."

This is in fact a theme-rheme pattern. The theme is a typical element used to begin a story. The verb-subject order within the rheme is beyond the scope of the present analysis.

In the following, a peddler had been driving a horse, although the horse itself is not mentioned. Mithun argues for the newsworthiness of the verb.

(13) U:ną haésną: [θahra*nù:ri*]$_{Rheme}$ ha*á:ha:θ.
     now then   again he drove   the horse
     "Now then he drove his horse again."

Again, this is not strictly rheme-first, and the constituent "the horse" is predictable from the context.

Mithun does not discuss the context for the following, but says that the focal point is "behind her."

(14) [ae*taéhsnakw]$_{Rheme}$ wahra*ná*nihr.
     behind her   he stood
     "He stood behind her."

The information "he stood" must be predictable. In the following, although "in front" is probably not completely predictable, it seems to have a low entropy readily inferrable from "behind."

(15) [*Yú:ʔnaeks*]_{Rheme} *uhá̧ʔna̧ʔ*.
    it burns   in front
    "A fire was burning before her."

Mithun cites the literature and observes that in spoken language, significant new ideas are introduced *one at a time* [8]. For the above example, we could even say that the story can continue by linking the rhemes (and the themes preceding the rhemes), but omitting the constituents after the rhemes. Thus, in these rheme-theme patterns, we can still see zero-entropy themes after the rhemes. This observation is consistent with the present hypothesis.

Why there are (more or less) rheme-first languages and why there are also so few are intriguing questions. As a cognitive motivation for the rheme-first pattern, Downing refers to *primacy effect* [24]. In addition, Mithun adds that the sentence-initial position has an advantage of being more prominent prosodically because of *downstepping* (gradually decreasing pitch) [8]. However, since even Iroquoian allows sentence-initial adverbials as a part of the theme, neither of these proposals seems convincing. Finally, Mithun points out that the arguably rheme-first languages are highly agglutinating with a small number of constituents in each utterance and that the development of affixes may have affected the different degree of rheme-first tendency in the Siouan, Caddoan, and Iroquoian languages [8]. Additional relevant data can also be found in the literature [25–28], which are left for future work.

4 On the Definition of Information Structure

In this section, we turn our attention to the definition of information structure. Although researchers have some general agreement about the notion of information structure, the precise definition is still a matter of controversy. This section adds yet another definition, because it is rather different from the previous ones and could provide a precise foundation for its predecessors.

4.1 Previous Definitions

The most common way of analyzing information structure is to use a *question test*, as already seen in example (1). We could even define information structure based on a question test. However, such a definition cannot be applied to analyze information structure in texts. Another popular definition by Halliday [4] is problematic, because it is limited to the theme-rheme order.

Lambrecht provides a more general definition as shown below [1].

> *That component of sentence grammar in which propositions as conceptual representations of states of affairs are paired with lexicogrammatical structures in accordance with the mental states of interlocutors who use and interpret these structures as units of information in given discourse contexts.*

This definition appears intuitive, but it still does not nail down the concept in a precise manner. In particular, its reference to mental states seems to leave room for further specification.

Although the referential status of the rheme can vary, there are certain restrictions on the referential status of the theme. Themes are in general *evoked* or *inferrable* in the sense of Prince [29]. However, it is extremely difficult to pinpoint to what extent we can actually infer a theme from the context. Any definition of information structure based on the referential status of the theme would face this problem.

4.2 Information-Theoretic Definition

One of our assumptions is that the theme has lower entropy than the rheme. In this section, we attempt to define information structure based on this idea. Here is our definition:[4]

Definition 2. *The information structure of an utterance is the linguistic realization of a binary partition (composition) of the semantic representation of the utterance between theme and rheme, such that the entropy of the rheme is greater than that of the theme.*

Let us examine some of the prominent features of this definition. First, it assumes a binary partition. We also assume that partitions are those grammatically feasible ones. For example, such a partition can be represented using Combinatory Categorial Grammar as discussed in Steedman [22].

Definition 2 refers to the entropies of the theme and the rheme only relatively and does not directly refer to absolute properties of the theme or the rheme. As mentioned in Section 2, the computation of entropy would eventually depend on the analysis of inference. Thus, various problems of dealing with inference will not go away. However, it seems advantageous to abstract away from the difficulty with inference, as we can leave it all in the computation of entropy.

Except for the binary partition requirement, Definition 2 does not refer to linguistic notions such as reference to a verb and argument-adjunct distinction (cf. [7, 1]). As a result, the definition can be applied robustly to any construction in any language.

Since Definition 2 is based on entropy that evaluates to a numeric value, it can be compared with our own occasionally grayish judgment about information structure. In many cases, it is difficult to analyze information structure, especially in a written text. A theory of information structure may actually need to fail gracefully if the situation is not clear-cut. Unlike previous definitions, the present approach accepts such a possibility. Furthermore, the use of probability distribution would still allow us to assign small probabilities to unexpected outcomes. This can be adopted to account for unexpected options and indirect responses to a question.

---

[4] This definition is not compatible with recursive analyses of information structure including [4]. More details on this point are available in [6].

5 Conclusion

This paper proposes a hypothesis that information structure is to even out the information load of the theme and the rheme (information balance). Assuming that the theme is the low-entropy component of an information structure, we show that placing the theme before the rheme is, in this respect, never worse than the other order. A natural consequence is the theme-first tendency. One interpretation is that information structure is a way to minimize the required channel capacity.

The rheme-first examples are analyzed as involving zero-entropy themes. Since the information balance is not affected by the position of such themes, these examples are still consistent with our proposal. The paper also discusses a new definition of information structure as informational contrast between theme and rheme, which can serve as the basis for the entire discussion of this paper.

The current proposal is to some extent consistent with many other proposals about the relation between word order and information structure. However, the proposal is novel in that it relates certain word-order phenomena directly with the notion of entropy, which is widely applied to various fields, including linguistics. This approach also introduces a possibility of applying psycholinguistic/cognitive techniques for further evaluation. The proposal is arguably the first to derive both theme-first tendency and seemingly exceptional cases from a single hypothesis. This is desirable as we can now view more diverse phenomena with fewer principles.

References

1. Knud Lambrecht. *Information Structure and Sentence Form: Topic, focus, and the mental representations of discourse referents.* Cambridge University Press, 1994.
2. Vilém Mathesius. *A Functional Analysis of Present Day English on a General Linguistic Basis, edited by Josef Vachek.* The Hague: Mouton, 1975.
3. Jan Firbas. On defining the theme in functional sentence analysis. *Travaux Linguistiques de Prague*, 1:267–280, 1964.
4. M. A. K. Halliday. *An Introduction to Functional Grammar.* London: Edward Arnold, 1985.
5. Nobo N. Komagata. *A Computational Analysis of Information Structure Using Parallel Expository Texts in English and Japanese.* PhD thesis, University of Pennsylvania, 1999.
6. Nobo Komagata. Information structure in subordinate and subordinate-like clauses in special issue on discourse and information structure (to appear). *Journal of Logic, Language and Information*, 12(3), 2003.
7. Petr Sgall, Eva Hajičová, and Jarmila Panevova. *The meaning of the sentence in its semantic and pragmatic aspects.* D. Reidel, 1986.
8. Marianne Mithun. Morphological and prosodic forces shaping word order. In Pamela Downing and Michael Noonan, editors, *Word Order in Discourse.* John Benjamins, 1995.
9. Doris L. Payne. Information structuring in papago narrative discourse. *Language*, 63(4):783–804, 1987.

10. Chet A. Creider and Jane T. Creider. Topic-comment relation in a verb-initial language. *J. African Languages and Linguistics*, 5:1–15, 1983.
11. Elena M. Leman. *Cheyenne Major Constituent Order*. Dallas, TX: Summer Institute of Linguistics, 1999.
12. Enric Vallduví. *The informational component*. PhD thesis, University of Pennsylvania, 1990.
13. Fred I. Dretske. *Knowledge and the Flow of Information (originally published in 1981 from the MIT Press)*. CSLI, 1999.
14. I. Kruijff-Korbayová, G.J.M. Kruijff, and John Bateman. Generation of contextually appropriate word order. In Kees van Deemter and Rodger Kibble, editors, *Information sharing*. CSLI, 2000.
15. Fazlollah M. Reza. *An Introduction to Information Theory (first published in 1961 by McGraw-Hill)*. Dover, 1994.
16. Christopher D. Manning and Hinrich Schütze. *Foundation of Statistical Natural Language Processing*. MIT Press, 1999.
17. Yehoshua Bar-Hillel. *Language and Information*. Addison-Wesley, 1964.
18. Frederick J. Crosson and Kenneth M. Sayre, editors. *Philosophy and Cybernetics*. University of Notre Dame, 1967.
19. Colin Cherry. *On human communication: a review, a survey, and a criticism*. MIT Press, 1978.
20. Angelica Kratzer. Stage-level and individual-level predicates. In Gregory N. Carlson and Francis Jeffry Pelletier, editors, *The Generic Book*, pages 125–175. University of Chicago Press, 1995.
21. Nomi Erteschik-Shir. The syntax-focus structure interface. In Peter W. Culicover and Louise McNally, editors, *Syntax and Semantics, Vol. 29: The limits of syntax*, pages 211–240. Academic Press, 1998.
22. Mark Steedman. Information structure and the syntax-phonology interface. *Linguistic Inquiry*, 31(4):649–689, 2000.
23. Doris L. Payne. Verb initial languages and information order. In Pamela Downing and Michael Noonan, editors, *Word Order in Discourse*. John Benjamins, 1995.
24. Pamela Downing. Word order in discourse: By way of introduction. In Pamela Downing and Michael Noonan, editors, *Word Order in Discourse*. John Benjamins, 1995.
25. Knud Lambrecht. On the status of SVO sentences in French discourse. In Russell S. Tomlin, editor, *Coherence and Grounding in Discourse*, pages 217–262. John Benjamins, 1987.
26. Marianne Mithun. Is basic word order universal? In Doris L. Payne, editor, *Pragmatics of Word Order Flexibility*, pages 15–62. John Benjamins, 1992.
27. Doris L. Payne. Nonidentifiable information and pragmatic order rules in 'o'odham. In Doris L. Payne, editor, *Pragmatics of Word Order Flexibility*, pages 137–166. John Benjamins, 1992.
28. Russell S. Tomlin and Richard Rhodes. Information distribution in Ojibwa. In Doris L. Payne, editor, *Pragmatics of Word Order Flexibility*, pages 117–136. John Benjamins, 1992.
29. Ellen F. Prince. Toward a taxonomy of given-new information. In Peter Cole, editor, *Radical Pragmatics*, pages 223–256. Academic Press, 1981.

A Generic Framework for Context-Based Distributed Authorizations

Ghita Kouadri Mostéfaoui[1] and Patrick Brézillon[2]

[1] Software Engineering Group
Department of Informatics - University of Fribourg
Rue Faucigny 2, CH-1700
Fribourg, Switzerland
Ghita.KouadriMostefaoui@unifr.ch
[2] LIP6, case 169
Université Paris 6
Rue du Capitaine Scott 8, 75015
Paris, France
Patrick.Brezillon@lip6.fr

Abstract. In conventional security systems, protected resources such as documents, hardware devices and software applications follow an On/Off access policy. On, allows to grant access and off for denying access. This access policy is principally based on the user's identity and is static over time. As applications become more pervasive, security policies must become more flexible in order to respond to these highly dynamic computing environments. That is, security infrastructures will need to be sensitive to context. In order to meet these requirements, we propose a conceptual model for context-based authorizations tuning. This model offers a fine-grained control over access on protected resources, based on a set of user's and environment state and information.

1 Introduction

Research in the security field covers many aspects such as the improvement of cryptographic algorithms in order to make them more resistant to hackers, implementation of new authentication methods and designing access control mechanisms, etc. In traditional security systems, the security policy is pre-configured to a static behavior and cannot be seamlessly adapted dynamically to new constraints. This situation is due to the lack of consideration for the context in existing security systems. As a consequence, there is a lack of clearly defined conceptual models of context and system software architectures.

The goal of our research is to develop a conceptual framework for context-based security systems. Context-based security aims at adapting the security policy depending on a set of relevant information collected from the dynamic environment. As the environment evolves, the context change, some contextual elements being integrated in the proceduralized context, others leaving the proceduralized context [1]. Thus, security policies dynamically change in order to cope with new requirements.

Our model is intended to handle all the components of a security system including authentication, privacy and authorization and may be easily extensible to include new security modules. However, this article discusses essentially resources access control or authorizations in the case of distributed systems, where a set of independent computers and devices communicate via a network in order to share data and services.

The structure of the paper is as follows. Section 2 discusses security issues in ubiquitous applications. The main contributions of the paper are presented in Section 3. The foundations for designing context-based authorizations frameworks are laid in Section 4. Section 5 concludes this paper with an outline of future research directions.

2 Security Issues in Ubiquitous Applications

The use of widely distributed resources provides a huge potential for expanding the way that people and businesses communicate and share data, provide services to clients and process information to increase their efficiency. This broad access has also brought with it new security vulnerabilities. Security systems developed now suppose a given and static framework, when attacks generally try to bypass these static contexts of effectiveness of security systems. Amazingly, security has often been the last requirement in designing such dynamic environments. This situation is due to the high cost of security infrastructures, export controls of cryptography technologies and the lack of experts in the security field for specific applications [2]. This is particularly true inside corporate networks where a firewall is assumed to keep all hackers out [3]. Firewalls are, however, not sufficient to protect shared resources. The main function of a firewall is to block unwanted traffic and hide vulnerable internal-network systems. It provides no data integrity and does not check traffic not sent through it, which means that it cannot protect the corporate network from internal attacks. People inside the network may maliciously or unintentionally reveal critical data to unauthorized users or disturb the well-working of the system. As a conclusion, firewalls should always be viewed as a supplement to a strong security policy.

According to Merriam-Webster a policy is *"a definite course or method of action selected from among alternatives and in light of given conditions to guide and determine present and future decisions."*

In the same spirit, we define a security policy as *"a set of rules that monitor all the security components behaviors acting on the framework to secure."* The security policy must be concise, descriptive and easily implementable. Security components consist of access control lists, cryptographic algorithms, and users authentication tools. They act over the following security levels: network and application levels.

2.1 The Network Level

Networks are all about the sharing of data and applications. In recent years, network security breeches have increased in occurrences and more importantly, in severity. Security in these environments is thus a perquisite. Actually, there is no single technique to ensure reliable networks but different technologies (firewalls, encryption, etc...) are combined together in order to face security attacks. They try to extend security frameworks by combining them, but staying in a static approach. The next generation of Internet protocol (IPv6) is intended to add new security features at the network level over its predecessor version.

2.2 The Application Level

Network-based security suffers from some limitations on the kind of security checks that can be performed. The reason is that network-based security systems do not operate on a high level of data abstraction and cannot interpret the content of the traffic. They only know about hosts, addresses, and network related concepts. Application-based security is in contrast intended to provide a security layer based on user roles and identity along with other high level concepts such as protected resources and access policies. Our context-based security model is intended to operate at the application level. This does not mean that there is no need for network-level security. The main reason is simply that it is much cheaper to reconfigure the security infrastructure at the application level than at low-level (network).

Other requirements must be taken into account regarding the security policy. Following the definition given above, the security policy is intended to manage all the security components of the distributed system. Namely, the authentication, authorization, integrity and confidentiality modules and must be easily extensible to manage newly integrated modules. Following the aim of our work, the security policy must also be **reconfigurable** depending on the user and application environment context. This leads to the definition of a **context-based** security policy. Due to the pervasive nature of recent distributed environments, an additional requirement is the definition of **shared** policies. These features will be detailed in the following section.

3 Research Aim and Scope

Works addressing security issues in pervasive computing, basically provide technical solutions such as authentication, access control, integrity and confidentiality, and the security models are generally static. That means that they are built according to already identified threats. The resulting infrastructures are thus, very difficult to adapt to new threats. This work, rather, focuses towards a new aspect of security. We believe that more secure systems can be achieved by adding to these systems the ability to automatically adapt their security policy depending on new constraints. These constraints are dictated by the user's and application environment. Figure 1 illustrates this idea. The distributed application is controlled with an initial security policy in an initial context. This context is continually changing in request to triggers (dynamic changes in the environment). The security policy must then adapt itself to the new context.

Our approach will thus combine the two fields of context-aware computing and security in pervasive computing in order to provide the foundations for "context-based security".

3.1 Related Work

Integrating security with context-aware environments is a recent research direction. Most of the efforts are directed to securing context-aware applications. In [4] and [5], Covington et al. explore new access control models and security policies to secure both information and resources in an intelligent home-environment. Their framework makes use of environment roles [6]. In the same direction, Masone designed and implemented RDL (Role-Definition Language), a simple programming language to describe roles in terms of context information [7]. There have also been similar initiatives in [8] and [9].

Interestingly, we observed that all previous work on combining security and context-aware computing follow the same pattern: using contextual information to enrich the access control model in order to secure context-aware applications with a focus on specific applications.

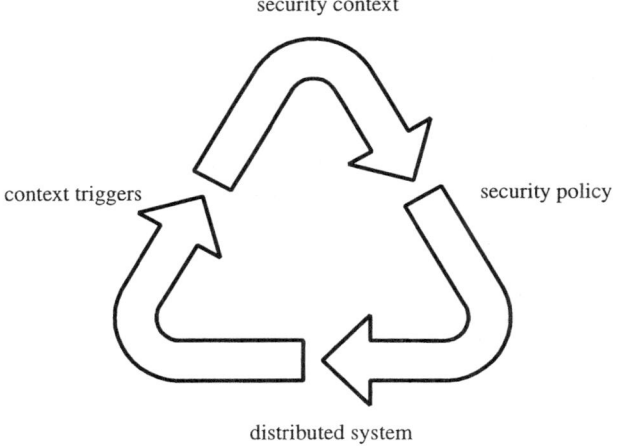

Fig. 1. Context-based security

3.2 Contributions of This Paper

By comparison with previous contributions discussed in Section 3.1, our work is about contextualizing security rather than securing context-aware applications. Even if the difference is not completely apparent actually where we begin by describing the overall architecture, fundamental differences will emerge in future

contributions when detailing each component. As a preliminary example, we cite the new concept of "security context" introduced in Section 4.3.
We summarize in the following the main contributions of our research:

1. To lay down the minimal foundations for a generic context-based security framework with a focus on the software architecture. Generic means to provide the minimal software architecture that can be easily extended to build more specific applications. In addition, context-related modules and resources are loosely coupled allowing the adding/removing of new resources and to modify their respective access policies in a transparent manner. This is an appropriate choice for highly dynamic environments.
2. The second main contribution is to provide a way for managing federations of resources following a specific global policy (an organization's policy for instance) where resources and services join and leave the federation in an ad hoc manner. That is why access policies are organized by resource type and their corresponding actions (see the example in Section 4.1).
3. The third main contribution is to require specific authentication methods depending on a partial context built from the state of the federation.

The resulting prototype will then be designed with the following requirements in mind:

- Provides a framework for the rapid prototyping of context-based security systems,
- Handles both simple and high-level contextual information related to security,
- Easily extensible to manage new protected resources,
- Easily reconfigurable to adapt to new access policies,
- Allows a customizable (context-based) method of authentication (username/password, certificates, etc) by requiring specific credentials depending on a partial context.
- Transparent to both resources and requesting clients; no need to an a priori knowledge of the federation policy.

The following section describes the overall architecture.

4 Context-Based Authorizations Tuning

The term context-aware computing was first introduced by Schilit et al in 1994 [10] as a software that " *adapts according to its location of use, the collection of nearby people and objects, as well as the changes to those objects over time*".
Another given by Dey in [11] states that " *A system is context-aware if it uses context to provide relevant information and/or services to the user, relevancy depends on the user's task.*"
Now context awareness is a well established community with conferences as ubiquitous computing, pervasive computing, etc.

In the field of context-aware computing, context is generally defined as physical parameters (location, temperature, time, etc.) obtained from sensors. However, the user is not really considered in these approaches. In this sense, context is generally managed as a layer between an application and the external world, a type of middleware. Conversely, there is also another approach in which the user (through his knowledge and reasoning) is central in the modelling of context. In this second area, knowledge and reasoning in the accomplishment of a task are described in a context-based formalism, i.e. inside the application itself (e.g. see the contextual graphs in [12]).

Our research covers both the first and second work with an application to the security infrastructure of ubiquitous applications. We concentrate on the main part of the security framework of a distributed system. Namely, access control to shared resources. The framework makes use of RBAC (Role-Based Access Control). Users are affected to roles based on their credentials and competencies [13]. Role-based access is more suitable for pervasive environments since it simplifies the administration of permissions; updating roles is easier then updating permissions for every user individually [14] [15].

4.1 A Case Study

To illustrate the main functionalities of the proposed architecture, we consider a simple example. This example will be developed along with the definition of each component.

We consider a protected document that offers the following operations: *read*, *write* and *delete*. Depending on their credentials and identity, requesting users are a priori affected to one of these two roles: *administrator* or *guest*. The document is available on the network and its access policy is defined and stored in the context engine. In our model, resources are managed by a specific access policy depending on their type; the type of service they provide.

The actual access policy is defined using a rule-based formalism with a simplified grammar (no explicit If/Then clauses). Rule-based reasoning is an area of artificial intelligence (AI) wherein people simulate human behaviors when presented with a new case requiring some action. This approach is used here to specify context-based access policies in order to grant or deny access to resources (see [16] and [17] for more information).

We consider that all protected resources are protected by default, thus, their corresponding policies express only cases when the access is granted (which justified the lack of If/Then clauses). This design choice aims at lightening the process of policies specification. Here is an example:

A Simple Access Policy

```
Resource_type = document;
Action = read ((Role = administrator) OR
        (Role = guest; (Date = Weekdays AND Time = between 8:00-18:00)));
        Authentication = username/password;
Action = write (Role = administrator);
```

```
            Authentication = username/password;
Action = delete (Role = administrator; Date = Weekends);
            Authentication = certificates;
```

Each shared resource defines access rules for each individual operation. The *authentication* tag is used to specify the authentication method required in the actual partial context. The access is granted only when the complete context is build; if the conditions are satisfied and the corresponding authentication phase succeeded. The pattern used above eases further updates of the access policy by adding or removing conditions on it. Defining access policies manually is a cumbersome task in complex real applications with complex relationships among roles. This process can, however, be performed visually using a graphical interface. In [4], Covington et al. propose a graphical policy editor for specifying available roles, their relationships, and policy rules.

Based on the above example, we present the main parts of the security architecture.

4.2 Protected Resource

We consider three types of resources: hardware devices, physical resources (documents, databases) and software resources (operations on a software object or data structure). In order to fit within our model, each resource must respect the following structure (see Fig. 2):

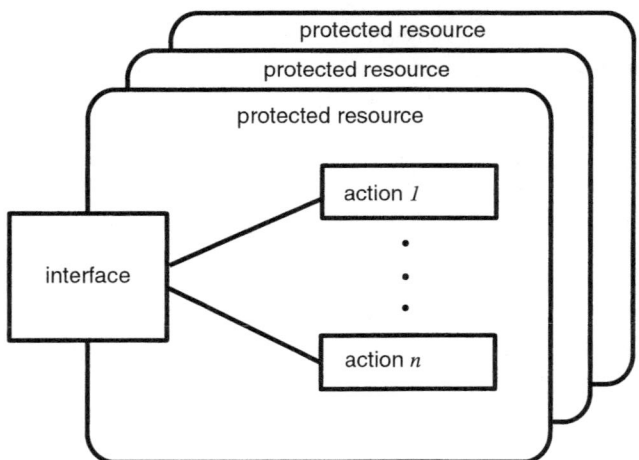

Fig. 2. Structure of protected resources

- Any interaction with the resource is performed via an interface that presents the set of all actions available for the resource,

- Resource actions contain an additional flag that accept only two states: true or false. This flag is used to allow or deny the operation on the resource,
- Each resource is protected by default (all of its corresponding actions have a flag value equal to false).

A user whose role is guest invokes the read action on the document via the interface attached to it. The protected resource identifies the request as coming from a client. It then, forwards it to the context engine.

4.3 The Security Context Engine

The context engine has two responsibilities: modelling contextual entries in order to build a security context and mapping between the security context and the corresponding authorizations on resources. Modelling context requires picking out the most relevant features to reduce it to a meaningful representation [18],[19]. We provide herein two types of classifications of contextual entries depending on their representation aspect and temporal aspect:

Representation Aspect

- *Simple:* The collected information is used in its original format. For example, it can represent the value of a parameter,
- *Interpreted:* The collected data cannot be used as it is but needs to be converted in a more meaningful format. For example, the contextual entry is "Sunday" that needs to be converted into "Weekday" or "Weekend",
- *Composite:* It is a set of simple and/or interpreted entries collected as a whole.

Temporal Aspect

- *Static:* It describes contextual information that is invariant, such as a person's date of birth,
- *Transient:* The value of a transient contextual entry is updated at run-time and does not need to store information about its past state. For example, time, date, etc,
- *Persistent:* Some entries must store historical data about their past state. Persistent contextual entries need to be marked with a time stamp.

Building a Security Context. Our model relies on a set of contextual information relevant to security. This set forms what we call a security context. Designing context in general is not easy and designing a security context suffers the same problem. We present herein an attempt definition of the security context.

> *A security context is a set of information collected from the user's environment and the application environment and that is relevant to the security infrastructure of both the user and the application.*

The word relevant means that has direct or indirect effect on the security policy. In the present work, we are dealing with authorizations, so the contextual information is more precisely relevant to authorizations on available resources. What is really relevant is not fully predictable in advance and depends on the application. This information may include, user identity, membership, resource location, date, time, the user's interaction history with the system, social situation, etc.

In our example, the context engine extracts the operation type and the client's role from the received message. It then builds a partial context based on the client role and the access policy of the requested resource. To build a partial context, the engine retrieves the suggested contextual information from the context bucket. Following the policy defined in section 4.1, requests to the bucket will ask for date and time. The resulting partial context requires a specific authentication method (a username/password method in our example). Thus, credentials provided by the user (username/password, certificates, etc) are additional sources of contextual information. Once the complete security context is built, final actions are performed; access to the document is granted (if authentication succeeded) or denied (if authentication failed). This process is practically equivalent to setting the read operation flag to true or false. Contextual data are received from the security bucket in a primitive type, and then interpreted at the context engine level. For example, date is represented as "*Monday*" and it is the responsibility of the context engine to interpret it as "*Weekday*". This design requirement eases the reuse of the context bucket by different applications with different interpretations of the same contextual data.

4.4 The Security Context Bucket

One of the main problems of context is how to store it and in a way that many applications can use it. This is true especially in distributed applications where both the contextual information and the applications that need it are naturally spread and shared [20]. In order to store contextual entries, we investigate a central point of fall. All the security contextual data are collected into a logical bucket; the security context bucket.

The security context bucket is a shared software data structure that offers the notion of container in order to handle the security contextual information.

A similar approach has been proposed in [21].

At first sight, this approach may seem not very suitable for distributed systems since components interested in context (subscribers) are distributed over different computing devices and developed by different programmers. This incompatibility leads to different interpretations of the same context data. For example, a user's location may be interpreted by one component as a relative distance (near, far) and by another component as an absolute location (using the coordinates). We argue that even if the storage medium is centralized, the interpretation of the selected entries is performed in a distributed manner, at the component level. In addition, the security bucket acts as a service that has the ability to retrieve a given contextual entry when requested. Thus, multiple security buckets with the same functionality may exist in the network in order to ensure availability and load balancing.

The main advantage of this approach is to ease the scale up of the system for a large number of contextual entries across a wide network. The second advantage is the robustness to failures by making the contextual data available from different places, and finally, to ease the protection of contextual data.

The security context bucket offers the same advantages as encapsulation; a key feature in the object-oriented paradigm. Object encapsulation is also known as data-hiding. It is a software mechanism that protects code and data from being accessed by everyone but only to the methods that need it. In the same manner, the context bucket hides its content, and access to it (read and write operations) is subject to a security policy that manages interactions with clients.

Context entries are collected from the distributed network by the mean of a group of agents. Gathering agents are mobile [22] and launched by the security bucket when requested. Their main role is to collect needed contextual data from their remote location, by requesting sensors, software applications and environment. The content of the context bucket is primordial in configuring the security policy of the system. This content must then be protected.

Protection and Privacy of the Security Context Bucket. Designing context-based security systems poses a kind of tricky issue. The more a context-based security system knows about the user's and the application environment, the more it can provide fine-grained access control to protected resources. On the other hand, it becomes easier for hackers (at least theoretically) to compromise the security of the system not directly (by attacking resources) but may do it indirectly by providing false contextual data to the bucket or by accessing critical users information contained in it. The first can be achieved by launching malicious gathering agents that provide corrupt data and the second can be achieved by accessing critical information from the context bucket, such as users' private data.

Thus, and in order to achieve protection and privacy of the security context, an additional component is then required in our architecture. The authentication module authenticates both entities that provide contextual entries (gathering agents) and entities that need access to the security context (context engines).

However, in case of the unavailability of a contextual entry, the system must be able to learn from previous experiences and propose an alternative.

Collected contextual information are used by the context engine (described in section 4.3) in order to build a security context and then to deduce the actions to perform. The security bucket requirements are summarized in the following:

- The security context bucket has the ability to create, manage and authenticate gathering agents,
- Contextual entries are sensed and stored in a primitive format that eases all possible interpretations,

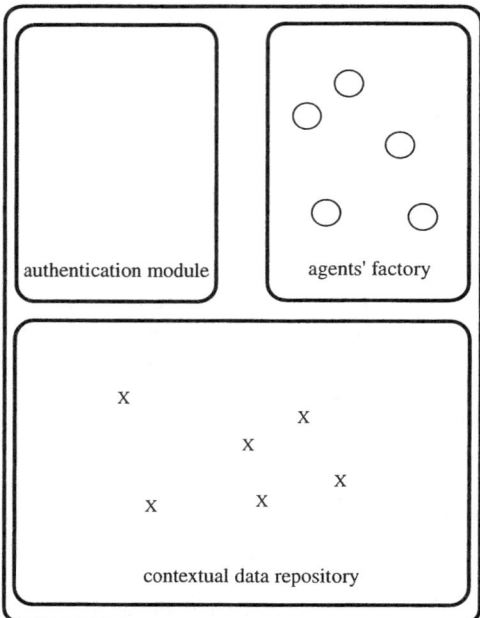

Fig. 3. The security context bucket

- Only authorized sources (gathering agents) are able to add/update data in the security context bucket,
- Only authorized destinations (security context engines) are able to read the bucket content,
- The security context bucket must maintain historical information at the finest level of detail possible of its content.

Figure 3 illustrates the main parts of the security context bucket:

1. The agents' factory produces gathering agents in order to collect contextual data upon request,
2. The contextual data repository is used to store gathered contextual data,
3. An authentication module is also needed in order to authenticate gathering agents and requesting context engines.

The following figure (Fig. 4) illustrates the overall structure of the framework and the relationships among the different components. Further changes in the access policy of protected resources can be transparently performed by updating the corresponding policy in the context engine. Resource can join or leave the distributed infrastructure without disturbing the security infrastructure.

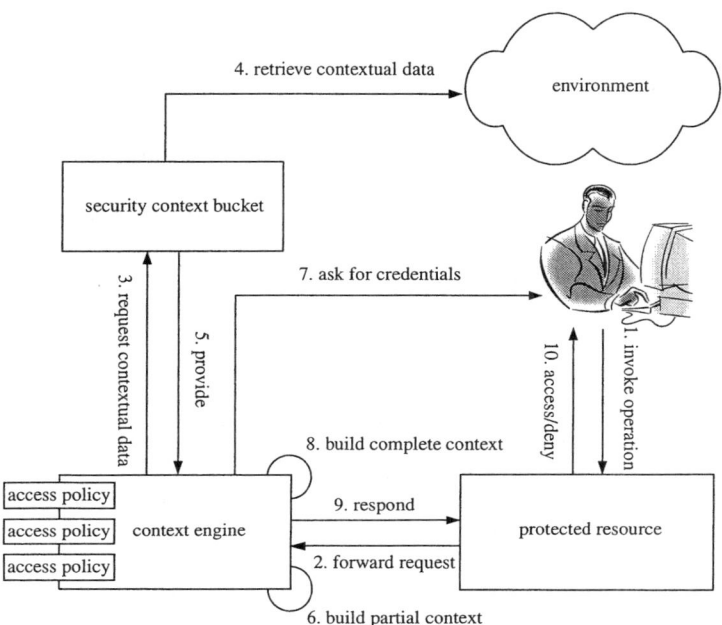

Fig. 4. Overall architecture

5 Conclusion and Future Work

Security systems for distributed infrastructure are generally bound to static access policies that make them very difficult to adapt to new threats. This situation is due to the lack of consideration for context in existing security systems. Context-based security is a recent research direction that aims at providing flexible security models for distributed infrastructures, where the user's and application environments are continually changing.

In this paper we have introduced a new model for context-based authorizations tuning in distributed systems. Much of this work is focused on providing a generic minimal architecture based on loosely-coupled components. The architecture provides tools for collecting and modelling security contextual data. We have introduced the concept of partial context and illustrate how it can be used to request specific authentication methods in order to control access to protected resources. In the near future, we intend to extend the proposed framework to handle inaccurate or unavailable contextual data, specify a registration protocol that allows adding context-based access policies in the context engine and investigate the case of complex relationships between user's roles. The use of contextual graphs [23] [24] is also a potential methodology for modelling the security context; in order to access a resource, the specification of an exhaustive graph may prevent frauds by including only "safe" cases.

We are actually investigating a test-bed application with a federation of services inside a university network. Both tutors and students provide plug and play services to be used in a safe environment. These services (or resources) include: course subscription services, online exercises, chat systems, printing service, etc. Our model provides an administrative tool to manage authorizations on these resources based on the user's role (regular student, auditor, tutor or guest) and the context of interaction (time, day, history of the user's use of the service, etc). We believe that this approach will prove to be an interesting starting point for further investigations of flexible security models for next-generation distributed authorizations frameworks.

References

1. Brézillon, P., Pomerol, J-Ch.: Contextual Knowledge Sharing and Cooperation in Intelligent Assistant Systems. Le Travail Humain 62 (3), PUF, Paris, (1999) 223–246
2. Balfanz, D., Dean, D., Spreitzer, M.: A Security Infrastructure for Distributed Java Applications. In Proceedings of the 2000 IEEE Symposium on Security and Privacy, Oakland, California, (2000) 15–26
3. Wack, J.: Guidelines on Firewalls and Firewall Policy. Computer Security Division. Information Technology Laboratory. National Institute of Standards and Technology
4. Covington, M.J., Ahamad, M., Srinivasan, S.: A Security Architecture for Context-Aware Applications. Technical Report GIT-CC-01-12, College of Computing, Georgia Institute of Technology, May 2001
5. Covington, M.J., Fogla, P., Zhan, Z., Ahamad, M.: A Context-Aware Security Architecture for Emerging Applications. In Proceedings of the Annual Computer Security Applications Conference (ACSAC), Las Vegas, Nevada, USA, December 2002
6. Covington, M.J., Long, W., Srinivasan, S., Dey, D., Ahamad, M., Abowd, A.: Securing Context-Aware Applications Using Environment Roles. In Proceedings of the 6th ACM Symposium on Access Control Models and Technologies (SACMAT '01), Chantilly, Virginia, USA, May 2001
7. Masone, C.: Role Definition Language (RDL): A Language to Describe Context-Aware Roles. Dartmouth College, Computer Science. Hanover, NH. TR2002-426. May 2002
8. Shankar, N., Balfanz, D.: Enabling Secure Ad-hoc Communication Using Context-Aware Security Services. Extended Abstract. In Proceedings of UBICOMP2002 - Workshop on Security in Ubiquitous Computing
9. Osbakk, P., Ryan, N.: Context Privacy, CC/PP, and P3P. In Proceedings of UBICOMP2002 - Workshop on Security in Ubiquitous Computing
10. Schilit, B.N and Theimer, M.M.: Disseminating Active Map Information to Mobile Hosts. IEEE Network, 8(5): (1994) 22–32
11. Dey, A.K.: Understanding and Using Context. Personal Ubi Comp 5 (2001) 1, 4–7
12. Brézillon, P.: Using Context for Supporting Users Efficiently. Proceedings of the 36th Hawaii International Conference on Systems Sciences, HICSS-36, Track "Emerging Technologies", R.H. Sprague (Ed.), Los Alamitos: IEEE, (2003)

13. Georgiadis C., Mavridis I., Pangalos G., Thomas, R.: Flexible Team-based Access Control Using Contexts. In Proceedings of the 6th ACM Symposium on Access Control Models and Technologies (SACMAT 2001) ACM SIGSAC, Chantilly, VA, U.S.A, May 2001
14. Ferraiolo, D.F. and Kuhn, D.R.: Role Based Access Control. In Proceedings of the 15th National Computer Security Conference (1992)
15. Sandhu R. S., Coyne E. J., Feinstein H. L., and Youman C. E.: Role-Based Access Control Models. IEEE Computer, Volume 29, Number 2, February 1996, 38–47
16. Clancey, W. J.: The Epistimology of a Rule-Based Expert System: A Framework for Explanation. Artificial Intelligence Journal, 20(3): (1983) 197–204
17. Clancey, W. J.: Notes on "Epistimology of a Rule-Based Expert system". Artificial Intelligence Journal, 59: (1993) 197–204
18. Kokinov, B.: A Dynamic Approach to Context Modeling. Proceedings of the IJCAI-95 Workshop on Modeling Context in Knowledge Representation and Reasoning. LAFORIA 95/11, (1995)
19. Bouquet, P., Serafini, L., Brezillon, P., Benerecetti, M., Castellani, F. (eds.): Modelling and Using Context. In Proceedings of the 2nd International and Interdisciplinary Conference, CONTEXT99, Lecture Notes in Artificial Intelligence, Vol. 1688, Springer Verlag, (1999)
20. Henricksen, K., Indulska, J., Rakotonirainy, A.: Modeling Context Information in Pervasive Computing Systems. In Proceedings of the 1st International Conference, Pervasive 2002 - Zurich August 2002, Lecture Notes in Computer Science, Vol. 2414, Springer Verlag, (2002) 167–180
21. Hong, J.I.: The Context Fabric: An Infrastructure for Context-Aware Computing. Extended Abstract. Conference on Human Factors and Computing Systems, (2002) 554–555
22. White, J.E.: Mobile Agents. In I. Bradshaw and M. Jeffrey, editors, Software Agents. MIT Press and American Association for Artificial Intelligence, (1997)
23. Brézillon, P., Pasquier, L., Pomerol J-Ch.: Reasoning with contextual graphs. European Journal of Operational Research, 136(2): (2002) 290–298
24. Brézillon, P., Cavalcanti, M., Naveiro, R., Pomerol J-Ch.: SART: An intelligent assistant for subway control. Pesquisa Operacional, Brazilian Operations Research Society, 20(2): (2002) 247–268

Unpacking Meaning from Words: A Context-Centered Approach to Computational Lexicon Design

Hugo Liu

MIT Media Laboratory
20 Ames St., E15-320D
Cambridge, MA 02139, USA
hugo@media.mit.edu

Abstract. The knowledge representation tradition in computational lexicon design represents words as static encapsulations of purely lexical knowledge. We suggest that this view poses certain limitations on the ability of the lexicon to generate nuance-laden and context-sensitive meanings, because word boundaries are obstructive, and the impact of non-lexical knowledge on meaning is unaccounted for. Hoping to address these problematics, we explore a context-centered approach to lexicon design called a Bubble Lexicon. Inspired by Ross Quillian's Semantic Memory System, we represent word-concepts as nodes on a symbolic-connectionist network. In a Bubble Lexicon, a word's meaning is defined by a dynamically grown context-sensitive bubble; thus giving a more natural account of systematic polysemy. Linguistic assembly tasks such as attribute attachment are made context-sensitive, and the incorporation of general world knowledge improves generative capability. Indicative trials over an implementation of the Bubble Lexicon lends support to our hypothesis that unpacking meaning from predefined word structures is a step toward a more natural handling of context in language.

1 Motivation

Packing meaning (semantic knowledge) into words (lexical items) has long been the knowledge representation tradition of lexical semantics. However, as the field of computational semantics becomes more mature, certain problematics of this paradigm are beginning to reveal themselves. Words, when computed as discrete and static encapsulations of meaning, cannot easily generate the range of nuance-laden and context-sensitive meanings that the human language faculty seems able to produce so effortlessly. Take one example: Miller and Fellbaum's popular machine-readable lexicon, WordNet [7], packages a small amount of dictionary-type knowledge into each *word sense*, which represents a specific meaning of a word. Word senses are partitioned *a priori*, and the lexicon does not provide an account of how senses are determined or how they may be systematically related, a phenomenon known as systematic polysemy. The result is a sometimes arbitrary partitioning of word meaning. For example, the WordNet entry for the noun form of "sleep" returns two senses, one which means "a slumber" (i.e. a long rest), and the other which means "a nap" (i.e. a brief rest). The systematic relation between these two senses is unaccounted for, and their classification as separate senses indistinguishable from homonyms give the false impression that there is a no-man's land of meaning in between each predefined word sense.

Hoping to address the inflexibility of lexicons like WordNet, Pustejovsky's Generative Lexicon Theory (GLT) [19] packs a great deal more meaning into a word entity, including knowledge about how a word participates in various semantic roles known as "qualia," which dates back to Aristotle. The hope is that a densely packed word-entity will be able to generate a fuller range of nuance-laden meaning. In this model, the generative ability of a word is a function of the type and quantity of knowledge encoded *inside* that word. For example, the lexical compound "good rock" only makes sense because one of the functions encoded into "rock" is "to climb on," and associated with "to climb on" is some notion of "goodness." GLT improves upon the sophistication of previous models; however, as with previous models, GLT represents words as discrete and pre-defined packages of meaning. We argue that this underlying word-as-prepackaged-meaning paradigm poses certain limitations on the generative power of the lexicon. We describe two problematics below:

1) **Artificial word boundary.** By representing words as discrete objects with predefined meaning boundaries, lexicon designers must make *a priori* and sometimes arbitrary decisions about how to partition word senses, what knowledge to encode into a word, and what to leave out. This is problematic because it would not be feasible (or efficient) to pack into a word all the knowledge that would be needed to anticipate all possible intended meanings of that word.

2) **Exclusion of non-lexical knowledge.** When representing a word as a predetermined, static encapsulation of meaning, it is common practice to encode only knowledge that formally characterizes the word, namely, *lexical knowledge* (e.g. the qualia structure of GLT). We suggest that non-lexical knowledge such as *general world knowledge* also shapes the generative power and meaning of words. General world knowledge differs from lexical knowledge in at least two ways:

    a) First, general world knowledge is largely concerned with **defeasible** knowledge, describing relationships between concepts that *can* hold true or *often* holds true (connotative). By comparison, lexical knowledge is usually a more formal characterization of a word and therefore describes relationships between concepts that *usually* holds true (denotative). But the generative power of words and richness of natural language may lie in defeasible knowledge. For example, in interpreting the phrase *"funny punch,"* it is helpful to know that *"fruit punch can sometimes be spiked with alcohol."* Defeasible knowledge is largely missing from WordNet, which knows that a *"cat"* is a *"feline"*, *"carnivore"*, and *"mammal"*, but does not know that *"a cat is often a pet."* While some defeasible knowledge has crept into the qualia structures of GLT (e.g. *"a rock is often used to climb on"*), most defeasible knowledge does not naturally fit into any of GLT's lexically oriented qualia roles.

    b) Second, lexical knowledge by its nature characterizes only word-level concepts (e.g. *"kick"*), whereas general world knowledge characterizes both word-level and **higher-order concepts** (e.g. *"kick someone"*). Higher-order concepts can also add meaning to the word-level concepts. For example, knowing that *"kicking someone may cause them to feel pain"* lends a particular interpretation to the phrase *"an evil kick."* WordNet and GLT do not address general world knowledge of higher-order concepts in the lexicon.

It is useful to think of the aforementioned problematics as issues of context. Word boundaries seem artificial because meaning lies either wholly inside the context of a word, or wholly outside. Non-lexical knowledge, defeasible and sometimes characterizing higher-order concepts, represents a context of connotation about a word, which serves to nuance the interpretation of words and lexical compounds. Considering these factors together, we suggest that a major weakness of the word-as-prepackaged-meaning paradigm lies in its inability to handle context gracefully.

Having posed the problematics of the word-as-prepackaged-meaning paradigm as an issue of context, we wonder how we might model the computational lexicon so that meaning contexts are more seamless and non-lexical knowledge participates in the meaning of words. We recognize that this is a difficult proposition with a scope extending beyond just lexicon design. The principle of modularity in computational structures has been so successful because encapsulations like frames and objects help researchers manage complexity when modeling problems. Removing word boundaries from the lexicon necessarily increases the complexity of the system. This notwithstanding, we adopt an experimental spirit and press on.

In this paper, we propose a context-centered model of the computational lexicon inspired by Ross Quillian's work on semantic memory [21], which we dub as a *Bubble Lexicon*. The Bubble Lexicon Architecture (BLA) is a symbolic connectionist network whose representation of meaning is distributed over nodes and edges. Nodes are labeled with a word-concept (our scheme does not consider certain classes of words such as, *inter alia,* determiners, prepositions and pronouns). Edges specify both the symbolic relation and connectionist strength of relation between nodes. A word-concept node has no internal meaning, and is simply meant as a reference point, or, *indexical feature,* (as Jackendoff would call it [9]) to which meaning is attached. Without formal word boundaries, the "meaning" of a word becomes the dynamically chosen, flexible context bubble (hence the lexicon's name) around that word's node. The size and shape of the bubble varies according to the strength of association of knowledge and the influence of active contexts; thus, meaning is nuanced and made context-sensitive. Defeasible knowledge can be represented in the graph with the help of the connectionist properties of the network. Non-lexical knowledge involving higher-order concepts (more than one word) are represented in the graph through special nodes called *encapsulations,* so that they may play a role in biasing meaning determination.

The nuanceful generative capability of the BLA is demonstrated through the linguistic assembly task of attribute attachment, which engages some simulation over the network. For example, determining the meaning of a lexical compound such as "fast car" involves the generation of possible interpretations of how the "fast" and "car" nodes are conceptually related through dependency paths, followed by a valuation of each generated interpretation with regard to its structural plausibility and contextual plausibility. The proposed Bubble Lexicon is not being presented here as a perfect or complete solution to computational lexicon design, but rather, as the implementation and indicative trials illustrate, we hope Bubble Lexicon is a step toward a more elegant solution to the problem of context in language.

The organization of the rest of this paper is as follows. First, we present a more detailed overview of the Bubble Lexicon Architecture, situating the representation in the literature. Second, we present mechanisms associated with this lexicon, such as context-sensitive interpretation of words and compounds. Third, we discuss an implementation of Bubble Lexicon and present some evaluation for the work through some indicative trials. Fourth, we briefly review related work. In our conclusion we return to revisit the bigger picture of the mental lexicon.

2 Bubble Lexicon Architecture

This section introduces the proposed Bubble Lexicon Architecture (BLA) (Fig. 1) through several subsections. We begin by situating the lexicon's knowledge representation in the literature of symbolic connectionist networks. Next, we enumerate some tenets and assumptions of the proposed architecture. Finally, we discuss the ontology of types for nodes, relations, and operators.

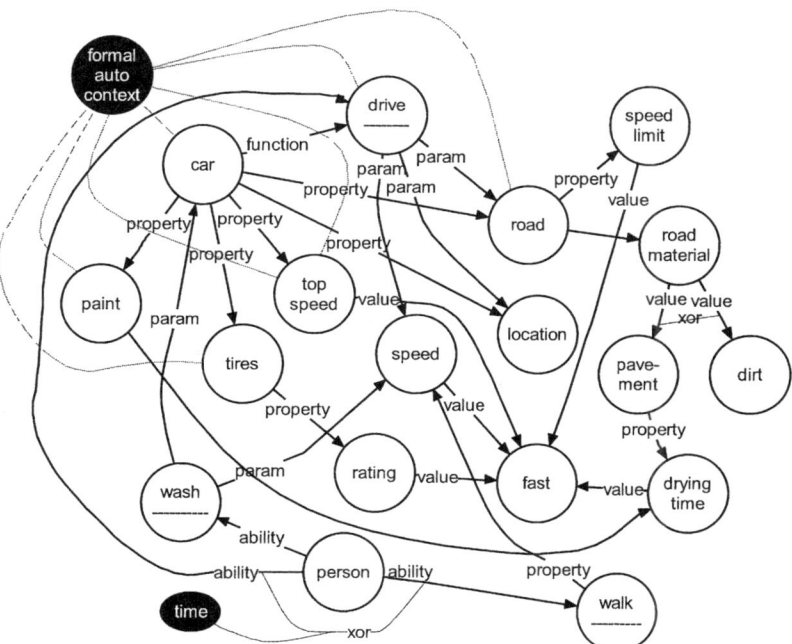

Fig. 1. A static snapshot of a Bubble Lexicon. We selectively depict some nodes and edges relevant to the lexical items "car", "road", and "fast". Edge weights are not shown. Nodes cleaved in half are causal trans-nodes. The black nodes are context-activation nodes.

2.1 Knowledge Representation Considerations

A Bubble Lexicon is represented by a symbolic-connectionist network specially purposed to serve as a computational lexicon. Nodes function as *indices* for words, lexical compounds (linguistic units larger than words, such as phrases), and formal contexts (e.g. a discourse topic). Edges are labeled dually with a minimal set of structural structural dependency relations to describe the relationships between nodes, and with a numerical weight. Operators are special relations which can hold between nodes, between edges, and between operator relations themselves; they introduce boolean logic and the notion of ordering, which is necessary to represent certain types of knowledge (e.g. ordering is needed to represent a sequence of events).

Because the meaning representation is distributed over the nodes and edges, words only have an *interpretive meaning*, arising out of some simulation of the graph. Spreading activation (cf. [5]) is ordinarily used in semantic networks to determine semantic proximity. We employ a version of spreading activation to dynamically create a context bubble of interpretive meaning for a word or lexical compound. In growing and shaping the bubble, our spreading activation algorithm tries to model the influence of active contexts (such as discourse topic), and of relevant non-lexical knowledge, both of which contribute to meaning.

Some properties of the representation are further discussed below.

Connectionist weights. Connectionism and lexicon design are not usually considered together because weights tend to introduce significant complexity to the lexicon. However, there are several reasons why connectionism is necessary to gracefully model the context problem in the lexicon.

First, not all knowledge contributes equally to a word's meaning, so we need numerical weights on edges as an indication of semantic relevance, and to distinguish between certain from defeasible knowledge. Defeasible knowledge may in most cases be less central to a word's meaning, but in certain contexts, their influence is felt.

Second, connectionist weights lend the semantic network notions of memory and learning, exemplified in [16], [17], and [22]. For the purposes of growing a computational lexicon, it may be desirable to perform supervised training on the lexicon to learn particular meaning bubbles for words, under certain contexts. Learning can also be useful when importing existing lexicons into a Bubble Lexicon through an exposure process similar to semantic priming [1].

Third, connectionism gives the graph *intrinsic semantics*, meaning that even without symbolic labels on nodes and edges, the graded inter-connectedness of nodes is meaningful. This is useful in conceptual analogy over Bubble Lexicons. Goldstone and Rogosky [8] have demonstrated that it is possible to identify conceptual correspondences across two connectionist webs without symbolic identity. If we are also given symbolic labels on relations, as we are in BLA, the structure-mapping analogy-making methodology described by Falkenhainer et al. [6] becomes possible.

Finally, although not the focus of this paper, a self-organizing connectionist lexicon would help to support lexicon evolution tasks such as lexical acquisition (new word meanings), generalization (merging meanings), and individuation (cleaving meanings). A discussion of this appears elsewhere [11].

Ontology of Conceptual Dependency Relations. In a Bubble Lexicon, edges are relations which hold between word, compound, and context nodes. In addition to having a numerical weight as discussed above, edges also have a symbolic label representing a dependency relation between the two words/concepts. The choice of the relational ontology represents an important tradeoff. Very relaxed ontologies that allow for arbitrary predicates like `bite(dog,mailman)` in Peirce's existential graphs [18] or node-specific predicates as in Brachman's *description logics* system [2] are not suitable for highly generalized reasoning. Efforts to engineer ontologies that enumerate *a priori* a complete set of primitive semantic relations, such as Ceccato's *correlational nets* [3], Masterman's primitive concept types [14], and Schank's Conceptual Dependency [23], show little agreement and are difficult to engineer. A small but insufficiently generic set of relations such as WordNet's nyms [7] could also severely curtail the expressive power of the lexicon.

Because lexicons emphasize *words*, we want to focus meaning around the word-concept nodes rather than on the edges. Thus we propose a small ontology of generic *structural* relations for the BLA. For example, instead of `grow(tree,fast)`, we have `ability(tree,grow)` and `parameter(grow,fast)`. These relations are meant as a more expressive set of those found in Quillian's original Semantic Memory System. These structural relations become useful to linguistic assembly tasks when building larger compound expressions from lexical items. They can be thought of as a sort of semantic grammar, dictating how concepts can assemble.

2.2 Tenets and Assumptions

Tenets. While the static graph of the BLA (Fig. 1) depicts the meaning representation, it is equally important to talk about the simulations over the graph, which are responsible for *meaning determination.* We give two tenets below:

1) **No coherent meaning without simulation.** In the Bubble Lexicon graph, different and possibly conflicting meanings can attach to each word-concept node; therefore, words hardly have any coherent meaning in the static view. We suggest that when human minds think about what a word or phrase means, meaning is always evaluated in some context. Similarly, a word only becomes coherently meaningful in a bubble lexicon as a result of simulation (graph traversal) via spreading activation (edges are weighted, though Fig. 1 does not show the weights) from the origin node, toward some destination. This helps to exclude meaning attachments which are irrelevant in the current context, to hammer down a more coherent meaning.

2) **Activated nodes in the context biases interpretation.** The meaning of a word or phrase is the collection of nodes and relations it has "harvested" along the path toward its destination. However, there may be multiple paths representing different interpretations, perhaps each representing one "word sense". In BLA, the relevance of each word sense path depends upon context biases near the path which may boost the acti-

vation energy of that path. Thus meaning is naturally influenced by context, as context nodes prefers certain interpretations by activating certain paths.

Assumptions. We have made the following assumptions about our representation:
1) Nodes in BLA are word-concepts. We do not give any account of words like determiners, pronouns, and prepositions.
2) Nodes may also be higher-order concepts like "fast car," constructed through encapsulation. In lexical evolution, intermediate transient nodes also exist.
3) In our examples, we show selected nodes and edges, although the success of such a lexicon design thrives on the network being sufficiently well-connected and dense.
4) Homonyms, which are non-systematic word senses (e.g. fast: not eat, vs. quick) are represented by different nodes. Only systematic polysemy shares the same node. We assume we can cleanly distinguish between these two classes of word senses.
5) Though not shown, relations are always numerically weighted between 0.0 and 1.0, in addition to the predicate label, and nodes also have a *stable activation energy*, which is a function of how often active a node is within the current discourse.

2.3 Ontology of Nodes, Relations, and Operators

We propose three types of nodes (Fig. 2). **Normal nodes** may be word-concepts, or larger encapsulated lexical expressions. However, some kinds of meaning i.e. actions, beliefs, implications are difficult to represent because they have some notion of syntax. Some semantic networks have overcome this problem by introducing a causal relation [22], [17]. We opted for a causal node called a **TransNode** because we feel that it offers a more precise account of causality as being *inherent* in some word-concepts, like actions. This also allows us to maintain a generic structural relation ontology. Because meaning determination is dynamic, TransNodes behave causally during simulation. TransNodes derive from Minsky's general interpretation [15] of Schankian transfer [23], and is explained more fully elsewhere [11].

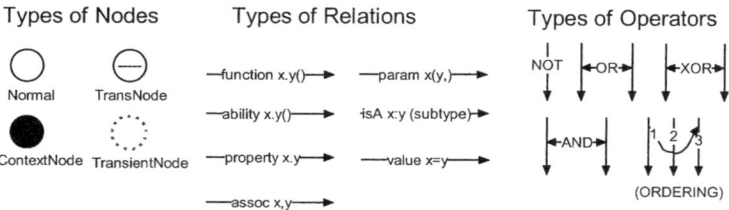

Fig. 2. Ontology of node, relation, and operator types.

While normal nodes can act as contexts when they are activated in the BLA, there is no formal definition to those groupings. We suggest that sometimes, human minds may employ more formal and explicit notions of context which define a topic or domain of discourse (e.g.: "automotives," "finance"). For example, the meaning of the formal context "finance" is somewhat different than the meaning that is attached to that word-concept node. For one, the formal context "finance" may be a well-defined term in the financial community. The external definition of certain concepts like formal contexts is supported by Putnam's idea of semantic externalism [20]. We introduce **ContextNodes** as an explicit representation of externally-defined formal contexts. ContextNodes use the assoc (generic association) relation, along with operators, to cause the network to be in some state when they are activated. They can be thought of as *externally grounded contexts*. Meta-level ContextNodes that control a layer of ContextNodes are also possible. In Figure 1, the "formal auto context" ContextNode is meant to represent formally the domain of automotives, to the best of a person's understanding of the community definition of that context.

Because ContextNodes help to group and organize nodes, they are also useful in representing perspectives, just as a semantic frame might. Let us consider again the example of a car, as depicted in Figure 1. A car can be thought of as an assembly of its individual parts, or it can be thought of functionally as something that is a type of transportation that people use to drive from point A to point B. We can use ContextNodes to separate these two perspectives of a car. After all, we can view a perspective as a type of packaged context.

So far we have only talked about nodes which are stable word-concepts and stable contexts in the lexicon. These can be thought of as being stable in memory, and changing slowly. However, it is also desirable to represent more temporary concepts, such as those used in thought. For example, to reason about "fast cars", one might encapsulate one particular sense path of fast car into a **TransientNode**. Or one can instantiate a concept and overload its meaning. TransientNodes explain how fleeting concepts in thought can be reconciled with the lexicon, which contains more *stable* elements. The interaction of concepts and ideas constructed out of them should not be a strange idea because in the human mental model, there is no line drawn between them. In the next section we illustrate the instantiation of a TransientNode.

We present a small ontology of structural **relations** to represent fairly generic structural relations between concepts. Object-oriented programming notation is useful shorthand because the process of meaning determination in the network engages in *structural marker passing* of *relations*, where symbol binding occurs. It is also important to remember, that each edge carries not only a named relation, but also a numerical weight. Connectionist weights are critically important in all processes of Bubble Lexicons, especially spreading activation and learning.

Operators put certain conditions on relations. In Figure 1, road material may only take on the value of *pavement* or *dirt,* and not both at once. Some operators will only hold in a certain instantiation or a certain context; so operators can be conditionally activated by a context or node. For example, a person can drive and walk, but under the time context, a person can only drive XOR walk.

3 Bubble Lexicon Mechanisms

We now explain the processes that are core themes of the Bubble Lexicon.

Meaning Determination. One of the important tenets of the lexicon's representation in Bubble Lexicons is that coherent meaning can only arise out of simulation. That is to say, out-of-context, word-concepts have so many possible meanings associated with each of them that we can only hope to make sense of a word by putting it into some context, be it a formal topic area (e.g. traversing from "car" toward the ContextNode of "transportation") or lexical context (e.g. traversing from "car" toward the normal node of "fast"). We motivate this meaning as simulation idea with the example of attribute attachment for "fast car", as depicted in Figure 1. Figure 3 shows some of the different interpretations generated for "fast car".

As illustrated in Figure 3, "fast car" produces many different interpretations given no other context. Novel to Bubble Lexicons, not only are numerical weights passed, structural messages are also passed. For example, in Figure 1, "drying time" will not always relate to "fast" in the same sense. It depends on whether or not pavement is drying or a washed car is drying. Therefore, *the history of traversal functions to nuance the meaning of the current node.* Unlike Charniak's earlier notion of marker passing [4] used to mark paths, structural marker passing in Bubble Lexicons is accretive, meaning that each node contributes to the message being passed.

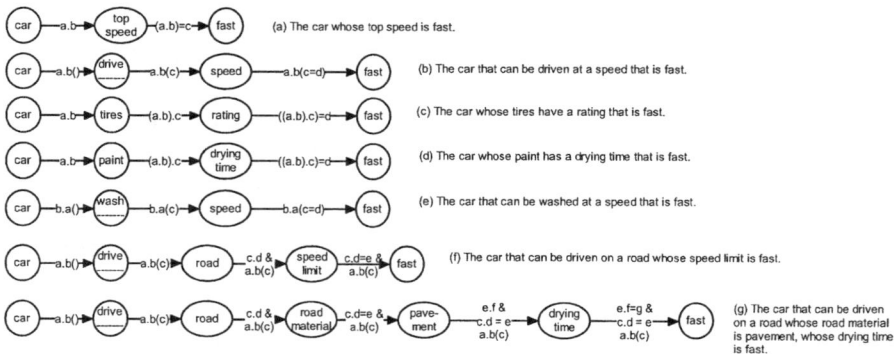

Fig. 3. Different meanings of "fast car," resulting from network traversal. Numerical weights and other context nodes are not shown. Edges are labeled with message passing, in OOP notation. The ith letter corresponds to the ith node in a traversal.

Although graph traversal produces many meanings for "fast car," most of the senses will not be very *energetic*, that is to say, they are not very plausible in most contexts. The senses given in Figure 3 are ordered by plausibility. Plausibility is determined by the activation energy of the traversal path. Spreading activation across a traversal path is different than classical spreading activation from the literature.

$$A_{ij_x} = \sum_{n=i}^{j} \omega_{n-1,n} \alpha_n \quad (1) \qquad A_{ij_x} = \sum_{n=i}^{j} \sum_{c}^{\text{active contexts}} \omega_{n-1,n} \pi_{M_{n,n+1}} \alpha_n A_{cn} \quad (2)$$

Equation (1) shows how a typical activation energy for the *x*th path between nodes *i* and *j* is calculated in classical spreading activation systems. It is the summation over all nodes in the path, of the product of the activation energy of each node *n* along the path, times the magnitude of the edge weight leading into node *n*. However, in a Bubble Lexicon, we would like to make use of extra information to arrive at a more precise evaluation of a path's activation energy, especially against all other paths between i and j. This can be thought of as meaning disambiguation, because in the end, we inhibit the incorrect paths which represent incorrect interpretations.

To perform this disambiguation, the influence of contexts that are active (i.e. other parts of the lexical expression, relevant and active non-lexical knowledge, discourse context, and topic ContextNodes), and the plausibility of the structural message being passed, are factored in. If we are evaluating a traversal path in a larger context, such as a part of a sentence or larger discourse structure, or some topic is active, then there will likely be a set of word-concept nodes and ContextNodes which have remained active. These contexts are factored into our spreading activation valuation function (2) as the sum over all active contexts c of all paths from c to n.

The plausibility of the structural message being passed $\pi_{M_{n,n+1}}$ is also important. Admittedly, for different linguistic assembly tasks, different heuristics will be needed. In attribute attachment (e.g. adj-noun compounds), the heuristic is fairly straightforward: The case in which the attribute characterizes the noun-concept directly is preferred, followed by the adjective characterizing the noun-concept's ability or use (e.g. Fig. 3(b)) or subpart (e.g. Fig. 3(a,c,d)), followed by the adjective characterizing some external manipulation of the noun-concept (e.g. Fig. 3(e)). What is not preferred is when the adjective characterizes another noun-concept that is a sister concept (e.g. Fig. 3(f,g)). Our spreading activation function (2) incorporates classic spreading activation considerations of node activation energy and edge weight, with context influence on every node in the path, and structural plausibility.

Recall that the plausibility ordering given in Figure 3 assumed no major active contexts. However, let's consider how the interpretation might change had the discourse context been a conversation at a car wash. In such a case, "car wash" might be an active ContextNode. So the meaning depicted in Fig. 3(e) would experience increased activation energy from the context term, $A_{"car-wash", wash}$. This boost makes (e) a plausible, if not the preferred, interpretation.

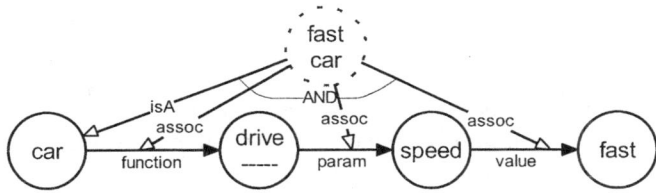

Fig. 4. Encapsulation. One meaning of "fast car" is encapsulated into a TransientNode, making it easy to reference and overload.

Encapsulation. Once a specific meaning is determined for a lexical compound, it may be desirable to refer to it, so, we assign to it a new index. This happens through a process called encapsulation, in which a specific traversal of the network is captured into a new TransientNode. (Of course, if the node is used enough, over time, it may become a stable node). The new node inherits just the specific relations present in the nodes along the traversal path. Figure 4 illustrates sense (b) of "fast car".

More than just lexical compounds can be encapsulated. For example, groupings of concepts (such as a group of specific cars) can be encapsulated, along with objects that share a set of properties or descriptive features (Jackendoff calls these *kinds* [9]), and even assertions and whole lines of reasoning can be encapsulated (with the help of the Boolean and ordering operators). And encapsulation is more than just a useful way of abstraction-making. Once a concept has been encapsulated, its meaning can be *overloaded*, evolving away from the original meaning. For example, we might instantiate "car" into "Mary's car," and then add a set of properties specific to Mary's car. We believe encapsulation, along with classical weight learning, supports accounts of lexical evolution, namely, it helps to explain how new concepts may be acquired, concepts may be generalized (concept intersection), or individuated (concept overloading). Lexical evolution mechanisms are discussed elsewhere [11].

Importing Existing Knowledge into the Bubble Lexicon. One question which may be looming in the reader's mind is how a Bubble Lexicon might be practically constructed. One practical solution is to bootstrap the network by learning frame knowledge from existing lexicons, such as GLT, or even Cyc [10], a database of lexical and non-lexical world knowledge. Taking the example of Cyc, we might map Cyc containers into nodes, predicates into TransNodes, and map micro-theories (Cyc's version of contexts) into ContextNodes, which activate concepts within each micro-theory. Assertional knowledge can be encapsulated into new nodes. To learn the intrinsic weights on edges, supervised learning can be used to semantically prime the network to the knowledge being imported. Cyc suffers from the problem of rigidity, especially contextual rigidity, as exhibited by microtheories which pre-fix context boundaries. However, once imported into a Bubble Lexicon, meaning determination may become more dynamic and context-sensitive. Contexts will evolve, based on the notion of lexical utility, not just on predefinition.

4 Implementation

To test the ideas put forth in this paper, we implemented a Bubble Lexicon over a adapted subset of the Open Mind Commonsense Semantic Network (OMCSNet) [13] based on the Open Mind Commonsense knowledge base [24]. We use the adaptive weight training algorithm developed for a Commonsense Robust Inference System (CRIS) [12]. OMCSNet is a large-scale semantic network of 140,000 items of general world knowledge including lexical and non-lexical, certain and defeasible. Its scale provides BLA with a rich basis from which meaning can be drawn.

With the goal of running trials, edge weights were assigned an *a priori* fixed value, based on the type of relation. The spreading activation evaluation function described in equation (2) was implemented. We also labeled three existing nodes in OMCSNet as ContextNodes and translated the nodes' *hasCollocate* relations, into the *assoc* relation in the Bubble Lexicon. Predefining nodes, while not generally necessary, was done in this case to make it easier to observe the effects of context bias in indicative trials. Trials were run over four lexical compounds, alternatingly activating each of these ContextNodes plus the null ContextNode. Context activations were set to a very high value to exaggerate, for illustrative purposes, the effect of context on meaning determination. Table 1 summarizes the results.

One discrepancy between the proposed and implemented systems is that assertional knowledge (e.g. *"Gas money to work can be cheap"*) in the implementation is allowed to be in the traversal path. Assertional knowledge is encapsulated as nodes.

The creative and nuanceful interpretations produced by the BLA demonstrate clearly the effects of active context on meaning determination. The incorporation of non-lexical knowledge into the phrasal meaning is visible (e.g. *"Horse that races, which wins money, is fast"*). By comparison, WordNet and GLT would not have produced the varied and informal interpretations produced by BLA.

Table 1. Results of trials illustrate effects of active context on attribute attachment.

Compound (context)	Top Interpretation (A_{ij_x} score in %)
Fast horse ( )	Horse that is fast. (30%)
Fast horse (money)	Horse that races, which wins money, is fast. (60%)
Fast horse (culture)	Horse that is fast (30%)
Fast horse (transportation)	Horse is used to ride, which can be fast. (55%)
Cheap apartment ( )	Apartment that has a cost which can be cheap. (22%)
Cheap apartment (money)	Apartment that has a cost which can be cheap. (80%)
Cheap apartment (culture)	Apartment is used for living, which is cheap in New York. (60%)
Cheap apartment (transportation)	Apartment that is near work; Gas money to work can be cheap (20%)
Love tree ( )	Tree is a part of nature, which can be loved (15%)
Love tree (money)	Buying a tree costs money; money is loved. (25%)
Love tree (culture)	People who are in love kiss under a tree. (25%)
Love tree (transportation)	Tree is a part of nature, which can be loved (20%)
Talk music ( )	Music is a language which has use to talk. (30%)
Talk music (money)	Music is used for advertisement, which is an ability of talk radio. (22%)
Talk music (culture)	Music that is classical is talked about by people. (30%)
Talk music (transportation)	Music is used in elevators where people talk. (30%)

However, the implementation also reveals some difficulties associated with the BLA. Meaning interpretation is very sensitive to the quality and signal-to-noise ratio of concepts/relations/knowledge present in the lexicon, which in our case, amounts to knowledge present in OMCSNet. For example, in the last example in Table 1, "talk music" in the transportation context was interpreted as "music is used in elevators, where people talk." This interpretation singled out elevators, even though music is played in buses, cars, planes, and elsewhere in transportation. This has to do with the sparseness of relations in OMCSNet. Although those other transportation concepts existed, they were not properly connected to "music". What this suggests is that meaning is not only influenced by what exists in the network, it is also heavily influenced by what is *absent*, such as the absence of a relation that should exist.

Also, judging the relevance of meaning relies largely on the evolution of good numerical weights on edges; but admittedly, learning the proper weights is a difficult proposition: Though we point out that even a rough estimate of weights (for example, separating lexical and non-lexical knowledge by 0.5), as was employed in our implementation, vastly improved the performance of meaning determination.

Though the complexity and knowledge requirements remain lingering challenges for the BLA, the implementation and indicative trials do seem to support our hypothesis that unpacking meaning from predefined word structures is a step toward a more natural handling of nuance and context in language.

5 Related Work

Ross Quillian's Semantic Memory System [21] was the initial inspiration for this work, as it was one of the first to explore meaning being distributed over a graph. In the semantic memory system, Quillian sought to demonstrate some basic semantic capabilities over a network of word-concepts, namely, comparing and contrasting words. The relations initially proposed represented minimal structural dependencies, only later to be augmented with some other relations including proximity, consequence, precedence, and similarity. The type of knowledge represented in the network was denotative and dictionary-like. With the Bubble Lexicon, we attempt to build on Quillian's work. We explain how such a semantic memory might be used to circumvent the limitations of traditional lexicons. We populate the network with lexical and non-lexical knowledge, and demonstrate their influences on meaning. We give an account of context-sensitive meaning determination by modifying spreading activation to account for contextual and structural plausibility; and introduce connectionism as a vehicle for conceptual analogy and learning.

6 Conclusion

In this paper, we examined certain limitations that the word-as-prepackaged-meaning paradigm imposes on the ability of the lexicon to generated highly nuanced interpretations. We formulated these problematics as issues of context, and hypothesized that a context-centered design of the computational lexicon would lend itself more to nuanced generation. We proposed a context-sensitive symbolic-connectionist network called a Bubble Lexicon. Rather than representing words as static encapsulations of meaning, the Bubble Lexicon dynamically generates context bubbles of meaning which vary based on active contexts. The inclusion of non-lexical knowledge such as defeasible and higher-order conceptual knowledge, along with intrinsic weights on relations, all serve to nuance to meaning determination. More than a static structure, the Bubble Lexicon is a platform for performing nuanceful linguistic assembly tasks such as context-sensitive attribute attachment (e.g. "fast car").

An implementation of the Bubble Lexicon over a large repository of commonsense called OMCSNet yielded some promising findings. In indicative trials, context had a very clear effect in nuancing the interpretation of phrases, lending support to our hypothesis. However, these findings also tell a cautionary tale. The accuracy of a semantic interpretation is heavily reliant on the concepts in the network being well-connected and densely packed, and the numerical weights being properly learned. The task of building lexicons over symbolic connectionist networks will necessarily have to meet these needs and manage a great deal of complexity. However, we are optimistic that the large repositories of world knowledge being gathered recently will serve as a well-populated foundation for such a lexicon. The research described in this paper explores lexicons that approach the generative power of the human language faculty. We cannot help but note that as such a lexicon theory grows toward its goal, it also approaches a comprehensive model of thought and semantic memory.

Acknowledgements. We are indebited to the following individuals for their helpful ideas and comments in the course of this work: Deb Roy, Push Singh, Andrea Lockerd, and Marvin Minsky.

References

1. Becker, S. et al.: Long-term semantic priming: A computational account and empirical evidence. J. of Exp. Psych: Learning, Memory, & Cognition, 23(5), (1997) 1059–1082
2. Brachman, R. J.: On the epistemological status of semantic networks. In: Findler, N. V., (ed.), Associative Networks, pages 3–50. Academic Press (1979)
3. Ceccato, S.: Linguistic Analysis and Programming for Mechanical Translation, Gordon and Breach, New York (1961)
4. Charniak, E.: A Neat Theory of Marker Passing. Proceedings of AAAI, (1986) 584–588
5. Collins, A. M., and Loftus, E. F.: A spreading-activation theory of semantic processing. Psychological Review, 82, (1975) 407–428
6. Falkenhainer, B., Forbus, K. D. and Gentner, D.: The structure-mapping engine: algorithm and examples. Artificial Intelligence, 41:1–63. (1990).
7. Fellbaum, C. (Ed.).: WordNet: An electronic lexical database. MIT Press (1998).
8. Goldstone, R. L., Rogosky, B. J.: Using relations within conceptual systems to translate across conceptual systems. Cognition 84, (2002) 295–320
9. Jackendoff, R.: Chapter 10, Foundations of Language. Oxford University Press (2002)
10. Lenat, D.B.: CYC: A large-scale investment in knowledge infrastructure. Communications of the ACM, 38(11) (1995).
11. Liu, H.: Bubble Networks: A Context-Sensitive Lexicon Representation. MIT Media Lab Technical Report SA02-02. (2002). Available at: web.media.mit.edu/~hugo/publications.
12. Liu, H. and Lieberman, H.: Robust photo retrieval using world semantics. Proceedings of LREC2002 Workshop: Using Semantics for IR, Canary Islands, (2002) 15–20
13. Liu, H. and Singh, P.: OMCSNet: A commonsense inference toolkit. *In Submission*. (2003) Available at: web.media.mit.edu/~hugo/publications
14. Masterman, M.: Semantic message detection for machine translation, using an interlingua. In: NPL (1961) 438–475
15. Minsky, M.: Society of Mind. Simon and Schuster, New York. (1986)
16. Minsky, M. and Papert, S.: Perceptrons. MIT Press (1969)
17. Pearl, J.: Probabilistic Reasoning in Intelligent Systems: Networks of Plausible Inference. Morgan Kaufmann, San Mateo, CA. (1988)

18. Peirce, C.S.: On the algebra of logic. American Journal of Math 7, (1885) 180–202
19. Pustejovsky, J.: The generative lexicon, Computational Linguistics, 17(4), (1991) 409–441
20. Putnam, H.: Representation and Reality, MIT Press (1988) 19–41
21. Quillian, M.: Semantic Memory. In: M. Minsky (Ed.): Semantic Information Processing, 216–270. MIT Press, Cambridge, MA (1968)
22. Rieger, C.: An organization of knowledge for problem solving and language comprehension. Artificial Intelligence 7:2, (1976) 89–127.
23. Schank, R.C., Tesler, L.G.: A Conceptual Parser for Natural Language. Proceedings of IJCAI, (1969) 569–578
24. Singh, P. et al.: Open Mind Common Sense: Knowledge acquisition from the general public. In Proceedings of ODBASE'02. LNCS. Heidelberg: Springer-Verlag (2002)

A Contextual Approach to the Logic of Fiction

Rolf Nossum

Agder University College, Norway

Abstract. An algebraic variant of multi-context logic is considered as an alternative to existing logical accounts of fictional discourse. An associative and idempotent operator on reified fictions supercedes Woods' *olim* modality. Soundness and completeness results are obtained for certain inter-fictional deductive rules relative to semantical conditions which respect the 'authorial say-so' criterion of fictional truth.

1 Introduction

Fictional discourse has been a research programme for logicians at least since John Woods' seminal investigations more than a quarter of a century ago [Woo69,Woo74]. The problems encountered when attempting to model sentences of fiction are very tough, and to this day the field lacks a robust and widely accepted logical foundation. This is, however, not for lack of vigorous efforts by many researchers during the intervening years, cfr i.a. [Col73], [Blo74], [Dev74], [Par75], [Sea75], [How76], [Lew78], [Par78], [Cas79], [Gab79], [Hei79], [Rou79], [Pav86], [Cur90], [LO94], [vI00].

Giving a logical account of fiction poses more problems than analyzing discourse about what is not the case in the real world. A logic of fiction must tackle thorny issues of representation and existence, simultaneous real and fictional reference, consistency, and nesting, to name a few.

The aim of this paper is to present a logical framework which lacks some of the weaknesses that have been identified in existing logical accounts of fiction.

The paper is structured as follows: In the next section we make some remarks intended to motivate the logical account of fiction that we shall give toward the end of this paper. Then we cite some intuitive benchmark properties that a logic of fiction should have, and review some existing systems in the literature, with special attention to the quantificational-substitutional account of Woods [Woo74]. Then we proceed to give our own positive account, which is an adaptation of the algebraic multi-context system of [NS02], and conclude by evaluating it against the intuitive benchmarks.

2 Remarks about Fiction

Here are some general remarks about fiction, intended to illuminate central features of our approach:

Fictions are typically told by someone, and are thus not given a priori. On the contrary, every fiction is given relative to the fiction-teller's point of view. A proper logical account of fiction should reflect this.

Fiction is not necessarily unrealistic, indeed some life-like stories have many readers and viewers. Vice versa, reality is itself sometimes portrayed as fiction (relative to some point of view); e.g. as a story (told by someone), or as a dream (dreamed by someone). Cfr. "We are such stuff as dreams are made of ..." [Sha02]. Logical accounts of fiction should therefore not grant too much of a special status to the real world.

Fictions can be nested inside each other, as in telling a story about an author writing a book. This involves several points of view; the fiction-teller's point of view x, the point of view y of the author-in-the-story, and the point of view z of the fiction contained in the book-in-the-story. The multi-level pattern of nesting suggested here will be a main feature of our system.

Stories can be mutually fictional relative to each other. In [Bab66] two stories develop in alternate chapters, each chapter spanning a day. Each story turns out to consist of the dreams dreamt at night in the other story, and the fictionological clou occurs when the protagonist of one story manages to kill that of the other. Here, the points of view β, γ of the protagonists of the two intertwined stories are related in ways that should be investigated.

A story about a time-traveller who accidentally prevents his father from meeting his mother should also be accomodated within a logic of fiction. It is, after all, a fiction.

Fiction frequently interacts with reality: The fictional detective Sherlock Holmes lived in London. This is the same non-fictional London that some of Conan Doyle's readers live in, and others visit from time to time.

Another example is Nicolas Bourbaki, a fictional mathematician whose work (by a collective using Bourbaki as a peudonym) is not fictional. It has had a great deal of influence on how mathematics is studied and taught in reality.

In modelling fiction, it seems appropriate to allow elements of the fiction to agree with elements from the point of view of the fiction-teller. In a first approximation, equal names may corefer, in an elaboration there can be an overlap of semantic domains, and in the account we shall give later, there will be a mapping of the language of the fiction into that of the fiction-teller.

It is tenable that the fiction-reader's point of view is augmented by what is read, resulting in a new point of view. If this is accepted, then the resulting fiction must be said to encompass both the reader and the story. The point of view of the reader is a non-trivial part of the fiction.

Proceeding along this path, one might be inclined to blur the distinctions between such categories as point of view, reality, and fiction. One person's reality is just a point of view to another. Holmes' reality is Conan Doyle's fiction. If Holmes' aide Dr. Watson tells a fairytale to his grandchildren, (Did he have any? Let me just say-so!) the fiction deepens further. All along, additional fictional components must be understood through previous points of view, or fictional layers.

On his view then, fictions are structured as sequences of fictional layers. For the time being, let us make the convenient assumption that adding a fictional layer is an associative operation.

As a preliminary observation, it seems that the space of fictions is not tree-structured by the add-a-fictional-layer operation. It also seems reasonable to say that discourse wholly within one and the same point of view does not add another fictional layer, which suggests that adding a fictional layer will be an idempotent operation.

3 Intuitive Benchmarks

Perhaps the most important criterion by which to judge a logic of fiction is the extent to which it obeys common-sense intuitions about fictional discourse. Let us borrow the following five intuitive benchmarks from [WA02]:

A. Reference is possible to fictional beings even though they do not exist.
B. Some sentences about fictional beings and events are true.
C. Some inferences about fictional beings and events are correct.
D. These three facts are made possible, in a central way, by virtue of the creative authority of the authors of fiction. Indeed, the primary and basic criterion of truth for fictional sentences is the author's say-so.
E. It is possible in a fictional truth to make reference to real things. For example, "Sherlock Holmes lived in London" is true and refers to the actual capital city of Britain.

4 Exisiting Accounts

Existing theories of fictionality vary greatly in their shape and form, but we may group them into broad categories according to their main focus.

One group of theories is represented by [Sea75,Wal78,Wal90] with their focus on speech acts, authorial pretense and make-believe analyses. In reference to the above intuitive axioms of fictionality, this may be said to focus on axiom D.

An ontological point of view is taken by the theories of [Par75,Rou79,Cas79]. These admit the Meinongian view that objecthood does not necessarily entail existence, thus meeting benchmark A about reference to fictional beings.

A third group of theories distinguishes between the logical forms of sentences in fictional and non-fictional discourse by direct or indirect use of some kind of fictionality operator. Some of these take a possible-world approach [Pla74,Kap73, Gab79,Lew78,How79], while the seminal work of John Woods [Woo69,Woo74] takes a substitutional and quantificational approach.

4.1 Woods' Account

The present work aims in part to ameliorate some perceived weaknesses in Woods' original framework, so we pause for a brief review of [Woo74] by Howell, as quoted in [WA02]:

Fiction [is represented in] a formal language containing the *olim* (once-upon-a-time) modal operator O. He proposes, roughly speaking, that this language be given a say-so semantics, coupled with a substitutional treatment of quantification, for references to the fictional, and a normal Tarskian semantics, with objectual quantification for references to the real. The (imagined) fictional claim "Holmes squared the circle" is represented by

$$O(\text{Holmes squared the circle}),$$

for example. This latter sentence contains but is not identical to a self-contradiction, and Woods notes that in affirming this sentence (and so in affirming the fictive claim which it represents) we are therefore not ourselves affirming a self-contradiction. Thus one of the problems created by fictional inconsistency is circumvented. Woods urges in conclusion that his *olim* language and its associated semantics, if developed in detail, will let us solve all the other logical and metaphysical issues about fiction. [How76]p.355

Some semantical rules quoted from Woods and Alward [WA02] give an impression of the technical aspects of Woods' account:

1. If ϕ represents the usual symbolization of a sentence that occurs in a work of fiction, or if ϕ logically follows from a consistent sentence of this sort, then $O(\phi)$ is true.
2. If $O(Fa)$ is true and if $(x)(Fx \supset \neg Gx)$ is also true (with the variable ranging over real objects), then so is $O(\neg Ga)$.
...
5. $O(\phi \& \psi)$ is true iff both $O(\phi)$ and $O(\psi)$ are true.
...

Quantification is handled as follows:

8. If Φ is $\exists v_i O(\Psi)$, then Ψ is true iff for every sequence S of the theory's objects at least one of two conditions is met:
    i. $O(\Psi)$ contains free occurrences of the variable v_i and v_i denotes the i-th element of some sequence s' differing from s in at most the i-th place, a is the name of that element and χ is a substitution instance of $O(\Psi)$ with respect to a, and χ meets the say-so condition.
    ii. If $O(\Psi)$ is $O(\chi(v_i, a))$, then v_i denotes the i-th element if some sequence s' differing from s in at most its i-th place; that element knows $O\exists v_k (v_k = a)$ to be true; the predicate χ is such that in general $\chi(v_j, v_h)$ is semantically equivalent to v_j believes that $\chi(v_j, v_h)$; and the element denoted by v_i believes that $\chi(v_i, a)$.
9. A further condition on quantifiers is if ϕ is $O(\exists v(\psi))$ then ϕ is true iff for some name or singular term a of L, free for a free variable in Ψ, $O(S_x^v(\Psi))$ is true.

While Woods' system was a great advance over previous semantical analyses, and has greatly stimulated and influenced the research community, it has also been subjected to strong criticism, as conceded in [WA02], from which we paraphrase:

[How76,Rou79] Howell and Routley independently observed that, in Woods' system, inconsistent fictions narrate everything. More specifically, if a fiction admits an inconsistency, it admits any sentence at all. Formally, if $O(\phi \& \neg \phi)$ is in a given story, then so is $O(\neg\neg\psi)$ and $O(\psi)$ for arbitrary ψ.

[Par78] Parsons observed an anomaly in the way Woods' system handles sentences which mix fictional and real-world references. The requirement of epistemic intensionality in part (ii) of semantic condition 8 seems to be the crux of this matter. Parsons enjoyably calls the sentence

"Some fictional detective is more famous than any real-world detective"

a "bet-sensitive, indeed winning, claim", but observes that Woods' system represents such sentences in the form $\exists v O(\phi(v, \alpha))$, and that nothing resembling an intensional verb such as required by clause 8(ii) is present.

Furthermore, Parsons criticizes the deductive part of Woods' system for being too unrestrictive. The deductive mechanisms fail to suitably restrict inference to sentences made true by the author's say-so. This latter point may be the most severe criticism against Woods' logic of fiction.

5 A Multi-context Logic of Fiction

Let us write
$$x : \phi \tag{1}$$
to mean that the statement ϕ is asserted in the fiction x. For example:

$$\text{Hound-of-B} : domicile(\text{Sherlock Holmes}, \text{London}) \tag{2}$$

We can explicate the meaning of terms used in fictions through a mapping into the language of the fiction-teller. Thus, we can map the term "London" in Conan Doyle's stories to "London" in reality as we know it, and establish the correct meaning of references to London in that way:

$$\mathcal{T}_{\text{Hound-of-B}}(\text{"London"}) = \text{"London"} \tag{3}$$

But what about Sherlock Holmes, who does not occur in reality?

It is plausible to think that a reader of fiction takes notice of each new character when they are first mentioned, and refers back to that notice upon subsequent mention.

What is, then, the notice arising from a first encounter with the term "Sherlock Holmes"? His name, and the fiction he occurs in, are all the information there is at that stage. Let us enrich the reader's vocabulary with a fresh atom, and map the fictional detective to that:

$$\mathcal{T}_{\text{Hound-of-B}}(\text{"Sherlock Holmes"}) = \text{"Hound-of-B-fictional-Sherlock-Holmes"} \tag{4}$$

Keeping languages countable at all points of view, we shall avoid technical problems here.

When there are implicit equalities in the language of the fiction, the mapping M must respect them:

$$\text{If } t = u \text{ then } \mathcal{T}_x(t) = \mathcal{T}_x(u) \tag{5}$$

In this way, "Sherlock", "Sherlock Holmes", and "Holmes", are all interpreted the same.

With this notation we can apply the results of [NS02], and adapt their multi-context algebraic systems to the logic of fiction.

The central two rules for reasoning in systems of fictions structured like the contexts of [NS02], are the following:

$$\frac{u : \mathcal{T}_x(\phi)}{ux : \phi} \; Rup_x \qquad \frac{ux : \phi}{u : \mathcal{T}_x(\phi)} \; Rdw_x \tag{6}$$

Here, ux is the fiction obtained by adding fictional component x to fiction u. The operation of adding a fictional component is fundamental to our system, and reality, fictions, contexts, points of view, are all lumped into one and the same category and are represented as sequences of fictional components strung together associatively.

Mathematically, this amounts to an associative algebra on terms denoting fictions. Technically speaking, we are structuring the space of fictions as a semigroup.

As observed earlier, addition of a fictional layer to itself is idempotent:

$$uu = u \tag{7}$$

As shown in [Nos02], idempotence is within the scope of the theory of [NS02], so we are now in a position to make direct use of the latter. It gives sound and complete deductive rules for a class of multi-context systems that the present one is a member of.

For simplicity of presentation we restrict ourselves to the case in which the language of each fiction is propositional, and leave the first order case out for now.

The main definitions are given below, but for proofs and other technical details we refer the reader to [NS02] and [Nos02].

5.1 Languages and Mappings

Let C be a countable set of contexts, or fictions, given a priori, and let $C^\star$ denote the set of finite sequences of elements of C, without adjacent identical pairs. The condition corresponds to the idempotence of fiction composition, as discussed above.

We use a, b, c, d, sometimes subscripted, to denote primitive fictions from C, while t, u, v, w, x, y, z and their subscripted variants are used liberally to denote primitive fictions from C, composite fictions from $C^\star$, or composite fictions from $C^\star$.

For each fiction $u \in C^\star$ let there be a propositional language L_u, that is used to express sentences in this fiction.

Definition 1 (Well formed formulae) *Well formed formulae are defined as follows (for all $u \in C^\star$). If ϕ is a propositional formula in L_u, then for all $y \in C^\star$ $y.\phi$ is a well formed formula, and ϕ is called a y-formula.*

Definition 2 (Language mapping) *For all $c \in C$, there is a partial recursive function $\mathcal{T}_c$ that maps uc-formulae into u-formulae for arbitrary $u \in C^\star$.*

Intuitively, a language mapping from uc to u states which part and how the content of the fiction uc is represented in the fiction u.

5.2 Local Model Semantics

Every equivalence class of fictions, i.e. each $u \in C^\star$, has its own formula language L_u. The semantical structure we are about to define, takes as its basic building blocks the local interpretations of each language L_u. We can identify interpretations with subsets of L_u, i.e. the true formulas in each interpretation.

The semantical structure for the entire system of languages reflects the way in which fictions are augmented by adding fictional components. We start by defining ground extensions of fiction terms:

Definition 3 (x-continuation) *Given $x \in C^\star$, an x-continuation is a fiction*

$$xc_1c_2\ldots c_h$$

where $0 \leq h$ and $c_i \in C$ for $1 \leq i \leq h$. When $h = 0$, this is just x.

Definition 4 (x-chain) *For $x \in C^\star$, an x-chain m is a function which maps every x-continuation y to a set m_y of interpretations of L_y (the local models of fiction y), such that for some x-continuation y, m_y is not empty, and for all x-continuations y, and fictions $c \in C$:*

1 $m \models y.\mathcal{T}_c(\phi)$ *if and only if* $m \models yc.\phi$
2 *the cardinality of m_y is at most 1.*

Definition 5 (Satisfiability) *An x-chain m satisfies a formula $y.\phi$ where y is an x-continuation, in symbols $m \models y.\phi$, if for any $s \in m_y$, $s \models \phi$ according to the definition of satisfiability for propositional formulae.*

Definition 6 (Logical consequence) *A formula $x.\phi$ is a logical consequence of a set of formulae Γ, in symbols $\Gamma \models x.\phi$ if, for any z-chain m, such that x is an z-continuation, if $m \models \{y.\gamma \in \Gamma | x \neq y$ and y is a z-continuation$\}$, then for all $s \in m_x$, $s \models \{\psi | z.\psi \in \Gamma, x = z\}$, implies that $s \models \phi$.*

5.3 Reasoning between Fictions

The notion of x-continuations induces a partial order among fictions, each fiction preceding its continuations. Let us see how one moves between composite fictions which are related in a partial order. We rely on a natural deduction calculus extended with indices as described in [GG01] and [SG01], plus the following inter-fictional deduction rules:

$$\frac{u.\mathcal{T}_c(\phi)}{uc.\phi} \; Rup_c \qquad \frac{uc.\phi}{u.\mathcal{T}_c(\phi)} \; Rdw_c \tag{8}$$

$$\frac{uc.\phi \leftrightarrow \psi}{u.\mathcal{T}_c(\phi) \leftrightarrow \mathcal{T}_c(\psi)} \; RRI \tag{9}$$

$$\frac{uu.\phi}{u.\phi} \; IDEM1 \qquad \frac{u.\phi}{uu.\phi} \; IDEM2 \tag{10}$$

We say that a formula $u.\phi$ is derivable from a set Γ of formulae, in symbols $\Gamma \vdash u.\phi$, if there is a deduction of $u.\phi$ from Γ, that uses u-rules (as defined in [SG01]) and the above inter-fictional deduction rules.

6 Soundness and Completeness

We the following soundness and completeness result:

Theorem 1 (Soundness and Completeness) $\Gamma \models u.\phi$ *if and only if* $\Gamma \vdash u.\phi$.

6.1 Soundness

Soundness of Rdw_c and Rup_c are direct from item 1 of the chain conditions, and RRI is sound by virtue of item 2 of the chain conditions. Soundness of $IDEM1$ and $IDEM2$ follows because any u-chain is also a uu-chain and vice versa.

6.2 Completeness

The completeness proof relies on canonical models which respect the bridge rules. The basic building blocks will be maximal consistent sets of well-formed formulae. Let us state the versions of consistency and maximality that we need.

Definition 7 (x-consistency) *A finite set Δ of well-formed formulae is said to be x-consistent iff $\Delta \not\vdash x.\bot$, and an infinite set is x-consistent iff every finite subset is x-consistent.*

Definition 8 (x-maximality) *A set Δ of well-formed formulae is said to be x-maximal iff Δ is x-consistent and for all well-formed labelled formulae $y.\delta$ such that $\Delta \cup \{y.\delta\}$ is x-consistent, $y.\delta \in \Delta$.*

Theorem 2 (Lindenbaum) *Any x-consistent set of wffs can be extended to an x-maximal set.*

Proof: see [NS02].

Canonical model. Now let us choose an arbitrary x-consistent wff $x.\delta$ and construct an x-chain for it. To begin with, we expand $\{x.\delta\}$ to an x-maximal set Δ by the construction in the previous lemma.

Definition 9 (Canonical model) *For all x-continuations $y = xc_1 \ldots c_h$*

- *let $\Delta_y = \{\lambda \mid x.\mathcal{T}_{c_1}(\ldots \mathcal{T}_{c_h}(\lambda)) \in \Delta\}$*
- *let S_y be the set of interpretations of the language L_y*
- *let $T_y = \{s \in S_y \mid s \models \Delta_y\}$ be the subset of interpretations that validate Δ_y*
- *and let m be the function that maps y to $\emptyset$ if $T_y = \emptyset$ and to $\{t\}$ otherwise, where t is some arbitrary member of T_y.*

Our canonical model is the x-chain m.

Δ_y is well-defined, so m is really an x-chain. To see this, we prove that

$$x.\mathcal{T}_{c_1}(\ldots \mathcal{T}_{c_h}(\lambda)) \in \Delta \quad \text{iff} \quad x.\mathcal{T}_{d_1}(\ldots \mathcal{T}_{d_k}(\lambda)) \in \Delta$$

whenever

$$xc_1 \ldots c_h = xd_1 \ldots d_k. \tag{11}$$

In fact,

$$x.\mathcal{T}_{c_1}(\ldots \mathcal{T}_{c_h}(\lambda)) \in \Delta$$

iff, by h applications of *Rup*,

$$((xc_1)\ldots c_h) : \lambda \in \Delta$$

iff, by associativity,

$$xc_1 \ldots c_h : \lambda \in \Delta$$

iff, by (11),

$$xd_1 \ldots d_k : \lambda \in \Delta$$

iff, by associativity,
$$((xd_1) \ldots d_k) : \lambda \in \Delta$$

iff, by k applications of Rdw,
$$x.\mathcal{T}_{d_1}(\ldots \mathcal{T}_{d_k}(\lambda)) \in \Delta.$$

As regards the model conditions, the m we have defined here trivially fulfills condition 2, and condition 1 is fulfilled because for $c \in C$ and an x-continuation $y = xc_1 \ldots c_h$, we have

$$y : \mathcal{T}_c(\lambda) \in \Delta \quad \text{iff} \quad yc : \lambda \in \Delta$$

by Rup, Rdw, and x-maximality of Δ.

The x-chain m satisfies the wff $x.\delta$ (take $h = 0$), so we have completeness.

7 Conclusions

Now, how does our system fare with respect to the intuitive criterions A-E of section 3? Quite well, we maintain:

A. Reference is possible to fictional beings even though they do not exist.
    Each fiction is endowed with its own local language, with unrestricted freedom of reference to local fictional entities. The interface with other fictions in general, and with the real world in particular, is through language mappings which explicate the meaning of imported references.
B. Some sentences about fictional beings and events are true.
    Each local language has its own truth assignment.
C. Some inferences about fictional beings and events are correct.
    Rules (8), (9), (10) tell us which ones are correct.
D. These three facts are made possible, in a central way, by virtue of the creative authority of the authors of fiction. Indeed, the primary and basic criterion of truth for fictional sentences is the author's say-so.
    Again, the local languages have full autonomy with respect to their assignment of truth and their consequence relation. There is nothing to prevent a fiction from having exactly the set of truths that correspond to an author's say-so.
E. It is possible in a fictional truth to make reference to real things. For example, "Sherlock Holmes lived in London" is true and refers to the actual capital city of Britain.
    The meaning of references which migrate between fictions and the real world is explicated through the language mappings $\mathcal{T}_c$. In the example, the name "London" in the language of the detective story would, when imported into the real world, be mapped to the name "London" in the language of the real world, and the reference to the capital of Britain would be secured.

We have presented a logic of fiction which does a straightforward application of the Multi-Context language and Local Models semantics of [GG01,SG01, NS02] and measures well against the above benchmarks. However, much is still left for future study: for instance, a more thorough comparison with existing logics of fiction should be made, and the applicability of analogy-related work to this line of inquiry remains to be investigated.

References

[Bab66] Mihály Babits. *The Nightmare*. Corvina Press, Budapest, 1966.

[Blo74] H. G. Blocker. The truth about fictional entities. *Philosophical Quarterly*, 24:27–36, 1974.

[Cas79] Hector-Neri Castaneda. Fiction and reality: their fundamental connections. *Poetics*, 8:311–62, 1979. "Formal semantics and literary theory", special issue edited by John Woods and Thomas Pavel.

[Col73] F. X. J. Coleman. A few observations about fictional discourse. In B. R. Tilghman, editor, *Aesthetics and language*, pages 31–42. University of Kansas Press, Lawrence, 1973.

[Cur90] Gregory Currie. *The nature of fiction*. Cambridge University Press, Cambridge, 1990.

[Dev74] P. E. Devine. The logic of fiction. *Philosophical studies*, 26:389–399, 1974.

[Gab79] Gottfried Gabriel. Fiction: a semantical approach. *Poetics*, 8:245–265, 1979. "Formal semantics and literary theory", special issue edited by John Woods and Thomas Pavel.

[GG01] C. Ghidini and F. Giunchiglia. Local models semantics, or contextual reasoning = locality + compatibility. *Artificial Intelligence*, 127(2):221–259, April 2001.

[Hei79] John Heintz. Reference and inference in fiction. *Poetics*, 8:85–99, 1979. "Formal semantics and literary theory", special issue edited by John Woods and Thomas Pavel.

[How76] Robert Howell. Review of John Woods: Logic of Fiction. *Journal of Aesthetics and Art Criticism*, 34:354–355, 1976.

[How79] Robert Howell. Fictional ojects: How they are and how they aren't. *Poetics*, 8:50–72, 1979. "Formal semantics and literary theory", special issue edited by John Woods and Thomas Pavel.

[Kap73] David Kaplan. Bob and Carol and Ted and Alice. In J. M. E. Moravcsik Jaakko Hintikka and P. Suppes, editors, *Approaches to natural language*, pages 490–518. Reidel, Dordrecht, 1973.

[Lew78] David Lewis. Truth in fiction. *American Philosophical Quarterly*, 15:37–46, 1978.

[LO94] P. Lamarque and S. Olsen. *Truth, Fiction and Literature: A Philosophical Perspective*. Oxford University Press, Oxford, 1994.

[Nos02] Rolf Nossum. Propositional logic for ground semigroups of context. *Logic Journal of the IGPL*, 10(3):273–297, May 2002.

[NS02] Rolf Nossum and Luciano Serafini. Multicontext logic for semigroups of contexts. In J. Calmet, B. Benhamou, O. Caprotti, L. Henoque, and V. Sorge, editors, *Artificial Intelligence, Automated Reasoning and Symbolic Computation*, volume 2385 of *LNAI*. Springer-Verlag, 2002.

[Par75] Terence Parsons. A Meinongean analysis of fictional objects. *Grazer Philosophische Studien*, pages 73–86, 1975.
[Par78] Terence Parsons. Review of John Woods: Logic of Fiction. *Synthese*, 39:155–164, 1978.
[Pav86] Thomas Pavel. *Fictional Worlds*. Harvard University Press, Cambridge, MA, 1986.
[Pla74] Alvin Plantinga. *The nature of necessity*. Oxford University Press, Oxford, 1974.
[Rou79] Richard Routley. The semantic structure of fictional discourse. *Poetics*, 8:3–30, 1979. "Formal semantics and literary theory", special issue edited by John Woods and Thomas Pavel.
[Sea75] John Searle. The logical status of fictional discourse. *New Literary History*, 6:319–332, 1975.
[SG01] Luciano Serafini and Fausto Giunchiglia. ML systems: A proof theory for contexts. *Journal of Logic, Language and Information*, July 2001.
[Sha02] William Shakespeare. *The Tempest*. The New Cambridge Shakespeare. Cambridge University Press, June 2002.
[vI00] Peter van Inwagen. Quantification and fictional discourse. In Anthony Everett and Thomas Hofweber, editors, *Empty Names: Fiction and the Puzzles of Non-existence*, pages 235–247. CSLI publications, Stanford, 2000.
[WA02] John Woods and Peter Alward. The logic of fiction. manuscript, 2002.
[Wal78] Kendall Walton. On fearing fictions. *Journal of Philosophy*, 75:5–27, 1978.
[Wal90] Kendall Walton. *Mimesis as make-believe*. Harvard University Press, Cambridge, MA, 1990.
[Woo69] John Woods. Fictionality and the logic of relations. *The Southern Journal of Philosophy*, 7:51–64, 1969.
[Woo74] John Woods. *The logic of fiction: Philosophical Soundings of Deviant Logic*. Mouton, The Hague and Paris, 1974.

Predictive Visual Context in Object Detection*

Lucas Paletta

JOANNEUM RESEARCH
Institute of Digital Image Processing
Wastiangasse 6, 8010 Graz, Austria
lucas.paletta@joanneum.at

Abstract. This work discriminates external and internal visual context according to a recently determined terminology in computer vision. It is conceptually based on psychological findings in human perception that stress the utility of visual context in object detection processes. The paper outlines a machine vision detection system that analyzes external context and thereby gains prospective information from rapid scene analysis in order to focus attention on promising object locations. A probabilistic framework is defined to predict the occurrence of object detection events in video in order to significantly reduce the computational complexity involved in extensive object search. Internal context is processed using an innovative method to identify the object's topology from local object features. The rationale behind this methodology is the development of a generic cognitive detection system that aims at more robust, rapid and accurate event detection from streaming video. Performance implications are analyzed with reference to the application of logo detection in sport broadcasts and provide evidence for the crucial improvements achieved from the usage of visual context information.

1 Introduction

In computer vision, we face the highly challenging object detection task to perform recognition of relevant events in outdoor environments. Changing illumination, different weather conditions, and noise in the imaging process are the most important issues that require a truly robust detection system. This paper considers exploitation of visual context information for the prediction of object location and identity, respectively, that would significantly improve the service of quality in real-time interpretation of image sequences.

Research on video analysis has recently been focussing on object based interpretation, e.g., to refine semantic interpretation for the precise indexing and sparse representation of immense amounts of image data [13,17]. Object detection in real-time, such as for video annotating and interactive television [1], imposes increased challenges on resource management to maintain sufficient quality of service, and requires careful design of the system architecture.

---

* This work is funded by the European Commission's IST project DETECT under grant number IST-2001-32157.

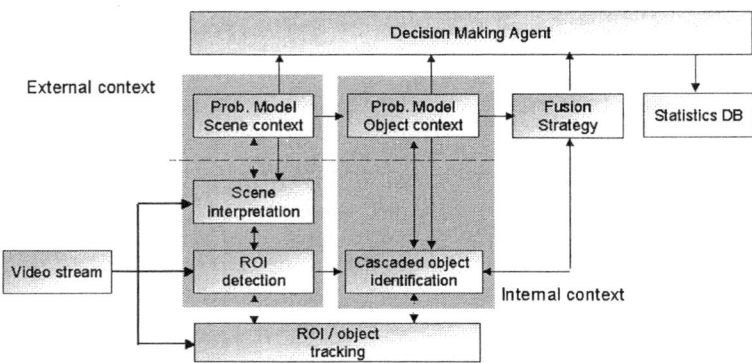

Fig. 1. External and internal visual context as a means to trigger discrimination processes for the purpose of object detection in video streams.

Fig. 1 illustrates how external and internal visual context are used to detect object information in a video stream. Rapid extraction of the context of a scene - the *global spatial context* with respect to object detection - might trigger early determination of regions of interest (ROI) and support careful usage of resources for more complex discrimination processes (Section 3.1). Within the ROI, object identification requires a grouping of local information [6]. In particular, the presented work describes how internal context from a configuration of object appearances - the *local spatial context* with respect to object detection - is exploited to distinguish collections of local measurements by means of their geometrical relations (Section 4). Finally, a federation of discriminatory processes is controlled by a supervising decision making agent [8,22] to feed object information into a database for statistical performance evaluations.

Recent work on real-time interpretation applies attentional mechanisms to coarsely analyze the external context from the complete video frame information in a first step, reject irrelevant hypotheses, and iteratively apply increasingly complex classifiers with appropriate level of detail [32,20]. In addition, context priming [31,21] makes sense out of globally defined environmental features to set priors on observable variables relevant for object detection. Investigations on the binding between scene recognition and object localization made in experimental psychology have produced clear evidence that highly local features play an important role to facilitate detection from predictive schemes [4,11,5]. In particular, the visual system infers knowledge about stimuli occurring in certain locations leading to expectancies regarding the most probable target in the different locations (*location-specific target expectancies*, [10]).

Extraction of internal object context often optimizes single stage mapping from local features to object hypotheses [16,25]. This requires either complex classifiers that suffer from the course of dimensionality and require prohibitive computing resources, or provides rapid simple classifiers with lack of specificity. Cascaded object detection [8,32] has been proposed to decompose the mapping into a set of classifiers that operate on a specific level of abstraction and focus on a restricted classification problem. We investigate the impact of local spatial context information on the performance of object detection processes, using a Bayesian method to extract *context from object geometry*.

The methodology to eploit visual context is embedded within a global framework on integrated evaluation of object and scene specific context (see [31], Section 2) with the rationale to develop a generic cognitive detection system that aims at more robust, rapid and accurate event detection from dynamic vision.

2 Visual Context in Object Detection Processes

In general, we understand *context* to be described in terms of *information that is necessary to be observed* and that can be *used to characterize situation* [7]. We refer to the ontology and the formalization that has been recently defined with reference to perceptual processes for the recognition of activity [6], and a Bayesian framework on context statistics [31], with particular reference to video based object detection processes.

In a probabilistic framework, object detection requires the evaluation of

$$p(\varphi, \sigma, \mathbf{x}, o_i | \mathbf{y}), \tag{1}$$

i.e., the probability density function of object o_i, at spatial location $\mathbf{x}$, with pose φ and size σ given image measurements $\mathbf{y}$. A common methodology is to search the complete video frame for object specific information. In cascaded object detection, search for simple features allows to give an initial partitioning into object relevant regions of interest (ROIs) and a background region.

The *visual context* is composed of a model of the *external context* of the embedding environment, plus a model of the object's *internal context*, i.e., *the object's topology characterized by geometric structure and associated local visual events* (e.g., local appearances) [31] so that local information becomes characterized with respect to the object's model (e.g., Fig. 5). Measurements $\mathbf{y}$ are separated into *local* object features representing object information $\mathbf{y}_L$ and the corresponding local visual *environment* represented by context features $\mathbf{y}_E$. Assuming that - given the presence of an object o_i at location $\mathbf{x}$ - features $\mathbf{y}_L$ and $\mathbf{y}_E$ are independent, we follow [31] to decompose Eq.1 into

$$p(\mathbf{y}|\varphi, \sigma, \mathbf{x}, o_i) = p(\mathbf{y}_L|\varphi, \sigma, \mathbf{x}, o_i) \cdot p(\mathbf{y}_E|\varphi, \sigma, \mathbf{x}, o_i). \tag{2}$$

Cascaded object detection leads to an architecture that processes from simple to complex visual information, and derives from global to local object hypotheses (e.g., [8,32]). Reasoning processes and learning might be involved to select the most appropriate information according to an objective function and learn to integrate complex relationships into simple mappings. They are characterized by tasks, goals, states defined with respect to a model of the process, and actions that enable transitions between states [22], much in the sense of a decision making agent controlling discriminatory processes to improve quality of service in object detection (Fig. 1).

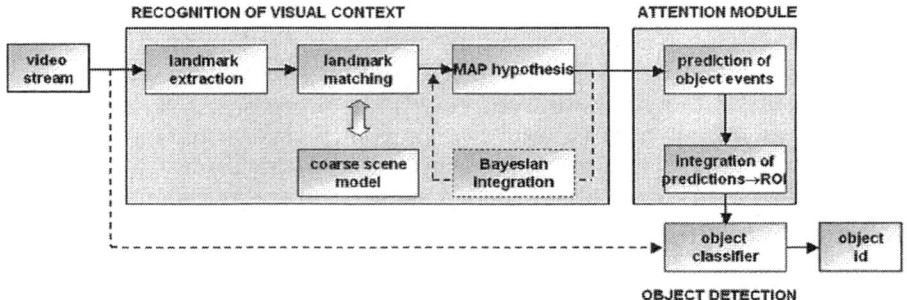

Fig. 2. Concept for recognition of external context for attention. Landmarks are extracted from the scene and matched towards a simlpe scene model. Bayesian recognition enables evidence integration over time and space. Attentive predictions on the location of embedded objects finally instantiate a complex object classifier ([21], Section 4) that verifies or rejects the object hypotheses.

3 External Context for ROI Detection

The concept is to propose attention from scene context using knowledge about forthcoming detection events that has been built up in repeated processing on the scene before. The knowledge which is derived from a simple scene model is activated from rapid feature extractions (e.g., using color regions) in order to operate only in those image regions where object detection events will most likely occur. The localization within an already modeled video scene is on the basis of a Bayesian prediction scheme. Recent investigations on human visual cognition give evidence for memory in visual search [12], underlining the assumption that already simple modeling mechanisms significantly support the quality of service in object detection.

3.1 Scene Representations from Landmarks

The basis for landmark based localization within a video scene is the extraction of discriminative and robustly re-locatable chunks of visual information in the scene. Landmarks have been efficiently defined on local greyvalue invariants [28], color and edge features [30], based on local appearance [29] and distinguished regions [18].

We apply an approach that rapidly extracts color and shape features but also considers the contrast of the extracted landmark region with respect to the corresponding features of its local neighborhood, being motivated by human perception, where, e.g., color is addressed by attentional mechanisms in terms of its diagnostic function [19]. Note that any other choice of local landmark representation would enable to pursue the methodology described in Sections 3.2, 3.3 as well.

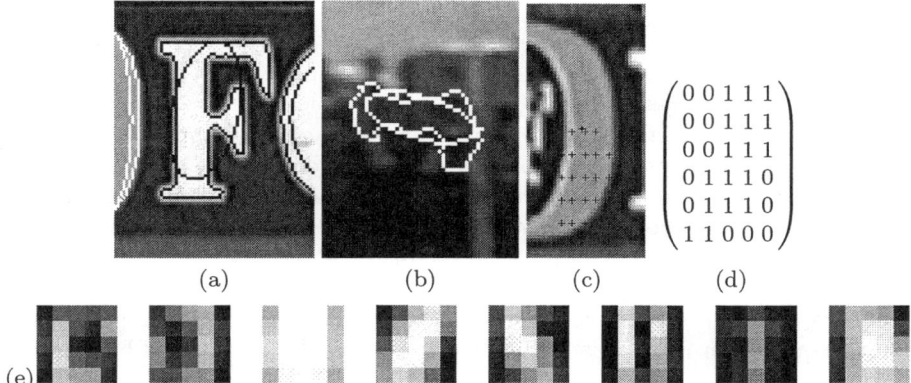

Fig. 3. Characteristic landmark features. (a,b,c) Color based ROIs for landmark definition, denoting the ROI border and the variance ellipsoid of the spatial distribution of ROI member pixels. (c,d) Class based extraction of shape: (c) Sampling (crosses) within the landmark region, (d) binary pattern received from color class based interpretation of the pixels sampled in (c), and attributed to class 4 in (e). (e) All prototypical patterns of shape to classify (d).

Fig. 4. Triple configuration of landmarks in a sample video frame using the landmark extraction described in Section 3.1.

In order to increase the discriminability of the locally extracted context, it is useful to combine landmarks into geometric configurations of 1-, 2-, and 3-tuples of landmarks (Fig. 4). Tuples of localized image properties own specific characteristics of scale invariance, ordering and topology [9] that make them attractive for landmark usage. Each single *landmark region* is encoded by a vector $\boldsymbol{\lambda}$ with landmark specific components $\boldsymbol{\nu}_i = (\mathbf{c}, \mathbf{n}, \mathbf{s}, \ldots)$, with features being vector-coded by color ($\mathbf{c}$), contrast ($\mathbf{n}$), and shape ($\mathbf{s}$). A 3-tuple *landmark configuration* denotes $\boldsymbol{\lambda} = [\boldsymbol{\nu}_1, \boldsymbol{\nu}_2, \boldsymbol{\nu}_3, \boldsymbol{\alpha}]^T$, where $\boldsymbol{\alpha}$ encodes the angles between landmarks $\boldsymbol{\nu}_i$.

3.2 Bayesian Scene Recognition

The goal of rapid scene modelling is to provide a simple and efficient encoding of the environment. The presented work is based on localization of a given landmark within a complete video sequence. The extracted landmark results in a hypothesis on representing a sample of a physical identity l_i of a landmark, i.e., the real landmark that generates a distribution of features from appearances to the observer. In our model, we pursue a framework of recognition and attribute each landmark sample λ to a physical landmark identity l_i and associated semantic blocks (frames) f_j in the reference (training) video sequence.

A simple scene model is rapidly generated from the frames of a video training sequence in terms of a list of landmark vectors $l_i \in \Lambda$ that can be matched against a currently extracted landmark sample λ_t. Scene recognition from interpretation of a landmark l^* is then computed via $l^* = \arg\min_{l_i} ||\lambda_t - \lambda(l_i)||$, which represents a nearest-neighbor matching to stored landmarks $\lambda(l_i)$ in 'λ-space'.

In order to represent the uncertainty in landmark classification, the landmark l_i specific sample distribution is modelled using an unimodal Gaussian, $N_{l_i}(\mu_\lambda, \Sigma_\lambda)$. The posterior interpretation of a landmark configuration λ is then outlined as follows,

$$P(l_i|\lambda) = \frac{p(\lambda|l)P(l_i)}{p(\lambda)} = \frac{p(\lambda|l_i)\sum_{j=1}^{F} P(l_i|f_j)P(f_j)}{p(\lambda)}, \quad (3)$$

where λ denotes a sample landmark extraction from a test image, $P(l_i|\lambda)$ is the posterior with respect to a corresponding physical identity of a landmark, $P(l_i|f_j)$ is the probability for observing a physical landmark given a specific frame of the video sequence. To be precise, we require f_j to partition the space of landmarks l_i, which is the case in video block segmentation.

3.3 Contextual Cueing to Predict Object Detection Events

Assuming that the scene has been repeatedly viewed and in a prevalent direction, each landmark configuration can be associated with a pointer to a succeeding object event that has been extracted before using any highly accurate, computationally expensive object identification method [32,21]. In the scene model, a directional information in terms of an angle interval ($\beta \pm \sigma$), is provided in which the object event is completely embedded; β is in the direction of the center of the predicted detection event, and $\pm \sigma$ designates an angle interval so that the detection event is completely embedded within. This interval $\pm \sigma$ defines the standard deviation with respect to a one-dimensional normal distribution, i.e., $N(\mu_\beta, \sigma)$, that is defined geometrically normal to the straight line originating in landmark l_i with angle β. In total, these operations will define a probability density function (PDF) on the image, $p(\mathbf{x}|\Omega, l_i)$, with image locations $\mathbf{x}$ carrying confidence information about the support for a local object detection event, out of the set of objects, i.e., Ω, and in terms of a landmark specific *confidence map* (Fig. 6). However, in real-time implementations, Monte-Carlo sampling [15] would be appropriate to approximate the estimated PDFs.

To increase the robustness of the approach, we integrate the confidences from those landmarks $l_k \in K$ that have been consecutively visited in an observation sequence and been selected as estimators for the forthcoming object location, e.g., simply using a naive Bayes estimator,

$$p(\mathbf{x}(\beta)|\Omega, l_1, l_2, \ldots, l_K) = \prod_{k=1}^{K} p(\mathbf{x}(\alpha)|\Omega, l_k), \quad (4)$$

and thereby receive an incremental fusion of individual confidence maps. Fusion might use all those predictions $N()$ that correspond to the selected l_i giving $P(l_i|\boldsymbol{\lambda})$ (Section 3.2), weighting individual contributions according to the confidences given in Eq. 3 (Fig. 6).

4 Internal Context from Probabilistic Structural Matching

Recently, the requirement to formulate object representations on the basis of local information has been broadly recognized [26,18]. Crucial benefits from decomposing the recognition of an object from global into local information are, increased tolerance to partial occlusion, improved accuracy of recognition (since only relevant - i.e., most discriminative - information is queried for classification) and genericity of local feature extraction that may index into high level object abstractions. In this paper we are using simple brightness information to define local appearances, but the proposed approach is general enough to allow any intermediate, locally generated information to be used as well, such as Gaussian filter banks [14], etc.

Context information can be interpreted from the relation between local object features [26] or within the temporal evolution of an object's appearance [2, 23]. Decomposing the complete object information into local features transforms $p(o_i|\mathbf{y}_L)$ into $p(o_i|\mathbf{y}_{L_1}, ..., \mathbf{y}_{L_N})$, N determines the size of the object specific environment. The *grouping* of conditionally observable variables to an *entity of semantic content*, i.e., a visual object, is an essential perceptual process [6].

The relevance of structural dependencies in object localization [26,27] has been stressed before, though the existing methodologies merely reflect co-location in the existence of local features. The presented work outlines full evaluation of geometrical relations in a framework of probabilistic structural matching using Bayesian conditional analysis of local appearances as follows.

Geometrical information is derived from the relation between the stored object model - the trajectory in feature space - and the actions (shift of the focus of attention) that are mapped to changes in the model parametrization (e.g., change in viewpoint, i.e., $\Delta \varphi_j$). Figure 5 illustrates the described concept in the reference frame of the local appearance based object model. The geometry between local appearances is now explicitly represented by the shift actions a_i (deterministically causing $\Delta \varphi_j$) that feed directly into Bayesian fusion [22] by

$$P(o_i, \varphi_j|\mathbf{y}_1, a_1, \mathbf{y}_2) = \alpha P(o_i, \varphi_j|\mathbf{y}_1, a_1) p(\mathbf{y}_2|o_i, \varphi_j, \mathbf{y}_1, a_1). \quad (5)$$

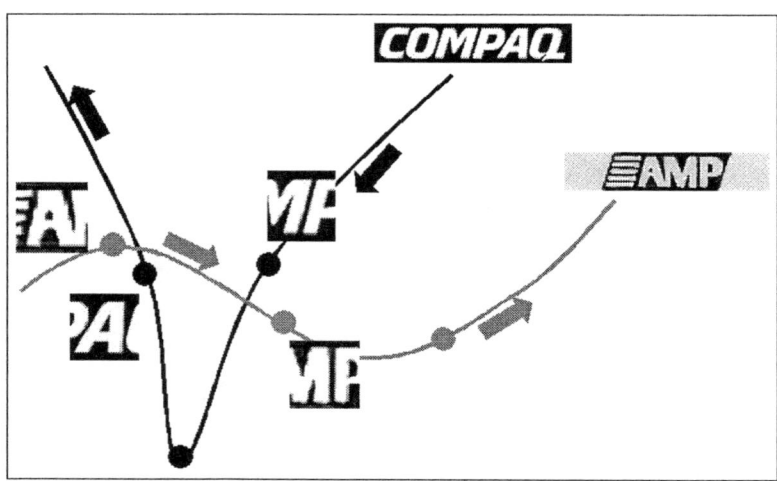

Fig. 5. Spatial context from the geometry of local information. A single appearance might give rise to evidence for multiple objects (at crossings of manifolds), even from a second measurement. The shift action to change a visual parameter of the manifold (e.g., a viewpoint change) and the associated appearances is then matched towards the manifold's trajectory in feature space to discriminate between object hypotheses.

Spatial context from probabilistic sturctural matching is now exploited using the conditional term $P(o_i, \varphi_j | \mathbf{y}_1, a_1)$: The probability for observing view (o_i, φ_j) as a consequence of deterministic action $a_1 = \Delta\varphi_1$ must be identical to the probability of having measured at the action's starting point before, i.e. at view $(o_i, \varphi_j - \Delta\varphi_1)$, thus $P(o_i, \varphi_j | \mathbf{y}_1, a_1) \equiv P(o_i, \varphi_j - \Delta\varphi_1 | \mathbf{y}_1)$. Note that this obviously does not represent a naive Bayes classifier since it explicitly represents the dependency between the observable variables $\mathbf{y}_i, a_i$.

Furthermore, the probability density of $\mathbf{y}_2$, given the knowledge of view (o_i, φ_j), is conditionally independent on previous observations and actions, and therefore $p(\mathbf{y}_2 | o_i, \varphi_j, \mathbf{y}_1, a_1) = p(\mathbf{y}_2 | o_i, \varphi_j)$. The recursive update rule for *conditionally dependent* observations accordingly becomes,

$$P(o_i, \varphi_j | \mathbf{y}_1, a_1, \ldots, a_{N-1}, \mathbf{y}_N) = \alpha p(\mathbf{y}_N | o_i, \varphi_j) P(o_i, \varphi_j - \Delta\varphi_{N-1} | \mathbf{y}_1, a_1, \ldots, \mathbf{y}_{N-1}) \quad (6)$$

and the posterior, using $\mathbf{Y}_N^a \equiv \{\mathbf{y}_1, a_1, \ldots, a_{N-1}, \mathbf{y}_N\}$, is then given by

$$P(o_i | \mathbf{Y}_N^a) = \sum_j P(o_i, \varphi_j | \mathbf{Y}_N^a). \quad (7)$$

The experimental results in Figures 8 and 9 demonstrate that context is crucial for rapid discrimination from local object information. The presented methodology assumes knowledge about (i) the scale of actions and of (ii) the directions with reference to the orientation of the logo, which can be gained by ROI analysis beforehand.

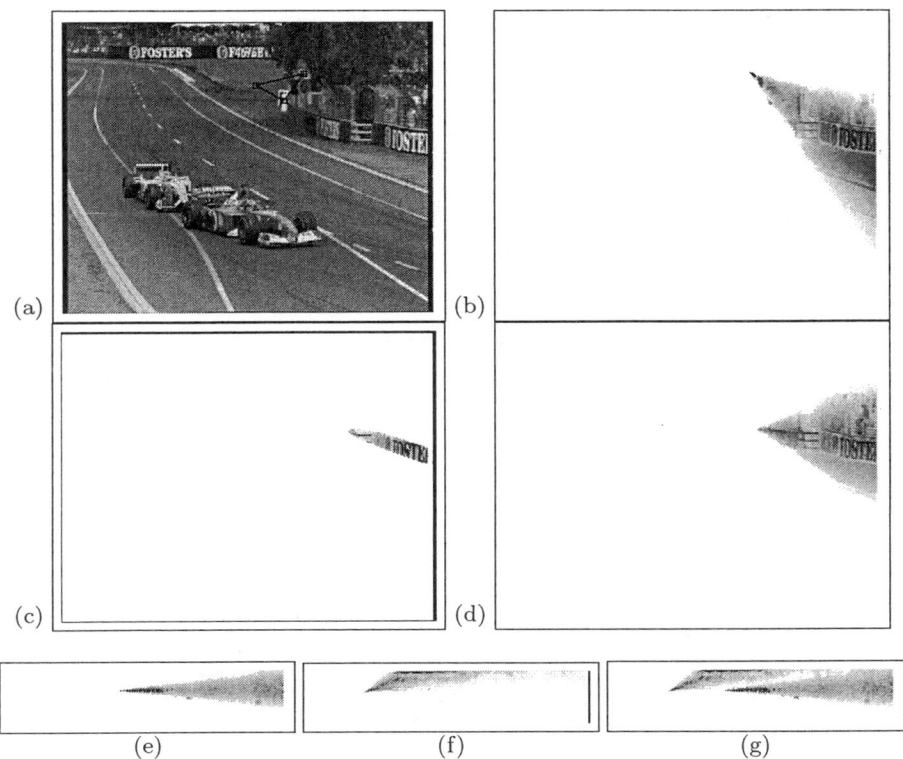

Fig. 6. Recursive contextual cueing and spatial attention for object detection. (a) Original frame with extracted landmark configurations. (b,d) Confidence maps derived from 2 individual landmarks. (c) Confidence map after Bayesian integration, depicting confidence beyond the threshold of $\Theta = 0.9$. (e) Accurate and (f) inaccurate predictive search regions, (g) integrated confidence map contains target. Since the target is represented in the fused confidence map, landmark 1 and 2 would impose object hits. In contrast, single landmark evaluation of 1 and 2 would produce one erroneous result.

5 Experimental Results

The object detection experiments were performed on 'Formula One' sport broadcast image sequences. The proposed object detection system first applies ROI detection based on contextual cueing from landmark configurations, supported by some color specific pattern classifiers [24,21]. Within these detection regions, it extracts internal context from local features. The following paragraphs describe the recognition performance from (i) external context and (ii) using probabilistic structural matching (internal context).

(i) Context from landmark configurations. The experiments were conducted on prediction of object detection in 'Formula One' broadcast videos. In

Table 1. Performance analysis of contextual cueing for attentive detection of objects of interest. Using extensive image analysis, ca. 73,1 % of the image had to be analyzed in order to detect a logo. In contrast, using CCA (contextual cueing for attention) analysis, only 46, 1% of the complete image had to be processed, resulting in a gain of 36,7 % of unprocessed image parts. CCA does not only provide impressive gains in speedup, but also in the statistically estimated accuracy of object detection as illustrated in Fig. 7b.

Extensive analysis %	CCA analysis %	unprocessed image part (CCA) %
73,1	46,1	36,7

particular, a video sequence of 71 frames (of 795 × 596 pixels) was used as training sequence and analysed to setup the scene model of the complete sequence, i.e., the interpretation of the landmark information, configurations, and the associated indexing and probabilistic interpretation for Bayesian scene recognition (Section 3.2).

The ROI color information was clustered into 12 Gaussian unimodal kernels via expectation maximization (EM) [3]. Shape patterns were clustered into 12 classes alike. The interpretation of this sequence resulted in 4351 n-tuple landmark registrations from 2123 physical landmark identities. The attribution to detection events was performed manually and under the assumption that this particular scene is captured by a specific camera motion (left to right) so that events are always encountered from one direction.

Via the localization of landmarks one can predict the successive detection event. The error in degree per single prediction is on average $2,6°$, $\pm 6,39°$ stdev). A direct hit rate of 93, 7% is achieved within the $2 \times \sigma_\beta$ interval (Fig. 7a). The resulting ROC curve (Fig. 7b) interprets the contextual cueing method in terms of a detection classifier, leading to excellent results with respect to its object detection performance. Finally, table 1 illustrates the gain in resources due to contextual cueing.

(ii) Context from geometry. Spatial context from geometry can be easily extracted based on a predetermined estimate on scale and orientation of the object of interest. This is computed (i) from the topology of the ROI, and from (ii) estimates on $p_\varphi(\varphi|\sigma, \mathbf{x}, o_i, \mathbf{y}_E)$ and $p_s(\sigma|\mathbf{x}, o_i, \mathbf{y}_E)$ from global image features [31]. We present a recognition experiment from spatial context on 3 selected logos (Figure 8(a)) with local appearance representation as described above, and a 3-dimensional eigenspace representation to model highly ambiguous visual information. Figure 9 (right) demonstrates the dramatic decrease of uncertainty in the pose information for object o_2, i.e., $p(o_2, \varphi_j|\mathbf{y})$, from several steps of information fusion according to Eq. 6. Figure 8(b) illustrates the original and final distribution for all objects, $o_1 - o_3$.

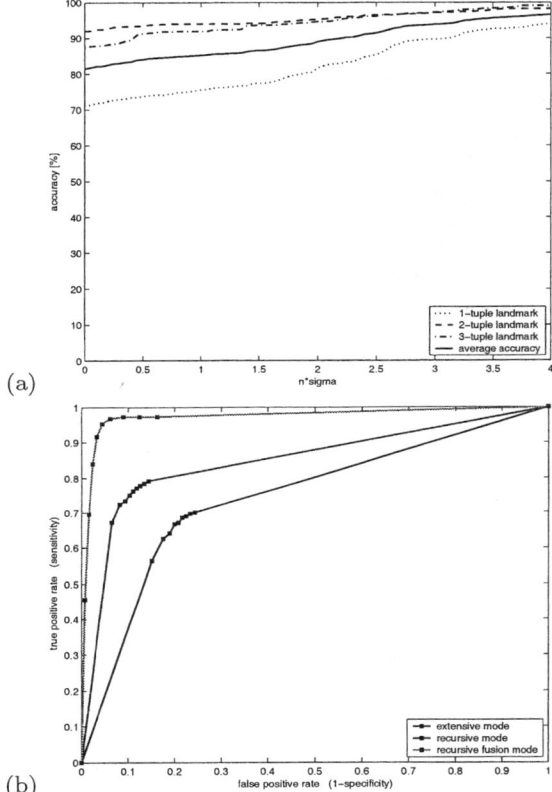

(a)

(b)

Fig. 7. Performance evaluation of the contextual cueing system. (a) Tolerated error and associated percentage of predictions (%) within this interval (for n-tuples of landmarks: point=1-tuple, dotdashed=2-tuple, dashed=3-tuple landmarks, line=avg.). (b) Receiver operator characteristic (ROC) curve demonstrating the high capabilities for object detection understanding the contextual cueing in terms of a detection system.

6 Conclusions

Context information contributes in several aspects to robust object detection from video. This work presents a predictive framework to focus attention on detection events instead of extensively searching the complete video frame for objects of interest.

Firstly, the probabilistic recognition of scenes from a landmark based description of the scene context are the innovative components that enable both rapid, predictable, and robust determination of relevant search regions. Secondly, grouping of local features can be rapidly applied and yields improved results. Additional computing derives the *context from the geometry of local features* which has been demonstrated to dramatically improve object recognition. Further ex-

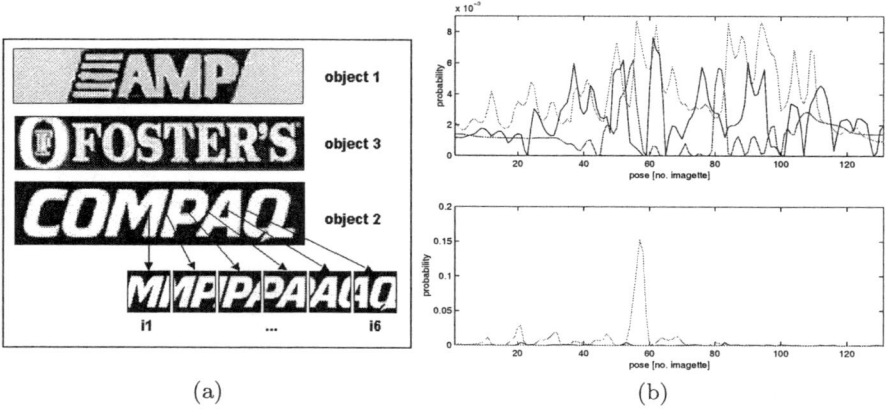

(a)   (b)

Fig. 8. (a) The logo object set and associated pattern test sequence. (b) Probability distribution on pose hypotheses w.r.t. all 3 logo objects from a single imagette interpretation (*top*) and after the 5th fusion of local evidences (*bottom*).

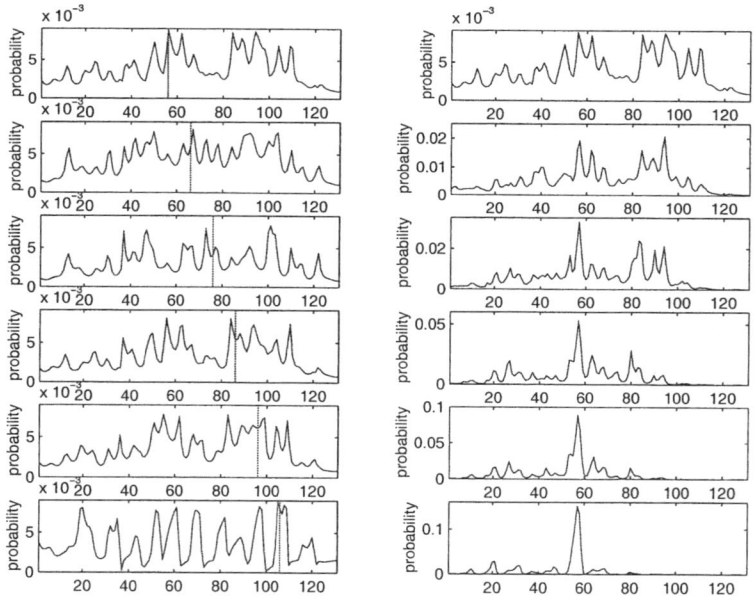

Fig. 9. *Left:* Probability distributions over pose hypotheses (imagette pose no.1-131 within logo) from individual test patterns no. 1-6, from top to bottom. *Right:* Corresponding fusion results using spatial context from geometry illustrating fusion steps no.1-5.

periments on contextual cueing demonstrate that prediction of object events from landmark based scene context can decisively determine an efficient focus of attention that would permit to save a substantial amount of computational resources from extensive processing.

Future work will focus on the extraction of local context from scene information in order to predict the future locations of detection events. We will consider the temporal context in the occurrence of landmark configurations and therefore most probably improve the landmark based scene recognition, together with the prediction performance as well.

References

1. J. Assfalg, M. Bertini, C. Colombo, and A. Del Bimbo. Semantic annotation of sports videos. *IEEE Multimedia*, 9(2):52–60, 2002.
2. S. Becker. Implicit learning in 3D object recognition: The importance of temporal context. *Neural Computation*, 11(2):347–374, 1999.
3. S. Belongie, C. Carson, H. Greenspan, and J. Malik. Color- and texture-based image segmentation using em and its applications to content-based image retrieval. In *Proc. International Conference on Computer Vision*, pages 675–682. Bombay, India, 1998.
4. I. Biederman, R.J. Mezzanotte, and J.C. Rabinowitz. Scene perception: Detecting and judging objects undergoing relational violations. *Cognitive Psychology*, 14:143–177, 1982.
5. M.M. Chun and Y. Jiang. Contextual cueing: Implicit learning and memory of visual context guides spatial attention. *Cognitive Psychology*, 36:28–71, 1998.
6. J. L. Crowley, J. Coutaz, G. Rey, and P. Reignier. Perceptual components for context aware computing. In *Proc. 4th International Conference on Ubiquitous Computing*, 2002.
7. A. K. Dey. Understanding and using context. In *Proc. 3rd International Conference on Ubiquitous Computing*, 2001.
8. B. A. Draper. Learning control strategies for object recognition. In K. Ikeuchi and M. Veloso, editors, *Symbolic Visual Learning*, chapter 3, pages 49–76. Oxford University Press, New York, 1997.
9. G.H. Granlund and A. Moe. Unrestricted recognition of 3-D objects using multi-level triplet invariants. In *Proc. Cognitive Vision Workshop*, Zürich, Switzerland, September 2002.
10. J. Hoffmann and W. Kunde. Location-specific target expectancies in visual search. *Journal of Experimental Psychology: Human Perception and Performance*, 25:1127–1141, 1999.
11. A. Hollingworth and J. Henderson. Does consistent scene context facilitate object perception. *Journal of Experimental Psychology: General*, 127:398–415, 1998.
12. A. Hollingworth and J.M. Henderson. Accurate visual memory for previously attended objects in natural scenes. *Journal of Experimental Psychology: Human Perception and Performance*, 28(1):113–136, 2002.
13. M. Irani and P. Anandan. Video indexing based on mosaic representation. *IEEE Transactions on Pattern Analysis and Machine Intelligence*, 86(5):905–921, 1998.
14. L. Itti, C. Koch, and E. Niebur. A model of saliency-based visual attention for rapid scene analysis. *IEEE Transactions on Pattern Analysis and Machine Intelligence*, 20(11):1254–1259, November 1998.
15. M.H. Kalos and P.A. Whitlock. *Monte Carlo Methods, I: Basics*. John Wiley & Sons, New York, NY, 1986.

16. A. Mohan, C. Papageorgiou, and T. Poggio. Example-based object detection in images by components. *IEEE Transactions on Pattern Analysis and Machine Intelligence*, 23(4):349–361, 2001.
17. M.R. Naphade and T.S. Huang. A probabilistic framework for semantic video indexing, filtering, and retrieval. *IEEE Transactions on Multimedia*, 3(1):141–151, 2001.
18. S. Obdrzalek and J. Matas. Object recognition using local affine frames on distinguished regions. In *Proc. British Machine Vision Conference*, 2002.
19. A. Oliva and P.G. Schyns. Diagnostic colors mediate scene recognition. *Cognitive Psychology*, 41:176–210, 2000.
20. L. Paletta, A. Goyal, and C. Greindl. Selective visual attention in object detection processes. In *Proc. Applications of Artificial Neural Networks in Image Processing VIII*. SPIE Electronic Imaging, Santa Clara, CA, in print, 2003.
21. L. Paletta and C. Greindl. Context based object detection from video. In *Proc. International Conference on Computer Vision Systems*, pages 502–512. Graz, Austria, 2003.
22. L. Paletta and A. Pinz. Active object recognition by view integration and reinforcement learning. *Robotics and Autonomous Systems*, 31(1-2):71–86, 2000.
23. L. Paletta, M. Prantl, and A. Pinz. Learning temporal context in active object recognition using Bayesian analysis. In *Proc. International Conference on Pattern Recognition*, pages 695–699, 2000.
24. F. Pelisson, D. Hall, O. Riff, and J.L. Crowley. Brand identification using Gaussian derivative histograms. In *Proc. International Conference on Computer Vision Systems*, pages 492–501. Graz, Austria, 2003.
25. F. Sadjadi. *Automatic Target Recognition XII*. Proc. of SPIE Vol. 4726, Aerosense 2002, Orlando, FL, 2002.
26. B. Schiele and J.L. Crowley. Recognition without correspondence using multidimensional receptive field histograms. *International Journal of Computer Vision*, pages 31–50, 2000.
27. C. Schmid. A structured probabilistic model for recognition. In *Proc. IEEE International Conference on Computer Vision*, 1999.
28. C. Schmid and R. Mohr. Combining greyvalue invariants with local constraints for object recognition. In *Proc. International Conference on Computer Vision*, pages 872–877. San Jose, Puerto Rico, 1996.
29. R. Sims and G. Dudek. Learning visual landmarks for pose estimation. In *Proc. International Conference on Robotics and Automation*, Detroit, MI, May 1999.
30. Y. Takeuchi and M. Hebert. Finding images of landmarks in video sequences. In *Proc. Conference on Computer Vision and Pattern Recognition*, 1998.
31. A. Torralba and P. Sinha. Statistical context priming for object detection. In *Proc. IEEE International Confernce on Computer Vision*, 2001.
32. P. Viola and M. Jones. Rapid object detection using a bossted cascade of simple features. In *Proc. IEEE Conference on Computer Vision and Pattern Recognition*, 2001.

Copular Questions and the Common Ground

Orin Percus

Università "Vita-Salute" San Raffaele, Via Olgettina 58, 20132 Milano, Italy
percus.orin@hsr.it

Abstract. Stalnaker ([9]) and much subsequent work argues that we should model the common ground as a set of possible worlds, but there are different ways in which one might imagine doing so. This paper asks which way is most appropriate when it comes to explaining how language works. The paper looks at a certain restriction on the use of copular questions, and suggests that accounting for this restriction is easier given one way of modeling the common ground than given others.

1 A Question about Language

This paper will revolve around a particular fact. To get a handle on this fact, consider the question in (1).

(1) Who do you think is John?

There are certain things that we *cannot* use this question to ask. In particular, we cannot use it to ask what function John performs. To see what I mean, imagine *Scenario I*. In this kind of scenario, I couldn't use the question in (1) to ask what instrument John plays in the trio. I couldn't use it, that is, to ask for information of the kind we would express with a sentence like (2). It is interesting to note that (1) contrasts in this respect with the sentence in (1'), which I *can* use to ask what instrument John plays in the trio[1] – so in some way this restriction on the use of (1) relates to the fact that *who* originates in subject position, and not in object position as in (1').

---

[1] For reasons not obvious to me, the corresponding which-question -- *Which (one) do you think John is ?* -- sounds awkward as a request for this kind of information. The awkwardness goes away when the "range" of *which one* is linguistically determined, as in *Which (one) do you think John is, the cellist or the violinist?* Incidentally, I suspect that one can make an argument similar to the one I make in this paper by considering sentences of the latter kind: relevant would be the contrast on Scenario II between *Which (one) do you think is John, the guy on the left or the guy on the right?* (good) and *Which (one) do you think is John, the violinist or the cellist?* (odd, in my judgment odder than *Which one do you think John is,...*).

> Scenario I. When we arrive at the piano trio concert, a friend of ours tells us that he is going to introduce us to a couple of the musicians. He brings over two men in tuxedos and introduces one as John (the one on the left), and the other (the one on the right) as Bill. At this point, we know that John is one of the musicians, but we don't know whether he is the cellist, violinist or pianist. Likewise for Bill.

(1') Who do you think John is?

(2) John is the violinist.

What *can* we use the question in (1) to ask? To see an example, consider a slight variation on our first scenario, which I have labelled *Scenario II*. On this scenario, while again it would be odd to use the question in (1) to ask what instrument John plays, I could use it to ask for information of the kind we would express with an answer like (3).[2] On Scenario I, this kind of question is infelicitous – the answer is already common knowledge – but on Scenario II, where the answer is not common knowledge, it is clear that the question can have this use.

> Scenario II: We weren't paying attention, and our backs were turned when the introductions were made. When we turn around, we see those two people in front of us, but we don't know which one got introduced to us as John and which one got introduced to us as Bill.

(3) John is the guy on the left.

With this in mind, here is the fact that the paper will revolve around. If we take a copular question of the form *Who is John?* where *who* originates in subject position ((4)), we cannot use it to ask what function John performs, but we can use it to ask which of the people in front of us got introduced as John.

(4) Who$_i$ is$_j$ [$_{IP}$ t$_i$ t$_j$ John]

(structure before wh-movement and auxiliary raising: [$_{IP}$ **who is John**])

To point you towards this fact, I had to ask you to think about the question *Who do you think is John?* rather than *Who is John?* This is because the main auxiliary moves above the subject in English questions, and so the question *Who is John?* is actually ambiguous between two different structures, one in which *who* originates in subject position and one in which *who* originates in object position. But the question *Who do you think is John?* is not ambiguous in this way, there it is clear that *who* originates in subject position, and so there we can see our fact: structures where *who* originates in subject position – structures like (4) – cannot be used to ask what function John performs. (To convince yourself of the same thing in another way, you could think about the contribution that the embedded interrogatives make to the sentences in (5). These too are cases where *who* clearly originates in subject position, since the embedded auxiliary doesn't move above the embedded subject.)

(5) a. Guess [who is John]   (cf.   a'. Guess [who John is]   )
    b. May I ask [who is John]   (cf.   b'. May I ask [who John is]   )

---

[2] Or alternatively the answer *The guy on the LEFT is John*. I think there is some variation as to which form of the answer is preferred.

One concern of this paper will be to ask: What is the right way to describe this restriction on the use of (4)? And why is the use of (4) restricted in this way?

2 A Question about Information States

That was a question about language: why can a structure of the kind in (4) convey some things but not others? Now here is a different kind of problem. I want to consider our first problem in the context of this second one.

Suppose we assume that the right way to model information states, and in particular common grounds, is as a set of possible worlds. That is, suppose we imagine a la Stalnaker that there is a "context set" that consists of the open candidates for the actual world as far as the participants in the conversation are concerned. Suppose we also assume (pace Lewis) that individuals exist across worlds. Now, with these assumptions in mind, recall our second scenario, the one in which our backs were turned at the point when introductions were made. The problem is: how should we model the common ground that the scenario leads to? At the point when I ask you the question in (1), what kinds of worlds are in our context set?

(The more general problem that this is getting at, of course, is: how should we model the common ground in cases where we would say that it is not yet part of our presumed common knowledge who is John.)

Here are two ideas that we might entertain.

2.1 Idea A

On Idea A, there are two kinds of worlds in our "context set." In some, on the left is a certain individual j who exists throughout our possible worlds, and on the right is a certain individual b who exists throughout our possible worlds. In others, it's the other way around. The information that we lack by virtue of having had our backs turned when introductions were made is the information that would allow us to exclude one of these bunches of worlds.

(6)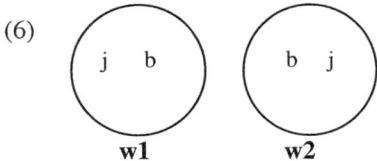

The information that we lack by having our backs turned is the information that someone could supply us with by telling us the sentence in (7). On Idea A, if we accept the sentence (7) as true, this is how we adjust our context set: we keep those worlds in which we find j on the left (the **w1** worlds), but throw out those worlds in which j is not on the left (the **w2** worlds). That is, on Idea A, to talk about John is to talk about j and to talk about Bill is to talk about b.

(7) The guy on the left is John.

2.2 Idea B

On Idea B, again there are two kinds of worlds in our "context set," and the information that we lack by virtue of having our backs turned – the information that (7) communicates – is the information that would allow us to exclude one of these kinds of worlds. But the distinction between the two groups of worlds is of a different nature.

On Idea B, unlike on Idea A, the same individual – call him x – is on the left in all worlds in the context set, and the same individual – call him y – is on the right in all worlds in the context set. Where the worlds differ has to do with the properties that x and y have. In some worlds, x has a property that for convenience we might call the "John" property, and y has a property that for convenience we might call the "Bill" property. In others, it's the other way around.

(8)

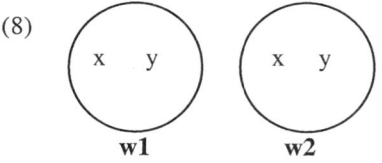

w1: x has the "John" property, y has the "Bill" property
w2: x has the "Bill" property, y has the "John" property

On Idea B, how do we adjust our context set when we receive the information that a sentence like (7) expresses? We keep those worlds in which the guy on the left has the "John" property – that is, those worlds in which x has the "John" property, the **w1** worlds – and we throw out those worlds in which the guy on the left does not have the "John" property – the **w2** worlds. So basically, on Idea B, to talk about John is to talk about the property we called the "John" property and to talk about Bill is to talk about the property we called the "Bill" property. These are properties that an individual may have in some worlds but not others.[3]

What I will argue now is that the idea we adopt about how to model the common ground can determine how we look at the problem we started out with – the problem of why a question of the form in (4) can convey some things but not others. Interestingly, one idea about how to model the common ground, because it encourages us to look at the problem in a particular way, could make a direction for solution more evident than would another idea. In that case, we would have a reason to favor one idea over the other. I will suggest that this is indeed the case.

(Incidentally: In the **Appendix** I mention a third idea about how to model the common ground that Scenario II leads to. I don't discuss it in the text because it departs from an assumption that I don't want to depart from here – the assumption that different

---

[3] But what kinds of properties are they? One way of taking the mentality behind Idea B is that these properties are the properties of occupying a particular position in the repertoire that we keep of individuals important to us. I will leave the possibilities open. One should certainly ask under what kinds of assumptions about the nature of worlds, individuals and properties, Idea B can be made sense of.

information states are to be modeled as different sets of worlds. I include it in the paper because it might be more intuitive than the two ideas I have set out here, and because I think the discussion in the text has some bearing on this idea too.)

3 Tackling These Questions

Let us now return to our initial fact: in a situation in which we have been introduced to someone named "John," (4) can ask which of the salient people is the one so introduced, but not which function that person performs. One aspect of this fact happens to be that, in the scenarios we considered, (4) does not request information of the kind (9a) would contribute, but does request information of the kind (9b) would contribute. Let us ask how the different ideas of how to model the common ground lead us to think about this fact. Importantly, the different ideas of how to model the common ground will mean different ways of characterizing the kind of information that (4) can and cannot request.

(9) a. # John is the violinist.   b. John is the guy on the left.

I will conduct this discussion from the standpoint of a few assumptions about how questions work. The background assumption is about declarative sentences: I assume that, in general, declarative sentences[4] express propositions – functions from worlds to truth values – and that, when we accept a sentence as true, we eliminate from the context set those worlds for which the proposition does not yield 1.   Questions, I assume, are instructions to express one proposition in a certain set that the question[5] makes relevant; a felicitous answer will then be a sentence whose denotation is contextually equivalent to one of these propositions[6]. (Propositions ϕ and ψ are contextually equivalent with respect to a context set C when, for every world w in C, ϕ is true in w iff ψ is true in w.)   Bearing this in mind, what do Ideas A and B suggest as far as the set of propositions that the question instructs us to choose among?

Let us start with Idea B, the idea under which there is a "John" property that potentially holds of different people in different worlds. On Idea B, we can entertain the hypothesis that the set (4) instructs us to choose among is a set of propositions that vary with respect to an <u>individual</u> – a subset of (10).

(10) { $\lambda w.\ n =$ the person who has the "John" property in w | n an individual }

some propositions in this set:
$\lambda w.\ x =$ the person who has the "John" property in w
$\lambda w.\ y =$ the person who has the "John" property in w

This hypothesis enables us to explain as follows why the question does not elicit responses like (9a) in the scenarios we considered. The idea would be that (9a) plausi-

---

[4] More precisely, the LF of a declarative sentence. Sentences that have different LFs are potentially ambiguous between different propositions.
[5] More precisely, the LF of a question. Questions that have different LFs are potentially ambiguous between different kinds of instructions. (See Section 4.)
[6] More precisely, a sentence whose LF has a denotation that is contextually equivalent to one of these propositions.

bly expresses the proposition in (11). Since on a scenario like ours it is not known which of the individuals in front of us plays the violin, (11) will not convey what any of the propositions in (10) conveys, and so (9a) will not be a felicitous answer. In a similar way, this hypothesis will enable us to explain why the question *does* elicit responses like (9b). Idea B suggests an analysis of (9b) along the lines of (12) (this should be clear from our exposition of Idea B). Given that in all of the worlds in our context set the guy on the left is the same individual x, (12) *is* contextually equivalent to one of the propositions in (10): the proposition $\lambda w.\ x =$ the person who has the "John" property in w.

(11) $\lambda w.$ the person who has the "John" property in w = the violinist in w

(12) $\lambda w.$ the person who has the "John" property in w = the guy on the left in w[7]

Now suppose instead we adopt Idea A. Unlike on Idea B, it doesn't look as though we are going to get anywhere by saying that (4) instructs us to choose from among a set of propositions that vary with respect to an individual. It seems more promising to say that the set is made up of propositions that vary with respect to an <u>individual concept</u> – a subset of (13). But there is an important difference between the approach that Idea B suggests and the approach that Idea A suggests. On Idea A, the subset must be restricted in such a way as to exclude concepts like *the violinist* (including concepts like *the guy on the left*). By contrast, on Idea B, where we dealt with a set of propositions that varied with respect to an individual, no individuals needed to be excluded.

(13) { $\lambda w.\ F(w) = j$ | F a function from worlds to individuals }
    some propositions in this set (abbreviated):
    $\lambda w.$ the person on the left in w = j
    $\lambda w.$ the person on the right in w = j

(14) $\lambda w.\ j =$ the violinist in w

To see that the subset must be restricted, note that, on Idea A, the natural way of analyzing our (9a) is as expressing the proposition in (14). If the subset were not restricted, it would include (14), and so we would wrongly predict (9a) to be a possible answer on a scenario like Scenario 1 or Scenario 2.

To summarize, when we ask what our ideas of how to model the common ground suggest about what the set of propositions is that (4) asks us to choose among, this is what we find: in a certain sense, the proposition set that Idea A points to is more constrained than the proposition set that Idea B points to. This suggests a moral. The tool for deciding between Idea A and Idea B should be the theory that we adopt of what propositions a question makes relevant. The deciding factor will be how easy it is to account for the different constraints on proposition sets that the different views force us to. If it is easy for our theory to account for the constraints on proposition sets that we are forced to once we adopt one of these views, that is a point in favor of the view; if hard, that is a point against.

---

[7] This is shorthand: to "be on the left" in w is roughly to be standing on the left of where we are in w.

Right now I see that there is a direction to pursue in order to account for Idea B's less constrained proposition sets – that is what I will outline next. I don't see any direction when it comes to Idea A's more constrained proposition sets. So it seems to me that the balance tips in favor of Idea B.

4 A Syntactic Puzzle

Our initial question was: why does (4) mean what it does? On Idea B, this question becomes: why does (4) makes relevant propositions that vary with respect to an individual? I will now pull another item out from our toolbox: a theory that links the propositions a question makes relevant to what the question's syntactic structure is. I will show that, if we take for granted a Heim and Kratzer ([6])-style theory of how syntactic structures are interpreted, then by assuming Idea B we can see our initial question as a syntactic puzzle. The puzzle is that there seem to be limitations on what syntactic structures we can generate for a question pronounced *Who is John* where *who* originates in subject position. The fact that we can transform our question into a syntactic puzzle means that we potentially open up areas for investigation.

Here is the theory I assume in what follows. The basic idea is that a question's LF determines what proposition set it makes relevant. Questions that are syntactically ambiguous in the sense that they admit different LFs are therefore also ambiguous in the sense that the different structures make different proposition sets relevant. How does the LF of a question determine a proposition set? LFs are interpreted along the lines given in Heim and Kratzer. In the case of questions, I assume, interpretation works in such a way that the denotation of a question's LF is a function from worlds to sets of propositions. At the same time, it is only licit to use a question when it yields the same proposition set for every world in the context set[8] -- so that is how the single set of propositions arises.

Given this background, imagine that we also assume Idea B. We can then explain why (4) means what it does if we can guarantee that (4) only admits LFs with a denotation like (15). (15) will lead to propositions that vary with respect to an individual.

(15) $\lambda u.$ { $\lambda w.$ n = the person who has the "John" property in w | n is a person in u }

With this in mind, here is the reasoning behind the claim that, if we assume Idea B, our initial question becomes a syntactic puzzle. (i) There is one LF (call it LF1) that it

---

[8] This is reminiscent of a principle that Stalnaker ([9]) argues for. A slightly distorted version of Stalnaker's view is: declarative sentences yield not propositions, but functions from worlds to propositions; a condition on use limits us to those sentences that take every world in the context set to the same proposition. As it happens, Stalnaker also argues that in cases where this condition is not met, speakers can make use of a recovery procedure – "diagonalization" -- that, out of the function that the sentence denotes, determines a single proposition. The idea is that, given the function p, the procedure yields $\lambda w.\ p(w)(w)$. Maybe, in cases where the condition of use for questions is not met, there is a parallel recovery procedure that, out of the function that the question denotes, yields a single set of propositions. Given the function Q, say, the procedure might first choose a set of (declarative-like) functions, Q', such that Q = $\lambda u.$ { $p(u)$ | $p \in Q'$ }, and then on this basis produce the set of propositions { $\lambda w.\ p(w)(w)$ | $p \in Q'$ }. However, I will ignore this possibility in what follows.

is plausible that we can generate for (4) and that has precisely the denotation in (15). (ii) If the LFs that we could build for (4) were exactly parallel to the LFs that we can build for some other questions, we might *also* expect to be able to generate a second LF (call it LF2), which does *not* have the denotation in (15). (iii) So to explain why (4) has the meaning it has, we have to explain why the syntax cannot generate LF2. In seeking to explain why the syntax cannot generate LF2, there are naturally different avenues to investigate, since the architecture of LF2 is not identical to the architecture of LF1.

In the ensuing subsections, I will flesh out some details of these steps of reasoning. Space limitations force me to be brief, though.

4.1 Step i: An LF with the Right Denotation That We Can Plausibly Generate for (4)

The simplified structure in (16) gives the general idea of one LF that we can plausibly generate for (4). (I assume here that the verb has lowered back into IP, and that *who* actually has a complex structure. Note too that the structure contains silent items that function as variables over possible worlds.) It would not be controversial to posit such an LF, and it gives us just the denotation in (15).[9]

(16)

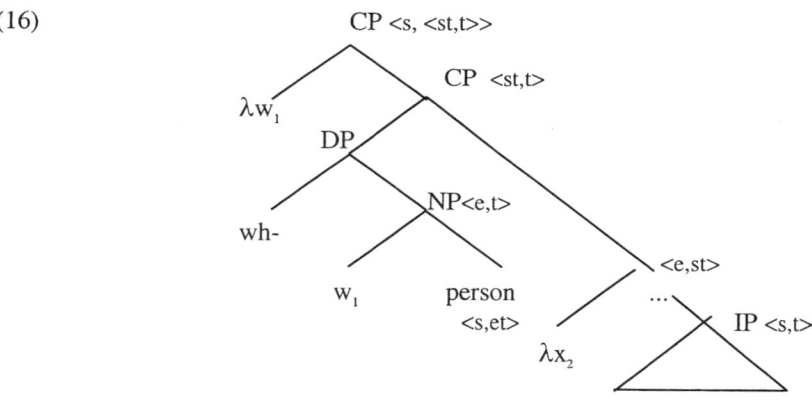

4.2 Step ii: Why We Might also Expect a Second LF with a Different Denotation

The reason why we cannot stop here and say that we have solved our problem is that it isn't obvious that an LF like (16) is the *only* LF the question has, and other potential

---

[9] To get (16) to yield the denotation in (15), we need these denotations for its parts: $[[\textbf{wh}]]^g = \lambda F_{<e,t>}.\lambda P_{<e,st>}.\ \{\ P(n)\ |\ F(n) = 1\ \}$; $[[\textbf{person}]]^g = \lambda w_s.\lambda n_e.$ n is a person in w ; $[[\ [_{IP} ...t_2\ \text{is John} ...\ ]]]^g = \lambda w_s.$ g(2) = the person who has the "John" property in w. Note that **wh** takes a predicate and a property (like the property of "being John"), and creates a set of propositions each of which says that the property holds of some individual the predicate characterizes.

LFs might *not* lead to a denotation of the kind we want. What other potential LFs do I have in mind? Some background is in order.

Basically since Engdahl ([2]), it has been clear that, once we say that questions like *(17)* have one LF that makes relevant propositions that vary with respect to an individual ((18)), then we must say they also have a second LF, which makes relevant different kinds of propositions. We must take this step because the questions license answers such as (19) even when such answers are not contextually equivalent to any of the propositions in (18) -- for example, when it isn't known who the candidates are. What kind of propositions does the second LF make relevant? A possible answer is: propositions that differ with respect to an individual concept ((20)). But now the worry should be clear: if (17) admits a second LF that makes relevant propositions that vary with respect to an individual concept, then (4) potentially might as well. And in that case, we might lose our account of the restriction on (4)'s interpretation.

(17) Who will win the next election?

(18) { λw. in w, individual n wins the next election | n an individual }

(19) The candidate with the biggest campaign budget [will be the winner].

(20) { λw. in w, F(w) wins the next election | F a function from worlds to individuals }

  (e.g., $\mathbf{F}_1$(w) = the election candidate in w with the biggest campaign budget in w)

Naturally, what additional LF we might expect (4) to admit depends on what we decide the second LF for (17) looks like. In (21) I have given one possibility (along the lines of the functional wh-analyses of [2], [1] and [5]). This LF has the same basic architecture as the earlier one in (16), but differs in two important ways: first, the wh-phrase contains a silent affix; second, movement leaves a complex trace (it consists of an item that functions as a variable over concepts together with an item that functions as a variable over worlds). The net effect of these differences is to give individual concepts the role that individuals play in (16): this LF will make relevant propositions that vary with respect to individual concepts rather than individuals.[10] If (4) admitted this LF, we would not be able to account for (4)'s meaning. Why is this? The denotation of (21) is (roughly) as in (22). This denotation means that (23) is among the propositions that the LF makes relevant. But Idea B suggests an analysis of *John is the violinist* (our (9a)) basically along these lines, so if (4) had this LF, we would expect *John is the violinist* to be a possible response – contrary to fact.

---

[10] Note the change in the order of *person*'s arguments, and the appearance of world variables in IP (I simplified away from these details in the discussion of the earlier LF). To get out of (21) a denotation that "ranges over" individual concepts, we will need the following denotations, among others: $[[\mathbf{AFF}]]^g$ ($P_{<e,st>}$) = $\lambda K_{<s,e>}$. for all w in dom(K), P(K(w))(w) = 1 ; $[[\mathbf{wh}]]^g$ = $\lambda \mathbf{F}_{<X,t>} \lambda \mathbf{P}_{<X,st>}$. { $\mathbf{P}$(N) | $\mathbf{F}$(N) = 1 } . (The new cross-categorial denotation for *wh-* enables it to range over individual concepts. The effect of AFF is that NP will be a predicate of individual concepts rather than of individuals. The effect of the "big trace" and the lambda that binds it will be to create out of the movement remnant too a function that takes individual concepts, rather than individuals, as arguments.)

(25)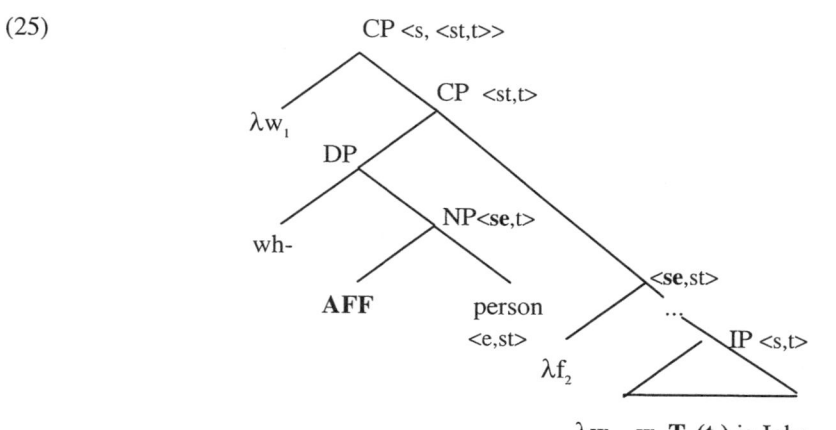

(22) λu. {λw. in w, K(w) has the "John" property | for all v, K(v) yields a person in v}
(23) λw. in w, the violinist in w has the "John" property

4.3 Step iii: The Syntax Must Not Generate the Second LF

So if we adopt Idea B, we arrive at a syntactic puzzle. We can account for the restriction on the use of (4) if we say that (4) *admits* the *first* kind of LF – the one in (16), the kind that makes relevant a set of propositions that differ with respect to an individual – but does *not* admit the *second* kind of LF – the one in (21), the one that makes relevant a set of propositions that differ with respect to an individual concept. Evidently, even if the syntax can generate LFs of the second kind for other sentences, like (17), it cannot generate this LF for (4). The puzzle is: why is this second LF excluded?

I don't know how tractable this problem is, but at least it is clear what the problem is that we have to solve. And there are places to start investigating. We noticed some ways in which the architecture of the second LF differs from that of the first. One is that the second LF contains a silent affix, another is that the second LF contains a complex trace. Might it be that the syntax is unable to generate a complex trace in the position where it appears in the second LF? These are the kinds of questions to ask.

5 Concluding Remarks

What I hope came out of the discussion is this. When we look at ways of combining assumptions about ontology, on the one hand, with ideas about interpretation, on the other, we find that some ways are more suited to describing natural language than other ways are. Therefore, if we take a stand on one of these things – in this case, a theory of how we interpret syntactic structures – we will be driven to conclusions about the other -- in this case, about what view of possible worlds is the right one for modeling common grounds like the one Scenario II leads to. In this way, theories of how we interpret syntactic structures can become tools for investigating how we represent common grounds.

The discussion so far suggested that we should adopt Idea B, on which it makes sense to talk about a "John" property that individuals have in some worlds but not others. This is interesting for a number of reasons. For one thing, it could help us to understand what articles in the copula literature mean when they say that name-like expressions can sometimes behave like "predicates" and sometimes like "referring expressions." On Idea B, there is a natural way of explicating this terminology: a name like *John* behaves like a "predicate" when the "John" property holds of different individuals in different worlds in the context set, and behaves like a "referring expression" when the "John" property holds of the same individual in all worlds in the context set. On Idea A, by contrast, explicating this terminology looks less straightforward.

Speculating, here is another advantage that Idea B might have. It might lead to insight as to why, unlike our friend (24a), parallel sentences with pronouns like (24b) seem bizarre in any context. When we ask how the denotations of questions like (24a) are built out of the parts of the sentence, it is a small step to say that the name *John* denotes an individual concept – the person who has the "John" property. It is another small step to say that in the structure for this sentence, *John* combines with a world variable, and the result occupies a slot reserved for an individual-denoting expression ((25a)). Now, the idea to pursue is that, while names denote properties, pronouns denote individuals. In that case, in the structure for (24b), the pronoun will occupy the position that the name-world variable complex occupies in the structure for (24a). If the pronoun in (24b) is a simple variable, the result will be that, when we compute the denotation for the question in (24b), and we look at the propositions the question will makes relevant, we will find only tautologies or contradictions.[11]

(24) a. Who is [t t John] ?   b. ?? Who is [t t him]?

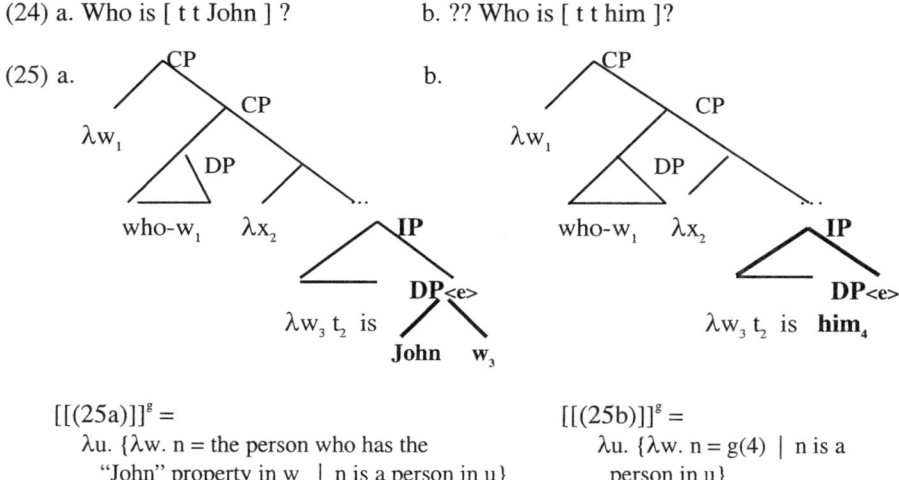

[[(25a)]]^g =
λu. {λw. n = the person who has the
"John" property in w | n is a person in u}

[[(25b)]]^g =
λu. {λw. n = g(4) | n is a
person in u}

---

[11] The explanation is not obviously so simple: questions like *Which one is t t him?* are fine (imagine that we are at the opera, and know that a mutual friend of ours is in the chorus), and this suggests that we might not *always* want to say that pronouns are simple individual variables.

Acknowledgments. For helpful comments, I am grateful to Carlo Cecchetto, Gennaro Chierchia, Bill Ladusaw, Line Mikkelsen, Francesca Panzeri, Barbara Partee, Michal Starke, Satoshi Tomioka, Anna-Lena Wiklund, and Roberto Zamparelli. I also thank the ZAS Cleft Workshop (November 1997) for indirectly getting me to think about these issues. Similar versions of this work were presented at the Penn Copula Workshop (November 2001), IGG 2002 (February 2002) and EDILOG 2002 (September 2002).

Data of the kind in this paper were originally brought to light (together with many others) in Higgins ([7]), ch. 5 (see esp. p. 154).

References

1. Chierchia, G.: Questions with Quantifiers. Natural Language Semantics 1 (1993) 181–234
2. Engdahl, E.: Constituent Questions. Kluwer, Dordrecht (1986)
3. Groenendijk, J., Stokhof, M., Veltman, F.: Coreference and Modality. In: Lappin, S. (ed.): The Handbook of Contemporary Semantic Theory. Blackwell, Oxford (1996) 179–213
4. Hamblin, C.L.: Questions in Montague Grammar. Foundations of Language 10 (1973) 41–53
5. Heim, I: unpublished lecture notes. MIT, Cambridge MA (1991)
6. Heim, I., Kratzer, A.: Semantics in Generative Grammar. Blackwell, Malden (1998)
7. Higgins, F.R.: The Pseudo-Cleft Construction in English. PhD dissertation. MIT, Cambridge MA (1973)
8. Karttunen, L.: Syntax and Semantics of Questions. Linguistics and Philosophy 1 (1977) 3–44
9. Stalnaker, R: Assertion. Syntax and Semantics 9 (1978) 315–332

Appendix: A Third Idea

Naturally, Idea A and Idea B are not the only views that one might entertain about how to model the common ground in Scenario II. Here is a sketch of a third idea. This third idea departs from our original assumptions in that it assumes that a set of possible worlds alone is not sufficient to describe an information state: in a certain sense, a description in terms of possible worlds has to be supplemented with a description of what is known about who is named what.

This third idea shares with Idea B the view that in all of our possible worlds the same individual – call him x – is on the left, and the same individual -- call him y -- is on the right. And it shares with Idea A the view that names are meant to talk about individuals. But there is an important difference between the third idea and our other ideas. Idea A and Idea B assume that the parties to conversation have decided what individual, or property, it is that a name like *John* evokes, and therefore that sentences with the name *John* impart some kind of information to them. The third idea assumes instead that the parties to conversation have not resolved what individual to associate with the name *John*, and if they were to hear a sentence like *John is happy*, they would not necessarily be able to determine a way in which to reduce the context set.

(26)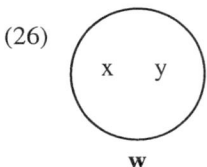

A proponent of this third view would look at our scenario as a case in which communication does not proceed quite as smoothly as it could. The idea would be roughly as follows. At the outset, since we do not know that to talk about John is to talk about x, there are a lot of sentences with the name *John* that we could not use to narrow down the worlds in our candidate set – including the phrase that accompanied the introduction, say *I present you John*. The principal effect of a sentence like *The guy on the left is John* ((7)) is then not to eliminate worlds from our candidate set, but just to let us know that there is a certain way of talking about the worlds in our candidate set – that to talk about John is to talk about x. Once we know this, we are then in a position to use the information provided by other phrases, like *I present you John*, and to narrow down our candidates to those worlds in which we have been introduced to x.

In other words, unlike on Idea A and Idea B, on the third idea there is no sense in which the worlds in the context set divide up into two classes according to who is John. And an assertion like *The guy on the left is John* is not designed to get us to eliminate worlds from our context set – in fact, it does not directly cause the elimination of any. Rather, it is designed to allow us to interpret previously uninterpretable messages.[12]

What consequences might this third idea have for an analysis of our fact about the use of questions like (4)? In fact, it is not transparent how to apply this idea, but here is an attempt.

Maybe one could maintain what might at first look like a non-starter -- that the question makes relevant a set of propositions that, in terms of the way they are built up, vary with respect to an individual, but that are either a tautology or a contradiction ((27)). However, we must also stipulate that the question requires its answer to take a certain form: to be of the form *X is John*. What will happen then? The idea is roughly as follows. Since one of the propositions is a tautology, the answerer will have the obligation to utter a proposition that is true in all worlds in the context. As soon as he answers a sentence of the form *The guy on the left is John*, the parties to conversation, who know that x is the guy on the left, will see that to talk about John is to talk about x. This, then, looks like another case where we would like to say that a question makes relevant a set of propositions that vary with respect to an individual. So at first glance at least, it looks as though, since we want to insure variation with respect to an individual, we are going to wind up with the same problems that we have to face with Idea B.

(27) { $\lambda w.\, x = j$,
       $\lambda w.\, y = j$,
       $\lambda w.\, z = j$,
           ...   }

---

[12] Line Mikkelsen pointed out to me that the proposals of Groenendijk, Stokhof and Veltman ([3]) are in this spirit.

Contextual Coherence in Natural Language Processing

Robert Porzel and Iryna Gurevych

European Media Laboratory, GmbH
Schloss-Wolfsbrunnenweg 31c
D-69118 Heidelberg, Germany
{porzel,gurevych@eml.org}

Abstract. Controlled and restricted dialogue systems are reliable enough to be deployed in various real world applications. The more conversational a dialogue system becomes, the more difficult and unreliable become recognition and processing. Numerous research projects are struggling to overcome the problems arising with more- or truly conversational dialogue system. We introduce a set of contextual coherence measurements that can improve the reliability of spoken dialogue systems, by including contextual knowledge at various stages in the natural language processing pipeline. We show that, situational knowledge can be successfully employed to resolve pragmatic ambiguities and that it can be coupled with ontological knowledge to resolve semantic ambiguities and to choose among competing automatic speech recognition hypotheses.

1 Introduction

Following Allen et al. (2001), we can differentiate between controlled and conversational dialogue systems. Since controlled and restricted interactions between the user and the system decrease recognition and understanding errors, such systems are reliable enough to be deployed in various real world applications, e.g. timetable or cinema information systems. The more conversational a dialogue system becomes, the less predictable are the users' utterances. Recognition and processing become increasingly difficult and unreliable. Research projects are struggling to overcome the problems arising with more- or truly conversational dialogue systems, e.g. Wahlster et al. (2001). Their goals are more intuitive and conversational natural language interfaces that can someday be used in real world applications. The work described herein is part of that larger undertaking: we view the handling of contextual - and therefore linguistically implicit - information as one of major challenges for understanding conversational utterances in complex dialogue systems.

In this paper we report on a set of research issues, solutions and results pertinent to the construction of mobile multi-domain spoken dialogue systems. These systems aim at providing conversational speech interfaces to complex and heterogeneous applications and their domains, e.g. touristic, spatio-geographic or entertainment information as well as various assistance domains such as planning, electronic communication or electronic commercial transactions. A common feature of the solutions, to be described below, is that they involved the inclusion of extra-linguistic contexts into the natural language processing (NLP) pipeline by applying contextual coherence measurements.

In this work, we will focus on two specific extra-linguistic knowledge stores - namely ontological- and situational knowledge - and introduce the corresponding ontological - and situational coherence measurements.[1] Ontological knowledge, for example, may assert that a bakery is a store and that it has specific properties, such as opening times, specific goods for sale etc. Situational knowledge, on the other hand, may assert that the bakery Seitz is located in a specific street and currently open. Given a user utterance such as: *Is there a bakery somewhere around here?*, we ultimately want an NLP system to understand that the user might want to go there in order to buy something to eat and supply corresponding spatial instructions - to the nearest bakery or other shop depending on what is actually open given the situation at hand - rather than answering the question solely with yes or no. While the ontologies employed herein model more or less static world, conceptual and common-sense knowledge concerning *types* and *roles* (Russell and Norvig, 1995) based on the standard combinations of frame- and description logics, situational knowledge is induced in specific *instances* and highly dynamic states of affairs.

Our overall goal is to produce reliable natural language understanding components that increase dialogue quality metrics,[2] by applying context sensitive analysis such as described below. After a brief outline of contextual processing in spoken dialogue systems in Sect. 2, we will introduce *situational coherence* and the resulting model, employing data, analyses and results from the domain of spatial information in Sect. 3. We will discuss data, results and model for *ontological coherence* scoring applied in automatic speech recognition and semantic interpretation in Sect. 4. A conclusion on contextual coherence scoring is given in Sect. 5.

2 Contextual Interpretation in NLP

Utterances in dialogues, whether in human-human interaction or human-computer interaction, occur in a specific situation that is composed of different types of contexts. A broad categorization of the types of context relevant to spoken dialogue systems, their content and respective knowledge stores is given in Table 1. Following the common distinction between linguistic and extra-linguistic context[3] our first category, i.e. the dialogical context, constitutes the linguistic context, encompassing both co-text as well as intertext.

---

[1] The role of linguistic- and user-context for NLP is included via discourse-, user- and belief-modeling (LuperFoy 1999, Paris 1993, Narayanan 1997).
[2] Measurable in the PARADISE evaluation framework (Walker et al. 2000).
[3] All extra-linguistic contexts are also often referred to as the *situational context* (Connolly, 2001). however, we adopt a finer categorization thereof.

Table 1. Contexts, content and knowledge sources

types of context	content	knowledge store
dialogical context	what has been said by whom	dialogue model
situational context	time, place, etc	situation model
interlocutionary context	properties of the interlocutors	user model
domain context	world/conceptual knowledge	ontology

In linguistics the study of the relations between linguistic phenomena and aspects of the context of language use is called *pragmatics*. Any theoretical or computational model dealing with reference resolution, e.g. anaphora- or bridging resolution, spatial- or temporal deixis, or non-literal meanings, requires taking the properties of the context into account. In current spoken dialogue systems *contextual interpretation* follows *semantic interpretation*, which follows automatic speech recognition (ASR) (additionally fused with other modality-specific information). That is, the modality-specific signals, (e.g. speech or gesture) are transfered into graphical representations (e.g. word- or gesture graphs) and then fused and mapped onto some meaning representation followed by contextual interpretation (Allen, 1987). Computationally, this implies that context-independent graphical and semantic representations can be computed and the context-dependent contributions are associated with the semantic interpretation thereafter, resulting in the final representation.

This so-called *modular* view supports a distinct study of meaning (corresponding to the semantic representation) without having to muck around in the mirky waters of language use. This view is supported by the claim that some semantic constraints seem to exist independent of context. In this work we propose a different view that also allows for context-independent constraints, but offers a less modular point of view of contextual interpretation. We will show that, given the notion of context introduced above, contextual analysis can be employed already at the level of speech recognition, during semantic interpretation and, of course, thereafter. The central claim is being made, that - as in human processing - contextual knowledge can be used successfully in a computational framework in all processing stages.[4] While most research in linguistics has consequently departed from this view, most computational approaches still feature a modular pipeline architecture in that respect.

In linguistics utterances which are context-dependent are called *indexical* utterances (Bunt, 2000). Computationally they exhibit a difference in their semantic and final representation. Indexical utterances are - by virtue of the pervasiveness of contextual knowledge - the norm in discourse, with linguistic estimations

---
[4] In recent times the so-called *modular* theory of cognition (Fodor, 1983) has been abandoned more or less completely. The so-called *new look* or modern cognitivist positions hold that nearly all cognitive processes are interconnected, and freely exchange information; e.g. influences of semantic and pragmatic features have been shown to arise already at the level of phonological processing (Bergen, 2001).

of declarative non-indexical utterances around 10% (Barr-Hillel, 1954). Without contextual knowledge utterances, or fragments thereof, become susceptible of interpretation in more than one way. Computer languages are designed to avoid anaphoric, syntactic, semantic and pragmatic ambiguity, but human languages seem to be riddled with situations where the listener has to choose between multiple interpretations. In these cases we say that the listener performs *pragmatic analysis*; corresponding to contextual interpretation on the computational side. For human beings the process of resolution is often unconscious, to the point that it is sometimes difficult even to recognize that there ever was any ambiguity.

The phenomenon that this process of resolution, frequently goes unnoticed is due to the fact that in many cases the ambiguity is only perceived if the contextual factors that allowed the listener interpret the utterance unambiguously are missing. For example, if shared ontological and situation-specific knowledge provided information that was elided in the utterance. These utterances/texts, therefore, become ambiguous only after they have been stripped of discourse-, situation-, domain- and speaker-context, and, for example, appeared as a text(-fragment) in a linguistics textbook. The problem for computational linguistics originates at least partially in the fact language understanding has to make do with exactly such a contextually and pragmatically impoverished input.

3 Situational Coherence

In this section we display findings from experiments tailored towards identifying and learning contextual factors relevant to understanding a user's utterance in an uncontrolled dialogue system. That system supplies touristic and spatial information (Porzel and Strube,). In this data we find many instances of phenomena usually labeled as *pragmatic ambiguity*. In our view these examples constitute *bona fide* cases for contextual interpretation after phonological and semantic processing has been concluded. We show how natural language analysis can employ models that incorporate specific situational factors, resulting in a context-dependent analysis of the given utterances, thereby increasing the conversational capabilites of dialogue systems.

Several NLP research efforts have adopted the tourism domain as a suitably complex challenge for an intuitive conversational natural language processing system (Johnston et al. 2002, Wahlster et al. 2001). Supplying spatial information, specifically spatial instructions and spatial descriptions, constitutes an integral part of the functionality of a mobile tourist information system. We regard a **spatial instruction** - e.g. *"In order to get to the castle you have to turn right and follow the path until you see the gate tower"* - as a felicitous response to a corresponding *instructional* request. A **spatial description** - e.g. *"The Cinema Gloria is near the marketplace on the Hauptstrasse "* - is appropriate for a *descriptive* request.

We can, therefore, say that a spatial instruction is an appropriate response to an instructional request and a spatial description, e.g. a localization, constitutes an appropriate response to a descriptive request. Responding with one to the other does not constitute a felicitous response, but can be deemed a misunderstanding of the questioner's intention, i.e. an intention misrecognition. In all dialogue systems intention misrecognitions decrease the overall evaluation scores, since they harm the dialogue efficiency metrics, as the user is required to paraphrase the question, resulting in additional dialogue turns. Furthermore, satisfaction measures decrease along with perceived task ease and expected system behavior.[5]

The Data: In an initial data collection (Porzel and Gurevych, 2002) we find 128 instances of instructional requests out of a total of roughly 500 requests from 49 subjects. The types and occurrences of these categories are in Table 2.

Table 2. Request types and occurrences

Type Example	#	%
(A) How interrogatives, e.g., *How do I get to the Fischergasse*	38	30%
(B) Where interrogatives, e.g., *Where is the Fischergasse*	37	29%
(C) What/which interrogatives, e.g., *What is the best way to the castle*	18	14%
(D) Imperatives, e.g., *Give me directions to the castle*	12	9.5%
(E) Declaratives, e.g., *I want to go to the castle*	12	9.5%
(F) Existential interrogatives, e.g., *Are there any toilets here*	8	6%
(G) Others, e.g., *I do not see any bus stops*	3	2%

While handling both instructional and descriptive requests for spatial information our parsers identify types A, C, D and E as instructional request. This corresponds to a baseline of recognizing roughly 63% of the instructional requests contained in our first data sample as such. Changing the grammars to treat type B and F as instructional request would consequently raise the coverage to 98%. However, **Where interrogatives** do not only occur as requests for spatial instructions but also as requests for spatial descriptions, i.e. localizations.[6] The problem is that the current parser grammars either interpret all **Where interrogatives** as descriptive requests or as instructional requests. This implies that both systems can either misinterpret 29% of the instructional request from our initial data as descriptive requests or misinterpret all descriptive request as instructional ones. In short, they lack a systematic way of asking which type of **Where interrogative** might be at hand.[7]

Resulting from these observations we conducted an experiment in which we ask people on the street always the same **Where interrogative**, i.e. *Excuse me, can you tell me where X is.* We logged several factors:

---
[5] Unfortunately in PARADISE dialogue quality metrics are not effected by intention misrecognitions, as they are not taken into account (Walker et al. 2000).
[6] Numerous instances of **Where interrogatives** requesting spatial localizations can be found also in other corpora such as the HCRC Map Task Corpus.
[7] As the data discussed herein show a simple approach to employ the system's class-based lexicon to make this decision hinge on the object-type, e.g. building or street, will not suffice to solve the problem completely.

- the goal object, i.e. either the castle, city hall, a specific school, a specific discotheque, a specific cinema, a bank (ATM) and a specific clothing store, all of which can be either open or closed depending on the time of day,
- the time of day (i.e. morning, afternoon, evening),
- the proximity to the goal object, i.e. near (< 5 minutes walk), medium (5 - 30 minutes walk) and far (> 30 minutes walk). - additionally we kept track of the approximate age group (young, middle, old) and gender of the subjects.

In this set of contextual features we find that the results of generating decision trees and rules applying a c4.5 learning algorithm (Winston, 1992), show that:
- if the object is currently closed, e.g. a discotheque or cinema in the morning, almost 90% of the *Where interrogatives* are answered by means of localizations, a few subjects asked whether we actually wanted to go there now, and one subject gave instructions.
- if the object is currently open, e.g. a store or ATM machine in the morning, people responded with instructions, unless - and this we did not expect - the goal object is near and can be localized by means of a reference object that is within line of sight.

Looking at the problem of analyzing Where interrogatives correctly, we can conclude already that, depending on the combination of at least two contextual features, accessibility and proximity, responses were either instructions, localizations or questions. The following sections will describe how we have chosen to incorporate findings such as the ones described above into the natural language understanding process.

Requirements for Contextual Analysis: We have noted above that current natural language understanding systems lack a systematic way of asking, for example, whether a given Where interrogative at hand is construed as an instructional or a descriptive request. Speakers habitually rely on situational and other contextual features to enable their interlocutor to resolve such construals appropriately. This is not at all surprising, since conversational dialogues - whether in human-human interaction or human-computer interaction - that occur in a specific context are consequently composed of utterances based upon specific knowledge of that context.

In order to capture the diverse kinds of contextual information, studies and experiments of the type described above need to be conducted, so that the individual factors and their influences for a set of additional construal resolutions can be identified and formalized. Looking at the domain of spatial information alone we find a multitude of additional decisions that need to be made in order to enable a dialogue system to produce felicitous responses. Next to the *instruction versus localization* decision, we find construal decisions, such as:

— does the user want to enter, view or just approach the goal object
— does the user want to take the shortest, fastest or nicest path
— does the user intend to walk there, drive or take public transportation

as relevant to answering instructional requests felicitously. In many cases, e.g. the ones noted above, construal resolution corresponds to an automatic context-dependent generation of paraphrases in the sense of Ebert et al. 2001. That is, to explicate information that was left linguistically implicit, e.g. to expand an utterance such as *How do I get to the castle* depending on the context into *How do a get to the castle by car on a scenic route.*

These decisions hinge on a number of contextual features much like the instruction versus localization decision discussed above.[8] In our minds a model resolving the construal of such questions has to satisfy the following demands:

- it has to model the data collected in the experiments, which provide the statistic likelihoods of the relevant factors, for example, the likelihood of a `Where interrogative` being construed as a descriptive or instructional request, given the accessibility of the goal object,
- it has to be able to combine the probabilistic observations from various heterogenous knowledge sources, e.g. what if the object is currently accessible, but too far away to reach within a given time period,
- it has to be robust against missing and uncertain information, as these contextual features may not always be observable, e.g. in case specific services of the system such as location modules (GPS) or weather information services are currently offline.

Applying the Contextual Analysis: As a first approach we have chosen Belief- or *Bayesian* networks employing a generalized version of the variable elimination algorithm, described in Cozman (2000), to represent the relations and conditional probabilities observed in the data and to compute the posterior probabilities of the decision at hand. Bayesian networks are well-suited for combining heterogeneous, independent and competing input to produce discrete decisions and can even be regarded as suitable mathematical abstractions over the cognitive processes underlying the way human speakers process natural language (Narayanan and Jurafsky, 1998). The simplest network possible, estimating the liklihood of a `Where interrogative` being construed as an instructional or descriptive request, needs only three *observation nodes*. These nodes observe whether a `Where interrogative` is at hand, the goal object is open or closed and its proximity to the user. The single *decision node* - whether a spatial location or instruction constitutes an appropriate response - is connected to the three observation nodes.

We have linked the network to interfaces providing that contextual information. For example within the SmartKom framework (Wahlster et al. 2001), a database called the *Tourist-Heidelberg-Content Base* supplies information about individual objects including their opening and closing times. A global positioning system built into the mobile device supplies the current location of the user. This is handed to the geographic information system that computes the respective distances and routes to the specific objects. It is important to note that this type of context monitoring is a necessary prerequisite for context-dependent

---

[8] Here also ontological factors, e.g. object type and role, additional situational factors, e.g. weather, discourse factors, e.g. referential status, as well as user-related factors, e.g. tourists or business travelers as questioners and their time constraints, constitute significant factors.

analysis. These technologies enable our model to make dynamic observations of the factors determined as relevant/significant by the data collected.

These observations, captured by the monitoring modules and converted into a context representation, and the given utterance at hand, i.e. the parser output, constitute the input into our belief network. The resulting output constitutes a measurement of the *situational coherence* of the possible alternative readings. In other words it represents a list of ranked construals, e.g. a ranked list of two decisions for a given `Where interrogative` with their corresponding situational coherence scores (e.g. (probability(instruct), 0.64223 p(true | evidence) 0.35777 p(false | evidence))). This can then be employed to interpret requests accordingly, i.e., the parser output is either converted into the system's representation of an instructional or localizational request.

Results: As we have seen the current baseline performance results in a misinterpretation rate of 37% of the instructional requests of our initial data set. More specifically, all requests of type B and E, will falsely be interpreted as localizational requests and type F is not recognized at all and causes the system to indicate non-understanding. The context-adaptive enhancement described herein, lowers the error rate to 8%, which, in our minds, constitutes a significant improvement. If additional data indicate that we can treat `Existential Interrogatives` in a similar fashion, this would result in an additional lowering by 6%, leaving only 2% of the initial data set as unanalyzable for the system.

4 Ontological Coherence

As we have seen above one of the fundamental issues concerning pragmatic ambiguity, is to enable dialogue systems to pick the most appropriate reading given the contextual factors at hand. This is equally true for ambiguities that arise during semantic interpretation and automatic speech recognition.

4.1 Speech Recognition Ambiguities: N-Best Lists

A common phenomena found in different fields of NLP, e.g. automatic speech recognition, information retrieval or question answering, is that current processing techniques seem to hit a ceiling of performance. In ASR systems have progressed to a level where they are close to extracting as much information as possible from the acoustic stream. Some context-dependent features have been added to handle dialectal- and speaker-adaptation and dynamic lexica, to handle novel input (Rapp et al. 2000). However, neither ontological nor situational knowledge is taken into account, which leaves the known problem of dealing with phonetically indistinguishable input, unresolved. The classic example in the community is, that a large vocabulary speech recognition (LVSR) system, as needed for more conversational dialogue systems, could hardly differentiate between homonymic utterances such as: "*it is hard to wreck a nice beach*" and "*it is hard to recognize speech*". Humans on the other hand *hear* either one or the other depending on the context.

Today's LVSR systems rarely feature simple one-best hypothesis as interface between ASR and NLU. While that may suffice for restricted dialogue systems, most systems either operate on n-best-lists as ASR output or convert ASR word graphs (Oerder and Ney 1993) into n-best lists, given the distribution of acoustic and language model scores (Schwartz 1990). In our data a user expressed in Example (1) the wish to see a specific city map again, leading to the top two speech recognition hypotheses (1a,1b). Annotators found that Example (1a) constituted a pretty much well formed representation of the utterance whereas Example (1b) constituted an inadequate representation thereof:

(1) Ich   würde   die   Karte   gerne   wiedersehen
    I   would   the   map   like   to see again

(1a) - Ich   würde   die   Karte   eine   wieder   sehen
     - I   would   the   map   one   again   see

(1b) - Ich   würde   die   Karte   eine   Wiedersehen
     - I   would   the   map   one   Good Bye

Facing multiple representations of a single utterance consequently poses the question which of the different hypotheses most likely corresponds to the user's utterance. Several ways of solving this problem have been proposed and implemented in various systems, i.e. to use scores provides by the ASR system, i.e. acoustic and language model probabilities or to use scores provided by the natural language understanding and discourse modeling components, c.f. Litman et al. (1999).

We claim that contextual extra-linguistic knowledge can as well be used at this point to provide further information and to help in solving this task, especially in those cases where ASR and semantic scores fail. In the following we will report on the experimental setup and evaluations of this claim, thereby introducing the central notion of *ontological coherence*.

The Data: An initial experiment was reported in Gurevych et al. (2002) where we tested, whether or not human annotators could reliably classify 2300 speech recognition hypotheses (SRH) in terms of their ontological coherence, i.e. whether or nor a given hypothesis constitutes an internally coherent utterance. On an additional corpus of 1400 hypotheses we showed in recent experiments that annotators could also reliably (>94%) identify the best hypothesis, given a transcribed utterance and the corresponding SRHs choices.

Requirements and Application: The corresponding contextual analysis, then, needs to provide a coherence score automatically, that can be employed by any NLU system to select the best hypothesis from the N-best list independently or in conjunction with acoustic or statistical scores. We employ the ONTOSCORE system described in Gurevych et al (2003): Given a frame- and description logic-based ontology - e.g. a semantics as defined in oil-rdfs, daml+oil, or owl -[9] we map *words* to *concepts* and compute the average path-length of the shortest

---
[9] See www.daml.org or www.w3c.org/rdf.

graph found connecting all concepts excluding *isa* relations in the individual path-length measures, that we fitted with a conceptual context addition.
Results: Using the SmartKom system and its pre-existing ontology, ONTOSCORE correctly assigns the highest score to over 84% of the best hypothesis as defined in the merged human *gold standard* (baseline 63.91%). This coherence measuring method has, therefore, been shown to exhibit much greater than baseline-performance in an additional task and performs better or equal compared to the alternative scoring methods.

4.2 Semantic Interpretation and Construal

In much the same way ontological coherence can be employed to disambiguate between multiple representations of a user's input. We will show how it can serve to assist in semantic interpretation, i.e. in resolving semantic construal, that underlies many non-syntactic phenomena involving *unconventional* meaning (Langacker, 1998). Employing simple examples from our tourism domain:

(2) Goethe often visited the historical museum.
(3) The Palatine museum was moved to a new location in 1951.
(4) The apothecary museum was renovated in 1983.
(5) In 1994 the museum bought a new Matisse.

we find four instances of noun phrases featuring the word *museum* as an argument of four main verbs: *visited, moved, renovated* and *bought*. Linguistic analyses may vary in their classifications, however, commonly, Example (2) would be regarded as pretty conventional, Examples (3) and (4) as polysemous and Example (5) as metonymical language use with respect to the word *museum*. In many cases we find lexical ambiguities as for *kommen* in the SRH Example (6) and Example (7):

(6) *was   für   Spielfilme   kommen   heute   Abend*
    what   for   films   come   today   evening

(7) *wie   kann   ich   mir   zur   Schloss   kommen*
    how   can   I   me   to   castle   come

Due to the persuasiveness of construal in natural language, a formal model thereof as well as an account of its mechanisms, constitutes an important part of any approach to natural language understanding.
Data: As shown in Poesio (2002) about 50% of all noun phrases in their corpora are discourse-new, anaphoric noun phrases make up 30% of their data. The remaining 20% are made up by so-called *associative* expressions. In an additional experiment annotators labeled correct word-senses for all cases (1415 markables) of multiple word to concept mappings. For example, in SRHs containing forms of the verb *kommen* (to come/showing on), a decision had to be made whether it is a `MotionDirectedTransliterated`, as in Example (6) or

WatchPerceptualProcess as in Example (7) or undecidable - which was only the case in non-best SRHs (see Sect. 4.1).

Requirements and Application: Previous work on resolving ambiguities, metonymic language use and other types of *associative* meaning (Poesio, 2002) exists that also employ various kinds of hierarchical knowledge bases, showing promising results in domain-specific settings. The actual content of an ontology depends on the specific modeling choices made while constructing the ontology. Due to individual differences and even internally heterogeneous modeling choices, we need flexible algorithms for retrieving the appropriate information from the knowledge base, unlike those employed in previous approaches.[10] Additionally, the semantic web projects bring forth a multitude of external ontologies, whose modeling choices need not be known beforehand. Yet, if dialogue systems intent to profit from this undertaking, they will need to be able to extract the necessary information without knowing the specific modeling choices. As proposed in Porzel and Bryant (2003) an extra-linguistic knowledge store - an ontology - can be employed to find sets of alternative readings by searching the conceptual graph in ways as permissive as radial categories suggest. These ontological substitutions constitute an addition to ambiguity mappings from the lexicon.

This has been interfaced to the ontological coherence scoring application, i.e. ONTOSCORE, to calculate how often contextual coherence picks the appropriate reading. In order to aid semantic interpretation by means of contextual knowledge we can apply the same algorithm employed to score sets of speech recognition hypotheses for scoring different potential trigger - target pairings with respect to their ontological coherence. For metonymy or bridging resolution, however, an initial processing step is needed to find sets of possible pairings, i.e. candidates that are potentially more ontologically coherent.

Results: As a result of measuring the ontological coherence of the conceptual representations we get a corresponding ranking for the alternative readings. Looking at the case of *kommen* as *showing (on TV)* versus *coming (to/from)*, given a pre-existing ontology we find 85 occurrences of this ambiguity in which the contextually enhanced ONTOSCORE picked in the correct reading in 72 cases, and not in 2 cases, and mixed in all 11 undecidable cases, which where not in the best SRH set. Baseline, given the majority distribution, was 56.5%.

The inclusion of such contextual interpretation during and before semantic interpretation can enable natural language understanding systems to become more conversational without loosing the reliability of restricted dialogue systems. Our work on combining situational coherence measures as reported in Sect. 3 with ontological coherence and discourse coherence has already shown an increase in performance on multiple tasks. We are, therefore, strongly encouraged by the results that this approach constitutes a suitable path towards making natural language processing more robust and human-like.

---

[10] While it is certainly feasible to limit bridging or metonymy resolution to a predefined set of ontological relations, such as *has-part* relations, if the ontology was especially crafted for that type of resolution (Poesio, 2002).

5 Contextual Coherence

Extra-linguistic factors relate not only to the situational context, but also to the other context stores, such as the discourse, interlocutionary and ontological context. For an integrated model of common sense-based contextual coherence, we have introduced a way of integrating diverse knowledge sources into belief networks by means of establishing a set of intermediate nodes that form a *decision panel*. In such a panel each weighable *expert node* votes on a common decision, e.g. the posterior probability of a Where interrogative being construed as a descriptive or instructional request, - or of the *museum* sense - as viewed from:
- a *situation expert* observing, e.g., time, date, proximity, accessibility
- a *user expert* observing, e.g., interests, transportation, thrift
- a *discourse expert* observing, e.g., referential status, discourse accessibility
- an *ontological expert* observing, e.g., object types and object roles

These weights and votes of the experts are, then, combined to achieve resulting posterior probabilities for the decision at hand that equal 1 in their sum.[11] In the simple case of a single decision (i.e. instructive versus descriptive requests) we have seen that the model is able to capture the data adequately and behaves accordingly. The full blown model features situational factors as introduced in Sect. 3 as well as ontological factors as input to the contextually enhanced ONTOSCORE system. It's integration into the SMARTKOM can be extended as collected data and monitoring capabilities, e.g. for the current weather conditions, become available. An additional reason for choosing these networks was that even if they become rather complex, they are naturally robust against missing and uncertain data, by relying on the priors in the absence of currently available topical data. This approach, therefore, offers a systematic and robust way of enabling natural language understanding modules to resolve different construals of conversational utterances via context-dependent analysis.

6 Conclusion

In this work we focus primarily on contextual interpretation that makes NLP applications more reliable and conversational. We rely on two primary contextual knowledge stores: world- and situational-knowledge captured, herein, by means of formal ontologies and belief networks. We argue that the addition of extra-linguistic knowledge, i.e. situational and ontological knowledge, can represent and integrate the diverse knowledge sources necessary for context-dependent natural language analysis. As a result we showed decreases in the amount of misinterpretations or intention misrecognitions applied at three stages in the processing pipeline of an implemented dialogue system. The application, thereby, increases the systems' performances on features crucial to user satisfaction evaluations, leading to measurable increases in evaluation criteria such as task ease,

---

[11] This addition offers a systematic way of combining evidences from independent factors in belief networks and shrinks the conditional probability tables.

expected behavior as well as dialogue metrics, due to a decrease in the number of turns necessary to achieve task completion.

We introduce contextual coherence measurements, i.e. the output of situational- and ontological coherence measurements. We employed these to find best-speech recognition hypotheses in n-best lists, rank ambiguous, polysemous and metonymical readings and resolve pragmatic ambiguities via inferences from knowledge- and belief-models - based on *common sense* knowledge. The general model introduced shows how such scores reflect a set of additional *common sense* constraints that can be applied as semantic- and pragmatic constraints next to phonological or morpho-syntactic constraints. We can, for example, consider the case of *where questions* as cases where all syntactic and semantic constraints are perfectly satisfied by a proposed filler, while pragmatic constraints concerning the accessibility of the goal object can be violated depending on the situational context.

Since the approach described herein results in ranked lists of possible construals for a given utterance, we can define a threshold for cases where the resulting scores can be considered too close. If, for example, the difference of the posterior probabilities of the `instruct` - `localize` decision is between 0.1 and -0.1, the system can respond by asking the user: *Do you want to go there or know where it is located?*, which incidentally is also a response we found in our initial experiments. This, in turn, would result in more mixed initiative of conversational dialogue systems next to increasing their understanding capabilities and robustness.

Acknowledgments. This work has been partially funded by the German Federal Ministry of Research and Technology (BMBF) as part of the SmartKom project under Grant 01 IL 905C/0 and by the Klaus Tschira Foundation.

References

Allen, James and Georga Ferguson and Amanda Stent: An architecture for more realistic conversational system. In Proc. of Intelligent User Interfaces. (2001) 1–8

Allen, James: Natural Language Understanding. Ben. Cummings (1987)

Barr-Hillel, Y.: Logical Syntax and Semantics. Language, 20 (1954) 230–237

Bergen, Ben: Of sound, mind, and body: neural explanations for non-categorical phonology. PhD Thesis UC Berkeley (2001)

Bunt, Harry: Dialogue pragmatics and context specification. Computational Pragmatics, Abduction, Belief and Context; Studies in Computational Pragmatics. John Benjamins (2000) 81–150

Connolly, John: Context in the Study of Human Languages and Computer Programming Languages: A Comparison. Modeling and Using Context, Springer, LNCS (2001) 116–128

Cozman, Fabio: Generalizing Variable Elimination in Bayesian Networks. In Proc. of the IBERAMIA Workshop on Probabilistic Reasoning in Artificial Intelligence (2000)

Ebert, Christian and Shalom Lappin and Howard Gregory and Nicolas Nicolov: Generating Full Paraphrases out of Fragments in a Dialogue Interpretation System. In Proc. 2nd SIGdial Workshop (2001) 58–67

Fodor, Jerry: The Modularity of Mind. MIT Press (1983)

Gurevych Iryna, Rainer Malaka, Robert Porzel and Hans-Peter Zorn: Semantic Coherence Scoring Using an Ontology. In Proc. of the HLT-NAACL Conference (2003)

Gurevych Iryna and Michael Strube and Robert Porzel: Automatic Classification of Speech Recognition Hypothesis. In Proc. of the 3rd SIGdial Workshop (2002) 90–95

Johnston, Michael and Bangalore, Srinivas and Vasireddy, Gunaranjan and Stent, Amanda and Ehlen, Patrick and Walker, Marilyn and Whittaker, Steve and Maloor, Preetam: MATCH: An Architecture for Multimodal Dialogue Systems. Proceedings of ACL'02. (2002) 376–383

Langacker, Ronald: Conceptualization, symbolization and grammar. Cognitive and functional approaches to language structure. Laurence Erlbaum (1998)

Litman, D. and Walker, M. and Kearns, M.: Automatic Detection of Poor speech recognition at the dialogue Level In Proc. of ACL'99 (1999)

LuperFoy, Susann: The representation of multimodal user interface dialogues using discourse pegs. In Proceedings of ACL'92 (1992) 22-31

Narayanan, Srini and Daniel Jurafsky: Bayesian Models of Human Sentence Processing. In Proc. 20th Cognitive Science Society Conference (1998) 84–90

Narayanan, Srinivas: KARMA: Knowledge-Based Active Representations for Metaphor and Aspect. PhD Thesis. UC Berkeley, (1997)

Oerder, Martin and Hermann Ney: Word Graphs: An Efficient Interface Between Continuous-Speech Recognition and Language Understanding. In Proc. of ICASSP Volume 2 (1993) 119–122

Paris, Cécile L.: User Modeling in Text Generation. Pinter (1993)

Poesio, Massimo: Scaling up Anahpora Resolution. In Proc. of the 1st Workshop on Scalable Natural Language Understanding (2002) 3–11

Porzel, Robert and Iryna Gurevych: Towards Context-adaptive Utterance Interpretation. In Proc. of the 3rd SIGdial Workshop (2002) 90–95

Porzel, Robert and John Bryant: Employing the Embodied Construction Grammar Formalism for Knowledge Represenation: The case of construal resolution. In Proc. of 8th International Conference on Cognitive Linguistic (2003)

Rapp, Stefan and Torge, Sunna and Goronzy, Silke and Kompe, Ralf: Dynamic speech interfaces In Proc. of 14th ECAI WS-AIMS (2000)

Russell, Stuart J. and Norvig, Peter: Artificial Intelligence. A Modern Approach. Prentice Hall (1995)

Robert Porzel and Michael Strube: Towards Context Adaptive Natural Language Processing Systems. In Proc. of the International Symposium: Computational Linguistics for the New Millenium (2000)

R. Schwartz and Y. Chow: The N-best Algorithm: an Efficient and Exact Procedure for Finding the N Most Likely Sentence Hypotheses. In Proc. of ICASSP'90 (1990)

Wolfgang Wahlster and Norbert Reithinger and Anselm Blocher: Smartkom: Multimodal Communication with a Life-Like Character. Proc. of the 7th Eurospeech (2001) 1547–1550

Marilyn A. Walker and Candace A. Kamm and Diane J. Litman: Towards Developing General Model of Usability with PARADISE. Natural Language Engeneering 6 (2000)

Patrick Henry Winston: Artificial Intelligence. Addison-Wesley (1992)

Local Relational Model: A Logical Formalization of Database Coordination

Luciano Serafini[1], Fausto Giunchiglia[1,2], John Mylopoulos[2,3], and Philip Bernstein[4]

[1] ITC-IRST, 38050 Povo, Trento, Italy
[2] University of Trento, 38050 Povo, Trento, Italy
[3] University of Toronto M5S 3H5, Toronto, Ontario, Canada
[4] Microsoft Corporation, One Microsoft Way, Redmond, WA

Abstract. We propose a new data model intended for peer-to-peer (P2P) databases. The model assumes that each peer has a (relational) database and exchanges data with other peers (its *acquaintances*). In this context, one needs a data model that views the space of available data within the P2P network as an open collection of possibly overlapping and inconsistent databases. Accordingly, the paper proposes the Local Relational Model, develops a semantics for coordination formulas. The main result of the paper generalizes Reiter's characterization of a relational database in terms of a first order theory [1], by providing a syntactic characterization of a relational space in terms of a multi-context system. This work extends earlier work by Giunchiglia and Ghidini on Local Model Semantics [2].

1 Introduction

Peer-to-peer (hereafter P2P) computing consists of an open-ended network of distributed computational *peers*, where each peer can exchange data and services with a set of other peers, called *acquaintances*. Peers are fully autonomous in choosing their acquaintances. Moreover, we assume that there is no global control in the form of a global registry, global services, or global resource management, nor a global schema or data repository of the data contained in the network. Systems such as Napster and Gnutella popularized the P2P paradigm as a version of distributed computing lying between traditional distributed systems and the web. The former is rich in services but requires considerable overhead to launch and has a relatively static, controlled architecture. The latter is a dynamic, anyone-to-anyone architecture with little startup costs but limited services. By contrast, P2P offers an evolving architecture where peers come and go, choose whom they deal with, and enjoy some traditional distributed services with less startup cost.

We are interested in data management issues raised by this paradigm. In particular, we assume that each peer has data to share with other nodes. To keep things simple, we further assume that these data are stored in a local relational database for each peer. Since the data residing in different databases may have semantic inter-dependencies, we require that peers can specify coordination rules which ensure that the contents of their respective databases remain "coordinated" as the databases evolve. For example, the patient database of a family doctor and that of a pharmacist may want to coordinate their information about a particular patient, the prescription she has been administered, the dates when these prescriptions were fulfilled and the like. Coordination may mean something as simple as propagating all updates to the PRESCRIPTION and MEDICATION relations, assumed to exist in both databases. In addition, we'd like to support query processing so that a query expressed with respect to one database fetches information from other relevant databases as well. To accomplish this, we expect the P2P data management system to use coordination rules as a basis for recursively decomposing the query into sub-queries which are translated and evaluated with respect to the databases of acquaintances.

Consider the patient databases example, again. There are several databases that store information about a particular patient (family doctor, pharmacist, hospitals, specialists.) These databases need to remain acquainted and coordinate their contents for every shared patient. Since patients come and go, coordination rules need to be dynamic and are introduced by mutual consent of the peers involved. Acquaintances are dynamic too. If a patient suffers an accident during a trip, new acquaintances will have to be introduced and will remain valid until the patient's emergency treatment is over.

In such a setting, we cannot assume the existence of a global schema for all the databases in a P2P network, or just those of acquainted databases. Firstly, it is not clear what a global schema means for the whole network, given that the network is open-ended and continuously evolves. Secondly, even if the scope of a global schema made sense, it would not be practical to build one (just think of the effort and time required.) Finally, building a global schema for every peer and her acquaintances isn't practical either, as acquaintances keep changing. This means that current approaches to information integration [3,4], are not applicable because they assume a global schema (and a global semantics) for the total data space represented by the set of peer databases.

Instead, the *Local Relational Model* (hereafter LRM) proposed here only assumes the existence of pairwise-defined *domain relations*, which relate synonymous data items, as well as *coordination formulas*, which define semantic dependencies among acquainted databases. Local relational model is an evolution of a first attempt in this direction presented in [5] which had the main limitation in the languages adopted to express peer's coordination. Among other things, LRM allows for inconsistent databases and supports semantic interoperability in a manner to be spelled out precisely herein. The main objective of this paper is to introduce the LRM, focusing on its formal semantics.

The LMS semantics presented in this paper are an extension of the *Local Model Semantics*, a new semantics motivated by the problem of formalizing contextual reasoning in AI [6], which was first introduced in [2].

2 A Motivating Scenario

Consider, again, the example of patient databases. Suppose that the Toronto General Hospital owns the Tgh database with schema:

```
Patient(TGH#, OHIP#, Name, Sex, Age, FamilyDr, PatRecord)
PatientInfo(OHIP#, Record)
Admission(AdmID, OHIP#, AdmDate, ProblemDesc, PhysID, DisDate)
Treatment(TreatID, TGH#, Date, TreatDesc, PhysID)
Medication(TGH#, Drug#, Dose, StartD, EndD)
```

The database identifies patients by their hospital ID and keeps track of admissions, patient information obtained from external sources, and all treatments and medications administered by the hospital staff. When a new patient is admitted, the hospital may want to establish immediately an acquaintance with her family doctor. Suppose the view exported by the family doctor DB (say, Davis) has schema:

> Patient(OHIP#, FName, LName, Phone#, Sex, PatRecord)
> Visit(OHIP#, Date, Purpose, Outcome)
> Prescription(OHIP#, Med#, Dose, Quantity, Date)
> Event(OHIP#, Date, Description)

Figuring out patient record correspondences (i.e., doing object identification) is achieved by using the patient's Ontario Health Insurance # (e.g., OHIP# = 1234). Initially, this acquaintance has exactly one coordination formula which states that if there is no patient record at the hospital for this patient, then the patient's record from Davis is added to Tgh in the PatientInfo relation, which can be expressed as:

$$\forall(\text{Davis}: fn, ln, pn, sex, pr).(\text{Davis}: \text{Patient}(1234, fn, ln, pn, sex, pr) \rightarrow \\ \text{Tgh}: \exists(tghid, n, a).(\text{Patient}(tghid, 1234, n, sex, a, \text{Davis}, pr) \land \\ n = concat(fn, ln))) \quad (1)$$

In the above formula the syntax "$\forall(\text{Davis}: fn, ln, pn, sex, pr)\ldots$" is a quantification of the variables fn, ln, pn, sex, pr in the domain of Davis; analogously the syntax Davis : Patient$(1234, fn, ln, pn, sex, pr)$ states the fact that the tuple $\langle 1234, fn, ln, pn, sex, pr \rangle$ belongs to the relation Patient of the database Davis.

When Tgh imports data from Davis, the existentially quantified variables $tghid$, n and a must be instantiated with some concrete elements of the domain of Tgh database, by generating a new TGH# for $tghid$, by inserting the Skolem constant <undef-age> for a and by instantiated n with the concatenation of fn (first name) and ln (last name) contained in Davis. Later, if patient 1234 is treated at the hospital for some time, another coordination formula might be set up that updates the Event relation for every treatment or medication she receives:

$$\forall(\text{Tgh}: d, desc).((\text{Tgh}: \exists(tid.tghid.pid.n.sex.a.pr). \\ (\text{Treatment}(tid, tghid, d, desc, pid) \land \\ \text{Patient}(tghid, 1234, n, sex, a, \text{Davis}, pr)) \rightarrow \\ \text{Davis}: \text{Event}(1234, d, desc))) \quad (2)$$

$$\forall(\text{Tgh}: tghid, drug, dose, sd, ed).(\\ \text{Tgh}: \text{Medication}(tghid, drug, dose, sd, ed) \land \\ \exists n, sex, a, p.\text{Patient}(tghid, 1234, n, sex, a, \text{Davis}, pr) \rightarrow \\ \text{Davis}: \forall d.(sd \leq d \leq ed \rightarrow \exists desc.(\text{Event}(1234, d, desc) \land \\ desc = concat(drug, dose, \text{"atTGHDB"})))) \quad (3)$$

This acquaintance is dropped once the patient's hospital treatment is over. Along similar lines, the patient's pharmacy may want to coordinate with Davis. This acquaintance is initiated by Davis when the patient tells Dr. Davis which pharmacy she uses. Once established, the patient's name and phone are used for identification. The pharmacy database (say, Allen) has the schema:

> Prescription(Prescr#, CustName, CustPhone#, DrugID, Dose, Repeats)
> Sales(CustName, CustPhone#, DrugID, Dose, Date, Amount)

Here, we want Allen to remain updated with respect to prescriptions in Davis:

$\forall$(Davis : $fn, ln, pn, med, dose, qt$)(
  Davis : $\exists ohip, date, sex, pr.$(Prescription($ohip, med, dose, qt, date$) $\wedge$
  Patient($ohip, fn, ln, pn, sex, pr$)) $\rightarrow$ (4)
  Allen : $\exists cn, amount.$(Prescription($cn, pn, med, qt, dose, amount$) $\wedge$
  $cn = concat(fn, ln)$))

Of course, this acquaintance is dropped when the patient tells her doctor that she changed pharmacy. Suppose the hospital has no information on its new patient with OHIP# 1234 and needs to find out if she is receiving any medication. Here, the hospital uses its acquaintance with Toronto pharmacies association, say TPhLtd. TPhLtd, is a peer that has acquaintances with most Toronto pharmacists and has a coordination formula that allows it to access prescription information in those pharmacists' databases. For example, if we assume that Tphh consists of a single relation

Prescription(Name, Phone#, DrugID, Dose, Repeats)

then the coordination formula between the two databases might be:

$\forall$(Davis : $fn, ln, pn, med, dose$).(
  Davis : $\exists ohip, qt, date, sex, pr.$(Prescription($ohip, med, dose, qt, date$)$\wedge$
  Patient($ohip, fn, ln, pn, sex, pr$)) $\rightarrow$ (5)
  Tphh : $\exists name, rep.$(Prescription($name, pn, med, dose, rep$) $\wedge$
  $name = concat(fn, ln)$))

Analogous formulas exist for every other pharmacy acquaintance of TPhLtd. Apart from serving as information brokers, interest groups also support mechanisms for generating coordination formulas from parameterized ones, given exported schema information for each pharmacy database. On the basis of this formula, a query such as "All prescriptions for patient with name N and phone# P" evaluated with respect to Tphh, will be translated into queries that are evaluated with respect to databases such as Allen. The acquaintance between the hospital and TPhLtd is more persistent than those mentioned earlier. However, this one too may evolve over time, depending on what pharmacy information becomes available to TPhLtd. Finally, suppose the patient in question takes a trip to Trento and suffers a skiing accident. Now the Trento Hospital database (TNgh) needs information about the patient from DavisDB. This is a transient acquaintance that only involves making the patient's record available to TNgh, and updating the Event relation in Davis.

3 Relational Spaces

Traditionally, federated and multi-database systems have been treated as extensions of conventional databases. Unfortunately, formalizations of the relational model (such as [1]) hardly apply to these extensions where there are multiple overlapping and heterogeneous databases, which may be inconsistent and may use different vocabularies and different domains. We launch the search for implementation solutions that address the scenario described in the previous section with a formalization of LRM.

The model-theoretic semantics for LRM is defined in terms of relational spaces each of which models the state of the databases in a P2P system. These are mathematical structures generalizing the model-theoretic semantics for the Relational Model, as defined by Reiter in [1]. Coordination between databases in a relational space is expressed in terms of coordination formulas that describe dependencies between a set of databases. Let us start by recalling Reiter's key concepts.

Definition 1 (Relational Language). *A relational language is first order language L with equality, a finite set of constants, denoted dom, no function symbols and finite set* **R** *of predicate symbols.*

The set *dom* of constants is called the domain and represents the total set of data contained in a database, while the predicates in **R** represent its relations. For instance, the language of Davis contains the constant symbol 1234, the relational symbols such as Patient, the unary predicates OHIP#, FName, LName, Phone#, Sex, and PatRecord; α(Patient) = $\langle$OHIP#, FName, LName, Phone#, Sex, *and*PatRecord$\rangle$.

We use the notation x for a sequence of variables $\langle x_1, \dots, x_n \rangle$ and d for a sequence of elements $\langle d_1, \dots, d_n \rangle$, each of which belongs to the domain *dom*; $\phi(x)$ is a formula with the free variable x, and $\phi(\mathbf{x})$ is a formula with free variables in x.

Definition 2 (Relational Database). *A relational database is a first order interpretation m of a relational language L on the set of constants dom, such that $m(d) = d$, for all constant d of L.*

Definition 2 does not properly represent partial databases, i.e., database that contain null values or partial tuples. Indeed, if m is a relational database, $m \models \phi$ or $m \models \neg\phi$ (where "$\models$" stands for "first order satisfiability"). In an incomplete database we would like to have for instance that neither ϕ not $\neg\phi$ are trie. A common approach is to model incomplete databases as a set of first order structures, also called a state of information. We follow this approach, and formalize an incomplete database on a relational language L as a set of relational databases on L. Notice that the set of relational databases corresponding to an incomplete database all share the same domain, consisting of the set of constants contained in the database. The partiality, therefore, concerns only the interpretation of the relational symbols. With this generalization we can capture inconsistent, complete, and incomplete databases. For instance, if db_a, db_b and db_c are three (partial) relational databases defined as

$$db_a = \{m_1\}, db_b = \{m_2, m_3\}, db_c = \emptyset$$

where $m_1, m_2,$ and m_3 are relational databases, we have that they are respectively, complete, incomplete, and inconsistent. Generally, db_i is complete if $|db_i| = 1$, incomplete if $|db_i| > 1$ and inconsistent if $db_i = \emptyset$.

Since we are interested in modelling P2P applications, we take a further step and consider, rather than a single database, a family (indexed with a set of peers I) of database. We call such of these databases a *local database* when we want to stress that it is a member of a set of (coordinated) databases.

When we consider a set of databases the same information could be represented twice in two databases. In this case we say that they *overlap*. Overlapping databases have nothing to do with the fact that the same symbols appear in both databases—the same constant can have completely different meanings in two databases—overlap occurs when the real world entities denoted by a symbol in different databases are somehow related. To represent the overlap of two local databases, one may use a global schema, with suitable mappings to/from each local database schema. As argued earlier, this is not feasible in a P2P setting. Instead, we adopt a localized solution to the overlap problem, defined in terms of pair-wise mappings from the elements of the domain of database i to elements of the domain of database j.

Definition 3 (Domain relation). *Let L_i and L_j be two relational languages, with domains dom_i and dom_j respectively; a* domain relation r_{ij} *from i to j is any subset of $dom_i \times dom_j$.*

The domain relation r_{ij} represents the ability of database j to import (and represent in its domain) the elements of the domain of database i. In symbols, $r_{ij}(d_i) = \{d_j | \langle d_i, d_j \rangle \in r_{ij}\}$ represents the set of elements in which j translates the constant d of i's domain. In many cases, domain relations are not, one to one, for instance if two databases represent a domain at a different level of details. Domain relation are not symmetric, for instance when r_{ij} represents a currency exchange, a rounding function, or a sampling function. In a P2P setting, domain relations need only be defined for acquainted pairs of peers. Domain relations between databases are conceptually analogous to *conversion functions* between semantic objects, as defined in [7]. The domain relation defined above formalizes the case where a single attribute of one database is mapped into single attribute of another database. It is often the case, however, that two (or more) attributes of a database correspond to a single attribute in another one. An obvious is when the attributes first-name and last-name in a database i are merged in the unique attribute name of a database j. Domain relation can be generalized to deal with these cases by allowing, for instance, a domain relation $r_{i:(\texttt{first-name},\texttt{last-name}),j:\texttt{name}}$ to be a subset of $dom_i^2 \times dom_j$.

Example 1. Let us consider how domain relations can represent different data integration scenaria. The situation where two databases have *different but equivalent representations of the same domain* can be represented by taking r_{ij} and r_{ji} as the translation function from dom_i to dom_j and vice-versa, namely $r_{ij} = r_{ji}^{-1}$. Likewise, disjoint domains can be represented by having $r_{ij} = r_{ji} = \emptyset$. Transitive mappings between the domains of three databases are represented by imposing $r_{13} = r_{12} \circ r_{23}$. Suppose instead that dom_i and dom_j are ordered according to two orders $<_i$ and $<_j$. A relation that satisfies the property: $\forall d_1, d_2 \in dom_i, d_1 <_i d_2 \Rightarrow \forall d_1' \in r_{ij}(d_1), \forall d_2' \in r_{ij}(d_2). d_1' <_j d_2'$ formalizes a mapping which preserves the orders, such as currency exchange. Finally, suppose that a peer with database i doesn't want to export any information about a certain object d_s in its database. To accomplish this, it is sufficient to ensure that the domain relations from i to any other database j, do not associate any element to d_s, namely $r_{ij}(d_s) = \emptyset$.

Definition 4 (Relational space). *A* relational space *is a pair $\langle db, r \rangle$, where db is a set of local relational databases on I and r is a function that associates to each $i, j \in I$, a domain relation r_{ij}.*

Example 2. A relational space modeling the states of the database described in Section 2, is a pair

$$\left\langle \begin{array}{l} db = \langle db_{\mathsf{Tgh}}, db_{\mathsf{Davis}}, db_{\mathsf{Allen}}, db_{\mathsf{Tphh}}, db_{\mathsf{TNgh}} \rangle \\ r = \langle r_{\mathsf{DavisTgh}}, r_{\mathsf{TghDavis}}, r_{\mathsf{DavisAllen}}, r_{\mathsf{DavisTphh}} \rangle \end{array} \right\rangle$$

where the first component, the local databases, contains five sets of interpretations of the relational languages associated to Tgh, Davis, Allen and Tphh and TNgh, respectively; and the second component, the domain relation, contains four domain relations between those databases which have to coordinate according to constraints (1–5).

The fact that $t = \langle 1234, \mathtt{Pippo}, \mathtt{Inzaghi}, 444, \mathtt{M}, \mathtt{Rec\_23} \rangle$ is a tuple of the relation PatRecord of the Davis database, if formalized by requiring $t \in m(\mathtt{PatRecord})$ for each interpretation $m \in db_{\mathsf{Davis}}$.

The fact that $t = \langle \mathtt{TG64}, 1234, "\mathtt{PippoInzaghi}", \mathtt{M}, \mathtt{<undef\text{-}age>}, \mathtt{Davis}, \mathtt{Rec\_23} \rangle$ is a tuple of the relation Patient of Tgh database, is represented by requiring that, for each natural number n, with $0 \leq n \leq \mathtt{MaxAge}$, db_{Tgh} contains a model a model m, with $t[\mathtt{<undef\text{-}age>}/n] \in m(\mathtt{Patient})$ ($t[\mathtt{<undef\text{-}age>}/n]$ is the result of substituting n for <undef-age> in t).

The fact that the TGH# 1234 uniquely identifies a patient in both Tgh and Davis, is represented by requiring $r_{\mathsf{DavisTgh}}(1234) = r_{\mathsf{TghDavis}}(1234) = \{1234\}$.

4 Coordination in Relational Spaces

Two (or more) peers who want to coordinates each other, need a language in which they can express the inter-dependencies between the information stored in their database. To this purpose, we define a declarative language by which it is possible to express semantic relations between local databases. The formulas of this language, called *coordination formulas* can be used to describe cross-database views and cross-databases constraints.

Definition 5 (Coordination formula). *The set of* coordination formulas CF *on the family of relational languages* $\{L_i\}_{i \in I}$ *is defined as follows for each* $i \in I$ *and each formula* ϕ *of* L_i[1].

$$CF ::= i : \phi \mid CF \rightarrow CF \mid CF \wedge CF \mid CF \vee CF \mid \exists i : x.CF \mid \forall i : x.CF$$

We use Greek letters ϕ, ψ, to denote formulas of any languages L_i $i \in I$, and Latin capital letters A, B, and C to denote coordination formulas. The basic building blocks of coordination formulas are expressions of the form $i : \phi$ and are called *atomic coordination formulas*. An occurrence of a variable x in a coordination formula is a *free occurrence*, if it is not in the scope of a quantifier. Examples of coordination formulas are shown in Section 2.

To give an interpretation of coordination formulas in relational spaces, let us start by considering Definition 5 in detail. Item 1 states that coordination formulas are defined on the basis of atomic formulas of the form $i : \phi$, where ϕ is any formula of L_i. $i : \phi$ intuitively means "ϕ is true in database i" and its interpretation follows the standard

---

[1] The following precedence rules apply: $i : \ldots$ has the highest precedence, followed by quantifiers, then $\wedge$, then $\vee$, and finally $\rightarrow$. For instance, $\forall i : x.i : \phi \wedge j : \psi \rightarrow k : \theta \vee h : \eta$, stands for: $((\forall (i : x).(i : \phi)) \wedge (j : \psi)) \rightarrow ((k : \theta) \vee (h : \eta))$.

rules of first order logic. Thus, in particular, if ϕ is of the form $\forall x.\psi(x)$ or of the form $\exists x.\psi(x)$ then its interpretation is given in terms of the possible assignments of x to elements of dom_i.

The crucial observation for the evaluation of quantified formulas is that a free occurrence of a variable can be quantified in four different ways: by $\forall x$, $\exists x$ within an atomic coordination formula (as from Item 1), and by $\forall i : x$ or $\exists i : x$, within a coordination formula. In the two latter cases the index i tells us the domain where we interpret x. Thus, the formula $\forall i : x.A(x)$ (where $A(x)$ is a coordination formula and not a formula!) must be read as "for all elements d of the domain dom_i, A is true for d". Likewise, $\exists i : x.A(x)$, must be read as "there is an element in the domain dom_i such that A is true". The trick is that A, being a coordination formula, may contain atomic coordination formulas of the form $j : \phi(x)$, with $j \neq i$. For instance in the coordination formula (5) the variables fn and ln occur free a coordination formula with index Tgh (the consequence of the implication), while they are bound by the quantifiers $\forall(\text{Davis} : fn, ln, \ldots)$.

The intuition underlying the interpretation of quantified indexed variables is that, if x is a variable being quantified with index i and occurring free in a coordination formula with index j, then we must find a way to relate the interpretation of x in dom_i to the interpretation of x in dom_j using the mapping defined by r_{ij}. More precisely, the coordination formula $\forall i : x.j : P(x)$, means, "for each object of dom_i, the *corresponding object* w.r.t. the domain relation r_{ij} in dom_j has the property P". Thus, for instance, in order to check whether the coordination formula

$$\forall i : x.(i : P(x) \to j : Q(x) \wedge k : R(x)) \qquad (6)$$

is true in a relational space, one has to consider all the assignments that associate to the occurrence of x in $i : P(x)$ any element of $d \in dom_i$, and to the occurrences of x in $j : Q(x)$ and $k : R(x)$ any element of $r_{ij}(d)$ and $r_{ik}(d)$, respectively. Dually, the coordination formula $\exists i : x.j : P(x)$, means "there is an element in dom_j that corresponds w.r.t. the domain relation r_{ji} to an element of dom_i with property P". Thus, for instance, in order to check whether the coordination formula

$$\exists i : x.(i : P(x) \wedge j : Q(x) \wedge k : R(x)) \qquad (7)$$

is true in a relational space, one has to find an assignment that associates to the occurrence of x in $i : P(x)$ an element d of dom_i, and to the occurrences of x in $j : Q(x)$ and $k : R(x)$ two elements $d' \in dom_j$ and $d'' \in dom_k$, respectively, such that $d \in r_{ji}(d')$ and $d \in r_{ki}(d'')$.

Notice that in our explanation of the universal quantification we used r_{ij}, while for existential quantification we used r_{ji}. This asymmetry is necessary to maintain the dual intuitive readings of existential and universal quantifiers. Indeed, the intuitive meaning of the formula $\forall i : x.j : P(x)$ is "*for all $d \in dom_i$, if $d' \in r_{ij}(d)$ then d' is in P*", which can be rephrased in its dual existential statement "*there does not exist any element $d' \in r_{ij}(d)$, which is not in P*". Notice that in this last sentence, the quantification is on the elements of dom_j, namely *on the elements in the codomain* of the domain relation r_{ij}, just like in the explanation of Equation (7) above.

To formalize the intuitions given above concerning the interpretation of coordination formulas, we need two notions. The first is *coordination space of a variable x in a coordination formula*. Intuitively this is the set of indexes of the atomic coordination formulas that contain a free occurrence of x. The coordination space is the set of domains where x must be interpreted. Thus, for instance, the coordination space of x in the $i : P(x) \wedge j : Q(x) \wedge k : R(x)$ is $\{i, j, k\}$.

Definition 6 (Coordination space). *The* coordination space *of a variable x in a coordination formula A is a set of indexes $J \subseteq I$, defined as follows:*

1. *the coordination space of x in $i : \phi$ is $\{i\}$, if x occurs free in ϕ according to the usual definition of free occurrence in a first order formula, and the empty set, otherwise;*
2. *the coordination space of x in $A \circ B$ (for any connective $\circ$) is the union of the coordination spaces of x in A and B;*
3. *the coordination space of x in $Qi : y.A$ (for any quantifier Q) is the empty set, if x is equal to y, and the coordination space of x in A, otherwise.*

The second notion is that of *assignment* for a free occurrence of a variable in a coordination formula. To evaluate a formula A quantified over x with index i, an assignment must consider dom_i but also all the domains in the coordination space. To understand how assignments work, look at Equations (6), (7). In Equation (6) we proceed "forward" from dom_i to reach dom_j and dom_k, by applying r_{ij} and r_{ik}. In this case we say that we have an i-to-$\{j, k\}$-assignment. Instead, in Equation (7), we proceed "backward" from dom_j and dom_k to reach dom_i by applying r_{ji} and r_{ki}. In this case we say that we have an i-from-$\{j, k\}$-assignment. If J is a coordination space, i-to-J-assignments take care of the assignments due to universal quantification, while i-from-J-assignments take care of those due to existential quantification.

Definition 7 (Assignment, x-variation i-to-J-assignment, i-from-J-assignment). *An assignment $a = \{a_i\}_{i \in J}$ is a family of functions a_i, where a_i assigns to any variable x an element of dom_i. An assignment a' is an x-variation of an assignment a, if a and a' differ only on the assignments to the variable x. Given a set $J \subseteq I$ and an index $i \in I$, an assignment a is an i-to-J-assignment of x if, for all $j \in J$ distinct from i, $\langle a_i(x), a_j(x) \rangle \in r_{ij}$. An assignment a is an i-from-J-assignment of x if, for all $j \in J$ distinct from i, $\langle a_j(x), a_i(x) \rangle \in r_{ji}$.*

Definition 8 (Satisfiability of coordination formulas). *The relational space $\langle db, r \rangle$ satisfies a coordination formula A under the assignment $a = \{a_i\}_{i \in J}$, in symbols $\langle db, r \rangle \models A[a]$, according to the following rules:*

1. $\langle db, r \rangle \models i : \phi[a]$, *if for each $m \in db_i$, $m \models \phi[a_i]$;*
2. $\langle db, r \rangle \models A \rightarrow B[a]$, *if $\langle db, r \rangle \models A[a]$ implies that $\langle db, r \rangle \models B[a]$;*
3. $\langle db, r \rangle \models A \wedge B[a]$, *if $\langle db, r \rangle \models A[a]$ and $\langle db, r \rangle \models B[a]$;*
4. $\langle db, r \rangle \models A \vee B[a]$, *if $\langle db, r \rangle \models A[a]$ or $\langle db, r \rangle \models B[a]$;*
5. $\langle db, r \rangle \models \forall i : x.A[a]$, *if $\langle db, r \rangle \models A[a']$ for all assignments a' that are x-variations of a and that are i-to-J-assignments on x, where J is the coordination space of x in A.*
6. $\langle db, r \rangle \models \exists i : x.A[a]$, *if $\langle db, r \rangle \models A[a']$ for some assignment a' that is an x-variation of a and that is an i-from-J-assignment on x, where J is the coordination space of x in A.*

A *coordination formula A is* valid *if it is true in all the relational spaces. A coordination formula A is a* logical consequence *of a set of coordination formulas Γ if, for any relational space $\langle db, r \rangle$ and for any assignment a, if $\langle db, r \rangle \models \Gamma[a]$ then $\langle db, r \rangle \models A[a]$.*

Item 1 states that an atomic coordination formula is satisfied (under the assignment a) if all the relational databases $m \in db_i$ satisfy it. Items 2–4 enforce the standard interpretation of the boolean connectives. Item 5 states that a universally quantified coordination formula is satisfied if all its instances, obtained by substituting the free occurrence of x in the atomic coordination formulas with index i with all the elements of dom_i, and the free occurrences of x in the atomic coordination formulas with index j different from i, with all the elements of dom_j, obtained by applying r_{ij} to the elements of dom_i, are satisfied. Item 6 has the dual interpretation.

Finally, notice that the language of coordination formulas does not include negation. The addition of negation with the canonical interpretation "$\neg A$ is true iff A is *not true*", implies the possibility to define the notion of "Global inconsistency", i.e., there are sets of inconsistent coordination formulas (e.g., $\{i : \phi, \neg i : \phi\}$). These sets are not satisfiable by any relational space. On the other hand, we have that the relational space composed of all inconsistent databases, is the "most inconsistent object that we can have (not allowing global inconsistency), we therefore should allow that this vacuous distributed interpretation satisfies any setxte of coordination formulas. Indeed we have that, in absence of negation, if $db_i^0 = \emptyset$ and $r_{ij}^0 = \emptyset$, $\langle db^0, r^0 \rangle \models A$ for any coordination formula A.

Coordination formulas can be used in two different ways. First, they can be used to define constraints that must be satisfied by a relational space. For instance, the formula $\forall 1 : x.(1 : p(x) \vee 2 : q(x))$ states that any object in database 1 either is in table p or its corresponding object in database 2 is in table q. This is a useful constraint when we want to declare that certain data are available in a set of databases, without declaring exactly where. As far as we know, other proposals in the literature for expressing inter-database constraints can be uniformly represented in terms of coordination formulas.

Coordination formulas can also be used to express queries. In this case, a coordination formula is interpreted as a deductive rule that derives new information based on information already present in other databases. For instance, a coordination formula $\forall i : x.(1 : \exists y.p(x, y) \rightarrow 2 : q(x))$ allows us to derive $q(b)$ in database 2, if $p(a, c)$ holds in database 1 for some c, and $b \in r_{12}(a)$.

Definition 9 (*i*-query). *An i-query on a family of relational languages $\{L_i\}_{i \in I}$, is a coordination formula of the form $A(\mathbf{x}) \rightarrow i : q(\mathbf{x})$, where $A(\mathbf{x})$ is a coordination formula, and q is a new n-ary predicate symbol of L_i and $\mathbf{x}$ contains n variables.*

Definition 10 (**Global answer to an *i*-query**). *Let $\langle db, r \rangle$ be a relational space on $\{L_i\}_{i \in I}$. The global answer of an i-query of the form $A(\mathbf{x}) \rightarrow i : q(\mathbf{x})$ in $\langle db, r \rangle$ is the set:*

$$\{\mathbf{d} \in dom_i^n | \langle db, r \rangle \models \exists i : \mathbf{x}.(A(\mathbf{x}) \wedge i : \mathbf{x} = \mathbf{d})\}$$

Notationally $\mathbf{x} = \mathbf{d}$ stands for $x_1 = d_1 \wedge \ldots \wedge x_n = d_n$, and $\exists i : \mathbf{x}$ stands for $\exists i : x_1 \ldots \exists i : x_n$. Intuitively, the global answer to an i-query is computed by locally evaluating in db_j all the atomic coordination formulas with index j contained in A, and by recursively composing and mapping (via the domain relations) these results according to the connectives and quantifiers that compose the coordination formula A. For instance to evaluate the query

$$(i : P(x) \vee j : Q(x)) \wedge k : R(x,y) \to h : q(x,y)$$

we separately evaluate $P(x)$, $Q(x)$ and $R(x,y)$ in i, j and k respectively, we map these results via r_{ih}, r_{jh}, and r_{kh} respectively obtaining three sets $s_i \subseteq dom_h$ $s_j \subseteq dom_h$ and $s_k \subseteq dom_h^2$. We then compose s_i, s_j and s_k following the connectives obtaining $(s_i \times s_j) \cap s_k$, which is the global answer of q.

Notice that the same query q has different answers depending on the database it is asked to (because of the quantification over $i : \mathbf{x}$). Notice also that Definition 10 reduces to the usual notion of answer to a query when A is an atomic coordination formula $i : \phi$ (case of a single database i). Finally, but most importantly, queries can be recursively composed. Indeed, a *recursive query* can be defined as a set of queries $\{q_h := A_h(\mathbf{x}_h) \to i_h : q_h(\mathbf{x}_h)\}_{1 \leq h \leq n}$ such that $A_h(\mathbf{x}_h)$ can contain of an atomic coordination formula $i_k : q_k(\mathbf{x}_k)$ for some $1 \leq k \leq n$. The evaluation of a query q_h in the i_h-th database is done by evaluating its body, i.e., the coordination formula A_h, which contains the query q_k. This forces the evaluation of the query q_k in the i_k-th database, and so in P2P network. We can prove the following theorem

Theorem 1. *Let $\langle db, r \rangle$ be a relational space and $rq = \{q_h := A_h(\mathbf{x}_h) \to i_h : q_h(\mathbf{x}_h)\}_{1 \leq h \leq n}$ be a recursive query. If $A(\mathbf{x})$ does not contain any $\to$ symbol, then there are n minimal sets $ans_1, \dots, ans_n$, such that each ans_h is the global answer of the query q_h, in the relational space $\langle db', r \rangle$, where db' is obtained by extending every relational database $m \in db_{i_k}$ with $m(q_k) = ans_k$, for each $k \neq h$.*

5 Representation Theorems

In this section we generalize Reiter's semantic characterization of relational databases to relational spaces. We start by recalling Reiter's result (in a slightly different, but equivalent, formulation).

Definition 11 (Generalized relational theory). *A theory T on the relational language L is a generalized relational theory if the following conditions hold.*

- *if $dom = \{d_1, \dots d_n\}$, $\forall x(x = d_1 \vee \dots \vee x = d_n) \in T$;*
- *for any $d, d' \in dom$, $d \neq d' \in T$;*
- *for any relational symbol $R \in \mathbf{R}$, there is a finite number of finite sets of tuples $E_R^1, \dots, E_R^n$ (the possible extensions of R) such that T contains the axiom:*

$$\bigvee\nolimits_{1 \leq k \leq n} \left(\forall \mathbf{x} \left(R(\mathbf{x}) \leftrightarrow \bigvee\nolimits_{\mathbf{d} \in E_R^k} \mathbf{x} = \mathbf{d} \right) \right)$$

Reiter proves that any partial relational database can be uniquely represented by a generalized relational theory. The generalization to the case of multiple partial databases models each of them as a generalized relational theory, and "coordinates" them using an appropriate coordination formula which axiomatizes the domain relation.

Definition 12 (Domain relation extension). *Let r_{ij} be a domain relation. The set of coordination formulas for the extension of r_{ij} is a the set R_{ij} that contains the coordination formula $\exists j : x.(i : x = d \wedge j : x = d')$ if $d' \in r_{ij}(d)$, and the coordination formula $\forall i : x.(i : x = d \to j : x \neq d')$ if $d' \notin r_{ij}(d)$.*

Theorem 2 (Characterization of domain relations). *Let R_{ij} be the set of coordination formulas for the extension of r_{ij}. For any relational space $\langle db, r' \rangle$ with db_i and db_j different from the empty set, $\langle db, r' \rangle \models R_{ij}$ if and only if $r_{ij} = r'_{ij}$.*

Theorem 2 states that, when db_i and db_j are consistent databases, the only domain relation from i to j that satisfies the coordination formulas for the extension of r_{ij} (i.e., R_{ij}) is r_{ij} itself. This means that R_{ij} uniquely characterizes r_{ij}. The characterization of a relational space (Theorem 3) is obtained by composing the characterization of local databases (Reiter's result) and the characterization of the domain relation (Theorem 2). A corollary of the relational space's characterization (Corollary 1) provides a characterization in terms of logical consequence of a global answer to a i-query.

Definition 13 (Relational multi-context system). *A relational multi-context system for a family of relational languages $\{L_i\}$ is a pair $\langle T, R \rangle$, where T is a function that associates to each i, a generalized relational theory T_i on the language L_i, and R is a set that contains all the coordination formulas for the extension of a domain relation from i to j for any $i, j \in I$.*

Theorem 3 (Representation of relational spaces). *For any relational multi-context system $\langle T, R \rangle$ there is a unique (up to isomorphism) relational space $\langle db, r \rangle$, with the following properties:*

1. *$\langle db, r \rangle \models i : T_i$ and $\langle db, r \rangle \models R$.*
2. *For each $i \in I$, db_i is different from the empty set.*
3. *$\langle db, r \rangle$ is maximal, i.e., for any other relational space $\langle db', r' \rangle$, satisfying condition 1 and 2, $db'_i \subseteq db_i$, and $r_{ij} = r'_{ij}$ for all $i, j \in I$.*

Vice-versa, for any relational space $\langle db, r \rangle$, there is a relational multi-context system $\langle T, R \rangle$ such that the maximal model of $\langle T, R \rangle$ is $\langle db, r \rangle$. We say that $\langle T, R \rangle$ is the multi-context system that represents $\langle db, r \rangle$.

Corollary 1 (Semantic characterization of queries). *Let $\langle T, R \rangle$ be the relational multi-context system that represents the relational space $\langle db, r \rangle$. for any i-query $q := A(\mathbf{x}) \to i : q(\mathbf{x})$, the n-tuple $\mathbf{d}$ belongs to the global answer of q, if and only if*

$$\{i : T_i\}_{i \in I}, R \models \exists i : \mathbf{x}(A(\mathbf{x}) \wedge i : \mathbf{x} = \mathbf{d})$$

Corollary 1 provides us with the basis for a correct and complete implementation of a query answering mechanism in a P2P environment.

6 Related Work

The formalism presented in this paper is an extension of the Distributed First Order Logics formalism proposed in [5]. The main improvements concern the language of the coordination formulas, their semantics and the calculus. In [5] indeed, relation between databases were expressed via *domain constraints* and *interpretation constraints*. These latter correspond to particular coordination formulas: namely domain constraints from

i to j corresponds to the coordination formulas $\forall i : x \exists j : yi : x = y$ and $\forall j : x \exists i : yi : x = y$, while interpretation constraints can be translated in the coordination formulas $\forall i : \mathbf{x}.(i : \phi(\mathbf{x}) \to j : \psi(\mathbf{x}))$. This limitation on the expressive power, does not allow to express in DFOL the fact that a table, say p, of a database i is the union of two tables, say p_1 and p_2 of two different databases j and k. This constraint can be easily expressed by the following coordination formula:

$$\forall i : x.(p(x) \leftrightarrow j : p_1(x) \lor k : p_2(x))$$

As far as the query language is concerned, our approach is similar in some ways to view-based data integration techniques, in the following sense. The process of translating a query against a local database into queries against an acquaintance would be driven by the coordination formulas that relate those two databases. If one thinks of our coordination formulas as view definitions, then the translation process is comparable to ones used for rewriting queries based view definitions in the local-as-view (LAV) and global-as-view (GAV) approaches ([8,9]. Although standard approaches cannot be applied directly to LRM, due to our use of domain relations and context-dependent coordination formulas, we expect it is possible to modify LAV/GAV query processing strategies for LRM. For example, one could define a sublanguage of LRM whose power is comparable to a tractable view definition language used for LAV/GAV query processing. One could then apply a modified LAV/GAV algorithm to that language. Or perhaps one could translate formulas and queries from the LRM sublanguage into a non-LRM (e.g., a Datalog dialect) and apply a conventional LAV/GAV query processing algorithm. If such a translation of formulas and queries proves to be feasible, then it would be important to compare the LRM notation to its translation in the non-LRM language, for example to determine their relative clarity and compactness.

Finally our approach provide a general theoretical reference framework where many forms of inter-schema constraints defined in the literature, such as [10,11,12,3,13,14]. For lack of space we briefly show only one case. Consider for instance directional existence dependences defined in [11]. Let $T_1[X_1, Y_1]$ and $T_2[X_2, Y_2]$ be two tables of a source database (let's say 1), and that $T[C_1, C_2, C_3]$ is a table of the target database (let say 2). An example of directional existence dependence is:

$$T.(C_1, C_2) \Leftarrow \text{select } X_1, X_2 \text{ from } T_1, T_2 \text{ where } T_1.X_1 \leq T_2.X_2 \qquad (8)$$

The informal semantics of (8) is that for each tuple of value $\langle V_1, V_2 \rangle$ produced by the RHS select statement, there is a tuple t in table T such that t projected on columns C_1, C_2 has the value $\langle V_1, V_2 \rangle$. The existence dependence (8), can be rewritten in terms of coordination formulas as

$$\forall 1 : x_1 x_2 (1 : \exists y_1 y_2 (T_1(x_1, y_1) \land T_2(x_2, y_2) \land x_1 \leq x_2) \to \\ \exists 2 : c_1 c_2 (1 : x_1 = c_1 \land x_2 = c_2 \land 2 : \exists c_3.T(c_1, c_2, c_3))) \qquad (9)$$

When the domain relation are identity functions, (9) capture the intuitive reading of (8).

7 Conclusion

We have argued that emerging computing paradigms, such as P2P computing, call for new data management mechanisms which do away with the global schema assumption inherent in current data models. Moreover, in a P2P setting the emphasis is on *coordinating* databases, rather than *integrating* them. This coordination is defined by an evolving set of coordination formulas which are used both for constraint enforcement and query processing. To meet these challenges, the paper proposes, the paper proposes the local relational model, LRM, where the data to be managed constitute a relational space, conceived as a collection of local databases inter-related through coordination formulas and domain relations. The main result of the paper is to define a model theory for the LRM. We use this semantics to generalize an earlier result due to Reiter which characterizes a relational space as a multi-context system. The results of this paper offer a sound springboard in launching a study of implementation techniques for the LRM, its query processing and constraint enforcement.

References

1. Reiter, R.: Towards a Logical Reconstruction of Relational Database Theory. In Brodie, M., Mylopoulos, J., Schmidt, J., eds.: On Conceptual Modelling. Springer-Verlag (1984) 191–233
2. Ghidini, C., Giunchiglia, F.: Local models semantics, or contextual reasoning = locality + compatibility. Artificial Intelligence **127** (2001) 221–259
3. Ullman, J.D.: Information Integration Using Logical Views. In: Proc. of the 6th ICDT (1997)
4. Florescu, D., Levy, A., Mendelzon, A.: Database techniques for the World-Wide Web: A survey. SIGMOD Record **27** (1998) 59–74
5. Ghidini, C., Serafini, L.: Distributed First Order Logics. In Gabbay, D., de Rijke, M., eds.: Frontiers Of Combining Systems 2. Studies in Logic and Computation. Research Studies Press (1998) 121–140
6. Giunchiglia, F.: Contextual reasoning. Epistemologia, special issue on I Linguaggi e le Macchine **XVI** (1993) 345–364 Short version in Proceedings IJCAI'93 Workshop on Using Knowledge in its Context, Chambery, France, 1993, pp. 39–49. Also IRST-Technical Report 9211-20, IRST, Trento, Italy.
7. Sciore, E., Siegel, M., Rosenthal, A.: Using semantic values to facilitate interoperability among heterogeneous information systems. ACM TODS **19** (1994) 254–290
8. Halevy, A.Y.: Answering queries using views: A survey. VLDB Journal **10** (2001)
9. Lenzerini, M.: Data integration: A theoretical perspective. In: PODS. (2002) 233–246
10. Carey, M., Haas, L., Schwarz, P., Arya, M., Cody, W., Fagin, R., Flickner, M., Luniewski, A., Niblack, W., Petkovic, D., II, J.T., Williams, J., Wimmers, E.: Towards heterogeneous multimedia information systems: The garlic approach. RIDE-DOM (1995) 124–131
11. Ceri, S., Widom, J.: Managing semantic heterogeneity with production rules and persistent queues. In Agrawal, R., Baker, S., Bell, D., eds.: 19th VLDB Conference 24–27, 1993, Dublin, Ireland, Proceedings, Morgan Kaufmann (1993) 108–119
12. Levy, A., Rajaraman, A., Ordille, J.: Querying Heterogeneous Information Sources Using Source Descriptions. In: Proceedings of the 22nd VLDB Conference, Bombay, India (1996)
13. Gupta, A., Widom, J.: Local verification of global integrity constraints in distributed databases. In: ACM SIGMOD International Conference on Management of Data. (1993) 49–58
14. Grefen, P., Widom, J.: Integrity constraint checking in federated databases. In: Proceedings 1st IFCIS International Conference on Cooperative Information Systems. (1996) 38–47

What to Say on What Is Said

Isidora Stojanovic

Stanford University, Dept. of Philosophy, Stanford, CA 94305-2155, USA
isidora@stanford.edu

Abstract. Discussions in philosophy of language, semantics, and pragmatics, often make crucial use of the notion of *what is said*. It is held that in order to account for our intuitions on what is said, we need a distinguished semantic level. A tripartite distinction is made among what the sentence means independently from the context of utterance, what it means (or "says") within the context, and what the speaker means (or "conveys"). I will challenge the need for that intermediate level of meaning, and argue that the enterprise of drawing a neat distinction between meaning and what is said is pretty hopeless. My main point is that our intuitions on what is said cannot be detached from the ways in which we talk about it, and from the semantics of speech-reports and attitude-reports in general.

1 Stirring Up Our Intuitions on What Is Said

What we say and what others say are things that undeniably play important roles in our lives. People get arrested for what they say, friendships break or come about because of what someone has said, and so on. There is little doubt that we have a certain intuitive notion of *what is said*, and attempts to account for it should be welcomed. However, this supposedly intuitive notion of what is said has also been used to draw the line between semantics and pragmatics, and to ground some substantial claims about semantics. It is widely held that, given some basic facts about the context in which a sentence is uttered, semantics, helped by the syntax of the sentence, allows us to figure out *what is said* by the utterance, which is what provides the utterance with a truth value, given further facts about the world. On the other hand, what the speaker is trying to achieve by means of the utterance most often goes beyond the reach of semantics. We need further facts about the context, concerning the speaker's beliefs and intentions, to figure out *what is conveyed* by the utterance. And the latter, it is held, is the realm of pragmatics.[1]

So far, so good. For, given any utterance u, the distinction is being made between the 'semantic' level of meaning, which may still be thought of as context-invariant and possessed by u in virtue of the sentence alone of which u is an utterance, and the 'pragmatic' level of the action performed by means of u.[2] Suppose that I say "It's

---

[1] I may be giving a caricature of the received wisdom, but the distinction drawn along these lines may be found all over the place. See e.g. [2], [7], or [10].
[2] Of course, what u, qua a mere string of sounds, has as its linguistic meaning, said to be 'context-invariant,' itself depends upon the context, since the context provides the language in which to interpret u. But once the language has been fixed, variations on other contextual features are not going to alter the meaning of u.

warm in here." Your linguistic knowledge, assuming that you are a competent English speaker, allows you to grasp the meaning of my sentence, and, tentatively, what I say. Yet, depending on the circumstances of my utterance, you may well start wondering what I mean. Perhaps I mean to ask you, indirectly, to open the windows. Perhaps I have just opened them, so I let you know why. Perhaps the windows are open, and I want you to close them and turn on the air-conditioning instead. Perhaps the heating is out of work, the temperature has dropped below zero, and I am just being ironic. Which among all those things I happen to be doing with my utterance is to be settled by considerations of the circumstances in which I made it, taking into account what I want, what I think, what I think that you think, and so on.

However, with this example already, we have hit upon a controversy. For, you could question my claim that your knowledge of English allows you to grasp what I said. Don't you need to know as well *where* I was when I said "It's warm in *here*"? If I say this in your living room, or in a van, or in a disco, will I be saying the same thing every time? Or will I be saying different things? This is, roughly, the point-break where our intuitions lose force and you can pull them one way or the other. Thus, I may insist, "yes, I am saying the same thing every time, namely, that it is warm in there." And you may either buy this, or protest, "no, you are saying different things. First, you say that it's warm in my living room. Then you say that it's warm in the van. Last, you say that it's warm in the disco." But now, how are we supposed to settle the question of what it is that I said?

Suppose, for the sake of the argument, that I agree with you on this much: when I say "It's warm in here" in your living room, I am saying that it is warm in your living room. Now, suppose that I say it at 6 p.m. I am warm, I would like you to open the windows, but you have not paid attention to what I said. So I repeat it ten minutes later, that is, at 6.10 p.m. The question becomes: do I now say the same thing as I did the first time, or do I say something different? Once again, I may insist that, yes, I said the same thing, namely that it was warm in there (in your living room), while you may disagree and say that what I first said is that it was warm in your living room as of 6 p.m., and what I second said is that it was warm in your living room as of 6.10 p.m. But who is to say whether you are right, or me, or neither, or both? Is there any matter of fact as to what it is that I said?

How do we identify, or individuate, what has been said? How do we decide when the relation of saying the same thing holds between any two utterances? There is a well-established tradition, from Frege, via Kaplan, to most contemporaries, to say that for every utterance there is a semantically relevant level of what is said, or *content*, or the proposition expressed, dependent on the context and distinct from the linguistic meaning of the sentence uttered. Contents are supposed to be specifiable beforehand, in the sense that there is a determinate method to figure out what the content is, given the sentence and the basic facts about the context of its utterance, like who the speaker is or what the time is. My aim is to argue against this tradition. I will point out a whole pattern of cases that seriously threaten the idea of isolating some determinate level of what is said, independently from any context in which the question of what was said has been raised.

2 Different Meanings, Different Contents, the Same Thing Said

I start with a few cases that motivate the distinction between meanings and contents. I then offer a few more cases that *dis*-motivate the distinction, so to speak. The lesson to draw is that our intuitions on what is said, on which the traditional view heavily relies, do not support any such neat distinction.

2.1 How You May Be Tempted to Appeal to Contents

The linguistic meaning of a sentence is a natural candidate to stand for what is said by an utterance of the sentence. But this suggestion has been widely rejected. Frege already wrote: "The sentence 'I am cold' expresses a different thought in the mouth of one person from what it expresses in the mouth of another."[3]

Suppose that you utter the following sentence twice, respectively in reference to Laura Bush and to Hilary Clinton:

She is arrogant. (1)

Your first utterance of (1) attributes arrogance to Mrs. Bush, whereas in your second utterance, you are saying of Mrs. Clinton that she is arrogant. The intuition that you have not said the same thing is supported by the fact that the truth of your first utterance depends on Laura Bush, while the truth of your second utterance depends on Hilary Clinton. It seems, then, that there may be something in what is said that does not come from the linguistic meaning of the words uttered. The two women seem to have gotten somehow into what you said on the two occasions. Certain items furnished by the context, like the person in reference to whom a personal pronoun has been used, seem necessary to the understanding of the utterance.

But if we are to take what is said to be simply the meaning *plus* whatever is required to understand the utterance, then it should work equally well to take what is said to be simply the meaning. The intuitive difference between your two utterances of (1) would be explained by the fact that you said the same thing *of* different people – first of Laura Bush, then of Hilary Clinton.

As this first case did not take us very far, suppose that I say to Laura Bush:

You are arrogant. (2)

Suppose that a guy called John, who happens to be standing nearby, mistakenly thinks that I was talking to him. Suppose also that I do not think that he is arrogant, nor does he think that he is arrogant, nor does Laura Bush think that she is arrogant. Intuitively,

---

[3] [1], p. 235. A "thought" in Frege's terminology is "what is said" in modern terminology. Kaplan's rejection of the suggestion at stake is even more explicit: "What is said in using a given indexical in different contexts may be different. Thus if I say, today, "I was insulted yesterday," and you utter the same words tomorrow, what is said is different. (...) There are possible circumstances in which what I said would be true but what you said would be false. Thus we say different things." [2], p. 500.

I disagree with Laura Bush, while I do not really disagree with John: there has simply been some misunderstanding between the two of us.[4]

If all aspects of the meaning were preserved in what is said, then what I said in (2) would be something like "x is arrogant and x is being talked to," and I would be saying it of Laura Bush. But then John and I would really disagree about the truth of what I said, since he thinks that it is him, and not Laura Bush, to whom I was talking. This clearly clashes with the intuition that he simply did not get what I said. In the light of this second case, it seems that some aspects of the meaning do not reach into what is said. The property of being talked to, carried by "you", seems to drop off what is said.[5]

Notwithstanding appearances, this case does not take us very far either. Consider it once again. It rests on the assumption that everything that can be evaluated for truth and is part of what is said is also something about which people may disagree. Now, the question of who is being talked to is not really open to disagreement, which further suggests that certain parts of the meaning do not reach to what is said. However, one may well question this suggestion, while preserving the intuition that it is not quite appropriate to talk of disagreement in the case of (2). John and I have conflicting opinions as to whether Laura Bush is being talked to or not. But how could I possibly go wrong on the issue of whom I was talking to, John or Mrs. Bush? As the speaker of (2), I am the best placed to know to whom I was talking. It is then inappropriate to say that John disagrees with me on something about which he knows that I cannot go wrong. I take it, therefore, that (2) no longer motivates having anything beyond meanings to stand for what is said.

So far, I have dealt away with two intuition-based arguments for a distinguished level of what is said. I now turn to the argument from same-saying, as Perry calls it.[6] It relies on utterances that intuitively say the same thing in spite of differences in meaning. Consider (1) and (2) together. Their meanings are clearly different, since "she" is to be used for the most salient female individual, while "you" is to be used for the addressee. Yet, it seems that what you say in (1), talking of Laura Bush, is the same thing as what I say in (2), something like *that Laura Bush is arrogant*. We both say of her that she is arrogant, only, I dare tell her straight. The insight again goes back to Frege, who further wrote: "It is not necessary that the person who feels cold should himself give utterance to the thought that he feels cold. Another person can do this by using a name to designate the one who feels cold."

2.2 A Dilemma

Suppose that your utterance of (1), in reference to Laura Bush, was made in June 2002. For the sake of the argument, assume that Laura Bush herself is part of what is

---

[4] A similar case is given by Stalnaker, who notes: "What one says (...) is itself something that might have been different if the facts had been different; and if one is mistaken about the truth value of an utterance, this is sometimes to be explained as a misunderstanding of what was said rather than as a mistake about the truth value of what was actually said." ([8] p. 279).

[5] As Recanati puts it: "the property of being the addressee is not a constituent of the proposition expressed [by an utterance in which 'you' occurs]: it is used only to help the hearer identify the reference, which is a constituent of the proposition expressed." ([7], p. 39)

[6] [6], p. 5.

said. Now, is the content of (1), what you said, temporally neutral? Does (1) merely attribute her arrogance, a property that she could possess at certain times and lack at others? Or is that content temporally specific? That is to say, does (1) attribute her the property of arrogance as of the time picked out by the present tense, June 2002?

How is this issue supposed to be settled, first of all? Presumably, we ought to consult our own intuitions on what is said. Suppose, then, that you utter the same sentence again in January 2003, referring once more to Laura Bush. You say:

She is arrogant. (3)

Will you say the same thing as you did in (1)? I can certainly reply, "hey, that is precisely what you said last June." And our judgments about the truth value of such a reply are that, in a suitable context, it comes out true. So, there is at least a sense in which what is said in (1) and what is said in (3) are one and the same thing. That is not sufficient evidence yet that the contents must be temporally neutral. For, suppose that you hold what seems to be the standard view, namely, that what is said must be a proposition, whose truth does not vary with time. The feeling that (1) and (3) in some sense say the same thing can be explained by their being utterances of one and the same sentence. More generally, the standard view will always have a choice between the context-independent meaning and the context-dependent content to account for what is said.

The dilemma is whether to think of contents as temporally neutral or temporally specific. The 'neutral' horn is motivated by the intuition that the same thing may be said by utterances that do not express the same proposition, as in the case of (1) and (3). The motivation is not conclusive, and is counterbalanced by the intuition that, if you were asked in January 2003 to repeat what you said in June 2002, you might well say something like:

In June 2002, Laura Bush was arrogant. (4)

There is little doubt that in a suitable context, on the basis of your utterance of (1), I can truly reply, "indeed, that is precisely what you said in June 2002." And if the same thing has been said by (1) and (4), it had better be temporally specific.

With the dilemma in mind, let us go back to (1) and (2) jointly considered. As before, (1) was uttered in reference to Laura Bush in June 2002. Suppose that I uttered (2) in January 2003, talking directly to Mrs. Bush. Your utterance of (1) and my utterance of (2) intuitively say the same thing. Or, more modestly, there is a sense of saying the same thing in which we are doing so, since both of us are attributing arrogance to Laura Bush.[7]

But here comes a problem. The pair of (1) and (4) strongly supports the idea that if there is a separate level of content to stand for what is said, it must consist of temporally specific contents. On the other hand, the pair of (1) and (2) strongly supports the idea that those contents must be temporally neutral. For, if what is said is something whose truth may vary with time, then (1) becomes a mere attribution of arrogance to Laura Bush, and so does (2), hence they "say the same thing." But if what is said is something whose truth cannot vary with time, then what you said in (1)

---

[7] Besides, note that our initial discussion of the intuition that (1) and (2) say the same thing was free of any assumptions about the times of our utterances.

will consist of her arrogance as of June 2002, while what I said in (2) will consist of her arrogance as of January 2003. If you opt for temporally specific contents, motivated, among other things, by (1) and (4), how can you account for the intuition that what you said in (1) is the same as what I said in (2)? You can no longer appeal to the sameness of the sentences uttered, or even to the sameness of meanings of the sentences uttered, given that "she" and "you," hence our sentences themselves, do not mean the same thing.

We have a dilemma, then, whose both horns are likely to leave us unhappy. But this is not the problem yet. For, one way of resolving the dilemma is to allow for a minor modification on the side of the 'neutral' horn. In rough lines, one might say that temporally specific contents could be subsumed under temporally neutral contents. Specific contents are a particular case of neutral contents, in the same way in which constant functions are a particular case of functions: they yield the same truth value, whatever temporal input you feed in. The next step, one might say, is to spot an ambiguity in any present-tensed sentence. Consider again (1), as uttered in June 2002. On the one reading, it expresses a temporally neutral content, namely, Laura Bush's arrogance, which obtains at some times and not at others, as her personality changes through time. On the other reading, it expresses a temporally specific content, namely, Laura Bush's arrogance as of June 2002. At last, when we get the intuition that (1) and (4) say the same thing, it must mean that (1) has been given a specific reading, while when we get the intuition that (1) and (2) say the same thing, both (1) and (2) assume neutral readings.

This way of resolving the dilemma is still unlikely to make us happy. For, it presupposes that the sentences uttered in (1) and (2) are ambiguous, which is very implausible. More plausibly, one could say that the sentences are not ambiguous, but unambiguously express *two* contents each: one neutral, one specific. The neutral content accounts for what is said by (1) in one sense (the temporally neutral sense), the specific content accounts for what is said by (1) in another sense (the temporally specific sense). So we can be happy now – if only *for* now.

2.3 The Problem

The argument from same-saying is a double-bladed sword. It was designed to defend the traditional view, but is can be equally well turned against it. In order to handle our intuitions on what is said, without giving up the very idea that there is some distinguished level of what is said, one seems forced into a position that differs from the received one on two crucial aspects: the notion of content becomes more flexible, since contents may change truth value through time, and given any utterance, instead of there being one content to stand for what is said, there may be several.

So far, the situation does not appear dramatic, since we have only inquired how time affects what is said. But the cases that pose problems for the traditional view, far from being confined to the issue of time, are pervasive. Consider the following scenario. In June 2002, it was particularly warm in San Francisco, which we both know. In July the same year, we happen to be in Chicago, and it is very warm. I say:

It's very warm, probably warmer than in San Francisco last month.   (5)

Later, in September, we are in Stanford, and the weather is again very warm. I say:

It's very warm, probably warmer than in the city last June. (6)

If my aim in making those utterances is simply to comment on the weather, then it should be not too hard to get the intuition that I am saying the same thing. What I am saying is, simply, that it's very warm, warmer than in June 2002 in San Francisco.[8] But the traditional view has no handle on such cases. The propositions expressed by (5) and (6) are different: one is true only if it is warmer in Chicago in July 2002 than in San Francisco in June, while the other is true only if it is warmer in Stanford in September than in San Francisco in June. And the linguistic meanings of the sentences are obviously different. Moreover, it is no longer enough to appeal to temporally neutral contents. We must allow for locationally neutral contents, too.

By way of a bonus example, suppose that we go with a friend to see Almodovar's latest movie, *Talk to Her*. Coming out of the movie theatre, I ask you "How did you like it?" Later on, I turn to our friend and I ask her "How did you like the movie?" The linguistic meanings of the questions that I asked are different (you may use the pronoun "it" for anything you wish, while you may use the description "the movie" only for the movies). At the same time, we are inclined to say that I asked the same question. I asked our friend what I asked you, namely, how she liked the movie in question, *Talk to Her*. If after a while I come upon someone who I know has seen *Talk to Her*, and I ask him "How did you like *Talk to Her*," I will be asking the same question again, namely, how he liked that movie. And if in a conversation about Almodovar someone asks me "How did you like his latest movie," then I will be asked the very question that I was previously asking, namely, how I liked the movie *Talk to Her*.

The case clearly poses a problem for the traditional view. Propositionally, the four questions are different: the first is how *you* liked *Talk to Her*, the second, how our friend liked it, the third, how that guy that I came upon liked the same movie, and the fourth, how *I* liked it. Temporally and locationally neutral contents are of little help. What you need is something like contents neutral with respect to the addressee. And it should take you little to think of a case that would call for contents neutral with respect to the speaker. But once you start making room for various sorts of contents, neutral with respect to various sorts of things, then the dichotomy between meaning and *what is said* is clearly lost.

Let us see what has been done so far. We were first led to accommodate contents whose truth varies with time. But once we allowed for temporally neutral contents, we realized that we could not limit ourselves to time. We had to allow for contents whose truth varies with locations, or, worse, with addressees. The problem is that we are likely to have to allow for contents neutral with respect to all sorts of things, contents whose truth may vary not only with places and people, but also with points of views, time zones, or situations in general. We are likely to have to allow for contents whose truth may vary with the context, contents that assume their truth value only relative to a context. But wait! What becomes of the difference, then, between meanings and contents? Sure, if you *define* meanings as functions from contexts to contents, the way

---

[8] It might be worth making it clear that when in Stanford one talks of "the city," one means San Francisco.

Kaplan does it, there is a difference. But if you simply conceive of meanings as encoding the conditions for the truth of the utterance, given the facts about its context, then the difference fades away. If you assume that the meaning of "she is nice" tells you that an utterance of that sentence is true if and only if the most salient female in the context of the utterance is nice in that context, then you have ipso facto a contextually neutral content, true in those and only those contexts in which the most salient female individual is nice.[9]

To bring the issue home, it turns out that when we start searching for the most appropriate way to conceive of what is said, we are bereft of motivations for any neat, binary distinction between what is said and meaning. We have not arrived yet at the result that what is said is simply the context-invariant meaning of the sentence uttered.[10] For the time being, we are willing to associate with any utterance several meanings, all of which are "said" in some sense. Before I outline how to retreat to the more austere, 'one meaning' position, I want to rebut what has been taken to be a knockdown argument against any such position, the argument from direct intuitions.

2.4 How Direct Are Our Intuitions on What Is Said?

The traditional view of what is said adopts the referential analysis of indexicals, by taking their contribution to what is said to consist of their reference rather than their meaning. It rejects the descriptive analysis, according to which indexicals do not differ from definite descriptions insofar as the nature of their contribution to what is said is concerned, but roughly behave like definite descriptions that take wide scope over any sort of operator. In the context of the present discussion, the distinguishing feature of indexicals would derive from the way in which they interact with devices used in reporting speech, such as "Bill said that..." or "that is what George said."

Kripke and Kaplan argued that the descriptive analysis was untenable.[11] Their argument uses simple sentences in which the indexical does not come embedded within any phrase out of whose scope it could leap. Imagine that we go through the following dialogue:

She is arrogant. – That's not the case, though it might have been. (7)

---

[9] You might object that the difference does not just "fade away," for meanings are admittedly nothing but contextually neutral contents, but there are other contents, – temporally neutral, locationally neutral, addressee-neutral, and so on. However, that would be a meagre objection, since it is easy to see that all those other contents can be subsumed under contextually neutral contents. Among the contents whose truth varies with the context, temporally neutral are those whose truth varies with the time of the context, addressee-neutral are those whose truth varies depending on who is being talked to in the context, and so on.

[10] To forestall a possible confusion, the meaning of a sentence is context-invariant in that it does not vary with the context, but at the same time, it is contextually neutral in that the truth value that it confers upon the sentence varies with the context.

[11] My source for this argument, which is a variant of Kripke's so-called modal argument, is [3], p. 10 ff. Kaplan seems to have a similar argument in mind in [2], p. 500.

Suppose that I am talking of Laura Bush. Our intuitions tell us that your reply will be true if and only if Laura Bush might have happened to be arrogant, but actually is not. Now, if the contribution of "she" were some general condition, like that of being the most salient female individual in the context of my utterance, then it seems that your reply in (7) is already true if someone actually arrogant other than Laura Bush might have happened to be the most salient woman in the context of my utterance. Our robust intuition that this is not enough for your reply to be true might lead you to conclude that the contribution of "she" to what I said must somehow involve Laura Bush herself.

Kripke used essentially the same example to ground his "main remark," namely, "that we have direct intuitions of truth conditions of particular sentences."[12] Adapted to our topic, Kripke's remark is that we have direct intuitions of what is said by particular sentences. However, I do not see that Kripke has given us any argument to the effect that we have such direct intuitions. What I do see in Kripke's discussion and similar ones of Kaplan, Recanati, Soames, and so on, is the use of propositional anaphora, that is, of pronouns like "that" in "that is not the case" and "it" in "it might have been." The same pronouns are systematically used in reported speech, like when we say: "*this* is what he said," "but I already said *that*," "he told *it* to me," etceatera. Now, *my* main remark is that dialogues such as (7) do not provide any conclusive evidence that we have direct intuitions either of truth conditions or of what is said. Such dialogues crucially involve phrases and constructions whose semantic behavior we do not fully understand yet. Why assume that "that" and "it" in (7) refer to what is said by the antecedent sentence? Why assume that we must consider those pronouns as referring at all? For one thing, "it might have been" is elliptical at least for "it might have been the case." But furthermore, it is not clear that "it" must be a pronoun anaphoric on the antecedent "that," rather than conceal a broader ellipsis, so that the last sentence of (7), with the ellipsis resolved, would amount to something like "it might have been the case that she were arrogant." But if (7) is to be analyzed along those lines, then we are far from being given anything that might undermine the descriptive analysis, even in its most radical form, in which it simply treats indexicals as definite descriptions that take wide scope.

Another reason not to give too much weight to dialogues such as (7) is that the occurrence of a definite description in the antecedent sentence gives rise to the usual scope ambiguities, in spite of the fact that the definite description does not come syntactically embedded within a phrase with whose scope it could interact:

> The President's wife, whoever she is, is surely arrogant. –   (8)
> That's not the case, though it might have been.

The wording makes it clear that the speaker in (8) has not used the definite description in order to refer to Laura Bush. In spite of that, the reply in (8) is multiply ambiguous. On one reading, it is true if it might have been the case that the person who were the President had a wife who were arrogant (e.g. the possibility of Clinton still being the President and Hilary being both his wife and arrogant would make this reading true). On another reading, the reply in (8) is true if the actual President, that is, George W. Bush, were married to someone who were arrogant (e.g. if we suppose that Laura

---

[12] [3], p. 14.

Bush is not arrogant, the fact that he could have been married to Hilary, supposedly arrogant, would make the reading true). Most importantly, there is also the reading on which the reply is true if Laura Bush, who happens to the wife of George W. Bush, who happens to be the President, might have been arrogant, but is not. The fact that this reading may obtain does not show that whenever this reading obtains, what the definite description "the President's wife" contributes to what is said must involve the actual President's actual wife herself. If there is a rendering of the last reading of (8) on which the contribution of the definite description to what is said is some descriptive condition, as Kripke and Kaplan would hold, this shows that there must be a rendering of (7) on which the contribution of "she" is some descriptive condition, like that of being the most salient female individual in the context of utterance. The claim, then, that "she" *must* contribute its reference to what is said is not as straightforward as it has been taken to be.[13]

3 An Alternative Approach to What Is Said

There has been a lot of discussion regarding the temporal character of contents. It seems to me, though, that the deeper issues and the generalizability to other features have not been fully seized. To my knowledge, the received wisdom on what is said has not been challenged with the sort cases that I have just offered. But it has been challenged, and there is a handful of philosophers overall doubtful that the notion of what is said makes sense on its own, if at all. Lewis wrote: "Unless we give it some special technical meaning, the locution 'what is said' is very far from univocal. It can mean the propositional content (...) It can mean the exact words. I suspect that it can mean almost anything in between."[14] A recent proposal by John Perry shares the spirit of Lewis' remark. I will briefly present Perry's proposal, apply it to the cases considered, explain what I find missing in it, and end with a suggestion regarding the sort of amendment that it needs in order to handle our intuitions on what is said.

3.1 Perry's 'Reflexive-Referential' Handle on What Is Said

Perry's main insight is that it is misleading to talk of *the* proposition expressed by an utterance, for there is a wide array of propositions connected with the utterance, all of which provide, in one way or another, a necessary and sufficient condition for the truth of the utterance. After noting that "the binary distinction (...) is too simple," he suggests: "An utterance has as wide a variety of contents as we may find useful to isolate, for particular purposes of description and explanation. We can say that in at

---

[13] One might insist that we can explain the *absence* of the other readings in (7) only if go referential. But that claim is not straightforward either. In the best case, we may hope that the absence of such readings may be explained by the same pragmatic mechanisms that are used wide and large to predict and explain the absence of some syntactically possible reading. In the worst case, we may assume that indexicals are conventionally tagged as expressions that always "take wide scope," in the widest sense of "scope."
[14] [4], p. 97.

least the vast majority of cases, the common sense concept of "what is said" (...) corresponds to content$_C$. This is a good reason for an account of content to recognize this concept, but not a good reason to expect it to be the only or even the most theoretically fruitful kind of content."[15]

By providing the utterance with many different, though interconnected contents, Perry gets a handle on the cases that pose problems for Kaplan. Suppose that John says:

$$\text{I find our President's wife attractive.} \quad (9)$$

Suppose that later on, I am talking of Laura Bush, and you tell me:

$$\text{I find her attractive.} \quad (10)$$

Then I may reply, "That is precisely what John said."[16] If John does not know you and could not have been talking of you, then I am clearly reporting John as having said that *he* finds Mrs. Bush attractive. Still, in some sense, you and John have said the same thing, namely, that you find Mrs. Bush attractive. This is typically a case that puts the traditional view in jeopardy. The two propositions are different – that you find Mrs. Bush attractive vs. that John finds her so –, but the meanings are different, too. In Perry's account, though, that dichotomy disappears. Instead, an utterance gets a *reflexive* content, which may be thought of as a general linguistically encoded condition on the utterance itself. The reflexive content of John's utterance u of (9) will roughly be:

$$\text{The wife of the President of the most salient group in the context of } u \text{ to which the speaker of } u \text{ belongs is} \quad (u1)$$
$$\text{such that the speaker of } u \text{ finds her attractive.}$$

Perry's strategy is to derive all the other relevant contents by increasing the reflexive content with the information available in the context of utterance. The information that we have in the case of (9) is: (i_1) that the speaker of u is John, and (i_2) that the wife of the President of the relevant group in u is Laura Bush.[17] Using simple inferential mechanisms, we can infer that u is true if and only if:

$$\text{John finds Laura Bush attractive.} \quad (u2)$$

($u2$) is indeed the referential content that the traditional view associates with (9).

More interestingly, we do not need to cash out at once all the information that we have about u. If we only use (i_1), then we can infer that u is true if and only if:

$$\text{The wife of the President of the most salient group in the context of } u \quad (u3)$$
$$\text{to which the speaker of } u \text{ belongs is such that John finds her attractive.}$$

If we only use (i_2), we can infer that u is true if and only if:

---

[15] [5], p. 17. Content$_C$, also called 'the official content,' corresponds to the traditional kind of Kaplanian contents.

[16] For the sake of clarity, I will be ignoring the issue of time in the rest of the discussion.

[17] Of course, (i_2) itself can be derived from the information that the relevant group in u is the American nation, the George W. Bush is their President, and that Laura is his wife.

> The speaker of *u* finds Laura Bush attractive. (*u*4)

As for your utterance *v* of (10), its reflexive content will roughly be:

> The speaker of *v* finds the most salient woman in the context of *v* attractive. (*v*1)

The information that Laura Bush is the most salient woman in the context of your utterance makes it possible to infer that your utterance is true if and only if:

> The speaker of *v* finds Laura Bush attractive. (*v*2)

Even though (*u*4) and (*v*2) are not exactly one and the same thing, since they are propositions about different particulars, viz. John's utterance and yours, they are still instances of one and the same condition on any arbitrary utterance *x*, namely:

> λ*x*: the speaker of *x* finds Laura Bush attractive. (*u-v*)

In general, suppose that in a content associated with an utterance we abstract on the utterance itself and obtain the same general condition that we obtain by similarly abstracting on the utterance in a content associated with another utterance. The two utterances may be said, then, to say the same thing.

3.2 A Shortcoming

While Kaplan allows for too little, Perry seems to allow for too much. For assume, with Perry, that for any utterance, there is an array of propositions, p_1 to p_n, obtained from the sentence uttered and the context in ways w_1 to w_n, and that any among those propositions specifies a condition for the truth of the utterance is the corresponding way. The idea that we can point in advance, à la Kaplan, to some particular way w_j such that the proposition obtained in that way is what is said by the utterance is, I have argued, hopeless, unless "what is said" is turned into a barren artificial notion. However, it will not work either to say that what is said may be *any* proposition you please among p_1 to p_n.[18] A follow-up on the last example reveals a problem. Suppose that in France, François, talking about John, says:

> He finds our President's wife attractive. (11)

Can I truly reply to François, on the basis of John's utterance of (9), "Indeed, that is precisely what he said"? Of course not. John never spoke of Bernadette Chirac, never said that he found her attractive. But if Perry's proposal were generally applicable, it would be unclear why we could not take François to be saying, in some sense, what John said. The reflexive content of François' utterance *w* of (11) is:

> The wife of the President of the most salient group in the context
> of *w* to which the speaker of *w* belongs is such that the most (*w*1)
> salient male individual in the context of *w* finds her attractive.

---

[18] To be sure, Perry does not quite say that. He suggests that what is said corresponds to the referential content, at least in the default case. He does not tell us, though, whether there are any constraints that other contents must meet in order to qualify as candidates for what is said.

Using the information that John is the most salient male in the context of w, we infer that w is true if and only if:

> The wife of the President of the most salient group in the context of w ($w2$)
> to which the speaker of w belongs is such that John finds her attractive.

Finally, if we abstract on the particular utterances u and w in ($u3$) and ($w2$), we are left with one and the same condition, namely:

> λx: the wife of the President of the most salient group in the context of x to   (v-w)
> which the speaker of x belongs is such that John finds her attractive.

The case is perfectly parallel to the previous one. But while we can easily hear (9) and (10) as saying the same thing, it is incredibly difficult, if possible at all, to hear (9) and (11) as saying the same thing, namely, that John finds the speaker's countrymen's President's wife attractive.

One might find the example unconvincing if one sees "our" in (9) as somehow anaphoric on "I." For, it would be all right to drop the anaphoric expression "our President's wife" and use "her" instead, but not to eliminate the anchor "I" and use "John" instead, thus forcing "our" to look for another anchor. If this kind of explanation is at least partly correct, then the challenge becomes to make room for it in Perry's account.[19] Presumably, the anaphoric linkage would be made apparent in the reflexive content of (9), which would then be:

> The speaker$_i$ of u is such that the wife$_j$ of the President of the contextually   ($u5$)
> relevant group to which he$_i$ belongs is such that he$_i$ finds her$_j$ attractive.

But even so, the fact remains that the usual inferential mechanisms, together with the information that John is the speaker of u, allow us to infer that u is true under condition ($u3$). It is unclear how we could deny that ($u3$) is a possible way of specifying the truth conditions of (9), and why would we? What seems to be missing in Perry's proposal is rather an account of further constraints that tell us which among the incremental contents may, under given circumstances, qualify as what has been said by the utterance.

4 Where to Look in Future

I have tried to make two points. First, the widespread idea that we have *direct* intuitions on what is said is ungrounded and probably mistaken. I am not saying that we have no intuitions at all on what is said. I am merely saying that our intuitions are indirect, mediated by our semantic intuitions on the conditions under which one may truly report what was said by a given utterance as being the same thing as what was

---

[19] I suspect that the explanation is only partly correct, since the same kind of problem occurs without any anaphora in the background. Suppose that John asks me "What do you think of our President?" What would it take for François, then, to ask *you* the question that John asked me? He might ask the same words, or he might ask "What do you think of George W. Bush?," or he might even ask "What does Isidora think of George W. Bush?," but there is no way that he could do so by asking "What does Isidora think of our President?"

said by some other utterance. Second, if our intuitions on what is said highly depend on our intuitions on the semantics of the locution "what is said" and of many other constructions used in reporting speech, then, given that those constructions can only be interpreted in a context, the question whether the same thing has been said by two reported utterances can only make sense in a context. We need to be given not only the contexts of the reported utterances, but also the context in which the report itself is taking place.

The traditional account of our intuitive notion of what is said leads to a dead end. If any alternative account is to lead anywhere worth going, then it had better proceed hand in hand with the syntactic and semantic accounts of indirect discourse, propositional anaphora, and all the other devices that we use when we talk about what is said. Those devices are highly contextually dependent, and it is still an open issue how their meanings exploit various contextual cues. But once we get a solid grip on their syntax and semantics, we may expect to get some valuable insight into the general constraints that carve up our intuitions about what is said.

References

1. Frege, G.: Logic, in Beaney, K. (ed.), *The Frege Reader*, Blackwell Oxford (1997)
2. Kaplan, D.: Demonstratives, in Almog, J. et al. (eds.), *Themes About Kaplan*, Oxford UP (1989)
3. Kripke, S.: *Naming and Necessity*, Oxford UP (1980)
4. Lewis, D.: Index, Context and Content, in Kanger, S., Ohman, S. (eds.), *Philosophy and Grammar*, Dordrecht, Reidel (1980)
5. Perry, J.: Reflexivity, Indexicality and Names, in Kunne, W., et al., *Direct Reference, Indexicality, and Propositional Attitudes*, CSLI Publications (1997)
6. Perry, J.: *Reference and Reflexivity*, CSLI Publication (2002)
7. Recanati, F.: *Direct Reference*, Blackwell Oxford (1993)
8. Stalnaker, R.: Assertion, in Cole, P. (ed.) Syntax and Semantics **9** (1978)
9. Taylor, K.: *Truth and Meaning*, Blackwell Oxford (1996)

Modelling "but" in Task-Oriented Dialogue

Kavita E. Thomas

School of Informatics, University of Edinburgh
kavitat@cogsci.ed.ac.uk

Abstract. We determine criteria for modelling plan-based "but" in task-oriented dialogue (TOD), following work by Lagerwerf [5] and focusing on cases in which it signals *denial of expectation* (DofE) and *concession*, to which end we propose a novel treatment of concession in TOD. We present initial considerations for an algorithm to address plan-based "but" in an Information State (IS) model of dialogue that predicts which interpretations (DofE and/or concessive) to generate. We motivate this work by showing how it updates beliefs in the PTT [6] model of dialogue and can be used to facilitate recognition of planning mismatches and more generally, discourse understanding and natural language generation (NLG).

1 Introduction

In this paper we try to model what's being communicated (both contrastively and otherwise) with the discourse marker "but" when it contrasts cross-speaker information in task-oriented dialogue (TOD), where speakers are planning and performing tasks cooperatively. We focus here on cases in which "but" signals denial of expectation (DofE) and concession where the contrast is made with respect to something in the preceding speaker turn. We aim to model these rhetorical relations across speakers in dialogue to clarify how speakers' turns are related, which helps establish the coherence of the dialogue. We extend the approach presented in Thomas and Matheson [3] to address cases in TOD which involve planning. Plan-based cases often don't involve DofE (in either the expectation-based or frustrated-plan senses presented in Knott [4], so predicting that a belief is defeated in these cases is misrepresentative.

Thomas and Matheson [3] argued that Lagerwerf's claim [5] that DofE involves causal presupposition can be extended to dialogue, as seen in the dialogue version (Ex. 1 below) of his example. [3] then presents an algorithm that predicts the defeated expectation in DofE cases from the hearer of the "but"'s perspective, and showed how this expectation facilitates both interpretation of "but" and generation of the DofE itself in the PTT framework. In the following example, their algorithm predicts that B has the expectation *beautiful(X) > married(X)*, (where > indicates defeasible implication). A's utterance triggers the denial of B's expectation, since B knows that Greta never married, resulting in B's contrastive (cued by "but") response.

Example 1:
A: Greta Garbo was called the yardstick of beauty.
B: But she never married.

The algorithm also predicts A's response to the DofE based on her private beliefs. E.g., if she disagrees with B's expectation but acknowledges that Greta never married, she might reply "Yeah, but beautiful people don't have to marry."

Plan-based "but" in TOD is often problematic in this algorithm, since it doesn't involve defeated expectations but rather signals a difference in the speaker's plan for the task at hand. Recognising this planning conflict is especially significant in TODs, since dialogue systems must be able to understand what expectations the user has in order to respond appropriately and collaboratively to her goals, constraints and beliefs. Consider the following example paraphrased from the TRAINS dialogues (d93-20.2 utt130-1):

Example 2:
A: so we should be (there) at eleven I think
B: eleven BUT that still doesn't give us enough time to get to Bath

In [3] the algorithm modelled B's expectation (in his private beliefs) as *be(ing) there at 11 > gives enough time to get to Bath*, which was odd, since it makes more sense to assume that B believes the more generalised rule that *effect of plan so far > preconditions for next goal will hold*. Generation of DofE only occurs in this algorithm if B has this expectation in his private beliefs and hears information that contradicts this expectation, which serves to constrain overgeneration somewhat. Here we'll argue that in many of these plan-based cases, it doesn't make sense to consider such highly specific and task-stage related expectations as part of the speaker's (static) private beliefs, and focus instead on a dynamic conceptualisation of these expectations in which they are launched at a specific stage in the task when A's utterance indicates a possible mismatch in the speakers' plans for their joint task.

In this paper, we will present examples of conflict in plan-based TOD signalled by "but" that don't involve causality at all, but often map onto planning operators in the task plan that are contingent upon one another, usually via satisfaction-precedence (s.p. $\prec$) or dominance relations. In our approach we consider whether these planning relations convey expectations in a similar way to the defeasible rules in DofE and concession and whether these expectations give useful information about dialogue participants' (DPs) task-related goals and beliefs. We motivate this work by arguing that information about contextually salient planning expectations enable DPs engaged in accomplishing a task together to detect mismatches in each others' plans at an early stage, thereby facilitating more collaborative behaviour. We extend Lagerwerf's monologue treatment of concession to present a novel treatment of concession across speakers in TOD, which gives salient information about speakers' attitudes towards contextually relevant questions and goals under discussion. Finally we determine considerations for an algorithm that predicts which interpretations (concessive and/or DofE) to generate, generates the appropriate expectations, and then updates the DPs' private beliefs accordingly. Social obligations [9] commit DPs to address expectations they infer promptly, facilitating more responsive generation which results in prompt detection of planning differences between the DPs.

316   K.E. Thomas

2   Modelling Plan-Based "but"

First we will present the different "but" cases contrasting planning operators and discuss what they communicate and how they differ from the expectation-based DofE cases discussed above (e.g., Ex.1).

The TRAINS example above is hard to model because it involves DPs planning their future actions rather than carrying out the actions they are discussing. We start by considering simpler examples involving agents carrying out actions rather than planning future ones:

Example 3:
A: Add the vinegar to the sauce.
B1:   (Yeah) But it's not tangy enough.
B2:   (Yeah) But we forgot to add the mushrooms.
B3:   (Yeah) But we still needs to add the salt.

In the above example, B2 and B3 refer to steps in B's plan for making sauce which haven't occurred (B2) or might not occur (B3) and B wants to indicate that this step is necessary in B's plan. B1 refers to B's judgement on the result of adding the vinegar upon tasting the sauce, where her perceived lack of tanginess triggers the "but". B1 involves causality, namely that adding vinegar makes things tangy, an expectation which is in her private beliefs and is violated by what she perceives to hold after the vinegar is added, so this is clearly a case of DofE, much like the Greta Garbo example.

However there is no causality between A and B2 or B3, so neither of these can be viewed as involving defeated expectations. B2 and B3 also involve actions in the plan rather than effects like B1. While it's natural to think of causality between actions and their effects in planning rules, the only structural contingency that can occur between actions in plans are (1) satisfaction precedence (s.p.) which partially orders them in terms of when they operate in the plan and (2) dominance relations which specify which actions need to be achieved to accomplish other actions. So distinguishing cases like B2 from B1 involves searching the plan for A and B and then determining where the conflict or defeated expectation arises.

We adhere to searching B's plan for the task rather than some objective third-party observer's record of Tom's actions or Tom's own plan (if it were accessible) since we claim that interpretation in dialogue is a subjective phenomenon, subject to the hearer based on their perception, rather than dependent on some objective model of the world.

Determining where these actions occur in B's plan and how they are related will indicate why B communicates contrast. In B2, B indicates a mismatch between what's been done already, and what should have been done according to her plan for making sauce. In B3 she indicates an action that still needs to be performed in response to A's implication that they have finished making the sauce after adding the vinegar. That is, B must believe that A implies that the task is finished in order to generate a contrastive reply in B3. In B's task plan adding vinegar is the last stage, so if there is no s.p. relation between adding salt and adding vinegar, and they have forgotten to add salt, it's reasonable for B to point this out to A, assuming that A might mistakenly think the task finished.

However the implication ([adding vinegar > sauce is done]) launched by A can't comfortably be viewed as an expectation stored in A's private beliefs as we've been assuming for DofE cases like Ex.1, since this implication is so closely tied to the stage of the task and the plan being discussed. I.e., it will not be salient at other stages of the task, unlike the expectation about beautiful people marrying in Ex.1 that is stored in B's private beliefs and which holds *before* A's utterance is heard. Expectations like the one in Ex.1 are *generalisations* that hold regardless of context. B discovers that this expectation doesn't hold due to some controversial fact she observes at any point in the dialogue, and the expectation is defeated. However in Ex.3 B3, it would be strange for B to have this expectation if they have just started making sauce, and began by adding vinegar. So it appears that these planning expectations are permissible only given a certain prior context and plan history.[1] Whether this in turn should be viewed as a special case of DofE is uncertain. In either case however, such an implication needs to be reached from A by B in order to license the "but" in these examples.

3 Concessive Interpretation

While *concession* differs from DofE by virtue of the argumentative force normally associated with the relation, it also involves defeasible expectations, following [5], and will possibly provide an alternative interpretation to DofE for many of these "but" cases.

The problem is, since concession has an argumentative function in which something is conceded in order to "win" the argument with an opposing point, in dialogue it doesn't make sense to consider A and B as giving the arguments of the concession relation directly, since a single speaker's intention is necessary to voice both conceded and victorious arguments. [3] present a way of getting around the difficulty of framing concession across speaker turns in dialogue by assuming that DPs minimally acknowledge the other person's utterance because of social obligations [9]. We view B's implicit acknowledgement of A's utterance as licensing the use of the relevant argument from A's turn (as B interprets it)[2] in the concession relation with the assertion in B's "but" turn containing the other argument.

Following [5], we view concession as giving rise to expectations that form defeasible rules which are presupposed by concessive discourse connectives like "but". According to Lagerwerf, concession requires a contextually available claim

---

[1] This is motivation for considering each speaker's evolving task-plan as part of their representation of context, along with dialogue history, since speakers need to access the current stage of the task in order to resolve many of these plan-based "but" cases.

[2] We assume for now that B simply accepts A's assertion as one of the arguments of the concession.

(or *tertium comparationis*, TC), for which both a positive and a negative argument are provided. The positive and negative arguments are presupposed as expectations launched by the relevant assertions in the adjacent speaker turns in these cross-speaker "but" cases with respect to the contextually relevant claim (TC). We determine whether the assertions favour the TC or disfavour it based on matching the assertions to planning operators in the plan and evaluating them with respect to the desired outcome at the current stage in the task (i.e., the TC[3]).

Unlike DofE, concession (or "concessive opposition") doesn't involve a causal relation between the clauses, and so might address cases like A followed by B2 or B3 in Ex.3 above.[4] Notice that for Ex.3, B1, both DofE and concessive interpretations work:

- DofE interpretation: *adding_vinegar > makes_sauce_tangy*
- Concessive interpretation:
  - *adding_vinegar > sauce_tastes_OK/better/ready*
  - *not_tangy_enough > sauce_doesn't_taste_OK/better/ready*
  - TC = Does the sauce taste OK?/ Is the sauce ready?

The causal link between adding vinegar and tangy taste might indicate that this example should be viewed as DofE rather than concession. However the concessive interpretation also seems reasonable, and adds further information with respect to the TC by arguing that the sauce doesn't taste OK. The DofE interpretation can be verified by finding an action-effect relationship in B's plan between adding vinegar and tanginess that will verify that this expectation is valid. Concessive interpretation relies on the TC being contextually available; the general TC of "Does the sauce taste OK" is valid at any stage, assuming that is a reasonable action to evaluate the sauce at that point. Asking whether the sauce is ready on the other hand is only valid if the task is finished, or at the final stage. So determining where in B's plan Tom's action falls will either rule out or allow this second possible TC.

Algorithm for generating concessive interpretation

1. Select most specific salient goal (which could be a desired effect at a given stage) from B's plan at current planning stage being described in dialogue. This is a good first guess at the TC.
2. Given this TC and B's plan for the task, determine whether A's and B's assertions favour the outcome of the TC or not by searching the plan, matching to planning operators, and evaluating w.r.t. the TC.
    a) If the outcome is favourable, then generate the expectation A > (TC = yes)
    b) Else generate the expectation A > (TC = No)
3. These two defeasible expectations form the concessive interpretation. Pass them both back to the main algorithm

---

[3] More on how we associate the TC with the desired outcome at a particular stage in the task plan is discussed in section 3.2.
[4] Concessive interpretation for Ex.3 B2 can be found on the next page.

3.1 Determining the TC

In order to generate concessive interpretations and then update the IS with these additional expectations, we will need to determine a relevant TC. We've discussed briefly why the given TCs were chosen in the examples above, but it will be necessary to be more precise about how the TC can be generated from the IS and how alternative possibilities for the TC will be considered in order to avoid overgenerating in the concessive case. In the examples above we saw that the TC was typically framed as "Is the sauce OK". Alternatively we had "Is the sauce ready" in Ex.3, B2, which only worked because the task was in its final stage. However in the final stage of the task, presumably both TCs are still valid, so the question arises as to which TC to prefer. Until evidence to the contrary arises, it seems reasonable to assume that a partial ordering of TCs with respect to specificity is possible. So in the final stage of the task, *Is the sauce ready* and *Is the sauce OK* should both be available, but the former should be preferred as it's more specific. In ambiguous cases with multiple possible TCs, it seems wiser to first pose expectations with respect to the more specific expectation before launching expectations based on the more general ones.

These TCs arise from the goals of the plan under discussion, and generally inquire about the desired effects of these goals or less specifically about the status of the plan. We can determine specificity of goals by ordering them based on dominance relationships, so that higher-level[5] goals are less specific than the goals they dominate. This means that the desired effect of the most specific task goal being achieved currently in the dialogue is the best candidate TC. In most cases, determining whether the assertion argues for or against this TC can be determined by searching the plan to check whether it is a favourable (or mentioned) operator in the plan. So it appears as if the plan itself provides the relevant TC in TOD. The relevant goals given the stage of planning involved will need to be updated in the IS in order to have a relevant TC available, so the planning module will need to update the IS with this set of partially ordered goals whenever a "but" occurs in the dialogue. Using this information to update the obligations[6] [6] might also prove useful.

3.2 Are Both Readings Always Reasonable?

However the question of whether the DofE and concessive readings are reasonable for all these TOD cases remains. For example, in Ex.3, B2 above, there is no causal relationship between adding vinegar and adding mushrooms, they simply describe successive operations in the plan for making sauce. Our algorithm would overgenerate in this sort of case and predict the DofE *adding_vinegar > have_already_added_mushrooms*. However, following the task plan, there is a contingency between these two events, namely the partial ordering in the plan. So B's expectation does hold in the context of her plan, even though there is no

---
[5] With respect to the hierarchical structure of goals in the plan.
[6] Or QUD, [2].

causality involved. B probably arrives at this expectation by searching her task plan rather than by relying on a stored expectation about the causality of these events as in the nontask-oriented dialogue (NTOD) Greta Garbo example. So the point here is that the way in which the expectation is generated (i.e., by searching a plan in TOD, and by recalling a stored expectation in NTOD) and also the causality vs. contingency aspects are novel in TOD. Concessive interpretation for this example also works and is possibly more natural in this case. The arguments for the concessive interpretation (given the TC "Is the sauce OK") for Ex.3 B2 are:

- *adding_vinegar > sauce_is_OK*
- *forgetting_to_add_mushrooms > sauce_isn't_OK*

While the expectation that "adding vinegar" implies that "the sauce is OK" (above) seems odd, it makes sense to conclude this given the context of evaluating the sauce at each stage. However, some cases do not seem to license concessive interpretation, as the second expectation *have_already_added_broccoli > sauce_isn't_OK* in Ex.4, B2 below shows (with the TC "Is the sauce OK").

Example 4:
A: Add the beans to the sauce.
B1:   (? OK) But we haven't finished sauteeing the onions.
B2:   (? OK) But we added the broccoli already, and the beans cook slower.
B3:   (OK) <adds beans> But now the sauce is too thick.

Here A presumably doesn't have the same task plan as B, whose plan precludes adding beans after broccoli (in B's plan: *adding_beans ≺ adding_broccoli*). This interpretation doesn't make sense unless we assume that both the agents are planning on adding beans and broccoli, which isn't explicitly present in the (negative) expectation launched from B2, so if these expectations are taken singly, they only make sense at the point of adding beans in the dialogue. Also, it seems odd to think of such specific expectations. It seems far more natural to think of this example involving the expectation that one should add broccoli after adding beans, as the DofE interpretation predicts, (since the beans cook slower, where this fact is part of B's knowledge about the world, relevant to making sauce). Presumably this ordering stems from B's task plan as a partial ordering between the two actions, which can be detected by searching the plan.

Notice that when DofE or concessive interpretations are evaluated in these cases, they are evaluated at the planning stage being described in the dialogue when the "but" utterance occurs. In this way we'll see whether specific expectations make sense at a given stage in planning and whether they are reasonable or help the planning process at all.

4 Searching the Task Plan

So far we've seen a variety of planning operator relationships between A and B's turns, and the particular combinations involved appear to shed some light on how the TC, expectations and contrast itself are resolved. In Ex.3 and 4 above,

A suggests an action that needs to be done (or has just been done) in the task at that point in the dialogue. B's response can suggest an alternative current action or mention a precondition to A's suggestion (e.g., Ex.4 B1), B can object to the current action on the grounds of some past action that prohibits A's suggestion (e.g., Ex.4 B2), or she can carry out the action and then find fault with the resulting effects (Ex.4 B3, Ex.3 B1). B can also mention a past action which doesn't match with her plan, as in Ex.3 B2. Finally, B can mention an action that still needs to be completed if B assumes that A is implying that the task is finished (Ex.3 B3). So if A opens with a suggested or described action, then B can respond with a goal that can't be met given this action, an alternative action, or can criticise an effect of the action. B can also object to the action on the grounds that some past action precludes A's suggested action from occurring.

We assume that effects don't generally get mentioned in TOD unless they are not as expected.[7] So if A opens with an effect, as in Ex.5 A1 below, B can (for example) express DofE that some preceding action didn't have the desired effect (as in B1), or introduce an action which needs to be carried out before the desired effect can be achieved (or not, in the case in which A mentions an undesired effect, as in A2). In A1 or A2 followed by B3 or B4 we see an effect followed by an effect. Generally here the effect asserted in B needs to be of a higher-level goal in the plan, since it doesn't make sense to criticise an effect by criticising or asserting a finer/lower level effect.[8] If B asserts an alternative effect (at the same level in the plan), then this effect must produce the opposite argument (e.g., B4).

Example 5:
A1:   The sauce isn't tangy enough.
A2:   The sauce is too tangy.
B1:   (* OK) But we just added vinegar.
B2:   (* OK) But we haven't added vinegar yet.
B3:   (? OK) But the sauce isn't ready.
B4:   (? OK) But the sauce (at least) has flavour.

Notice that for A2, B1 doesn't make sense as a possible response and wouldn't occur in a nonironic sense since it redundantly gives the expected action that results in A's effect, while B2 is OK and expresses DofE. However A1 followed by B2 can't express DofE in the typical way (our algorithm would predict $sauce\_isn't\_tangy > have\_added\_vinegar$ because our algorithm for DofE interpretation relies on actions preceding their effects in the dialogue. A1 followed by B2 makes sense because B is explaining why A's desired effect hasn't been achieved by alluding to the expectation that $adding\_vinegar > makes\_things\_tangy$, which is a case we will leave for future work.

To avoid overgenerating concessive interpretations in cases in which inappropriate expectations are generated, we will need to check whether the expectation generated from B2 (given the one generated from A) makes sense given the plan.

---
[7] Following Grice's Maxims in his Cooperative Principle.
[8] Unless introducing contrast at the SA level, e.g., with "even". We will consider SA interaction in future work.

In either A1 or A2 followed by B2, the expectation *haven't_added_vinegar* > *sauce_is_OK* doesn't make sense, since the sauce won't be OK (or ready) until all the ingredients (including vinegar) have been added. For A1 followed by B3, since the effects of A and B correspond to different levels of granularity in the plan, they should not be interpreted concessively or in terms of DofE.

To prevent overgenerating DofE in cases like A1 followed by B2, we will check whether the expectations generated match the predicted effects from the planning rules. Since B is describing the action that resulted in the effect described by A, the converse holds, and we can check that the converse of the rule from the plan (i.e., *sauce_is_tangy* > *have_added_vinegar*) isn't violated, which in this case it is, so the algorithm should not predict DofE in such a case.

4.1 Annotation of TODs

Since our goal is to determine what information about the contrasted speaker turns allows us to distinguish between DofE and concession, we first consider previous annotation schemes for TODs to determine whether they contain information that will be relevant for distinguishing between these relations. The DAMSL scheme [1] annotates dialogue acts (DAs) in several layers, accounting for (among other things) forward and backward looking functions that a given utterance unit may perform and describes how one distinguishes between these functions. For example, this scheme allows one to annotate an utterance unit (within a speaker turn) as both an assertion or information-request (forward looking functions) and also an agreement with partial acceptance (backward looking functions). Sikorski and Allen [8] propose an additional scheme to mark higher-level problem solving actions in TOD.

However, neither of these schemes annotates utterance units with lower-level planning operations to note whether they describe preconditions, effects, actions, or goals in the joint-task-plan being formed in the dialogue. We hypothesise that this information about what planning operations are being communicated might help distinguish which relation/s to generate in a given situation.

To this end, we performed an initial pilot study which involved annotating 20 examples each of both the MAPTASK and TRAINS TODs.[9] We searched for cross-speaker "but" examples which contrasted with material in the preceding speaker turn (ignoring occasional intervening turns involving back-channeling and utterances that signalled understanding). We then annotated both the relevant preceding utterance unit and the utterance unit containing the "but" with DAMSL tags, problem solving actions [8], and the planning operators involved. Annotation of planning operators involved (for each utterance unit) noting (1) the mood of the unit (i.e., is it a command, question, or assertion; while this does not involve planning operators, it was annotated because it reflects syntactic considerations which don't appear in either of the other two annotation

---

[9] MAPTASK dialogues can be browsed interactively at www.ltg.ed.ac.uk/ãmyi/maptask/demo.html and TRAINS dialogues can be found at www.cs.rochester/research/trains/.

schemes used), (2) whether the unit describes an action, effect, goal, precondition or constraint in the evolving joint-plan, and (3) whether this planning operation occurs in the present, past or future. Since this was an initial pilot study, only one annotator marked the two dialogues and no measures of intercoder agreement are available as yet (this will be done in future work). Of the 20 examples in each corpus (chosen only if they clearly involved cross-speaker contrast), Table 1 shows the distribution of cases.

Table 1. Classification of examples (20 examples from each corpus)

Corpus	Omitted	Clarification	Concession	DofE	Concession & DofE
MAPTASK	7	3	0	9	1
TRAINS	2	2	12	1	3

The corpora differ interestingly in the DofE:Concession ratio, with many more cases of DofE than concession in MAPTASK and vice versa for TRAINS, possibly because of the difference in corpora; TRAINS involves agents determining the optimal plan for actions that will subsequently be carried out while MAPTASK involves agents interleaving their speech and task-related actions rather than planning future ones. The fact that at least half these cross-speaker relations are either cases of concession or DofE is a good motivation for modelling these relations.

We then examined the annotated examples of concession and DofE to see if there are any annotation trends we might infer from DAMSL tags which might help us to predict which relation to generate, shown in Table 2 below, where "T1" below refers to the preceding speaker turn and "T2" to the turn containing "but" which responds to T1. The tags in the table are abbreviated from the DAMSL scheme, so Commit commits the speaker to some future action, while AD (action directive) and OO (open option) influence the listener's future actions. Reject and Accept can be partial, and IR is an information request. For now we simply consider individual tags rather than combinations within or across turns; this is left for future work when more examples have been annotated. Highly infrequent tags have also been omitted from the table, and only counts for cases of DofE and concession are shown.

One easily drawn observation is that the two corpora seem to indicate completely dissimilar trends for the same relations, which might be due to the differences between the corpora, or might indicate that the DAMSL tags don't provide enough discriminating information to predict relations from these tags alone. Possibly combinations of tags both (or either) within and across turns might be necessary to predict relations.

The next question to ask is whether our hypothesis that more planning-operator-specific information can help predict relations. We count the number of operator pairs for each relation ("Conc." stands for concession) and corpus

Table 2. DAMSL annotations: Counts correspond to DofE, Concession classification

Corpus & Turn	Assert	Commit	AD	OO	Reject	Accept	IR	Answer
TRAINS T1	4;9	2;1	0;2	0;2				
TRAINS T2	4;15			0;1	0;4	1;8	0;1	0;1
MAPTASK T1	2;1		1;0				7;0	
MAPTASK T2	4;1	1;0	3;0			0;1	1;0	7;0

in Table 3 below. The abbreviations correspond to Action, Effect, Goal and Precondition. Since we only consider 20 examples in each corpus, not all of which are classified as either DofE or concession, the data is sparse. For that reason, operator-pair counts for TRAINS DofE and MAPTASK concession should be ignored since there were so few examples (likewise for Table 2).

Table 3. Counts for planning operator sequences for TRAINS, MAPTASK

Relation	A,E	A,P	G,G	E,P	E,E	A,A	P,P	P,G	P,A
Conc.	1;0	4;0	1;0	4;0	1;0	1;1		1;0	1;0
DofE		1;1		2;1		0;3	1;3		0;1

However for the more numerous corpus-relation pairs, it appears as if concession seems to involve action–precondition or effect–precondition sequences more frequently, and DofE seems to involve action–action or precondition–precondition pairs. While these inferences are highly speculative since this is only a pilot study, they seem somewhat reasonable. For example, in concession, the second speaker (henceforth abbreviated as B) often points out a precondition for some future action that won't be met if a certain effect of some preceding action holds; i.e., B gives an opposing argument for the success of the given plan or action (which is the TC). Similarly if A (first speaker) proposes an action, B can concede the action but argue against it in the plan by pointing out a subsequent precondition failure. It also makes sense that sequences of actions (or preconditions for sequential actions) in the plan can function as DofEs, since the first action can indicate an expectation about subsequent actions which is then denied (e.g., if B knows information that A doesn't).

While many other planning operator sequences seem to lend themselves well to concession and DofE in similar ways, we will need to determine in further research whether some combination of planning operator sequence and DAMSL tags help predict relations somewhat. It might also be the case that we need to fine-tune these operator sequences and annotate whether they describe operators related by (for example) dominance or satisfaction-precedence, whether one planning operation precludes another one, or where they fall within the plan itself (e.g., relative temporal order, granularity, etc.).

4.2 Framework for an Algorithm for Cross-Speaker "But" in TOD

Our goal is to formulate an algorithm that generates the appropriate interpretations and bars the inappropriate ones based on how A's and B's adjacent turns match their evolving task-plans. Clearly determining which interpretations to generate in given situations is the tough part, and one which we hope will become feasible given a better characterisation of the specific situations involved (which is what the annotation work attempts to do). Given this, we would then generate the appropriate concessive and/or DofE interpretations and then update the private beliefs of the DP whose turn it is with the appropriate rhetorical rule/s and expectations, so the framework for an algorithm which would generate the appropriate interpretations would go as follows:

1. Initiate the update process with the two propositions related by "but"; for now we assume that these are just A and B's adjacent assertions.
2. Determine which interpretations (DofE, concessive, both or neither) to generate.
3. In the update module
   a) If concessive interpretation was generated, add both defeasible rules and rhetorical relation to B's private beliefs (in the IS).
   b) If DofE was generated, add DofE relation and defeasible rule to B's private beliefs.
4. Determine which interpretations for A to assume about B (after hearing the "but") by following step 2 with A's plan. Depending on what expectations A generates, she will respond accordingly, and potential mismatches will promptly be detected.

One last point to note is that these expectations are salient with respect to a given planning stage or task goal to be achieved; once the goal has been achieved, the expectations are no longer contextually relevant and should be discarded. The reason for this is the specificity of expectations formed by both the DofE and concessive interpretations in TOD and their close connection to planning stages. They are intended to aid local resolution of planning conflicts, but will not remain valid after the goal that they apply to has been resolved. So while the concessive and DofE interpretations remain in the dialogue history, the expectations themselves must be removed from the DPs' private beliefs after the goals they are relevant for have been resolved.

4.3 How This Helps Understanding and Response Generation

This approach attempts to predict which interpretations (i.e., DofE, concessive, both or neither) B might generate given how the assertion in A's turn fits into B's plan. Our goal is to describe conditions which would result in the formation of defeasible expectations that license the use of "but". Since we consider cases in which some form of contrast, conflict or defeated expectation is indicated via "but", we are focusing on how B detects and indicates (1) a potential planning mismatch as in Ex.4 B2, (2) a forgotten action (e.g., Ex.3 B2), or (3) an undesirable effect (e.g., Ex.3 B1). Since the plan being searched to determine these interpretations is B's plan, we ascribe these expectations to B's private beliefs after B's utterance occurs. Inferring these expectations will affect B's subsequent behaviour in the dialogue while the goals they are relevant for are salient.

We also determine interpretations from A and B's assertions from A's perspective (i.e., with respect to A's plan) following B's "but" turn. A searches her plan and determines which interpretations (DofE and/or concessive) B communicated according to the predictions of the algorithm above applied to her plan, and then updates her private beliefs accordingly. If there is a mismatch in their plans, A will not generate the same expectations as B. Assuming that A has a social obligation to respond to all interpretations (including rhetorical relations) she assumes are conveyed by B when her turn arrives, B should hopefully be able to detect a mismatch (if one occurs) based on A's response. For example suppose that after hearing B2 in Ex.4, A applies the algorithm, which predicts that both DofE and concessive interpretations should be generated (after searching A's plan). A then has an obligation to respond to these relations and expectations. Depending on her private beliefs, she might respond (A[3]) "I agree that if we've added broccoli we shouldn't add beans. I also think the sauce isn't going to turn out OK if we've already added broccoli." If B has the same expectations, A's explicit acknowledgement grounds them; otherwise if B doesn't have the same expectations, this enables B to detect this mismatch in beliefs and respond. For example, if B only intended to communicate the DofE interpretation, he might respond to A's acknowledgement with (B[4]) "But I thought we were making vegetable sauce which doesn't have beans."

5 Conclusions and Further Work

In this paper we address the oddness of plan-based "but" in TOD and show how it can't be treated as simple DofE in many cases. We present a novel treatment of concession in dialogue as a possible alternative and propose an approach for determining the concessive expectations and TC in plan-based TOD. We then informally survey some TOD annotation schemes in order to determine whether additional information about planning operators involved in A and B's turns might help distinguish which interpretations are licensed depending on how the operators they describe fit into each other's plans. We lay the framework for an algorithm that predicts which interpretations to generate that will hopefully constrain some of the overgeneration that might otherwise occur. Our approach follows a perspectivised treatment of dialogue that enables detection of mismatches in planning, whether with beliefs, goals, or the structure of the plan itself from a given speaker's perspective (i.e., with respect to their plan). The DP's private beliefs are then updated with these interpretations and contextually relevant expectations. Social obligations which require DPs to respond to these inferred interpretations ensure that they are communicated, enabling the other DP the chance to detect a planning mismatch and respond promptly.

In future work we hope to extend this treatment to cover more plan-based TOD cases and will conduct a more formal study of annotation schemes to determine what information helps distinguish which relations to predict. We will then incorporate these considerations into a distinguishing algorithm and evaluate it on a range of TOD corpora.

References

1. Allen, J., Core, M.: DAMSL: Dialog Act Markup in Several Layers. Available from http://www.cs.rochester.edu/research/cisd/resources/damsl/ (1997)
2. Ginzburg, J.: Interrogatives: Questions, Facts and Dialogue. Blackwell, Oxford, U.K. (1996)
3. Thomas, K., Matheson, C.: Modelling Denial of Expectation in Dialogue: Issues in Interpretation and Generation. In Proc. of the Sixth Annual CLUK Research Colloquium, Edinburgh, Scotland (2003)
4. Knott, A.: Discourse Relations as Descriptions of an Algorithm for Perception, Action and Theorem-proving. In Proc. of the International Workshop on Levels of Representation in Discourse (1999)
5. Lagerwerf, L.: Causal Connectives Have Presuppositions. PhD Thesis, Catholic Univ. of Brabant, Holland Academic Graphics, The Hague, The Netherlands (1998)
6. Matheson, C., Poesio, M., Traum, D.: Modelling Grounding and Discourse Obligations Using Update Rules. Proceedings of the North American Association for Computational Linguistics (2000)
7. Poesio, M., Traum, D.: Towards an Axiomatization of Dialogue Acts. Proceedings of Twente Workshop (1998)
8. Sikorski, T., Allen, J.: A Scheme for Annotating Problem Solving Actions in Dialogue. Working Notes of AAAI Fall Symposium on Communicative Actions in Humans and Machines (1997)
9. Traum, D., Allen, J.: Discourse Obligations in Dialogue Processing. Proceedings of the Association for Computational Linguistics (1994)

Dynamic Contextual Intensional Logic: Logical Foundations and an Application

Richmond H. Thomason

Philosophy Department
University of Michigan
Ann Arbor, MI 48109-2110, USA
rich@thomason.org

Abstract. This paper develops further aspects of Contextual Intensional Logic, a type-theoretic logic intended as a general foundation for reasoning about context. I motivate and formulate the dynamic version of the logic, prove some results about it, and show how it can be used to specify inter-contextual reasoning.

1 Introduction

Since 1997, I have advocated a type-theoretic approach to the logical formalization of context, which incorporates ideas of Alonzo Church, Richard Montague, and David Kaplan. The most complete available presentation of the core formalism, Contextual Intensional Logic (CIL), can be found in [1], along with references to the earlier work.

As soon as you have the idea of context, you need the idea of context dynamics. This need shows up in its purest form in the semantics of programming languages, where the context is an assignment of values to variables, and where the languages contain expressions (like '$x := 2$', i.e., 'Let x be 2') that change the assignment. Since all that a program can do between receiving an input and producing an output is to change variable assignments, there could be no nontrivial programs without context-changing operators.

Such phenomena are pervasive in natural language as well: an utterance can change the context that then figures in the interpretation of subsequent utterances. In retrospect, you can see this as a central insight of speech act theory; [2] contains one of the first explicit formulations of the point in the philosophical literature. Of course, this idea led to the enormous amount of work that has been invested in developing the various dynamic approaches to natural language semantics. And in computational applications of context, you want reasoning not only to depend on context, but to manipulate it.

2 Static CIL

CIL is a static logic; but one of the lessons of dynamic semantics is that a more or less standard logic can be dynamicized; see, for instance, [3]. In particular, dynamic versions of Intensional Logic have been produced as part of the project of making Montague grammar dynamic; see [4]. Accounting for context-shifting effects by making Contextual Intensional Logic dynamic is therefore a very natural and attractive program.

This program is sketched in [1], which contains a formulation of the logic DCIL. Here, the logic is formulated in greater detail, illustrated with an extended example, and is generalized to incorporate a record of the series of contexts that a reasoner has traversed; this generalization is needed to accommodate McCarthy's "exit" operation.

CIL is based on the architectural principles that underlie Montague's Intensional Logic [5], and is in fact an extension of IL. The syntax and model theory of (static) IL can be summarized as follows.

2.1 Types

(2.1.1) e is a type.
(2.1.2) t is a type.
(2.1.3) i is a type.
(2.1.4) If σ and τ are types, so are $\langle \sigma, \tau \rangle$ and $\langle w, \tau \rangle$.

Here, e is the type of individuals, t is the type of truth-values, and i is the type of indices. Indices are simultaneous disambiguations of every linguistic expression whose interpretation depends on context, including indexical expressions and lexically ambiguous expressions. Thus, indices are similar to "contexts", as David Kaplan uses the term. A type of the form $\langle w, \tau \rangle$ represents functions from possible worlds to objects of type τ. In this formulation there is no primitive type of possible worlds, although worlds can enter into complex types.

2.2 Syntax of CIL

A CIL language L is a function from the set of CIL types to nonempty sets of expressions; where τ is a type, L_τ is the set of basic constants of type τ. For each type τ, we postulate a denumerable set Var_τ of variables of type τ. (This set is the same for all CIL languages.) $L_\tau \cup Var_\tau$ is the set of basic expressions of type τ.

The following recursion extends L to a function L^* taking each type into the set of (basic or complex) expressions of that type.

Basic expressions: $L_\tau \cup Var_\tau \subseteq L_\tau^*$.

Identity: If ζ and $\xi \in L_\tau^*$, then $\zeta = \xi \in L_t^*$.

Functional application: If $\zeta \in L_{\langle \sigma, \tau \rangle}^*$ and $\xi \in L_\sigma^*$, then $\zeta(\xi) \in L_\tau^*$.

Lambda abstraction: If $\zeta \in L_\tau^*$, then $\lambda x_\sigma \zeta \in L_{\langle \sigma, \tau \rangle}^*$.

Intension: If $\zeta \in L_\tau^*$, then $^\wedge \zeta \in L_{\langle w, \tau \rangle}^*$.

Extension: If $\zeta \in L_{\langle w, \tau \rangle}^*$, then $^\vee \zeta \in L_\tau^*$.

Character formation: If $\zeta \in L_\tau^*$ then $^\sqcap \zeta \in L_{\langle i, \tau \rangle}^*$.

Content determination: If $\zeta \in L_{i,\tau}^*$ then $^\sqcup \zeta \in L_\tau^*$.

2.3 Domains of Arbitrary Type

Let D be a function taking the basic types e, w, and i into nonempty sets, or *domains*. The following recursive definition extends D to an assignment of a domain to arbitrary types. In this definition, $\top$ and $\bot$ are arbitrary different individuals, taken to stand for truth and falsity; I will assume that these two elements belong only to the domain D_t.

(2.3.1) $D_t = \{\top, \bot\}$.
(2.3.2) $D_{\langle \sigma, \tau \rangle} = D_\tau^{D_\sigma}$.

The organization of type theory around a family of domains that are constructed using the operation in (2.3.2) of forming a domain by taking the set of functions from one domain to another is due to [6]. Functions prove to be a powerful and useful organizing idea for such a logic. Montague extended Church-Henkin type theory by adding a domain of possible worlds; CIL extends Montague's logic by adding a domain of indices. In each case, the model theory is a straightforward extension of the one provided by Church. (Also, see [7].)

2.4 Models

A (static) model $\mathcal{M}$ of L on a domain assignment D is an assignment of a member $[\![x_\tau]\!]_\mathcal{M}$ of D_τ to each variable x_τ in Var_τ, and of a member $[\![\xi]\!]_{\mathcal{M},i,w}$ of D_τ to each basic constant in L_τ, for each $i \in D_i$ and $w \in D_w$. (Note that, unlike the values of constants, the values of variables are world- and index-independent.) In the rule that interprets lambda abstraction (clause (2.4.3), below), we appeal to the following device of semantic substitution for the variable x_τ: where $d \in D_\tau$, $\mathcal{M}^{d/x}$ is the model $\mathcal{M}'$ that is like $\mathcal{M}$ except that $[\![x]\!]_{\mathcal{M}'} = d$.

Such a model assigns a value $[\![\xi]\!]_{\mathcal{M},i,w} \in \tau$ to each expression $\xi \in L_\tau^*$, for each $i \in D_i$ and $w \in D_w$. This assignment conforms to the following rules of semantic interpretation, which uniquely determine the values assigned to complex expressions, given an assignment of values to basic expressions.

(2.4.0) For $x_\tau \in Var_\tau$, $[\![x_\tau]\!]_{\mathcal{M},i,w} = [\![x_\tau]\!]_\mathcal{M}$.
(2.4.1) $[\![\zeta = \xi]\!]_{\mathcal{M},i,w} = \top$ if $[\![\zeta]\!]_{\mathcal{M},i,w} = [\![\xi]\!]_{\mathcal{M},i,w}$. Otherwise $[\![\zeta = \xi]\!]_{\mathcal{M},i,w} = \bot$.
(2.4.2) $[\![\zeta(\xi)]\!]_{\mathcal{M},i,w} = [\![\zeta]\!]_{\mathcal{M},i,w}([\![(\xi)]\!]_{\mathcal{M},i,w})$.
(2.4.3) Where $\zeta \in L_\tau^*$ and $x_\sigma \in Var_\sigma$, $[\![\lambda x_\sigma \zeta]\!]_{\mathcal{M},i,w}$ = the function f from D_σ to D_τ such that $f(d) = [\![\zeta]\!]_{\mathcal{M}^{d/x},i,w}$.
(2.4.4) Where $\zeta \in L_\tau^*$, $[\![\wedge \zeta]\!]_{\mathcal{M},i,w}$ = the function f from D_w to D_τ such that $f(w') = [\![\zeta]\!]_{\mathcal{M},i,w'}$.
(2.4.5) Where $\zeta \in L_{\langle w,\tau \rangle}^*$, $[\![\vee \zeta]\!]_{\mathcal{M},i,w} = [\![\zeta]\!]_{\mathcal{M},i,w}(w)$.
(2.4.6) Where $\zeta \in L_\tau^*$, $[\![\ulcorner \zeta]\!]_{\mathcal{M},i,w}$ = the function f from D_i to D_τ such that $f(i') = [\![\zeta]\!]_{\mathcal{M},i',w}$.
(2.4.7) Where $\zeta \in L_{i,\tau}^*$, $[\![\cup \zeta]\!]_{\mathcal{M},i,w} = [\![\zeta]\!]_{\mathcal{M},i,w}(i)$.

In CIL, contexts are considered to involve two separate components: (1) a modal component, which has type $\tau_M = \langle\langle w,t\rangle,\langle w,t\rangle\rangle$ and (2) an indexical component, which has type i. According to McCarthy's account of contexts, a context c performs two distinct functions: it provides a source of specialized knowledge, and it serves to disambiguate the meanings of expressions. The two components of CIL contexts represent these two functions. The first component is a function that inputs a proposition and outputs a proposition; given a proposition p, this function returns the proposition that p is known in the context. The second component is an index; indices represent disambiguation policies. Separating these two functions of context is enforced by the type-theoretic framework of CIL. But the distinction also seems to be a highly useful representational and conceptual device.

Definition 2.1. *Contexts and their components.*
A *context*, relative to a domain assignment D, is a pair $\langle M,i\rangle$, where $M \in D_{\tau_M}$ and $i \in D_i$.[1] The modal component M of a context $c = \langle M,i\rangle$ is denoted by c_M; the indexical component i is denoted by c_i. (Therefore, $c = \langle c_M, c_i\rangle$.)

3 Dynamic CIL

Many natural forms of contextual reasoning involve navigating contextual space. If someone standing on the opposite side of the table tells us to move a pitcher on the table to the right, we may have to imagine things from their perspective to execute the instruction properly. [8] suggest that operations of entering and exiting contexts may be useful in many forms of contextual reasoning.

Dynamic logic (see, for instance, [9]) provides a natural way to integrate change of context into a logical formalism, and, in particular, to provide a logical semantics for context-changing operators. In typical formalizations of dynamic logic, the context is taken to be an assignment of values to variables; but it is perfectly possible to apply the ideas of dynamic logic to the contexts of CIL. That is the project that we will undertake here.

The project of modifying CIL to obtain Dynamic Contextual Intensional Logic (DCIL) is unlike Montague's project of extending the Church-Henkin type theory to IL, and my similar expansion of IL to CIL. In the latter cases, the basic model theoretic techniques are carried over without much change, though the ontology of the models is expanded by adding new primitive domains, and some constructions may be added to the language.

On the other hand, DCIL uses the same types and domains as CIL, and adds no new constructions to the syntax. But there are additions to the parameters on which denotations depend which result in rather fundamental changes to the logic.

We can best explain the nature of the change by concentrating on sentence-like formulas—expressions of type t. A model $\mathcal{M}$ of CIL assigns a denotation

---
[1] Recall that τ_M is $\langle\langle w,t\rangle,\langle w,t\rangle\rangle$.

$[\![\phi]\!]_{\mathcal{M},i,w}$ in $\{\top, \bot\}$ to a formula $\phi \in L_t^*$. Abstracting on the parameters i and w on which the truth-value depends we can repackage the interpretation of ϕ by saying that $\mathcal{M}$ assigns a set of index-world pairs to ϕ. Or, fixing world parameter w, CIL assigns a set S of indices to ϕ relative to w.

Generalizing this last idea, let's think of the interpretation S of ϕ at w as a set of contexts. Recall that, for us, contexts are modality-index pairs; so that we are simply adding another parameter to the interpretation, which represents a modality. In CIL, this addition is vacuous, because there are no formulas whose interpretation depends on the modality component. Nevertheless, the generalization provides a useful bridge to the dynamic logic.

According to this way of looking at the static logic, then, a sentential formula (relative to a model and a world) corresponds to the set of contexts in which it is true.

In dynamic logic, we are interested in expressions like 'enter context c' that enforce a change of context. We can interpret such expressions as sets of changes of context, where a change of context is simply a pair $\langle c, c' \rangle$ of contexts. This means that in dynamic CIL we want models to assign values to expressions relative to a world and *a pair* of contexts: $[\![\zeta]\!]_{\mathcal{M},c,c',w}$ is the value assigned to ζ relative to a world w and the contexts c and c'.

The intuitions guiding the extension of a static logic to one that is dynamic are not always as robust as one would like, especially when we are dealing with moderately powerful logics. In the case of DCIL, intuitions about the dynamic interpretation of higher types are not always entirely clear, and the fact that we are working without truth-value gaps also creates some difficulties. But the core of the following development of DCIL is, I believe, quite plausible. In explaining the logic I will concentrate on this core, and will try to show in terms of an example that the logic is potentially useful in specifying certain forms of reasoning about context.

3.1 Types and Syntax of DCIL

The types and the syntax of DCIL are the same as in CIL; there are no new syntactic constructions.

The complex expressions of DCIL are formed as in CIL. We add two basic expressions with fixed model-theoretic interpretations: @$_i$, which has type i, and @$_M$, which has type τ_M.[2] @$_i$ denotes the index of the current context, and @$_M$ denotes the modality of the current context.

These two new constants are (in a sense to be explained) static. Later, we will add other constants which are nontrivially dynamic.

---

[2] The symbol '@' is sometimes used in modal logic to denote the actual world ; this is a generalization of that notation to other types of semantic parameters.

3.2 Model Theory of DCIL

Definition 3.2. *Dynamic models.*
A *dynamic model* $\mathcal{M}$ of L on a triple $\langle D_e, D_w, D_i \rangle$ of domains is an assignment of a member $[\![\xi]\!]_{\mathcal{M},c,c',w} \in D_\tau$ to each basic expression $\xi \in L_\tau$, for each $c, c' \in D_{\tau_M} \times D_i$ and each $w \in D_w$.

This definition allows expressions of any type to denote nontrivial sets of context changes. For instance, a model $\mathcal{M}$ could assign a constant a of type e the following value.

$[\![a]\!]_{\mathcal{M},c,c',w} = a$ if and only if $c' = \langle M, i \rangle$, and
$[\![a]\!]_{\mathcal{M},c,c',w} = a'$ if and only if $c' = \langle M', i' \rangle$.

I can't think of any useful purpose that such a constant could serve. It seems that our core intuitions concerning dynamic interpretations only concern expressions of type t. No doubt this is related to the fact that it is utterances that that have the potential to change the context, and utterances are sentential.

In any case, the only plausible dynamic constructions—and the only ones with which we will be concerned here—have the sentential type t or, like dynamic conjunction, have a functional type that produces type t values.

We want the other expressions of DCIL to be vacuously or trivially dynamic; that is, we want these expressions to be static. In dynamic logic, it would be most natural to treat a static expression as one that is undefined except at identity transitions (transitions from c to c). But in the version of DCIL with which we are working, there are no "denotation gaps"; in every model $\mathcal{M}$, every expression ζ receives a denotation at every world, for every transition from c to c'. This is a logic development policy; adding denotation gaps is a complex matter that I would prefer to address in a separate paper, or series of papers. For the present, then, we have to find a solution to the problem of characterizing vacuously dynamic expressions in a gapless logic.

In the case of expressions of type t, it is natural to say that a static formula is false except at identity transitions. This definition must be generalized to expressions of arbitrary type. For this purpose, for each type τ we need to define a "null denotation" of this type. The following definition accomplishes this.

Definition 3.3. *Null value* $\perp_{\mathcal{M},\tau}$.
The null value of type τ is defined by the following recursion.

Basis. $\perp_{\mathcal{M},t} = \perp$; $\perp_{\mathcal{M},e} \in \text{Dom}_{\mathcal{M}}(e)$; $\perp_{\mathcal{M},w} \in \text{Dom}_{\mathcal{M}}(w)$; $\perp_{\mathcal{M},i} \in \text{Dom}_{\mathcal{M}}(i)$.

Induction. $\perp_{\mathcal{M},\langle\sigma,\tau\rangle}$ is the constant function f from $\text{Dom}_{\mathcal{M}}(\sigma)$ to $\text{Dom}_{\mathcal{M}}(\tau)$ such that $f(d) = \perp_{\mathcal{M},\tau}$ for all $d \in \text{Dom}_{\mathcal{M}}(\sigma)$.

Definition 3.4. *Static expression.*
An expression ζ is *static* relative to a model $\mathcal{M}$ iff for all c, c', w, if $c \neq c'$ then $[\![\zeta]\!]_{M,c,c',w} = \perp_{\mathcal{M},\tau}$.

The semantic rules for the basic expressions $@_M$ and $@_i$ are as follows.

If $c = c'$ then $[\![@_M]\!]_{\mathcal{M},c,c',w} = c_M$; otherwise, $[\![@_M]\!]_{\mathcal{M},c,c',w} = \bot_{\mathcal{M},\tau_M}$.
If $c = c'$ then $[\![@_i]\!]_{\mathcal{M},c,c',w} = c_i$; otherwise, $[\![@_i]\!]_{\mathcal{M},c,c',w} = \bot_{\mathcal{M},i}$.

Note that according to these rules, $@_M$ and $@_i$ are interpreted as static expressions.

The denotations of complex expressions in DCIL for the most part are characterized by clauses that look just like the corresponding clauses for CIL. The only exceptions are clause (3.2.2) for identity, and clause (3.2.6), for character formation. Both clauses render their constructions static. These restrictions are hard to motivate using direct intuitions, but they simplify things, and they are crucial for the proof of Theorem 3.1.

The semantic rules are as follows.

(3.2.0) For $x_\tau \in Var_\tau$, $[\![x_\tau]\!]_{\mathcal{M},c,c',w} = [\![x_\tau]\!]_{\mathcal{M}}$.
(3.2.1) Where ζ and ξ are expressions in L^*_τ, $[\![\zeta = \xi]\!]_{\mathcal{M},c,c',w} = \top$ iff $c = c'$ and $[\![\zeta]\!]_{\mathcal{M},c,c',w} = [\![\xi]\!]_{\mathcal{M},c,c',w}$.
(3.2.2) Where $\zeta \in L^*_{\langle\sigma,\tau\rangle}$ and $\xi \in L^*_\sigma$,
$[\![\zeta(\xi)]\!]_{\mathcal{M},c,c',w} = [\![\zeta]\!]_{\mathcal{M},c,c',w}([\![(\xi)]\!]_{\mathcal{M},c,c',w})$.
(3.2.3) Where $\zeta \in L^*_\tau$, $[\![\lambda x_\sigma \zeta]\!]_{\mathcal{M},c,c',w}$ = the function f from D_σ to D_τ such that $f(d) = [\![\zeta]\!]_{\mathcal{M},c,c',w}\mathcal{M}^{d/x}$.
(3.2.4) Where $\zeta \in L^*_\tau$, $[\![\char`\^\zeta]\!]_{\mathcal{M},c,c',w}$ = the function f from D_w to D_τ such that $f(w') = [\![\zeta]\!]_{\mathcal{M},c,c',w'}$.
(3.2.5) Where $\zeta \in L^*_{\langle w,\tau\rangle}$, $[\![\char`\v\zeta]\!]_{\mathcal{M},c,c',w} = [\![\zeta]\!]_{\mathcal{M},c,c',w}(w)$.
(3.2.6) Where $\zeta \in L^*_\tau$, if $c = c'$ then $[\![\ulcorner\zeta]\!]_{\mathcal{M},c,c',w}$ = the function f from D_i to D_τ such that $f(i) = [\![\zeta]\!]_{\mathcal{M},\langle c_M,i\rangle,\langle c_M,i\rangle,w}$. If $c \neq c'$ then $f = \bot_{\mathcal{M},\langle i,\tau\rangle}$.
(3.2.7) Where $\zeta \in L^*_{\langle i,\tau\rangle}$, then $[\![\llcorner\zeta]\!]_{\mathcal{M},c,c',w} = [\![\zeta]\!]_{\mathcal{M},c,c',w}(c_i)$.

The following result shows that DCIL is, in a sense, a conservative extension of CIL.

Definition 3.5. *Converting static to dynamic models.*
Let $\mathcal{M}$ be a static model of a language L (without $@_M$ or $@_i$).[3] The *dynamic equivalent* $dyn(\mathcal{M})$ of $\mathcal{M}$ is the dynamic model $\mathcal{M}'$ defined as follows:

If x_τ is a variable of type τ, then $[\![x_\tau]\!]_{\mathcal{M}'}$ is $[\![x_\tau]\!]_{\mathcal{M}}$.
If $\zeta_\tau \in L^*_\tau$ is a constant of type τ, then $[\![\zeta]\!]_{\mathcal{M}',c,c',w}$ is $[\![\zeta]\!]_{\mathcal{M}',c_i,w}$ if $c = c'$, and is $\bot_{\mathcal{M},\tau}$ otherwise.

Definition 3.6. $cl(\mathcal{M})$.
Let $\mathcal{M}$ be a (static or dynamic) model. Then $cl(\mathcal{M})$ is the smallest set of models containing $\mathcal{M}$ and such that if $\mathcal{M}' \in cl(\mathcal{M})$, $x \in Var_\tau$, and $d \in D_\tau$, then $\mathcal{M}'^{d/x} \in cl(\mathcal{M})$.

Theorem 3.0. For any static model $\mathcal{M}$ of L, the dynamic equivalent $dyn(\mathcal{M})$ of $\mathcal{M}$ is equivalent to $\mathcal{M}$, in the sense that for any type τ and any $\zeta \in L^*_\tau$, if $[\![\zeta]\!]_{dyn(\mathcal{M}),c,c',w} \neq \bot_{\mathcal{M},\tau}$ then $c = c'$ and $[\![\zeta]\!]_{dyn(\mathcal{M}),c,c',w} = [\![\zeta]\!]_{\mathcal{M},c_i,w}$.

---
[3] I have left out $@_M$ and $@_i$ to simplify things; the theorems apply if they are present, but CIL would need to be reformulated to accommodate these constants.

Proof. Let $\mathcal{M}_1$ be a static model. The proof is a straightforward induction on the syntactic complexity of expressions ζ. The hypothesis of induction is that the following holds for all $\zeta \in L_\tau^*$:

For all $\mathcal{M}_1' \in cl(\mathcal{M}_1)$, if $c = c'$ then
$[\![\zeta]\!]_{dyn(\mathcal{M}_1'),c,c',w} = [\![\zeta]\!]_{\mathcal{M}_1',c_i,w}$, and if $c \neq c'$
then $[\![\zeta]\!]_{dyn(\mathcal{M}_1'),c,c',w} = \perp_{\mathcal{M}',\tau}$.

The proof of the theorem consists in verifying the hypothesis of induction for each type of expression; each of the cases is straightforward.

A vacuous dynamic model is one in which the modal constituent of a context has no effect on semantic values, and in which all constants are interpreted as tests.

Definition 3.7. *Vacuous dynamic models.*

A dynamic model $\mathcal{M}$ is *vacuous* in case (1) for all τ and all $\zeta \in L_\tau$, $[\![\zeta]\!]_{\mathcal{M},\langle M,i\rangle,\langle M,i\rangle,w} = [\![\zeta]\!]_{\mathcal{M},\langle M',i\rangle,\langle M',i\rangle,w}$ for all modalities $M, M' \in D_{\tau_M}$ and (2) for all constants $\zeta \in L_\tau$, $[\![\zeta]\!]_{\mathcal{M},c,c',w} = \perp_{\mathcal{M},\tau}$ if $c \neq c'$.

Theorem 3.0. Every vacuous dynamic model is generated by a static model.

Proof. If $\mathcal{M}$ is vacuous, let $\mathcal{M}'$ be the static model that assigns variables the same values as $\mathcal{M}$, and that assigns constants ζ in L_τ the value $[\![\zeta]\!]_{\mathcal{M}',i,w} = [\![\zeta]\!]_{\mathcal{M},\langle M,i\rangle,\langle M,i\rangle,w}$, where M is an arbitrary modality in D_{τ_M}. It is not difficult to see that $\mathcal{M} = dyn(\mathcal{M}')$.

Putting Theorems 3.1 and 3.2 together, it follows that every vacuous dynamic model is equivalent to a static model.

4 Adding Some Dynamic Basic Expressions

For the dynamics of DCIL to be at all useful, we clearly must add some expressive power. We do this by adding nontrivially dynamic basic expressions.

4.1 Static Boolean Connectives

Church-style type theories in general, including CIL, provide powerful definitional mechanisms: in particular, boolean operations and quantifiers of arbitrary type can be defined using only identity and lambda abstraction. The following definitions are taken from [10]. The definition for $\wedge$ can be a bit puzzling; what it says is that Boolean conjunction is the function that gives $\top$ to arguments x and y if and only if every function that gives value y to x also gives $\top$ to $\top$.

$$\top =_{df} \lambda x_t\, x = \lambda x_t\, x$$
$$\perp =_{df} \lambda x_t\, x_t = \lambda x_t\, \top]$$
$$\neg =_{df} \lambda x_t\, [x_t = \perp]$$
$$\wedge =_{df} \lambda x_t \lambda y_t\, [\lambda z_{\langle t,t\rangle}\, [z(x) = y] = \lambda z_{\langle t,t\rangle}\, [z(\top)]]$$
$$\forall_\tau =_{df} \lambda x_{\langle \tau,t\rangle}\, [x = \lambda y_\tau \top]$$

These definitions were designed for a static logic; when incorporated into a dynamic type theory like CIL, they produce static or "test" operations, which produce expressions ζ such that if $c \neq c'$ then $[\![\zeta]\!]_{M,c,c'}w = \perp_{\mathcal{M},\tau}$.

4.2 Nontrivial Dynamicism

To begin with, I will introduce three new constants: ;, M-ENTER, and I-ENTER. ; is dynamic 'and'; its definition is standard. McCarthy proposed a context-shifting operator that allowed one to enter a designated context. Since in DCIL, contexts are divided into two components, the process of entering a context is divided between two dynamic operators. M-ENTER changes the modality of the current context, and I-ENTER changes its index.

The new constants and their formal interpretations are summarized as follows.

;, a basic expression of type $\langle t, \langle t, t \rangle \rangle$.[4]
$[\![;(\phi)(\psi)]\!]_{\mathcal{M},c,c',w} = \top$ iff there is a c'' such that $[\![\phi]\!]_{\mathcal{M},c,c'',w}$ and $[\![\psi]\!]_{\mathcal{M},c'',c',w}$.

M-ENTER, a basic expression of type $\langle \tau_M, t \rangle$.
Where $\zeta \in L^*\tau_M$ and $c_i = i$, $[\![\text{M-ENTER}(\zeta)]\!]_{\mathcal{M},c,c',w} = \top$ iff $c' = \langle [\![\zeta]\!]_{\mathcal{M},c,c',w}, i \rangle$.

I-ENTER, a basic expression of type $\langle i, t \rangle$.
Where $\zeta \in L^*\tau_i$ and $c_M = M$, $[\![\text{I-ENTER}(\zeta)]\!]_{\mathcal{M},c,c',w} = \top$ iff $c' = \langle M, [\![\zeta]\!]_{\mathcal{M},c,c',w} \rangle$.

5 A Database Example

I will use the database example from [1] to illustrate how DCIL can be used in the specification of applications. In this example, Ann, Bob and Charlie use personal databases to manage their calendars. The databases use the first-person pronoun to refer to the database user. They also have constants referring to Ann, Bob and Charlie. The databases contain information about meetings; it is important for the databases to agree on scheduled meetings. The databases communicate in order to ensure this. The purpose of the example is to make a convincing case that DCIL could be useful in specifying this communication process.

5.1 Database-Level Formalizations

In [1], I distinguished three levels at which the databases can be formalized. The *database format level* uses a representation that is close to the one actually manipulated by the databases. At this level, for instance, Ann's database might represent a meeting with Bob at 9 o'clock in the form of a triple $\langle I, b, 9 \rangle$, and represent the fact that Bob is aware of this meeting as a quadruple $\langle b, I, a, 9 \rangle$. (This second triple says that Bob's database contains a triple $\langle I, a, 9 \rangle$; so here, I refers to Bob.)

---

[4] We use the more familiar notation '$\phi;\psi$' in place of ';$(\phi)(\psi)$'.

5.2 Knowledge-Level Formalizations

The *knowledge level* describes the databases in much the same way that indirect discourse is used in natural language. The reference of I is fixed by the initial context, and modalities are used to characterize what various databases know. If we describe the databases at the knowledge level from a neutral context that is free of indexicals, all occurrences of I will have to be replaced by nonindexical equivalents. Using an expression MEET of type $\langle e, \langle e, \langle e, t \rangle \rangle \rangle$ for the 3-place meeting relation recorded by the databases, and $\Box a$ for the modality associated with Ann's database, the knowledge-level formulation of Ann's entry $\langle I, b, 9 \rangle$ is ⌜a⌝($^\wedge$MEET($a_e, b_e, 9$)). And the knowledge-level formulation of Ann's entry $\langle b, I, a, 9 \rangle$ is ⌜a⌝(⌜b⌝($^\wedge$MEET($a_e, b_e, 9$))).

5.3 The *ist* Operator and Context-Level Formalizations

The *context level* formalization of Ann's database, uses *ist*. For McCarthy, *ist* is a relation between contexts and sentences (or perhaps their meanings), and keeps track of the sentences that hold in contexts. In DCIL, *ist* is a function that inputs a modality and an index (the components of a context) and a propositional character (that is, a function from indices to propositions), and that outputs a proposition.[5] Therefore, *ist* has the type

$$\langle i, \langle \tau_M, \langle \tau\text{-Char-Prop}, \tau\text{-Prop} \rangle \rangle \rangle,$$

where τ-Prop $= \langle w, t \rangle$ and τ-Char-Prop $= \langle i, \tau\text{-Prop} \rangle$. *ist* can be defined as follows, using lambda abstraction.

(5.3.1) $ist = \lambda u_i \lambda x_{\tau_M} \lambda p_{\tau\text{-Char-Prop}}\, x(p(u))$.

Definition (5.3.1) gives rise to the following rule of *ist*-conversion.

(5.3.2) $ist(\eta)(\zeta)(\phi)$ is equivalent to $\zeta(\phi(\eta))$.

At the knowledge level, the formulation of Ann's entry $\langle I, b, 9 \rangle$ is ⌜a⌝MEET($a_e, b_e, 9$). The context-level formulation of this same entry $\langle I, b, 9 \rangle$ is $^\vee ist(a_i, $⌜$a$⌝$, ^{\ulcorner\wedge}$MEET$(I, b_e, 9))$, and the context-level formulation of $\langle b, I, a, 9 \rangle$ is $^\vee ist(a_i, $⌜$a$⌝$, ^\ulcorner ist(b_i, $⌜$b$⌝$, ^{\ulcorner\wedge}$MEET$(I, a_e, 9)))$.

DCIL provides another level—the *procedural level*—at which protocols can be formalized and proved correct. To illustrate this use of DCIL, suppose that Ann's database maintains its information about Bob's database by regularly communicating with Bob's database and fetching records. For instance, this procedure, on finding $\langle I, a, 9 \rangle$ in Bob's database, will record an entry $\langle b, I, a, 9 \rangle$ in Ann's database.

---

[5] In the example that I will develop, for instance, *ist* would input the modality of knowledge relative to Ann's database, the indices that disambiguates expressions using the policies of Ann's database, and the character associated with the sentence 'I have a meeting with Bob at 9', and would return the proposition that Ann has a meeting with Bob at 9.

Suppose that this procedure revises Ann's records about Bob's database by replacing them with a copy of Bob's meeting calendar. Then for all formulas MEET(I, ζ, ξ), where ζ and ξ are constants of type τ-Char-Prop $= \langle i, \tau\text{-Prop}\rangle$ (these are the formulas that could express the characters of propositions about Bob's meetings in his calendar), the following will hold immediately after the procedure has been applied. The sense of (5.3.3) is that in the context of Ann's database, Ann's database knows about Bob's entry for a meeting if and only if visiting Bob's database produces this entry.

(5.3.3)   [@$_M$ = [a] $\wedge$ @$_i = a_i$] $\rightarrow$ % Start in Ann's DB
       [$^\vee ist(a_i,$ [a] $, ^\sqcap ist(b_i,$ [b] $, ^{\sqcap \wedge}$MEET$(I, \zeta, \xi))) \leftrightarrow$
       M-ENTER([b]); I-ENTER(b_i);   % Enter Bob's DB
       $^\vee$@$_M$([$^{\sqcap \wedge}$MEET(I, ζ, ξ)](@$_i$));   % Test for MEET(I, ζ, ξ)
       M-ENTER([a]); I-ENTER(a_i)]   % Re-enter Ann's DB

As the annotations to the right side of (5.3.3) indicate, this dynamic formula corresponds to an algorithm that realizes the fetching procedure. The correspondence to such an algorithm is fairly coarse. To a large extent, this coarseness is due to the fact that the target procedure involves changes other than changes of context. For instance, the changes to Ann's database can't be formalized in DCIL; nor can the procedure of scanning all the records of Bob's database, which involves dynamic quantification. Also, we have left all temporal considerations out of the picture. By using a more powerful dynamic logic, it should be possible to have a more systematic and compositional relationship between algorithms and formulas of the logic.

The purpose of fetching records from Bob's database is to provide information about Bob's calendar in Ann's database. This purpose can be specified in DCIL by the following scheme, which says that what Ann knows (in her terms) about Bob's knowledge of his meetings is exactly what Bob knows (in his terms) about his meetings.

(5.3.4) $^\vee ist(a_i,$ [a] $, ^\sqcap ist(b_i,$ [b] $, ^{\sqcap \wedge}$MEET$(I, \zeta, \xi)))$
       $\leftrightarrow$ $^\vee ist(b_i,$ [b] $, ^{\sqcap \wedge}$MEET$(I, \zeta, \xi))$

Theorem 5.0. (5.3.4) follows in DCIL from (5.3.3); that is, every model of DCIL that satisfies (5.3.3) also satisfies every instance of (5.3.4), where ζ and ξ are constants in L_e^*.

Proof sketch. We assume that the interpretations of the constants ζ, ξ, and MEET—like most expressions of DCIL[6]—do not depend on the modal component of context. That is,
$[\![\zeta]\!]_{\mathcal{M}, \langle \text{M}, \text{i}\rangle, \langle \text{M}, \text{i}\rangle, w} = [\![\zeta]\!]_{\mathcal{M}, \langle \text{M}', \text{i}\rangle, \langle \text{M}', \text{i}\rangle, w}$, and similarly for ξ and MEET.
We also assume that these expressions are static. Furthermore, $D_i = \{\text{a}, \text{b}, \text{c}\}$ and the interpretations of I, b_i, [b], are static and are fixed as follows, where K_b is a subset of D_w:

---

[6] Cases like @$_M$ are an exception.

$[\![I]\!]_{\mathcal{M},\langle M,i\rangle,\langle M,i\rangle,w} = i$, for $i \in \{a, b, c\}$.
$[\![b_i]\!]_{\mathcal{M},\langle M,i\rangle,\langle M,i\rangle,w} = b$.
For $p \in D_{\langle w,t\rangle}$, $[\![[b]]\!]_{\mathcal{M},\langle M,i\rangle,\langle M,i\rangle,w}(p)$ is the constant function from D_w to $\top$ if for all w', w' $\in K_b$ if p(w') = $\top$; and otherwise is the constant function from D_w to $\bot$.

It follows that $^{\vee}ist(b_i, [b], ^{\sqcap\wedge}\text{MEET}(I, \zeta, \xi))$ is static and context-independent;

$[\![^{\vee}ist(b_i, [b], ^{\sqcap\wedge}\text{MEET}(I, \zeta, \xi))]\!]_{\mathcal{M},c,c,w} =$
$[\![^{\vee}ist(b_i, [b], ^{\sqcap\wedge}\text{MEET}(I, \zeta, \xi))]\!]_{\mathcal{M},c',c',w}$

for all c, c'. By similar reasoning, $^{\vee}ist(a_i, [a], ^{\sqcap}ist(b_i, [b], ^{\sqcap\wedge}\text{MEET}(I, \zeta, \xi)))$ is also static and context-independent. Now, suppose that $\mathcal{M}$ satisfies (5.3.3) at c, c', w. Since (5.3.3) is static, c = c'. Because (5.3.3) is context-independent, $\mathcal{M}$ satisfies (5.3.3) at $\langle M_a, a\rangle, \langle M_a, a\rangle, w$, where M_a is the denotation in $\mathcal{M}$ of the modality $[a]$. Further, suppose that $\mathcal{M}$ satisfies $^{\vee}ist(a_i, [a], ^{\sqcap}ist(b_i, [b], ^{\sqcap\wedge}\text{MEET}(I, \zeta, \xi)))$ at c, c, w. By context-independence of this formula, $\mathcal{M}$ satisfies $^{\vee}ist(a_i, [a], ^{\sqcap}ist(b_i, [b], ^{\sqcap\wedge}\text{MEET}(I, \zeta, \xi)))$ at $\langle M_a, a\rangle, \langle M_a, a\rangle, w$. Therefore $\mathcal{M}$ satisfies the right side of (5.3.3) at $\langle M_a, a\rangle, \langle M_a, a\rangle, w$. But this means that $\mathcal{M}$ satisfies $^{\vee}@_M(^{\sqcap\wedge}\text{MEET}(I, \zeta, \xi)$ at $\langle M_b, b\rangle, \langle M_b, b\rangle, w$. By context-independence, $\mathcal{M}$ must satisfy $^{\vee}ist(b_i, [b], ^{\sqcap\wedge}\text{MEET}(I, \zeta, \xi))$ at c, c, w.
The converse is proved in similar fashion.

6 The Exit Operation and Stack-Structured Memory

The space limitations for this paper don't allow for an extended presentation of how to treat the operation of exiting a context. This section is merely a brief sketch of the topic.

Where κ and κ' are nonempty sequences of contexts and $\phi \in L_t^*$, $[\![\phi]\!]_{\mathcal{M},\kappa,\kappa',w}$ represents a transition from a history in which the sequence κ is remembered to one in which κ' is remembered. We call an evaluation of an expression that uses context histories in this fashion a *historical* evaluation.[7]

In the following definition, $\kappa \frown c$ is the sequence resulting from appending c to κ.

Definition 6.8. *Locality.*

An expression ζ is *local* with respect to a historical interpretation iff
$[\![\zeta]\!]_{\mathcal{M},\kappa\frown c,\kappa'\frown c',w} = [\![\zeta]\!]_{\mathcal{M},\langle c\rangle,\langle c'\rangle,w}.$

An expression is local, then, in case its interpretation involves only the current context. Nonlocal expressions will be exceptional in historical DCIL.

---
[7] A sequence κ represents only the part of the history of contexts traversed in an inquiry that is remembered. Some things may be forgotten; see the rule for exiting for an example.

The clauses defining historical interpretations for the static part of DCIL are straightforward adaptations of the clauses from Section 3.2. For instance, the clause for functional application goes as follows.

(6.2) Where $\zeta \in L^*_{\langle\sigma,\tau\rangle}$ and $\xi \in L^*_\sigma$, $[\![\zeta(\xi)]\!]_{\mathcal{M},\kappa,\kappa',\mathrm{w}} = [\![\zeta]\!]_{\mathcal{M},\kappa,\kappa',\mathrm{w}}([\![(\xi)]\!]_{\mathcal{M},\kappa,\kappa',\mathrm{w}})$.

Dynamic operations are more complicated, but still straightforward; this is illustrated by the clause for dynamic 'and'.

$[\![;(\phi)(\psi)]\!]_{\mathcal{M},\kappa,\kappa',\mathrm{w}} = \top$ iff there is a κ'' such that $[\![\phi]\!]_{\mathcal{M},k,k'',\mathrm{w}}$ and $[\![\psi]\!]_{\mathcal{M},k'',k',\mathrm{w}}$.

I'll replace $@_i$ and $@_M$ by expressions CURRENT$_i$ of type $\langle i,t\rangle$ and CURRENT$_M$ of type $\langle\tau_M,t\rangle$. CURRENT$_i$ holds of an index i iff i is the index of the context currently being visited; similarly for CURRENT$_M$.

If $\kappa = \kappa'$, $\kappa = \kappa_1{}^\frown c$, and ζ is a constant of type τ_M, then
$[\![\mathrm{CURRENT}_M(\zeta)]\!]_{\mathcal{M},\kappa,\kappa',\mathrm{w}} = \top$ iff $c_M = [\![\zeta]\!]_{\mathcal{M},\kappa,\kappa',\mathrm{w}}$;
otherwise, $[\![\mathrm{CURRENT}_M(\zeta)]\!]_{\mathcal{M},\kappa,\kappa',\mathrm{w}} = \bot$.

If $\kappa = \kappa'$, $\kappa = \kappa_1{}^\frown c$, and ζ is a constant of type τ_i, then
$[\![\mathrm{CURRENT}_i(\zeta)]\!]_{\mathcal{M},\kappa,\kappa',\mathrm{w}} = \top$ iff $c_i = [\![\zeta]\!]_{\mathcal{M},\kappa,\kappa',\mathrm{w}}$;
otherwise, $[\![\mathrm{CURRENT}_i(\zeta)]\!]_{\mathcal{M},\kappa,\kappa',\mathrm{w}} = \bot$.

The clause for EXIT simply pops the current context off of the history, returning the value $\bot$ if there is no such context.

Where $\zeta \in L^*\tau_M$ and $c_i = \mathrm{i}$, $[\![\mathrm{EXIT}(\zeta)]\!]_{\mathcal{M},\kappa,\kappa',\mathrm{w}} = \top$ iff $\kappa = \kappa_1{}^\frown c_1$ for some c_1 (where κ_1 is non empty) and $\kappa' = \kappa_1$.

The following variation on (5.3.3) uses historical DCIL to characterize a procedure that uses the exit operation to update Ann's database with information from Bob's database.

(6.3) [CURRENT$_M$([a]) $\wedge$ CURRENT$_i(a_i)$] $\rightarrow$ %Start in Ann's DB
[$^\vee ist(a_i, [a], ^\neg ist(b_i, [b], ^{\neg\wedge}\mathrm{MEET}(I, \zeta, \xi))) \leftrightarrow$
M-ENTER([b]); I-ENTER(b_i);   %Enter Bob's DB
$^\vee @_M([^{\neg\wedge}\mathrm{MEET}(I,\zeta,\xi)](@_i))$;   %Test for MEET$(I,\zeta,\xi)$
EXIT; EXIT]   %Re-enter Ann's DB

In the last step, we have to exit twice because Bob's database was entered in two operations.

7 Conclusion

The application described in Section 5 of DCIL is too simple to do more than indicate a possible use of the logic in specifying tasks involving reasoning about context and proving correct the algorithms that do this reasoning. It remains to be seen if the logic is actually useful in formalizing domains that are complex enough to be realistic and useful. The success of this program depends on the development of generally applicable techniques for formalizing domains in DCIL. Also, as was mentioned in the discussion of the correspondence between (5.3.3) and the implemented algorithm, it might be useful to extend the dynamic logic so that the correspondence between the dynamic constructions and actual algorithms for reasoning about context is more systematic.

At several places, I indicated ways in with this approach to contextual reasoning could be extended and developed. For reasons that became evident when we introduced the null value in Section 3.2, the logic needs to be made partial. And of course, a nonmonotonic version of the logic remains to be developed. So there still remains much work to do on this approach to contextual logic, both to ensure a better relation to applications and a better match to general requirements on a fully adequate logic of context.

Acknowledgements. I owe thanks to the two referees who commented on this paper for Context'03.

References

1. Thomason, R.H.: Contextual intensional logic: Type-theoretic and dynamic considerations. In Bouquet, P., Serafini, L., eds.: Perspectives on context. CSLI Publications, Stanford, California (2003)
2. Lewis, D.K.: Scorekeeping in a language game. Journal of Philosophical Logic **8** (1979) 339–359
3. van Benthem, J.: Exploring Logical Dynamics. CSLI Publications, Stanford, California (1996)
4. Groenendijk, J., Stokhof, M.: Dynamic Montague grammar. Technical Report LP–90–02, Institute for Language, Logic and Information, University of Amsterdam, Faculty of Mathematics and Computer Science, Roeterssraat 15, 1018WB Amsterdam, Holland (1990)
5. Montague, R.: Pragmatics and intensional logic. Synthese **22** (1970) 68–94 Reprinted in *Formal Philosophy*, by Richard Montague, Yale University Press, New Haven, CT, 1974, pp. 119–147.
6. Church, A.: A formulation of the simple theory of types. Journal of Symbolic Logic **5** (1940) 56–68
7. Henkin, L.: Completeness in the theory of types. Journal of Symbolic logic **15** (1950) 81–91
8. McCarthy, J., Buvač, S.: Formalizing context (expanded notes). In Aliseda, A., van Glabbeek, R., Westerståhl, D., eds.: Computing Natural Language. CSLI Publications, Stanford, California (1998) 13–50
9. Harel, D.: Dynamic logic. In Gabbay, D., Guenther, F., eds.: Handbook of Philosophical Logic, Volume II: Extensions of Classical Logic. Volume 2. D. Reidel Publishing Co., Dordrecht (1984) 497–604
10. Gallin, D.: Intensional and Higher-Order Logic. North-Holland Publishing Company, Amsterdam (1975)

Comparatively True Types: A Set-Free Ontological Model of Interpretation and Evaluation Contexts

Martin Trautwein

Institute for Formal Ontology and Medical Information Science (IFOMIS),
University of Leipzig, Härtelstr. 16-18, 04107 Leipzig, Germany
martin.trautwein@ifomis.uni-leipzig.de

Abstract. The reality of linguistic forms (i.e. the language context) is connected to the world of prelinguistic and precognitive reality (i.e. the evaluation context) by the functions of truth and reference. Nevertheless, both truth and reference are relative. They are mediated by the cognitive module that controls all kinds of act of typing, type knowledge, or the lexicalization of these types (i.e. the interpretation context). This module has a socio-psychological reality and, in particular, has its special way of sorting and ordering the world from different perspectives. The paper argues that these perspectives bring about taxonomical orders, and granular, aspectual, and selective views of reality. Based on the Theory of Granular Partitions, we outline a formal-ontological approach to modeling both the evaluation and the interpretation context, bringing together the cognitive perspective and the actual counterparts of reference and truth.

1 Introduction

The entire system of natural language is based on cognitive acts of typing and on the representation of type knowledge. In our everyday usage of language, however, we are mostly referring to concrete objects and situations. Thus it is the abstraction power of types which enables language to generalize over an unlimited number of contexts and which enable us to speak or write about an infinite universe using a comparatively restricted amount of linguistic means. As a consequence, we cannot simply equate semantic content with the entities to which we refer. Such a view – let me call it the 'naïve extensionalist view' – has already silently adopted and presupposed the activeness of the type level of mental representation and linguistic knowledge which mediates between concrete linguistic forms and concrete situations in reality. In order to judge on semantic content, we need to know about the ontological structure of the world to which we refer, but also about the contributions of mental abstraction and our intentional directedness.

Types are not only mental doublings of the intrinsic structure of the world. They do not function like mirrors but rather like filters, micro- or telescopes, reliefs or black-and-white drawings, and so on. Thus all acts of typing, the usage of types in form of lexicalizations, and the interpretation of lexicalized types are context-dependent: they depend on the respective *cognitive* context. Inspired by Seuren et al. (2001), the model in Fig. 1 construes the language-world relation by dividing the system of reference and truth into types and tokens.

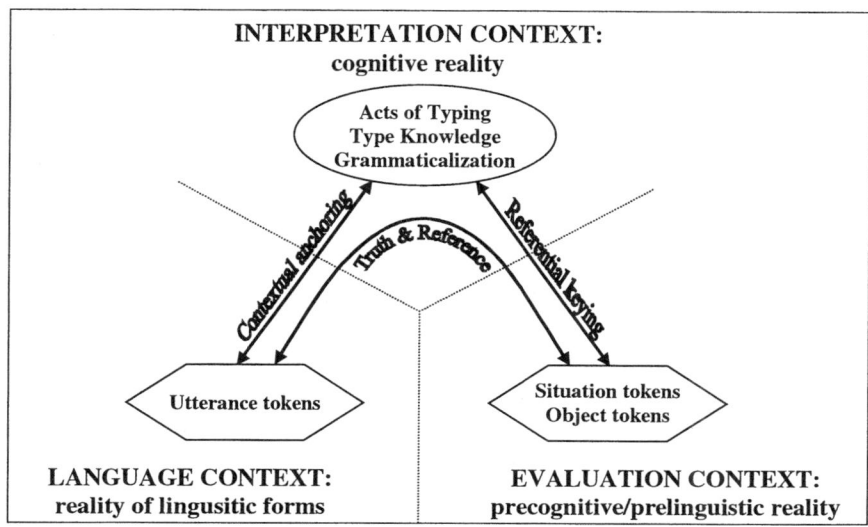

Fig. 1. A model of the system of reference and truth, based on the type-token distinction

Nominal and sentential descriptions are types of grammatical units out of the *interpretation context* (IC) that are complete and expressive enough for being used to refer to the *evaluation context* of truth- and reference-making reality (EC). Linguistic descriptions gain their expressiveness from a type expression and from a referential marker which indicates that the expression has to be linked to EC. If linguistic descriptions are performed in an utterance, they are realized as concrete instances of the *language context* (LC). Nouns denote information about a certain type of spatial object. The reference marker (e.g. the definite or indefinite article) indicates that an object of the kind specified can be found in a salient evaluation context, that is a context of reference-making objects shared by producer and recipient. Only an utterance token in a concrete discourse situation, however, is suitable to single out the portion of evaluation context which a recipient needs to determine in order to establish reference.

(1) [*the* [*dog*]]
     determiner   sortal noun
     = referential marker   = type expression

In this point, sentences do not differ from nominal descriptions. The core proposition of a sentence specifies a type of event or situation. The syntactic category of sentence mood anchors the description in EC such that, for instance, an declarative sentence asserts the existence of a salient truth-making event.

(2) [$e_{declarative}$ [*Peter is watching a dog*]]
     sentence mood   core proposition
     = referential marker   = type expression

As before, this context can only be determined in a concrete situation of utterance.

This paper's aim is to show how the way in which type expressions in linguistic descriptions are integrated in IC determines our capability of referring to and asserting on reality. The dependency of lexicalized types on IC suggests that the relationship between linguistic form and the structure of the world remains opaque as long as we try to explain reference and truth only in terms of Grammar Theory and an extensional theory of reality. The paper outlines an ontological understanding of cognitive reality instead in order to deal with the autonomous contribution of IC.

2 The Context-Dependency of Typing and Types

There are three exemplary problems that continuously recur in theories of truth and reference: (i) taxonomical hierarchies and the choice of the appropriate level of typing, (ii) cognitive perspectives such as granularity, aspectuality, or selectivity, and (iii) negative sentences as descriptions of absence or non-existence.

2.1 Taxonomical Hierarchies of Types

From an ontological view, the lexicalization of natural language is messed up since the mappings from word forms to word meanings are many to many. Polysemous word forms, for instance, are associated to more than one word meaning while two or more synonymous word forms are mapped onto one and the same word meaning. Nevertheless, both polysemy and synonymy are determined in language, interpretation, and evaluation contexts. It seems that the disambiguating contexts of the distinctive word meanings of a polysemous form reflect different ontological domains of EC.

(3) Word meanings of *grasp*[1]

1. **grasp**, grip, hold – (hold firmly)
2. get the picture, comprehend, savvy, dig, **grasp**, compass, apprehend – (get the meaning of something; "Do you comprehend the meaning of this letter?")

The first meanings of *grasp*, as shown in (3), for instance, belongs to the ontological domain of physical reality whereas the second meaning belongs to a socio-psychological context. It is obvious, therefore, that the IC of typing and lexicalization is able to unite portions of EC from different ontological domains by using similarity, analogy, metaphoric connections, or language change. Bundling subtypes in a super-type and bundling different word meanings in the semantic of one lexical item is part of the abstractive power of the typing module. The IC thereby builds taxonomies of types across ontological categories and domains. Moreover, it is interesting how different languages selectively realize the nodes of such a taxonomy.

---

[1] Source: WordNet 1.7.1 online version, http://www.cogsci.princeton.edu/cgi-bin/webwn1.7.1), query for 'grasp'

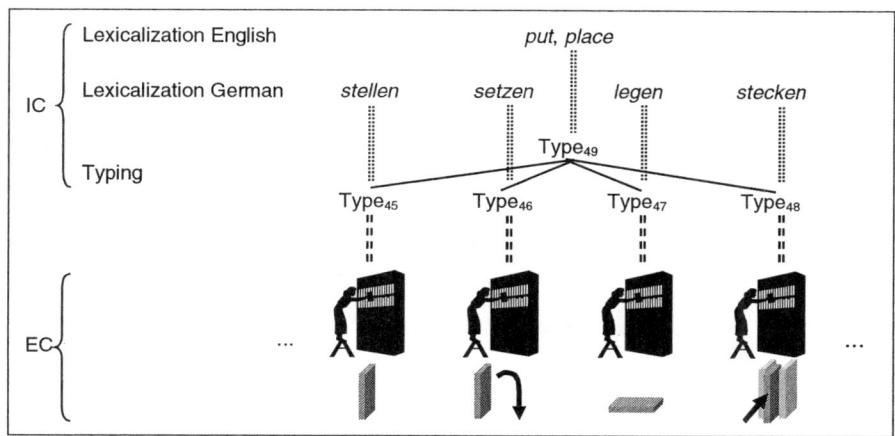

Fig. 2. Compensatory heteronymous lexicalization at different levels of the taxonomical hierarchy in English and German. Types abstract from EC (*dashed lines*) and build taxonomical hierarchies. Languages lexicalize the resulting types in different ways (*dotted lines*)

While English lexicalizes the supertype of PUT in the verbs *put* and *place*, German lexicalizes the respective subtypes which distinguish the situations with respect to the resulting orientation of the moved object. The types not realized by a language are regarded as lexical gaps. Lexical gaps, however, do mostly not decrease the expressiveness of a language since language compensates this lack by heteronymous forms of another taxonomical level. Beyond this, taxonomies unify types of different levels of abstraction in taxonomical granularity levels. One of these levels is the basic level that is used in a 'neutral' IC.

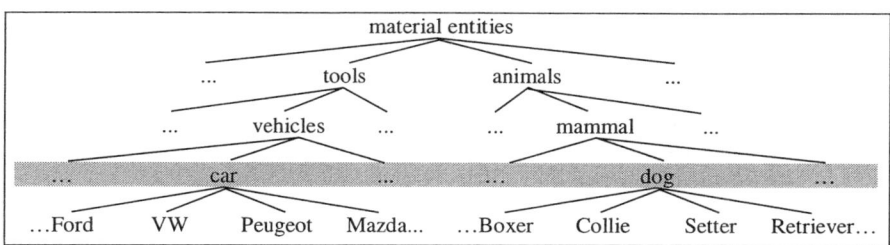

Fig. 3. Typing is subject to taxonomic granularity and comprises a basic level (*shaded*)

The taxonomical graduation of types and their lexicalization demonstrates that the contribution of IC goes far beyond the structural distinctions that the precognitive reality of EC suggests. IC creates a view of the world whose autonomous status lies in the particular cognitive capacity of typing.

2.2 The Context-Dependency of 'Perspection'

Besides taxonomical granularity, IC involves further cognitive perspectives. We do not only perceive the world, we also 'perspect' it, that is, we integrate the content of our perception into our knowledge by sorting it into relevant or irrelevant, salient or non-salient, typical or non-typical, or marked or unmarked structural aspects. The internal organization of our type knowledge plays a crucial role in perspection and is determined by three factors that influence language production and comprehension orthogonal to one another: **ontological granularity, aspectuality**, and **selectivity**.

2.2.1 Ontological Granularity

Any kind of typing is granular with respect to space, time, function, and complexity. This ontological granularity is part of EC and thus is precognitive and prelinguistic, but nevertheless IC is free in choosing a certain granularity level as its focus. Semantics tends to keep one and the same granular focus within one description. The spatial granularity levels lexicalized by verbal semantics, for instance, impose combinatorial restrictions on the semantic selection of locative modifiers.

Table 1. The granular compatibility of verbs and spatial modifiers

	TOO SMALL	FITTING	TOO LARGE
The people migrated	to the door to the park	to Egypt	
Sarah took a stroll	to the door	to the park	to Egypt
Sarah stepped		to the door	the park to Egypt

As well, the temporal granularity levels lexicalized with verbs impose some restrictions on selecting temporal modifiers.

Table 2. The granular compatibility of verbs and temporal modifiers

	TOO SMALL	FITTING	TOO LARGE
I educated my son	a few minutes ago last week	the last 5 years	
I taught my son to swim	a few minutes ago	last week	the last 5 years
I surprised my son		a few minutes ago	last week the last 5 years

Dealing with everyday reality, cognition and language commonly use the basic mesoscopic level. For instance, when we see a person crossing the street, we normally neither think of her as an aggregate of molecules (at a microscopic level) nor as a grain of dust in the infinity of the universe (at a macroscopic level). It is obvious, furthermore, that the mesoscopic level is predetermined by the thresholds of perception.

2.2.2 Aspectuality

Aspectuality reflects scales of spatial and temporal size at a single level of ontological granularity. Even if we judge on entities that we classify as events at the mesoscopic level, for example, IC may focus on the whole course of an action or activity, or just on one of its spatial and/or temporal parts. Fig. 4 shows how the linguistic descriptions chose an aspectual sentence type in order to refer to spatiotemporal segments of Sarah's walking to the Kilimanjaro.

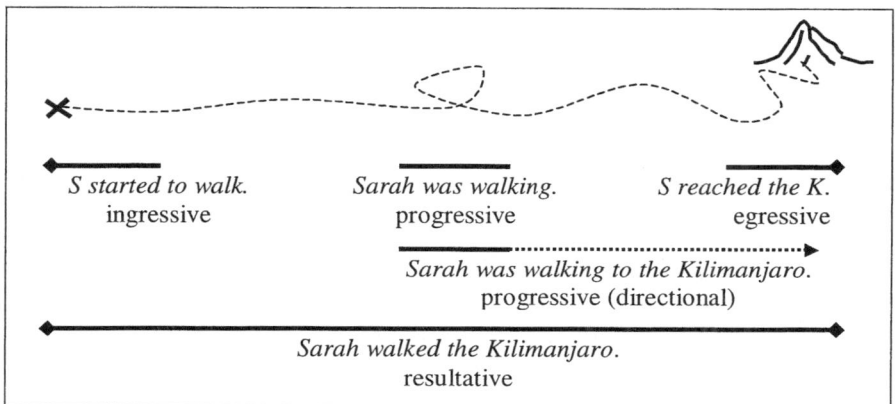

Fig. 4. The equivalence between the aspectual class of sentences and the spatiotemporal range of their referents. *Diamonds* mark natural spatiotemporal boundaries of the event course

Any kind of perspective of IC that allows pars-pro-toto reference, that is, it allows that a spatiotemporal part stands for a whole in a LC, can be called an aspect. But we can also use a term which types a spatiotemporal whole to refer, totus pro parte, to one of its spatial or temporal parts. If we speak about very large objects or long events, we often refer only to one of their parts that is located within our discourse situation. It is reasonable, for instance, that not all spatial segments of the road in (4) and not all temporal parts of the ceremony in (5) are involved in the situation.

(4) Thousands of potholes are scattered across the road.
(5) The shouting of the demonstrators disturbed the ceremony.

2.2.3 Selectivity

Finally, IC is selective. This feature is almost identical with what Talmy (1996) calls the 'windowing of attention'. IC puts relevant, salient, typical, or marked properties or aspects of a complex real structure to the foreground, and irrelevant, non-salient, untypical, or unmarked properties or aspects to the background of a verbal or preverbal message. The selective lexicalization of manner in the semantic field SPEAK shows that grammaticalized types may specify a conspicuous characteristic of a process if this characteristic is relevant in IC.

Table 3. The semantic field of SPEAK

Disregarding manner	*speak, talk, utter, verbalize, vocalize, voice*
Involving manner of articulating	*drawl, gabble, gasp, gibber, jabber, lisp, maunder, mispronounce, misspeak, mouth, mumble, murmur, mutter, shout, splutter, spout, whisper*
Involving manner of producing text	*descant, dilate on, expatiate, perorate*
Involving manner of communicating	*converse, discourse, allege, assert, aver, convey, declare, tell*

Nevertheless, grammaticalization choses integrative type concepts. This is signified by the fact that there is no single English verb that grammaticalizes two kinds of manner aspects (such as 'speaking quietly and long-winded' or 'saying something briefly with the intention of convincing'). There are no verbs which grammaticalize incidental or rare circumstances (such as 'articulating something with a hair between one's teeth' or 'telling something with the intention of offending the recipient's mother'). Moreover, verbal lexicalization maps the variety of relations of the common-sense world onto a very restricted set of thematic relations (at the level of semantics) and argument roles (at the syntactic level). The integrative power of types and of their grammatical realizations thus provides further indication for the economy of IC.

2.2.4 Coherence of Perspective

The IC shows a tendency to keep one perspective within one linguistic (e.g. sentential) context. Imagine that sentence (6) properly describes that a person had been taking a drug for a certain time, and that the therapy finally succeeded.

(6)   *The medication gave her some relief.*

Switching between different levels of granularity, aspectuality, or selectivity within one sentence, as in (7) to (10), complicates understanding and provokes metaphorical readings since we have to infer additional knowledge about causal connections in order to reconstruct how the situations at the different levels interrelate.

(7)   *The suppression of the enzyme H^+/K^+-ATPase gave her some relief.*
(8)   *Taking a tablet gave her some relief.*
(9)   *Choking the drug down gave her some relief.*
(10) *Drinking water gave her some relief.*

Sentence (7) confuses two levels of granularity, a biochemical and a mesoscopic one, (8) disregards the long-term medication as a whole and thereby ignores the aspectual perspective of the entire causal nexus. (9) superfluously expresses a certain manner which is irrelevant for the success of the therapy. Sentence (10), finally, lacks both aspectual and selective coherence, since *drinking water* is only one aspect and – even worse – an irrelevant aspect of taking a drug.

2.3 The Context-Dependency of Absence

Negative descriptions, that is descriptions of absence, might be regarded as the most striking argument for a division of mental context and world context.

(11) *There is no dog.*
(12) *Peter isn't laughing.*
(13) *Peter isn't watching a dog.*

It seems that these sentences refer to things that do not exist or are absent. EC does not contain 'negative entities' though. The world is the world of the existing. Reality does not miss something when we think or say that something is not there. 'Absence' is mainly a linguistic matter and thus a problem of IC. The difficulty of negative descriptions is that they do not refer to a transparent spatiotemporal constellation in EC. Positive descriptions, i.e. descriptions of existing entities, always at least implicitly express how the referent or truth-maker is integrated in the network of the world's relations.

(14) *There is a dog.*
(15) *Peter is laughing.*
(16) *Peter is watching a dog.*

Each of the sentences in (14) to (16) asserts or presupposes the existence of objects or events, and the IC involved in the descriptions give us a hint how these entities are integrated in EC. Thus we know that a dog and a person called Peter are situated in a space surrounding them, that *watching the dog* requires a certain distance and a period of time, and so on. The reference- and truth-makers of these sentences bring their own scenario, and, by describing the objects, persons, and events, we are also describing the scenario. The world context itself suggests certain options of describing it. Then the IC, which guides our linguistic descriptions, is almost self-explaining in view of reality.

Many negative descriptions, in contrast, do not point to such a scenario and thus do not reveal what 'is missing' in EC.

(17) *The physician A didn't administer the correct dose of the drug X to patient B.*
  → *The drug was the right one but the dose was wrong.*
  or *It was not the drug X but the drug Y.*
  or *A did not administer the drug but stopped the medication.*
  or *It was not the physician A but the physician C.*
  or *It was not patient B but patient D.*
  or *A didn't do anything with/to B.*
  or *There is no patient B.*
  or *Nothing happened at all.*
  or ...

Nonetheless, negative descriptions are always bound to a certain, clearly delineated, portion of EC since, otherwise, sentences such as (11) to (13) would not make sense. Using these negative descriptions, we do not want to assert that, in the whole universe, there are no dogs, that no person named Peter is laughing, or that there are no events wherein a person named Peter is watching a dog. Thus it is IC that delineates the contexts. It is our knowledge of comparable contexts or our comparison with 'what could be' that determines that portion of the world to which our description applies.[2]

---

[2] For a more detailed discussion of the ontological basis of negative descriptions, cf. Späth and Trautwein (2002), and Trautwein (2002).

3 The Misery of Extensionalism

Many formal-ontological theories about EC and its connection to the semantics of natural language are extensional.[3] Extensional models comprehend types as classes in terms of Set Theory. However, this view only describes already existing types of IC and common-sense entities of EC that are already typed in IC. Extensionalism is not able to say anything about the interrelation between both kinds of context. If we have a collection of entities, we receive the respective sets for free.

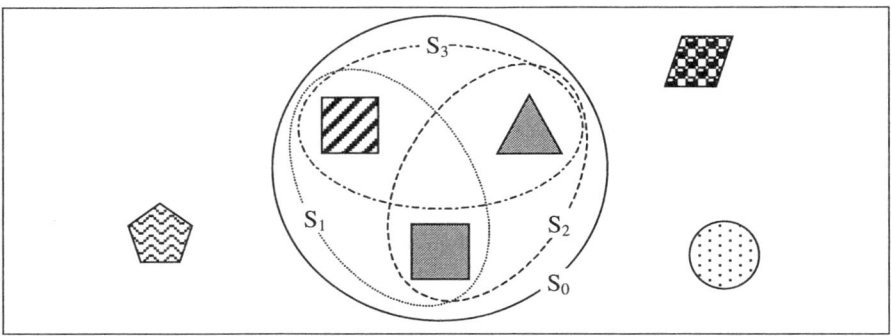

Fig. 5. A collection of entities and some of the resulting sets (distinguished by *line style*)

Set-theory, however, is not able to explain why S_1 and S_2 are relevant sets that could be seen by a type or to which we could refer by a linguistic description, whereas S_3 and S_0 only represent purely logical possibilities of segmenting the constellation which will never be lexicalized in a natural language. Obviously, the elements of S_3 do not have enough features in common, and the elements of S_0 are incomplete in order to reflect a supertype for all the entities. One of the problems is that sets, by their nature, are not integrated in orders, hierarchies, or contextual dependencies. This is not to say that set-theory is not expressive. It represents as much information about types and the underlying ontological structure as we are able to express. The theories remain purely descriptive though.

4 The Theory of Granular Partitions

The goal of the Theory of Granular Partitions (developed by Bittner/Smith (2001), Bittner/Smith/Donnelly (to appear)) is to provide a tool for representing all kinds of human listing, sorting, cataloguing, categorizing, and mapping activities. It is expressive enough to explain the selectivity of these cognitive activities, and thus provides an alternative to set theory. Granular Partition Theory (GPT in the following) is based on two subtheories: (i) a theory of tree-like cell structures, and (ii) a theory of how cell

---

[3] An overview on current top-level ontologies can be found in Degen et al. (2001)

structures are mapped onto a domain of entities. A system of cell structure is defined as CS = $\langle \mathbf{Z}, \subseteq \rangle$ with $\mathbf{Z} = \{z_1, z_2, \ldots\}$ being the domain of cells, and $\subseteq$ being a partial ordering on $\mathbf{Z}$. Every cell structures has a unique maximal cell (the root), such that every cell of the cell structures is a subcell of the maximal cell. Accordingly, each cell structures can be represented as a tree (i.e. a directed graph with a root and no cycles), as shown in Fig. 6.

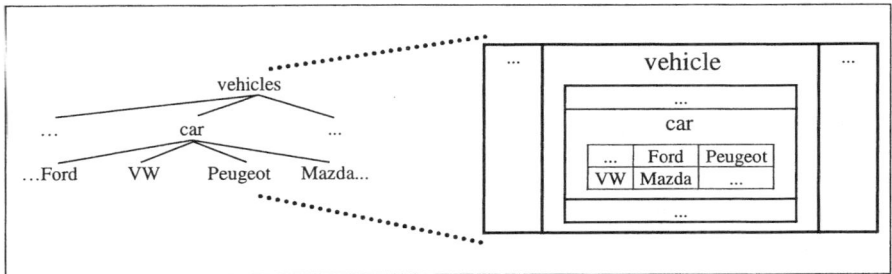

Fig. 6. Cell structures can be represented as trees and vice versa

A granular partition then is defined as P = $\langle CS, \Delta, l, p \rangle$, with CS being a cell structure and Δ being the domain we want to partition. The core idea of GPT is to complement the element relation of set theory by introducing two distinct relations. The location relation, l, replaces elementhood, defined as a mapping $l: \Delta \to CS$. The expression $l(a_j, z_i)$ can be read as 'the entity a_j is located in the cell z_i. The projection relation, p, is a mapping $p: CS \to \Delta$, saying that, if $l(z_i, a_j)$, the cell z_i projects onto the entity a_j. Then our understanding of the fact that a type of the interpretation context recognizes an entity of the evaluation context is represent in GPT if the respective cell projects onto the entity and if the entity is located in the cell.

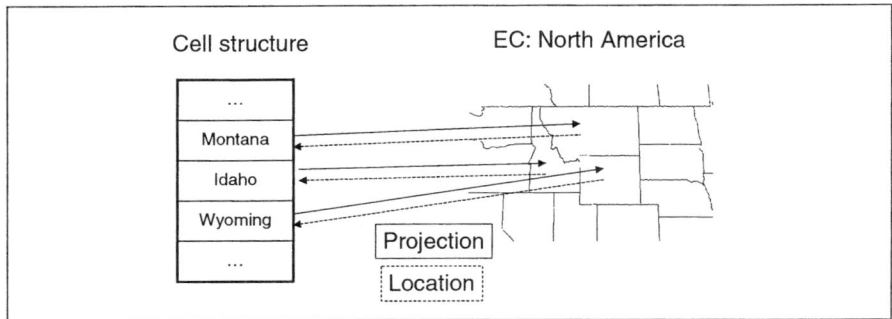

Fig. 7. Location and projection are the two sides of partitioning

GPT is able to express thereby that the type-token matching fails. A type represented by a cell, for instance, might project onto an entity, but, since the type does not comprise it, the entity is not located within the cell. Hence GPT includes an extensional and an intentional side.

4.1 Granular Partitions as Contexts

A partition may reflect an aspect of the ontological structure of its projection domain, but it might also be the case that the internal organization of the partition cannot bring about any typical characteristics of the domain if it is messed up. Thus the theory is completely neutral with respect to the question whether a type system results in a 'good' or a 'bad' division of the world.

GPT is a suitable tool for modeling both IC and EC, and thus enables us to solve the problems discussed in the first part of this paper. The advantage of the GPT is that partitions integrate types in intentional contexts. Each cell is embedded in a system of cells and is not just an arbitrary and isolated set of something. Partitions realize taxonomical hierarchies as they map complete tree structures. If they are 'good' partitions of reality, they also take perspection into account, such that each partition just involves only one level of ontological granularity, one level of aspectuality, and one grade of selectivity. This explains why linguistic descriptions with a coherent perspective are easier to understand than descriptions with changing perspectives. If we switch between different interpretation perspectives, our typing of reality analogously switches between different ways of partitioning the world. Thus we have to reconstruct how these different partitions used by the linguistic expression are connected, in order to receive a coherent interpretation and, finally, a referential or truth value.

The modeling of absence and negative descriptions, furthermore, is almost self-explaining in terms of GPT. Since a partition is a cell structure projected onto a domain, it is possible that certain cells remain empty. Now, however, we have the full required context for this kind of emptiness, since *we* are projecting a cell onto the world with the result that it does not recognize anything. Thus the cell is our window to a portion of the world. The cell, in turn, is nested in a structural context (i.e. the cell structure), and is predetermined by the perspective of this context. Negative descriptions seem to tell us: 'If you apply the type x in this interpretation context y to this special segment of the world z, you will see nothing.'

5 Modeling Lexicalizations Using Schemas

Lexicalizations, however, are much more complicated. Attempting to model the complex hierarchies and unifications of types that are hidden in polysemous lexical items, we have to face two problems: (i) superconcepts cross the boundaries of ontological categories and domains, and also the boundaries of levels of perspective. Superconcepts (ii) unify several subtypes, but only a few of these subtypes are actualized in a concrete utterance token. In a concrete utterance, we have to match the lexicalized meaning of one term with the other terms in the LC, with the intended IC of the message, and with the concrete portion of EC which provides the relevant reference- and truth-makers. Systems of multiple partitions provide a solution for these problems. In reality, entities are linked by an unlimited number of relations. Schema Theory demonstrates, however, that types can be linked, too (cf. D'Andrade (1995)). Schemas are constellation of place-holders for a certain type of entity. The place-holders are linked by thematic relations which abstract over the actual ontological relations of the world.

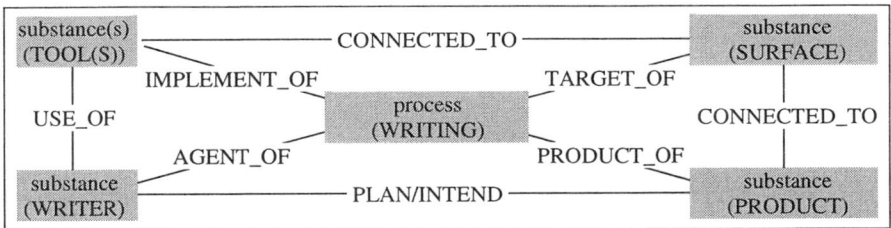

Fig. 8. A schema for WRITE. The place-holders marked for a certain type (*shaded*) are linked by thematic relations (indicated by *lines*)

Schemas are more than just tuples of types and relations, however. If a linguistic context fills in for some of the slots, it often restricts the contextual possibilities of neighboring slots and thereby predetermines the compatibility of further contextual tokens.
(18) *Peter is writing in the sand.*
Sentence (18), for instance, can be interpreted in the WRITE schema as illustrated by Fig. 9.

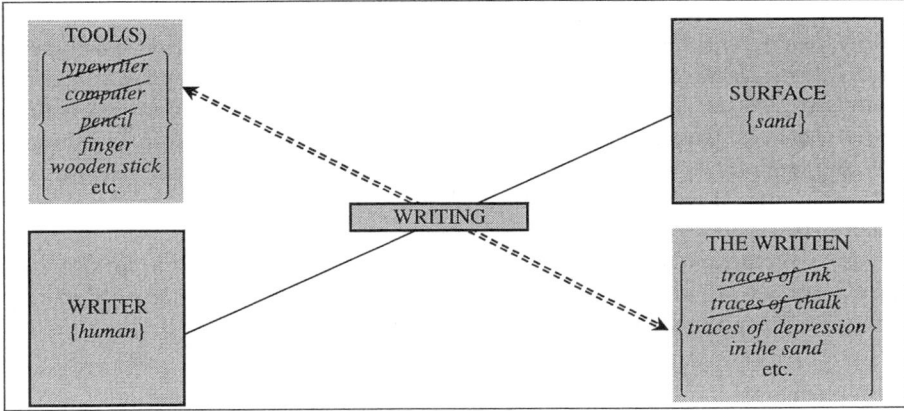

Fig. 9. Some of the place-holders of a schema are filled in by the context (*framed shaded boxes*) and predetermine how the neighboring place-holders (*frameless, shaded boxes*) may be replaced

This suggests that schemas are part of IC and can be defined in systems of multiple partitions, such that the typed place-holders are represented by the cells of the partitions and the thematic relations are represented as links between cells.

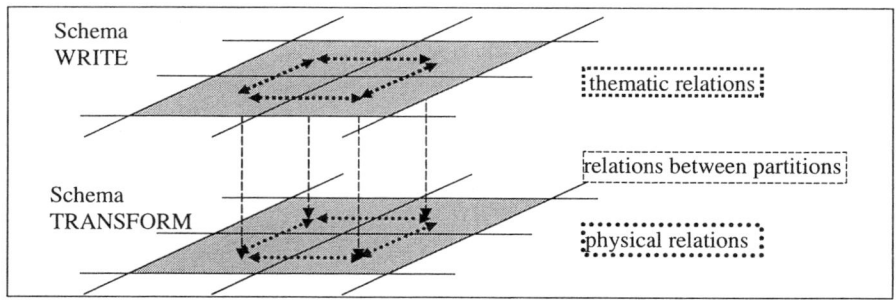

Fig. 10. Complex schemas can be represented by layered models of partitions

Such a model comprises different layers of partitions of different ontological domains, relations between neighboring cells, and relations between cells of different partition layers.

6 Conclusions

The discussion shows that, whenever we attempt to comprehend how an utterance relates to a portion of the world, we have to consider our cognitive capability of typing reality, storing type knowledge, and associating this knowledge to lexical forms. This capability constitutes the context which determines how we are able to speak about the world and understand linguistic descriptions. The ontological approach to both the world context to which we refer and the cognitive context of linguistic encoding and decoding outlines how formal semantics or lexical databases can tackle these problems. Partition Theory has been designed as a formal-ontological alternative to Set Theory. Being applied to the notion of typing, Partition Theory proves suitable to represent interpretation contexts and their mental background not just as neutral media for the transfer of information from the linguistic side to the world and back, but as a piece of independent reality that influences everything we communicate.

References

1. Bittner, T., Smith, B.: A Taxonomy of Granular Partitions. In: Montello, D. R. (ed): Spatial Information Theory: Foundations of Geographic Information Science. Lecture Notes in Computer Science, Vol. 2205. Springer-Verlag, Berlin Heidelberg New York (2001)
2. Bittner, Th., Smith, B., Donnelly, M.: A Theory of Granular Partitions. In: Duckham, M., Goodchild, M. F., Worboys, M. F. (eds.): Foundations of Geographic Information Science. Taylor & Francis Books, London (to appear)
3. D'Andrade, R. G.: The Development of Cognitive Anthropology. Cambridge University Press, New York (1995).
4. Degen, W., Heller, B., Herre, H.: Contributions to the Ontological Foundation of Knowledge Modelling. Report No. 2/2001. Department of Computer Science, University of Leipzig (2001). http://www.ontology.uni-leipzig.de/Publications/Paper-Ontologie-Degen-Heller-Herre-2001.pdf

5. Seuren, P. A. M., Capretta, V., Geuvers, H.: The Logic and Mathematics of Occasion Sentences. Journal of Linguistics and Philosophy 24 (2001) 531–595
6. Späth, A., Trautwein, M.: Events under Negation. In: Sinn und Bedeutung VII, Proceedings of the 7th Annual Meeting of the German Society of Semantics. Konstanz Linguistics Working Papers. University of Konstanz (to appear)
7. Talmy, L.: The Windowing of Attention in Language. In: Shibatani, M., Thompson, S. A. (eds.): Grammatical Constructions – Their Form and Meaning. Clarendon Press, Oxford (1996)
8. Trautwein, M.: The Time Window of Language: The Interaction between Linguistic and Non-Linguistic Knowledge in the Temproal Interpretation of German and English Texts. Doctoral thesis. University of Leipzig (2002)

Discourse Context and Indexicality

Mark Whitsey

University of Nottingham, UK

Abstract. Problems related to indexicality, such as Sidelle's Answering Machine Paradox [10], have been problematic for direct reference theorists. The solutions suggested to date are not wholly satisfactory. I suggest that the correct solution requires an account of context shifting in natural language. However, existing context shifting accounts assume that context shifting is a purely semantic, operator-governed mechanism. This view has trouble dealing with so-called 'free shifts', such as the answering machine problem. I discuss these approaches and sketch a new account in terms of *discourse context*, which should be viewed as a pragmatic feature.

Keywords: Natural language semantics, Natural language pragmatics, Natural language understanding

Introduction

When we report what someone thought or believed, the context of our report is often different from the context in which the thought or belief was entertained. What we report (in context c) must maintain the semantic or truth-conditional content of the thought or belief being reported. We must transpose any indexicals from the context in which the thought or belief was entertained to the context of the report in order to meet this condition, for indexicals are intrinsically tied to context. They are parasitic upon contextual parameters and take a semantic value only when values are available from the context. David Kaplan [5] provides a semantic framework according to which the semantic value of an uttered indexical is taken from the context in which it is uttered, where a context consists of an AGENT (the speaker or writer), a time, a place and a world. The Character of 'I', for example, which we can state as a mapping from contexts to contents, enables a speaker to refer to herself by uttering the first person pronoun. Similarly, she can refer to her location using 'here' and the time of her utterance using 'now'. Thus, according to Kaplan's framework, an utterance of "I am here now" will come out true, whatever context it is uttered in. Kaplan says that it is "universally true", for "[o]ne needs only understand the meaning of ["I am here now"] to know that it cannot be uttered falsely." ([6], p.509).

However, as Sidelle [10] has pointed out,[1] we often communicate using the utterance "I am not here now". We record such messages on answer machines and, when a similar message is played back to us over the telephone, we have no

---

[1] See also Kaplan [6], 491*ff.* and Vision [12].

trouble in understanding what it means. It means that the person being called isn't there to answer the phone (or at least doesn't want to answer the call; but in either case the *assertion* is that no one is in). How can Kaplan's framework accommodate this successful communication? Sidelle proposes that an utterance can be deferred [10]. However, the notion of a deferred utterance seems unable to explain, in a systematic way, similar examples, such as post-it notes [2]. If reference is secured by the indexicals 'I', 'here' and 'now' (and it seems that this is the case, for we understand what is said perfectly), then 'now' must refer to the time of the call, 'here' to the location of the recipient's telephone and 'I' to the absent recipient.

1 The Problem

The answering machine problem arises because the parameter AGENT becomes decoupled from the real-world speaker/writer. Several solutions have been proposed, notably in Corazza, Gorvett and Fish [2] and Predelli [7]. Both assume that a token of 'I' will always pick out the AGENT of the context of utterance, yet the AGENT *qua* abstract feature of our semantic framework need not be identical to the real-world speaker or writer, as Kaplan assumed. In Corazza, Gorvett and Fish, for example, *convention* is called upon to be the link between the AGENT and the world. They note that "it is a matter of the setting or broad context whether a given word, for instance, is used with such or such a conventional meaning [2]. The relevant convention in the case of the answering machine sentence is that the AGENT is whoever convention links to that machine, i.e. by ownership. We can thus view the AGENT in this particular case as the owner of the answering machine.

A different approach, taken by Predelli [7], identifies the linking factor with the communicative intentions of the speaker/writer. On such accounts, the intention of the owner of the machine, i.e. the intention to inform his callers that he is unable to take the call, automatically identifies himself as the AGENT, regardless of other contextual factors. Both approaches have their merits, and I do not intend to discuss and evaluate their relative appeal here. In my view, both approaches have erroneous views of what a context is and how contextual features relate to the context itself. Once this misunderstanding is repaired, we arrive at a position (which extends either the conventional or the intentional account) which has interesting consequences for the semantics of indexicality in general.

To begin with, let us isolate the common features of the two accounts and see how these allow us to solve the problem. Both the conventional and intention-based accounts view the context in which the answering machine sentence is uttered as having a vacant place for an AGENT. They view this parameter as a placeholder for a semantic value, which they seek to fill with the relevant person (i.e. the owner of the machine). Both accounts agree on these points; they only differ on how to make the link. On one view, the two are mediated by conventions and, on the other, by intentions. If this reading is faithful, it uncovers a misunderstanding held by both views, namely that a context is the kind of entity which can have a placeholder. It can't, for a context is *defined* by its constituents, just as a set is. When we write that a context c is defined as: $c = \langle a, t, p, w \rangle$ (where a is the AGENT, t a time, p a location and w a world), we use this as a *schema* to indicate that c contains an agent, a time, a place and a world. A placeholder for a person is not the same thing as a person, and so an entity containing a placeholder for a person does not fit into our schema. Because a context is defined by its constituents, we find that, when we alter a constituent, we change the context. This is intuitively correct, for when a speaker changes whom she talks with, or moves from place to place, we say her words must be put into a new context. The contextual parameters that form the tuple $\langle a, t, p, w \rangle$ must be restricted, such that, for any context c, c_p (i.e. the value of p in context c) must denote the location of c_a at c_t in c_w. If this condition were not met, the contextual values would be unrestricted and *any* combination of an AGENT, time, place and world would form a context, making indexical reference a non-starter. The idea that an individual could be conventionally/intentionally added to a context clearly does not respect this restriction. However, it is precisely this restriction that makes the answering machine sentence so puzzling.

The problem is that neither convention nor intention can tie an indexical utterance to the relevant context, precisely because there *is no relevant context* in the case of "I'm not here now." We cannot say that the relevant context contains an argument place for the relevant individual, because such an entity wouldn't be a context. The answering machine sentence can only obtain a semantic content by picking values from *distinct* contexts. In other words, a context shift occurs. But what is a context shift, and how does one occur in natural language? These questions will be addressed in the following section.

2 Context Shifting

From the standpoint of Kaplan's account of indexicality, context shifting appears *prima facie* to be a non-event. Consider the character of 'I': it is a function from a context to a particular person. In speaking or writing 'I', the speaker or writer appears to automatically become the agent of the relevant context. We have a strong case for asserting that one cannot use the first person pronoun in sincere, factual discourse without referring to oneself. Of course, there are ways in which 'I' may refer to another; in writing or acting fiction, for instance, or when impersonating someone. Dually, we can quote the words of another, as in:

(1)   Anyone who says "I am beautiful" must be vain

in which case indexicals are not transposed from that context to this, yet retain their original semantic value. (1) expresses the same proposition as:

(2)   Anyone who says that they are beautiful must be vain

One way of analysing 'I' in (1) is to treat it as any other token of the first person pronoun, picking out the agent of the relevant context. In the case of (1), however, the relevant context is not the context in which (1) itself is uttered, but instead a context in which "I am beautiful" is (non-quotationally) uttered. If we pursue this line of reasoning, we can interpret the quotation marks around "I am beautiful" as context shifting operators, capable of shifting the context relevant to indexical determination from the context of utterance to one which satisfies the relevant criteria.

What about the cases that are clearly non-quotational and nonfictional? Could (1) be uttered without the quotation marks, and yet not refer to the speaker? It certainly appears not, at first. But Philippe Schlenker [9] discusses data which suggests that, *pace* Kaplan, indexicals can select a value from a context other than the context of utterance. The most striking example provided is from Amheric, a West African language (D. Petros, quoted in [9], example (52)):

(3)   jon   ĵegna   n∂-ññ   yil-all
        John   hero   be$_{1s}$ say
        John says that he is a hero

where the verb modifier '-ññ' places the Amheric verb 'to be' in the 1st person. However, information from Amheric tells us that the embedded clause refers to *John*, i.e. the speaker in the reported context. This suggests that the context with which the embedded clause is interpreted with respect to is not the context of utterance, but instead to the reported context. If so, a context shift must have occurred. We also have temporal examples from closer to home:

(4)   John has told me repeatedly over the years:
        (a)   "I was sick two days ago"
        (b)   that he was sick two days ago

(a) and (b) are equivalent; the that- clause in (b) can only be interpreted *de se*, i.e. as denoting the time two day's before *John's* utterance. Clearly, 'two days ago' could not refer to a time two days previous to an utterance of (b), for John could not possibly have said many years ago that he was sick on *that* day! So we must treat 'two days ago' as referring to times relative to John's utterances, the very sentence we report in (b). It appears that in this case, too, context sensitive elements appearing within the scope of an attitude verb can be evaluated with respect to the reported context(s).

A natural way to incorporate this data is to claim that attitude verbs act as modal operators, capable of shifting the context with respect to which indexicals within their scope are interpreted. Thus we could write:

(5)   <c'> I am a hero   [uttered in context c]

to mean that 'I' should be interpreted with respect to a context c', the reported context, and not the context of utterance. (5) is thus true iff 'I am a hero' is true (with respect to a truth assignment s) when the context of utterance c is replaced by the reported context c' (with respect to s):

(6) $[[<c'> \phi]]^{c,s} = 1$ iff $[[\phi]]^{c',s} = 1$

where the double square brackets indicate semantic values, and we identify the bit 1 with the value *true*. However, Schlenker sets up his account in terms of *quantification* over contexts. An attitude verb introduces an argument place for contexts c_i. If we take a psychological verb (such as 'say', for instance), we can state this formally:

Definition 1 (Syntax). *If ϕ is a well-formed formula, if c_i is a context variable and if α', β', γ' are an individual, a time and a world term respectively, then $\ulcorner SAY_{<\alpha',\beta',\gamma'>} c_i \phi \urcorner$ is a well-formed formula.*

Definition 2 (Semantics). $[[SAY_{<\alpha',\beta',\gamma'>} c_i \phi]]^{c,s} = 1$ *iff for all c' compatible with the claim made by $[[\alpha']]^{c,s}$ at time $[[\beta']]^{c,s}$ in world $[[\gamma']]^{c,s}$: $[[\phi]]^{c,s[c'/c_i]} = 1$*

where the notation $[c'/c_i]$ indicates that the semantic value enclosed in the square brackets has had contextual items in c_i replaced by those of all compatible contexts c'. A claim is *compatible* with another if both denote the same entities, either in the actual context or in the reported context. On this view, attitude verbs *always* quantify over alternative contexts, not just when a context shift occurs. Schlenker thus suggests that indexical terms have a lexical specification, either '+actual' or '-actual', which determines whether the term selects its value from the actual or shifted context.

3 Free Context Shifts

However, there appears to be no such operator in "I'm not here now". We can also give further examples of context shifts that require no operator. Imagine a replay of a Manchester United football match, shown the day after the game took place, during which the following sentences are uttered by a viewer:

(7)   United are taking control *now!*
(8)   *Now* you see why United won the match!

The two tokens of 'now' are not co-referential, for the former token refers to a moment during the match (say, to a time at which the decisive goal was scored), whereas the latter token refers to a time *after* the replay. The latter sentence is to an extent self-referential, for it denotes a state of affairs only possible in light of the replay (or some representation of the game). The former sentence, however, does not require the replay to have taken place, for it could have been uttered at any point during the match.

Could we argue that such operators are found at the level of logical form, i.e. they are what Perry terms *unarticulated constituents* of the sentence? If the answering machine sentence is to have a content at all, the contexts with respect to which each indexical token is interpreted cannot be identical. The context from which 'I' takes its value cannot be that from which 'here' and 'now' obtain theirs.

Thus, even if we posit some implicit unarticulated modality to shift the context with respect to which 'I' is interpreted, what delimits its scope? In *oratio obliqua* the answer is obvious: an attitude verb shifts the context with respect to which the embedded clause is evaluated. Indexicals occurring outside of the embedded clause thus occur outside the scope of the modality. Since there is no operator to act as the shifting modality in "I'm not here now" and, furthermore, nothing to mark scope if there were, we can reject the modal shifting account. These considerations pose a problem for Schlenker's quantificational account, too. Recall that, according to Schlenker, attitude verbs act as quantifiers over contexts.

If we treat the answering machine sentence in a similar way to the examples from Amheric, or the temporal examples introduced above, we are faced with two questions. Firstly, what introduces the appropriate context and, secondly, what quantifies over them? In the case of attitude verbs, Schlenker claims that the appropriate context (the reported context) is introduced conventionally.

In the *oratio recta* case, there is no reported context, although the appropriate context could be reached by considering the conventions involved in using answering machines, or the intentions of whoever set up the playback message. In other words, the considerations of the conventional/intentional accounts discussed above might help us here. Even so, we must quantify over any context introduced and, as we have just seen, there is no explicit lexical item in the answering machine sentence to do the job for us. Positing an implicit or unarticulated quantifier at the level of logical form is not an attractive move, for we would then face a similar objection to the modal shifting account and the conventional/intentional accounts above: namely, that whatever considerations apply to the answering machine sentence must also apply to "I am here now".

However, even if these objections could be overcome, the account would still be flawed. Recall that the quantificational account requires the lexical entry for a term to specify *which* context a token of the term should be interpreted with respect to. The binary feature '±actual speech act' is added to the lexical entry of a type to do just this. However, we previously identified '+actual speech act' with English-type indexicals such as 'I', '±actual speech act' with Amheric type indexicals and '-actual speech act' with logophoric pronouns and quasi-indicators. However, in light of the data presented above, we need to modify the entry for 'I', which can be shifted in *oratio recta*.

If we consider example (7) above in terms of context shifting, it seems likely that every essential indexical (and hence all those which can be defined in terms of indexicals) must have the lexical feature '±actual speech act'. But if an indexical token can refer to either the context of utterance or some other context, irrespective of whether the token occurs in *oratio recta* or in *oratio obliqua*, what determines which context is used in a particular case? What filters the '+actual speech act' cases from the '-actual speech act' cases, given that the selection is underdetermined by the lexical specification of the term?

It appears that none of these attempts can rescue the quantification over contexts account. The problem is that the context with respect to which an uttered sentence is interpreted can be shifted *externally* to that sentence and even *externally to the language*. In the case of the answering machine sentence, the most important factor in determining the context with respect to which we interpret 'I' is not the sentence itself, but the fact that it is a recording, played back by a machine. Had the call been answered by a human, we would be involved in a very different situation from the one entered when an answering machine takes a call. They are two very different *language games*, and what makes them different isn't anything to do with the language used. Thus, the account in terms of quantification over contexts fails, precisely because no articulated (or even unarticulated) term exists to quantify over introduced contexts. Where a context shift occurs, but no operator exists to shift the context (or to quantify over possible contexts), we can say that a free shift occurs.

4 Types of Context

We have come to a view according to which context shifts are possible. *Pace* Kaplan, an indexical such as 'I' does not always refer to the speaker, for its semantic value, the relevant contextual parameter AGENT, need not be identified with the speaker of the utterance. In cases for which the context in which an utterance is made and the context from which its indexicals obtain their references are distinct, the only explanation is that the context is shifted from the former to the latter context. However, this by no means implies that the shift is affected by an operator contained within the utterance or in a preceding utterance. In fact, the direct discourse example of the answering machine message above shows that a context can shift when there is no suitable operator present at all. So how does the context shift? In the case of the answering machine example, it seems evident that various factors affect the shift. The fact that an answer machine is involved, for example, seems essential to the shift. Such factors are often termed "contextual". So, before we attempt the question of context shifting, we need to examine precisely what we mean by "context" and "contextual factors".

Kaplan's notion of context is a highly abstract one, which should be distinguished from the normal (and perhaps from the general philosophical) use of the term 'context'. Think of the politician persistently claiming that he was taken "out of context", or someone claiming that the joke was funny only in the right context—"you had to be there!" This is a much more plastic notion than that used by Kaplan in explaining the content of indexical utterances. To distinguish this notion of context from Kaplan's, I shall henceforth use 'Kaplanian context', or simply 'K-context' for short, to denote the apparatus used by linguists and philosophers of language (which has simply been called 'context' above). I will also use 'd-context' (discourse context) to mean 'context' it its more popular usage and 'd-components' to denote those extra-semantic factors, the components of d-contexts, which affect our linguistic interchanges. I will also refer to free K-context shifts (such as the shift that occurs in "I'm not here now") as *free K-shifts*.

It has become traditional to refer to the K-context of utterance, that it, the one which contains the speaker / writer, her location and the time of the utterance, as the *actual* context, as opposed to the shifted context. However, this is a bad use of terminology, as the suggested parallel with world shifts in modal logic is unwarranted. Sentences which require a K-shift in order to be understood are no more hypothetical or counterfactual that those which require no such shift. K-contexts, whether shifted or not, are very actual, although abstract, features of our world. So let us instead refer simply to *K-contexts of utterance* and *alternative K-contexts* (although we may drop the 'K-', since only one d-context can be associated with any discourse, and so the distinction is not meaningful in their case).

It might then be thought that K-contextual parameters are simply a subset of the d-components relevant to a particular discourse. If this were the case, d-context must be fundamentally constituted in the same way as K-contexts, only with more parameters. That is, they must comprise various indexed parameters, one for each d-component relevant to a particular discourse situation, including the K-contextual parameters a, t, p and w. Clearly, no one set of parameters will suffice here; just consider what might be needed to disambiguate the 'bank' utterances above. Our notion of d-content must be an extremely plastic one, because those d-components that are relevant in a discourse situation are themselves contextually determined. Thus, precisely which parameters would be required in a particular d-context, construed as an indexed set in the manner of K-contexts, themselves depend on contextual factors. If relevant d-components are specified contextually by a d-context c_1, what specifies *this* context? It appears that, whatever d-context might consist in, we cannot simply represent it as K-context with extra parameters.

This is well and good for our account of context shifting. If it were the case that d-contexts are simply extensions of K-contexts, then each K-context shift would be accompanied by a corresponding d-context shift. However, it seems clear that the information required to interpret utterances within a discourse (d-context) remains precisely the same during utterances such as "I'm not here now". All that changes is the relevant K-context. I therefore want to offer a brief sketch of d-context according to which information relevant to the structure of the discourse as a whole is tracked and stored, and becomes available to use in interpreting various sections of the discourse. Central to this picture is the idea that information about the discourse as a whole cannot be viewed merely as a concatenation of information encoded within (or localisable to) individual utterances. In order to interpret a section of a discourse, it is often necessary to refer to the relation that section bears to the discourse as a whole. I will further suggest that one such piece of information required to interpret sections of a discourse is information relating to how participants in the discourse should interpret indexicals, i.e. information about the available K-context(s). My aim is not to offer an exhaustive account of discourse context, but only to argue that, whatever such context consists in, it must contain information relating to both discourse structure and available K-contexts. I also hope to show that these to factors are not as disparate as might at first be thought.

5 Discourse Structure

The suggestion that information relevant to interpreting a section of a discourse is not always encoded in the sentences that make up that section certainly seems *prima facia* to be counterintuitive. However, Michael Glanzburg [4] makes the point explicitly and convincingly. His primary aim is to characterise information that is localisable to individual utterances and to show that, in order to evaluate certain types of utterance, such information will not suffice. Even though he treats the notion of context in a narrow sense, such that

$$\text{sentence} + \text{context} \rightarrow \text{truth conditions}$$

there must be a component of context which contains information that cannot be localised to individual utterances, the so-called global discourse component, in order to provide truth-conditions for certain types of sentence.

If this is the case, Glanzburg's notion of a global discourse component should certainly feature in d-contexts. We should therefore spend some time reviewing Glanzburg's arguments. So as not to make the argument concerning the non-utterance localisability of certain types of d-component too trivial, we can grant a generous notion of utterance localisable information. For example, the interpretation of cross-sentential anaphora requires that information be passed from one utterance to another, possibly at long distance:

(9)   Russell was a great philosopher. He also wrote on many social issues.

We need the content of the first sentence to interpret 'He' in the second, for 'He' is co-indexed (and thus co-referential) with 'Russell'. However, we can regard this passing of information as utterance localisable, for consultation of information encoded within the first sentence is wholly sufficient to interpret the second.

The problem for wholly utterance localisable accounts of d-context comes from the notion of discourse structure. Discourses cannot be viewed simply as linear sequences of utterances. Even monologues, which remove the need to interpret the turn-talking phenomenon, are highly structured, with distinct sections of a discourse forming a hierarchical structure. If we consider encyclopedia entries, for example, we see that they are usually headed by a summary of the discourse to follow.

Below this, we may find adjacent paragraphs which bear no coherent relation to one another, taken in isolation, but make sense when each is related to the discourse header. Each stands in an elaboration relation to the header of the discourse, which may operate long-distance, whilst standing in no interesting interpretative relation to adjacent sections in the discourse. The elaboration allows us to nest sections within ones that already stand in their own elaboration relation; any section may elaborate a preceding one and still require further elaboration later in the discourse. The point is that, at any point, we may jump out of a higher-level elaboration and start a totally new section, elaborating a section from above, which bears no interesting relations to the preceding section. In these cases, it is not the information provided by individual utterances that allows us to understand the new section. Rather, it is the relation this section bears to the structure of the discourse itself, which cannot be represented merely as a linear sequence of utterances.

Thus, to interpret utterances within a section of a discourse, we may well require global discourse information, i.e. a d-component that cannot be obtained merely by consulting individual utterances. We can picture discourse structure as a directed graph or a tree, with each node representing a sequence of sentences, and each edge representing the discourse relation a group of sentences bears to those above. Each node thus represents a section of the discourse.

6   A Contextual Framework

Since the relevant topic may differ from section to section of the discourse, our notion of d-context should not only contain the structure of the tree, in some form, but also information about the relevant topic at each node. There are two distinct notions of topic which to consider: *sentence topic* and *discourse topic*. Since we are dealing with entire sections of a discourse at particular nodes, we should work with a notion of discourse topic, or at least discourse section topic. However, for the sake of simplicity, I am going to somewhat blur this distinction, and take discourse topics to be, in some way which I leave unspecified, abstractions from sentence topics.

Reinhart [8] argues that (sentence) topics are simply the entities which sentences are about, where *aboutness* is treated within Stalnaker's [11] framework of *dynamic interpretation*. Stalnaker takes the current shared knowledge of speaker and audience to be expressed by a proposition. The propositional content of further utterances, if accepted by the audience, is added to this information. Reinhart proposes that, instead of simply adding propositions to this set as they are expressed in a discourse, we have to group propositions around relevant objects, i.e. we form ordered pairs $\langle \alpha, \Phi \rangle$, where α is an object and Φ is the set of propositions about that object.

This notion of context (the "context set") makes no reference to discourse structure. We can, however, integrate Reinhart's idea into our account of d-context by adding ordered pairs of the form $\langle \alpha, \Phi \rangle$ at each node. But if we consider fictional discourses, utterances have no propositional content, for characters such as Hamlet and Sherlock Holmes do not exist.[2] Are we then to say that fictions are about nothing—that they have no topic? We clearly require contextual information to interpret fictional discourse just as much as we do in the case of factual discourse.

However, if we remove the object-centered aspect of this account, we can retain much of Stalnaker's idea. Instead of grouping propositional information around objects, we use a set of propositions to keep track of the speaker's and audiences' *presuppositions*. In the case of fiction, for example, it is not the objects referred to that is important in understanding the discourse (for no such objects exist); what is crucial is the shared presupposition that the discourse is fictional. This notion of topic comes closer to the notion of discourse (section) topic that we require. So, we can represent d-context as a tree, where each node is an ordered tuple $\langle \Phi, \Sigma \rangle$, where Φ is the set of propositions expressed by the speaker's and

---

[2] Here, I simply assume the non-existence of fictional characters or *ficta*. For a defense of this view, see [3]. The assumption does not affect the argument greatly, as objects could be added to the account given here without altering it significantly.

audiences' presuppositions at that stage of the discourse, and Σ is the sequence of sentences occurring within that section of the discourse.[3]

A work of fiction invites us to engage in a pretense, within which the characters depicted in the fiction act as the referents of names and indexicals. In the latter case, we can consider fictional characters or *ficta* to be agents of K-contexts, as long as we remain within the pretense. We do not take a token of 'I' within a fiction, for example, to refer to the author or the actor who produces the token, but instead to the character depicted by the fiction. Thus, moving within a pretense from outside invokes a K-context shift, from the K-context of utterance to one appropriate to the world depicted in the fiction.[4] Let us refer to such K-contexts as *fictional contexts*.

We notice immediately that sentences within a work of fiction appear to be just the same as those used in factual discourse; in fact, if they differed substantially, authors would not be able to generate the pretense necessary for an audience to engage with a work of fiction. There is thus no operator present in individual sentences within a fictional discourse capable of making the required K-context shift. However, there is one large difference between fictional and factual discourse, and that is the general context within which each is interpreted. The former, but not the latter, needs to be interpreted within a pretense, which is entered at the start of the discourse. In order to engage with each subsequent section of the fictional discourse, one must first grasp its relation to the initial move to within the pretense; one must know, in other words, that a particular sentence is indeed part of the fictional discourse. Thus, a proposition to the effect that a pretense is presupposed by both speaker and audience should appear in Φ at each subsequent node of the fictional discourse.

Now suppose that a fiction features a narrator, recounting a further fiction. We are then faced with two distinct discourse sections (or even two distinct discourses), one (the narration) in which the narrator may refer to himself using 'I', and one (the story told by the narrator) in which characters in the story will be referents of 'I'. So, in order to grasp to whom a token of 'I' refers, we need to know the relation that a particular sentence in which 'I' occurs bears to those preceding it. Discourse structure, and hence d-context, plays a crucial rôle here. Usually, a work of fiction does not have to explicitly signal the change from one discourse section to another; by simply shifting the topic at any point, we, as the audience, grasp that the fiction or the setting has changed. So our notion of d-context can help explain our grasp of shifts within a work of fiction. Similarly, it can explain our ability to seamlessly move in and out of a particular pretense. For example, we do not enter a pretense with an author's preface to her novel, for the preface does not bear the appropriate relation to the fictional discourse.

---

[3] I include the relevant sentences in this representation of d-context simply to mark the correspondence between nodes and sections of the discourse, i.e. to make explicit which node relates to which section

[4] Assuming, of course, that *ficta* do not really exist, the shift is to a null K-context (or at least one with a null AGENT parameter). If this assumption is mistaken, then fictional contexts are just like any other, only containing abstract entities. So this assumption does not affect this account. See [3] for an exhaustive discussion of this issue.

Once we enter the pretence, we remain within it whenever a section of discourse bears the appropriate relation to other discourse sections within the fiction.

When we enter a pretence, the K-context available to indexical content determination shifts from the context of utterance to the fictional K-context. Indexical tokens within the fictional discourse obtain their content from a fictional K-context, which must be introduced at the same time as the jump into the pretence is made. In other words, the shift in K-context occurs at the root node of the tree that encodes the structure of the discourse. It thus seems natural to claim that each node encodes not only information about presuppositions relevant to that section of the discourse, but also about the relevant K-context. We can thus expand our tuple representation of d-context nodes to become $\langle K, \Phi, \Sigma \rangle$ where K is a (possibly singleton) set of K-contexts, one of which will be used to determine the content of indexicals occurring within Σ. Thus, d-context and K-context are related in that the former provides information about which K-context is at play during various sections of a discourse.

Can the same picture be applied in the case of the answering machine sentence? We can consider the sentence "I'm not here now" to be its own discourse section, i.e. we have a d-context node at which the only sentence in Φ is "I'm not here now" and where K includes both the K-context of utterance and the K-context in which the owner of the answering machine is the agent. However, this only answers the question of how a K-context other than that of utterance can provide indexical content. We still have to answer the question, what actually makes the selection between the available contexts? How is it, in other words, that 'I' selects its content from a different K-context than 'here' and 'now'?

There are two possible ways in which the above framework can accommodate such cases. Firstly, we could simply stipulate that each node in the discourse tree can contain at most one K-context. Thus, no choice need be made between K-contexts at each node. If we are to accommodate examples such as the answering machine sentence, we would then have to represent it as two separate nodes, one for the first person pronoun, one for the remaining part of the sentence. However, this seems to be a rather artificial suggestion with little independent motivation. In the case of the answering machine sentence, the fact that the discourse revolves around an answering machine is surely essential to any successful communication. After all, if one did not know that one was listening to a recording, an utterance of "I'm not here now" would appear self-contradictory. Thus, in order to understand this sentence, we need to relate it not to other sections of the discourse, but to the fact that we are dealing with an answering machine. In other words, believing that one is dealing with a recording device is a precondition of understanding sentences such as "I'm not here now". One must presuppose this information in order to grasp what is said, and so this presupposition should appear within Φ at the node containing "I'm not here now".

Within this framework, it is the information about our presuppositions concerning answering machines, plus the fact that we are dealing with an answering machine, that shifts the context half-way through "I'm not here now". In other words, the presupposition that recording devices do not feature as agents in discourses, plus the presupposition that we are dealing with a recording device (both of which are propositions stored at the relevant d-context node) entail that 'I' cannot take its semantic value from the context of utterance. The caller needs to presuppose this information in order to grasp the message played back to her by the answering machine. Thus, we can claim that information encoded within Φ at a particular node can affect the K-context available to sentences in Σ, from a choice of those K-contexts in K at each d-context node. The exact mechanism, however, is a task for future research to spell out.

7 Conclusion

I have appealed to two distinct notions of context. Firstly, we have the notion attached to the everyday use of 'context', which is those collection of features of the world required to give determinate truth-conditions to our utterances. I termed this *discourse context*, or d-context. On the other hand, we have K-contexts, which are tuples consisting of an AGENT, a time, a place and a world. It is the latter that are specifically required for indexicals to have a semantic value. An indexical takes its value from the available K-context, according to Kaplan's character-content framework. However, the available K-context need not be that of the utterance (the speakers, time, place and world of the utterance). Discourse contextual information can direct us to another K-context, using information about the speaker's and audiences' shared presuppositions. In this case, an alternative K-context becomes available to indexical reference. When values are picked out from this alternative K-context, we say a context shift has taken place.

References

1. Almog, J., Perry, J. and Wettstein, H. (eds.): *Themes from Kaplan*, Oxford: Oxford University Press (1989)
2. Corazza, E., Fish, W & Gorvett, J.: "Who is I?", in Philosophical Studies, Vol. 107, No. 1 (2002) 1–21
3. Corazza, E. & Whitsey, M.: "Fiction, Ficta and Indexicals", in *Dialectica*, forthcoming (2003)
4. Glanzburg, M.: "Context and Discourse", in Mind & Language, Vol. 17, No. 4 (2002) 333–375
5. Kaplan, D.: "Demonstratives", in Almog et el (1989) 481–463
6. Kaplan, D.: "Afterthoughts", in Almog et el (1989) 565–614
7. Predelli, S.: "I am not Here Now", in Analysis, vol. 58, no. 2 (1998) 107–15
8. Reinhart, T.: "Pragmatics and Linguistics: An Analysis of Sentence Topics", *Philosophica* 27 (1981) 53–94
9. Schlenker, P.: "A Plea for Monsters", in Linguistics & Philosophy, vol. 26, no. 1 (2003) 29–120
10. Sidelle, A.: "The Answering Machine Paradox", in Canadian Journal of Philosophy, vol. 81, no. 4 (1991) 525–539
11. Stalnaker, R.: "Assertion", in Syntax and Semantics 9 (1978) 315–332
12. Vision, G.: "I am Here Now", in Analysis, vol. 45, no. 4 (1985) 198–199

A Mathematical Model for Context and Word-Meaning

Dominic Widdows

Center for the Study of Language and Information,
Stanford University
dwiddows@csli.stanford.edu*

Abstract. Context is vital for deciding which of the possible senses of a word is being used in a particular situation, a task known as disambiguation. Motivated by a survey of disambiguation techniques in natural language processing, this paper presents a mathematical model describing the relationship between words, meanings and contexts, giving examples of how context-groups can be used to distinguish different senses of ambiguous words. Many aspects of this model have interesting similarities with quantum theory.

1 Introduction

Context plays a key role in determining the meaning of words — in some contexts the word *suit* will refer to an item of clothing, in others a legal action, and so on. In the past decade, the challenge of incorporating contextual models into the way information is described has become very immediate and practical, in the wake of rapid technological advances. To compare and combine information which is readily available but varies across languages, domains of expertise and media, it is important to have some way of expressing what that information actually means in a common and flexible framework. Context can be very useful here — if someone is trying to buy a new computer they will be much more interested in the term *PC* if it occurs in a magazine called *'Computing Today'* than if it occurs in the *'Political Activism Quarterly'*. A word in one language can often have several possible translations in another language depending on which meaning is appropriate (for example, English *drugs* can translate both to *drogen*='narcotics' and to *medikamente*='medicines' in German), and the correct translation can only be determined using context.

However, it is much easier to give examples of what context is used for and why it is important than it is to give a proper account what context is and how it is used to determine meaning. This paper attempts to bring some light on these issues, by describing techniques for resolving ambiguity in natural language

---

* This research was supported in part by the Research Collaboration between the NTT Communication Science Laboratories, Nippon Telegraph and Telephone Corporation and CSLI, Stanford University, and by EC/NSF grant IST-1999-11438 for the MUCHMORE project.

processing within a particular mathematical framework. This investigation leads to a concrete definition of context, and to promising similarities with traditional branches of mathematics and physics which are much more precisely understood than our current knowledge of the way context is used to determine meaning.

This paper proceeds as follows. Section 2 introduces the field of Word-Sense Disambiguation (WSD), the branch of natural language processing which is concerned with finding the correct meaning for a word used in a particular situation. Section 3 presents a mathematical model which can be used to define and describe WSD and other linguistic phenomena within a unified framework based upon the three spaces $\mathcal{W}$ ('words'), $\mathcal{L}$ (a lexicon of 'meanings'), $\mathcal{C}$ ('contexts') and mappings between them, in particular 'sense-mappings' which map a word w in a context c to a particular lexical meaning l. Section 4 examines the space $\mathcal{C}$ in the light of these mappings, which suggests an approach to defining context itself. More exactly, we define groups of contexts, two linguistic situations being in the same 'context group' with respect to a particular word if that word gets mapped to the same meaning in both contexts. Examples are given of how homonyms (unrelated senses of ambiguous words) give rise to disjoint context groups, and systematically related senses give rise to overlapping context groups. Finally, Section 5 describes some notable if surprising similarities between our model for disambiguation and the process of making an observation in quantum theory. It is hoped that these analogies will be of interest to many researchers, scientists and even interested layfolk.

Context is of importance in linguistics, philosophy, sociology and many other disciplines, as well as to natural language processing. The approach taken in this paper is to cast the NLP problem in a mathematical setting, and through this to enable the use of ideas from this very traditional discipline. Many other more established areas of scholarship, way outside this author's expertise, could almost certainly contribute fruitful ideas to NLP if researchers become more aware of the problems and possibilities therein. While this paper attempts to present a precise model of meaning and context, the main goal is to stimulate inquiry and exchange of ideas between disciplines.

2 Word-Sense Disambiguation

There are several situations in information technology, and particularly in natural language processing, where it is necessary or at least very desirable to know what a particular word or expression is being used to mean. This begins in very simple ways in situations with which most of us will be familiar. If you're booking an airline ticket online and say you want to travel to London, you will probably get a response such as:

> More than one airport matches 'london'. Please type a new
> airport name or code, or select an airport from the following list:
>
> London, England   LGW - Gatwick
> London, England   LHR - Heathrow
> London, England   LON - all airports

London, England   STN - Stansted
London, OA, Canada   YXU - London

The system which your browser is sending information to has a list of possible airports all over the world, each of which has a 3 letter code which means that airport and no other. A passenger requesting a flight to *London* could wish to travel to any of these airports. Effectively, each of these 3 letter code is a possible 'sense' of the word *London*, and to interact effectively the system must know which of these meanings are acceptable to the user. These meanings are not all created equal — *LHR*, *LGW* and *STN* are collectively referred to by *LON* [1] and this broader classification might be specific enough for some situations, so long as we know that *YXU* is completely different and must be distinguished. The user can see these options and choose the appropriate one for their travel needs — and if the system had more information about these travel needs it should be able to make the correct decision for itself. The word *London* is thus ambiguous (more particularly *polysemous*, meaning "many senses") and the process of deciding which meaning is appropriate is known as 'ambiguity resolution' or 'disambiguation'.

Most word-meanings in natural language have a much more complex structure than airport codes. There is a finite number of distinct airports and the system can rely on the user to say unequivocally which one is the correct choice. When a new airport is built, there is a recognised process for giving it its own unique code, the codes are universally recognised, and people do not start using a new and different code to refer to an airport which already exists. In natural language, people break all of these rules, and a careful analysis reveals all sorts of persistent difficulties in deciding what counts as a word-sense and which sense is being referred to in a particular instance [2].

Research on word-sense disambiguation (WSD) in natural language processing goes back several decades, and a historical survey of much of this work can be found in [3]. The central theme is that we should be able to work out which sense of a word is being used by examining the context in which a word is written (almost all work in WSD has been on text rather than speech). Initial work focussed on laboriously building 'expert' classifiers which would enumerate the different contexts in which a word might appear, with enough information to work out which sense was being used. Later on, as machine readable dictionaries (MRD's) became available, they were used automatically to provide information for disambiguation. An early and quite representative approach was that of Lesk [4], who used the words found in definitions from a MRD as clues that might indicate that one sense was being used rather than another. For example, one sense of the word *ash* is defined as

1) any of a genus (Fraxinus) of trees of the olive family with pinnate leaves, thin furrowed bark, and gray branchlets (Webster 1997)

---

[1] In a hierarchy of concepts, we might say that *LON* is a 'hypernym' [1, p 25] or 'parent node' of *LHR*, *LGW* and *STN*.

In a sentence, one might very well encounter the words *tree* or *leaves* near the word *ash*, and use that as evidence that this sense of *ash* is being used. The problem here is that the information provided is very sparse, and often very different in register from the words one encounters in more normal text. For example, the definition of the other major sense of *ash* includes the words *residue* and *combustible* and makes no reference to the words *cigarette* or *dust*, which might be much better contextual clues. Lesk's solution to this mismatch was to use not only words occurring in the definition of *ash*, but words whose definitions share words with that of *ash*. For example, *dust* is related to the 'residue from combustion' sense of *ash* and both definitions include the word *particles*.

A method of relating "definitions" which more clearly reflect actual usage is to use the "one-sense per collocation" assumption [5]. This works upon the premise that idiomatic phrases can be used as 'seed-examples' of particular senses with very high reliability. For example, Yarowsky distinguishes between the 'living thing' and 'factory' sense of the word *plant* by assuming that almost every instance of the collocation *plant life* uses *plant* to mean a living thing, and almost every instance of *manufacturing plant* uses plant to mean a factory. These examples can then be used to find other words which indicate one sense or the other, and so gradually extend coverage [6].

Such a method involves taking a few labelled instances of one sense or another, examining their surrounding contexts, and extrapolating to achieve a similar labelling in other similar contexts. To get a more representative sample of instances of different senses, the initial labelling phase can be carried out (at some cost) by human annotators. This gives rise to the process of *supervised* word-sense disambiguation, where statistical tendencies observed in labelled examples are used to classify new unseen examples [3, §2.4.2]. The most standard model used is naive-Bayes, where "indicative words" are extracted from the training examples and used as evidence of the sense they occured with in training if they appear within a distance of n words from the target word.

By the late 1990's, WSD had become a recognised field within natural language processing, with its own internally defined standards and SENSEVAL evaluation contests [7]. These contests provide manually annotated training data and held-out evaluation data to compare the performance of different systems. This framework encourages supervised methods, which perform understandably better in such situations, though it is unlikely that enough hand-labelled data will ever be available to use such disambiguation techniques in real situations with many thousands of different words.

Even after much progress in the last ten years or so, one of the clearest results is that the 'success' of a disambiguation system depends critically on whether it is evaluated on easy or difficult ambiguities. Comparing decisions made by different human judges shows that people often disagree with one another in labelling a particular usage with a particular sense [3, §3.3], and these discrepancies cast doubt on the significance of results obtained by comparing a word-sense disambiguation system against 'gold-standard' human judgments.

A major question is granularity: what distinctions in meanings are regarded as important enough to be considered different senses. Sometimes two senses are clearly different, such as the 'commercial-bank' and 'river-bank' senses of the word *bank*. This is an example of 'homonymy' or 'accidental polysemy', where two unrelated meanings share the same surface word-form almost accidentally. But many cases are more subtle — *bank* in the commercial sense is used to mean a financial institution and a building where that institution does business. Sometimes this distinction is not important — in the sentence

The bank closed at 5pm.

it probably means that the institution stopped doing business and the doors to the building itself were shut, so both senses are in use simultaneously. However, if one heard that

The bank collapsed at 5pm.

one would need to know whether it was the institution or the building which collapsed to determine whether to call in liquidators or rescue-workers. This sort of ambiguity is called 'regular polysemy' or 'systematic polysemy' — two senses are logically related, or one sense has different aspects which might be called upon in different situations. It is increasingly apparent that word-senses are not discrete, distinct units but adaptable and generative [8]. Rather than viewing ambiguity as a problem, one approach to creating lexicons is to use ambiguity as a guiding principle in defining systematically polysemous categories [9]. It is possible that the very choice of the term 'disambiguation' to describe the process of mapping word-usages to word-senses has led to a more divisive view of word-senses than is healthy.

Whether or not it is important to distinguish between senses depends very much on the task at hand. For translation, it is only important insofar as it determines the correct translation of a given word — if a single word in English with two possible senses is translated to the same word in French irrespective of which sense is being used, then the distinction does not matter for determining the correct translation. In information retrieval (IR), on the other hand, the distinction between two senses is important if and only if knowing the distinction could affect whether a particular document is relevant to a user. One solution to this problem in IR is to let the document collection itself determine which sense-distinctions are important. This entirely unsupervised approach to WSD was taken by Schütze [10,11], finding different 'senses' of particular words by identifying distinct clusters of 'context-vectors'. For example, the word *suit* often occurs with the words *court, judge* and *attorney*, or with the words *jacket, tie* and *dress*, and words from each of these groups occur far more often with one another than with words from the other group. These two groups fall into very distinct 'clusters', and once this has been noted the words in the different clusters can be used as clues that one sense or another is being used. This is one of the few methods for WSD that was shown to provide a reliable enhancement for information retrieval [12], possibly because it reflects the way words and senses are used in a particular document collection rather than how a dictionary says they should be used in general.

All of these methods for WSD share that feature that in some way, instances of words are being mapped to possible senses based upon information in a surrounding context. Unfortunately, the way 'context' is defined and used is often unstructured and inflexible. There are many kinds of context which might be relevant to determining meaning (for example, [3, §3.1] discusses *microcontext*, *topical context* and *domain context*), but much of this information is often neglected.

The practical challenge of combining information from different knowledge sources into a single system for disambiguation has been addressed with considerable success [13,14], and in combining different knowledge sources these methods implicitly combine the different contextual information that the different knowledge sources rely upon. A theoretical challenge, which is of possible interest to a broad group of people, is to give an account for words and their meaning in context which can recognise and explain the wide variations we encounter in language.

3 A Mathematical Model for Words, Meaning, and Context

The techniques for word sense disambiguation described in the previous section have several factors in common. These similarities encourage the development of a more structured mathematical model which we now present, beginning by introducing three important sets which we need in order to phrase these questions mathematically. These are the spaces of words, of meanings, and of contexts.

$\mathcal{W}$	Words	Primitive units of expression
		Single words
		Parts of compound words (eg. *houseboat*)
		Independent multiword expressions
$\mathcal{L}$	Lexicon	The available meanings to which signs refer
		Traditional Dictionaries
		Ontologies, Taxonomies
		Meanings collected from training examples
$\mathcal{C}$	Contexts	Pieces of linguistic data in which signs are observed
		Sentences
		Immediate collocations (eg. blood *vessel*)
		Whole domains have conventions (eg. acronyms)

The main goal of this paper is to explore the structure of $\mathcal{C}$, the set of possible contexts, and we will do this in the following section. Before this, we devote a few remarks to the sets $\mathcal{W}$ and $\mathcal{L}$, and to some important mappings between these spaces [2].

In discussing a space of words or expressions, we are assuming that a sentence is somehow decomposable into basic units. These will not necessarily be handy strings of letters with spaces in between them — some words can be identified as compounds of more primitive units (such as with the word *fishbone* = *fish* + *bone*), and some multiword expressions or 'collocations' have a more or less independent meaning of their own (such as *United States of America*). Sometimes it is not clear whether something is really a multiword expression or two primitive units or both (such as *Catholic Church*. While it is possible (and depressingly easy) to use such objections to claim that 'words' don't really exist at all, even very simple computational and statistical techniques can analyse a corpus of English and identify words and collocations which most of us would agree are a reasonable choice of 'primitive units' [15, Ch 5].

The structure of the lexicon $\mathcal{L}$ is much more subtle and has been the subject of considerable research [16]. The lexicon describes the senses used in the domain in question, for example a traditional dictionary or thesaurus, an ontology or a taxonomy. If the domain is computer manufacturing, the lexicon $\mathcal{L}$ might be an ontology describing parts of computers, and information about where they go, what their purpose is, how they relate to other parts, etc. More generally, lexical resources such as WordNet [1] (which is domain-general) try to locate concepts by introducing broad, general categories such as *event* or *artifact* and giving relations between the entries such as IS_A, PART_OF, etc. A standard objection is that all these 'lexicons' are finite and discrete, but it is impossible to enumerate all the nuances of meaning a word can have in advance because new words are always being coined and old words being used with new meanings which are intuitively related to their old ones (examples include words which have acquired a technological meaning such as *web* and *mouse*). An alternative to the enumerative approach is provided by the theory of generative lexicons in which a core meaning can be systematically extended to new situations [8].

Another way to build a 'lexicon' is to take labelled training data. Instead of a definition being a statement about what a word means or what other words it is related to, concepts are defined by a list of contexts in which they are represented by a given word. This way of defining a lexicon has some drawbacks in common with those above. It is costly and static: once the labelled instances have been used, you can't then ask the lexicon for more information, or information about a different domain, hence the difficulty in applying WSD methods that require training data to any domain other than that for which the training data was designed.

---

[2] Note that authors sometimes use the term 'lexicon' to refer not only to the meanings and the way they are organised, but also to the words and the various possible mappings from word-forms to lexical entries, effectively amalgamating the spaces $\mathcal{L}$ and $\mathcal{W}$. The main reason here for using a notation where these spaces are kept separate it to be able to describe the process of assigning word-senses to word-usages as a mapping from $\mathcal{W}$ into $\mathcal{L}$.

Mappings between $\mathcal{L}$, $\mathcal{W}$ and $\mathcal{C}$

Many important linguistic phenomena can be defined as mappings between these spaces or products thereof.

Traditional WSD. The traditional WSD problem is as follows. Take a predefined collection of meanings $\mathcal{L}$ and a collection of pairs of words and contexts $(w, c) \in \mathcal{W} \times \mathcal{C}$. Produce a mapping $\phi : (w, c) \to \mathcal{L}$ and compare the results with a 'gold-standard' of human judgments. We will refer to such mappings ϕ as 'sense-mappings'. The possible images of this map for a given word w,

$$S(w) = \{\phi(w, c) \text{ for any } c \in \mathcal{C}\} \subset \mathcal{L},$$

can be referred to collectively as the senses of w.

The WSD problem can be posed using any one of the lexicon types described above. One significant difference between hand-coded dictionary or thesaurus definitions and definitions extracted from training data is that for the latter we do have a collection of examples of the form $\phi(w, c) = l$. (In fact, this is all we have.) The traditional WSD problem is to extrapolate ϕ to other pairs of symbols and contexts. In practice, WSD algorithms are only evaluated on how well they generalise to other contexts: to extrapolate from known instances $\phi(w, c_1)$ to unknown instances $\phi(w, c_2)$. This goes some way to explaining to the new reader how narrow much work in WSD has been, which must surely suggest partial answers to questions such as "why has WSD not proven very useful for information retrieval?".

Synonymy. The words $w_1, w_2 \in \mathcal{W}$ are said to be *synonymous* in the context of $c \in \mathcal{C}$ if $\phi(w_1, c) = \phi(w_2, c)$. Synonymy is the name given to the phenomenon that mapping from $\mathcal{W}$ into $\mathcal{L}$ will not in real life be injective. Synonymy in one particular context could accurately be called 'partial synonymy'. Two symbols are said to be *totally synonymous* if they have the same meaning in *all* contexts, so that $\phi(w_1, c) = \phi(w_2, c) \; \forall \; c \in \mathcal{C}$. It is known that total synonymy is very rare — there is nearly always some context in which two different words will have (at least slightly) different meanings.

WSD with learning or 'eavesdropping'. Introducing some learning operation is likely to help WSD. Along with the dictionary and the local context, the system is encouraged to look at a range of unlabelled data that has been collected along the way to see if it gives any further clues for accurate sense-mapping. Effectively, our system can eavesdrop on lots of conversations to see if it gets any further clues about what the symbols are used to mean.

Thus our sense-mapping is a function not just using one context $c \in \mathcal{C}$, but using any subset of contexts in $\mathcal{C}$. If we let $\mathcal{C}_s$ denote those contexts that are potentially relevant to the symbol w, our sense-mapping takes the form $\phi : (w, c, \mathcal{C}_s) \to \mathcal{L}$.

This method uses our initial definitions $\mathcal{L}$ as 'seeds of meaning' and allows those seeds to sprout, gathering more information, before we expect the seeds to accurately catch the correct meaning of all the symbols which refer to them. One way of doing this is to use a corpus to extract semantic similarity scores between pairs of words and to use these similarities for smoothing (assigning a non-zero probability to events that were not observed in the training corpus but are possible nonetheless) to improve disambiguation [17].

4 Context Groups

In this, the most important section of this paper, we use the mappings from the previous section to define the notion of context groups. This essentially topological notion can be used to address the question of how much context is appropriate in a particular situation for accurately determining meaning.

The structure of the set $\mathcal{C}$ is of particular interest and flexibility. A context, on the global level, might be any subset of the total 'universe of discourse', and on the local level is some piece of language containing the word w which we want to map into the lexicon. The elements of $\mathcal{C}$ might take many different (and nested) forms from a single word to a whole domain. How much context we need to distinguish meanings is an important question for determining a suitable 'granularity' for $\mathcal{C}$, and as we have stressed in this paper, the answer to this question will vary considerably from case to case.

Many approaches to disambiguation (such as naive-Bayes and Schütze's vector models [11]) have assumed a model where a 'context' c is simply a list of words $(w_1, \ldots, w_n)$ (often without taking the order of words into account). These disambiguation systems therefore provide a mapping from $\mathcal{W} \times \ldots \times \mathcal{W} = \mathcal{W}^n \to \mathcal{C}$. However, it is clear that the space $\mathcal{C}$ also contains broader 'meta' information which, though it can be described in words, is essentially not wordlike. We would agree with the statement "in the context of medicine, *operation* usually means a surgical process (rather than a mathematical or a military operation". But this contextual information is certainly different from the usage of the single word *medicine* — we have an understanding of contexts as something that are often broader in scope than the individual words we use to name them.

One clear observation is that the relationship between granularity of contexts and granularity of senses is monotonic. That is to say that if two senses are to be clearly distinguished, then the contexts that include them must also be clearly distinguished and if the distinction between senses is relaxed, then the distinction between contexts can be relaxed. It follows that any measure of similarity or distance between senses will be mirrored in any corresponding measure of similarity or distance between contexts. This observation points to a way of defining the way context and meaning relate to one another without saying exhaustively what each *individual* context can consist of. This is accomplished by defining 'context groups' [3].

---

[3] This terminology is not intended to imply a group structure in the algebraic sense [18, Ch 10], since we have not yet defined a binary operation upon contexts, though

The context group of a word w with meaning l consists of precisely those linguistic situations under which that particular word will have that particular meaning. Phrases like 'in the legal context, *suit* means the same thing as *lawsuit*' are then given the following interpretation. Suppose the lexicon gives two meanings for the word *suit*, the legal meaning and the clothing meaning. Then as far as we are concerned, the 'legal context' $\mathcal{C}_{\text{legal}}$ is precisely those situations in which *suit* has the same sense as *lawsuit*,

$$\mathcal{C}_{\text{legal}} = \{c \in \mathcal{C} | \phi(suit, c) = l\},$$

where l is the 'lawsuit' meaning of suit.

This definition of context as an inverse image of the sense-mapping ϕ is essentially topological. Words appear in a 'neighbourhood' of surrounding information, and it is these neighbourhoods which are used to resolve any ambiguity [4]. Other contextual information may well be available but often unnecessary — if we know for sure from the topic of one article, or from a single sentence, that *suit* is being used to mean lawsuit, then all other observations in (say) the context of that publication can safely be treated as factors under which the meaning of suit remains constant.

In this localised version of events, the context group can be reduced to a local vector space of contextual symmetries [20, Ch 8]. Placing a word in a particular context is then conceptually similar to placing a ball on a sloping surface. It is the local structure of the surface, in particular the plane tangential to the surface at that point, which determines the direction in which the ball will roll. These 'local surfaces' will be orthogonal (or at least transverse) for semantically unrelated homonyms, and will have some intersection or overlap for senses which are more closely related through systematic polysemy.

Figures 1 and 2 give 2-dimensional projections of this information, derived automatically by analysing local collocations such as "arms and legs" in the British National Corpus. (Details of the corpus processing used to obtain these graphs are given in [21].) Two disjoint senses have been found for *arms* ('part of the body' and 'weapons'), whereas several systematically related senses have been found for *wing*, including the wing of an aeroplane, the wing of a bird, the wing of a building and a wing of an organisation such as a political party. These examples show clearly how different senses have made themselves apparent by appearing with different context groups.

This theoretical analysis suggests a simple order in which contextual information for disambiguation should be sought. Extremely local information such as collocational usage should be prior, followed by local syntactic information, broader coccurrence information and finally broad discourse and domain infor-

---

this is an open and promising question. Any reasonable combination of contexts from the context group of a particular word-to-sense mapping would be expected to preserve that mapping, so closure would certainly be satisfied.

[4] This analysis suggests some similarity between the idea of contexts which are in the same context group and possible worlds which are accessible from one another in intensional logic [19].

A Mathematical Model for Context and Word-Meaning   379

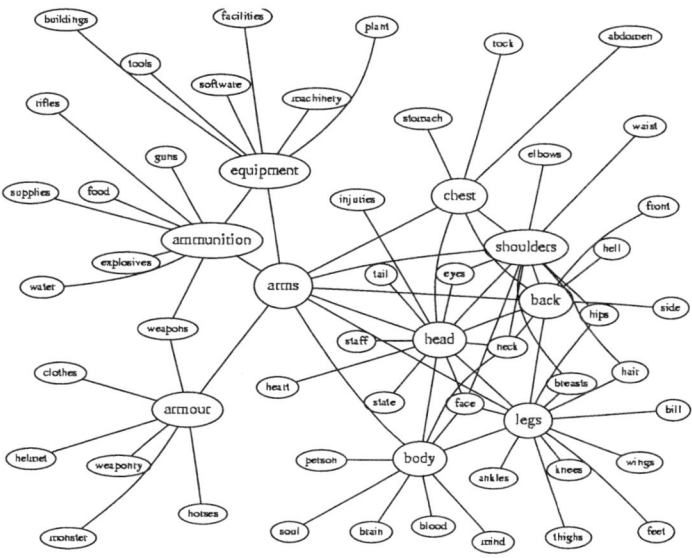

Fig. 1. Words related to different senses of *arms*

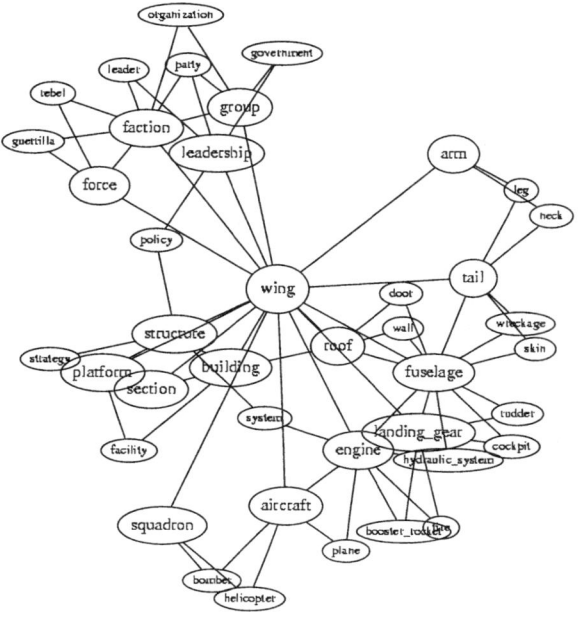

Fig. 2. Words related to different senses of *wing*

mation. The error made by statistical techniques [10, p 103] such as naive-Bayes for the sentence

> Sharpton said, "I have been on the attorney general's case, and I will be on his assistants like a suit jacket throughout the arraignment and the trial." (New York Times)

would be avoided by such a theory. Considering a context group as a topological neighbourhood around a particular word gives the simplest possible answer to the question "How much contextual information should be used to choose the correct sense?" — "Whatever is necessary".

5 The Parallel with Quantum Theory

Consider an electron e orbiting the nucleus of a hydrogen atom. As is well-known, the electron may be in several different 'states'. Its spin may be 'up' or 'down' and it may be in a variety of different energy states [22], and the possible energy states are predicted by the localised action of a particular group of symmetries [20, Ch 11]. Without further observations, it is impossible to say for sure which of these energy or spin states the electron is in — all we know is the prior probabilities of the different states. To determine the 'actual' energy or spin state of the electron, a measurement of some sort must be made by a conscious observer.

Of course, there are many particles other than electrons, and many situations other than this simplest case of the single electron orbiting a hydrogen nucleus. With many electrons orbiting bigger nuclei, knowing the states of some electrons gives a lot of information about the rest. Many forces work in different ways on different particles — the protons and neutrons together in the nucleus are bound together by the strong nuclear force which is very powerful over a small locality, the electrons are chiefly affected by electromagnetic forces which are broader ranging, weak nuclear forces can cause radioactive decay, and all the particles influence one another by the force of gravity which has a weaker global effect.

This situation is curiously reminiscent of the relationship between words, senses and contexts described in this paper. A word $w \in \mathcal{W}$ may have several possible senses $l \in \mathcal{L}$, some of which are more common than others. Once the word is observed in some actual situation, it is possible to ask which sense is being used. The answer will depend on a number of forces — collocational, syntactic, broad cooccurrence and domain — with different strengths and ranges of influence. The best way to understand the possible states of a particle is to understand the group of symmetries which preserves each state, just as we have suggested that context groups give more insight on word-senses than individual contexts. Assigning meaning is an interactive, not an isolated process — knowing the senses of some words can greatly assist us in determining the senses of others, and this process is often mutual. There are many common characteristics — the ambitious hope is that in years to come, scientific progress will provide theories which enable us to describe meaning with an accuracy similar to the accuracy with which quantum theory describes electrons.

A remaining difficulty is that the structure of word-senses is much more subtle and difficult to predict than the structure of energy-levels — at least for our current theories. However, this does not imply that meaning can only be written down exhaustively, guessed by statistical methods (both of which have already made great contributions) or left a mystery. The proposal in this paper is that careful scientific thought, together with empirical investigation, can provide models in which context and meaning are described clearly, flexibly, and accurately.

References

[1] Christiane Fellbaum, editor. *WordNet: An electronic lexical database*. MIT press, Cambridge MA, 1998.
[2] Adam Kilgarriff. Dictionary word sense distinctions: An enquiry into their nature. *Computers and the Humanities*, 26(1–2):365–387, 1993.
[3] Nancy Ide and Jean Véronis. Introduction to the special issue on word sense disambiguation: The state of the art. *Computational Linguistics*, 24(1):1–40, March 1998.
[4] M. E. Lesk. Automated sense disambiguation using machine-readable dictionaries: How to tell a pine cone from an ice cream cone. In *Proceedings of the SIGDOC conference*. ACM, 1986.
[5] David Yarowsky. One sense per collocation. In *ARPA Human Language Technology Workshop*, pages 266–271, Princeton, NJ, 1993.
[6] David Yarowsky. Unsupervised word sense disambiguation rivaling supervised methods. In *Proceedings of the 33rd Annual Meeting of the Association for Computational Linguistics*, pages 189–196, 1995.
[7] A. Kilgarriff and J. Rosenzweig. Framework and results for english senseval. *Computers and the Humanities*, 34(1–2):15–48, April 2000.
[8] James Pustejovsky. *The Generative Lexicon*. MIT press, Cambridge, MA, 1995.
[9] P. Buitelaar. *CoreLex: Systematic Polysemy and Underspecification*. PhD thesis, Computer Science Department, Brandeis University, 1998.
[10] Hinrich Schütze. *Ambiguity resolution in language learning*. CSLI Publications, Stanford CA, 1997.
[11] Hinrich Schütze. Automatic word sense discrimination. *Computational Linguistics*, 24(1):97–124, 1998.
[12] H. Schutze and J. Pedersen. Information retrieval based on word senses. In *Proceedings of the 4th Annual Symposium on Document Analysis and Information Retrieval*, pages 161–175, Las Vegas, 1995.
[13] Mark Stevenson and Yorick Wilks. The interaction of knowledge sources for word sense disambiguation. *Computational Linguistics*, 27(3):321–349, 2001.
[14] Mark Stevenson. *Word Sense Disambiguation: The Case for Combining Knowledge Sources*. CSLI Publications, Stanford, CA, 2003.
[15] Christopher D. Manning and Hinrich Schütze. *Foundations of Statistical Natural Language Processing*. The MIT Press, Cambridge, Massachusetts, 1999.
[16] L Guthrie, J Pustejovsky, Y Wilks, and B Slator. The role of lexicons in natural language processing. *Communications of the ACM*, 39(1):63–72, 1996.

[17] Ido Dagan, Lilian Lee, and Fernando C. N. Pereira. Similarity-based models of word cooccurrence probabilities. *Machine Learning*, 34(1–3):43–69, 1999.
[18] Barbara H. Partee, Alice ter Meulen, and Robert E. Wall. *Mathematical Methods in Linguistics*. Kluwer, 1993.
[19] L.T.F. Gamut. *Logic, Language, and Meaning*, volume II (Intensional Logic and Logical Grammar). University of Chicago Press, 1991.
[20] William Fulton and Joe Harris. *Representation theory – a first course*. Number 129 in Graduate texts in mathematics. Springer-Verlag, 1991.
[21] Dominic Widdows and Beate Dorow. A graph model for unsupervised lexical acquisition. In *19th International Conference on Computational Linguistics*, pages 1093–1099, Taipei, Taiwan, August 2002.
[22] H. F. Jones. *Groups, Representations and Physics*. Adam Hilger/IOP Publishing, 1990.

Demonstratives, Reference, and Perception

R.A. Young

Philosophy Department, University of Dundee, Scotland
r.a.young@dundee.ac.uk

Abstract. An account of reference which treats the role of perceptual demonstratives as fundamental is explored. Campbell's[1] account of reference and perceptual demonstratives is discussed, but it is argued that his account of perception is unsatisfactory, and an alternative account is proposed. This account is derived, albeit with substantial modifications, from Lewis[6] and Chisholm[2].

1 Introduction

When we consider the reference of demonstratives in philosophy of language, it is riddled with problems. Are these problems simply resolved in situations in which we can point to or grasp objects? It does not seem so. I point at my house and say to my artist friend, 'Isn't *that* beautiful, please paint it', and, perhaps understandably, she sees an ephemeral play of the light upon my house caused at just that minute of that hour of that day in that season with just those clouds dappling the light. To what else could I have been referring? So I feel that the Monet-like picture she paints might as well have been of a haystack. Perhaps I should have used the word 'house' in my specification of the picture I wanted, but then there can be many problems in the interpretation of even such a humble word. As with pointing, so with grasping. My wife stands on a stool with her arms aloft. She is lifting down a box that has been perched above the books on her high bookshelf. I grasp her fondly (probably to her annoyance at such a moment) and say to my sculptor friend, 'Please make me a sculpture of just *this* female form that I embrace'. What he sees is the waist and trunk I grasp, and he understands how it might be a fragment of a caryatid (such as might be left over from the destruction of a temple), and how just such a fragment could be sculpted out of the odd left-over mass of marble that clutters a corner of his studio. Despite such problems, perhaps pointing and grasping, and its context, can help us to understand how we cope with everyday communication. However, when I point, how do you know what I am pointing at; when I grasp, how do you know what I am grasping? Perhaps we can start by asking how *I know* what I am pointing at or what I am grasping.

John Campbell[1] has an account of how I typically know which thing, out of all the things I perceive at a given moment, is my object of reference when I am using a perceptual demonstrative. For our purposes in this paper, I employ a perceptual demonstrative when I employ a demonstrative term (e.g. 'this', 'that') and the reference of that term is understood by me through my perception of some object. On Campbell's account, my object of reference for a perceptual demonstrative is that object to which I am consciously attending when I make the reference. It is my conscious attention that distinguishes it out of all the objects in my perceptual field. Campbell holds that perceptual demonstratives and therefore conscious attention are fundamental to our understanding of reference. A worry here is that consciousness is at least as problematic as reference. One might feel that one could make an artificial device that communicated and referred[16], but could one make artificial consciousness? Yet, on Campbell's account genuinely to achieve the former we would need to accomplish the latter. Campbell offers us a reasonably detailed account of perception, conscious attention and perceptual demonstratives, and of their underpinnings in neural information processing, so it is unreasonable to dismiss his account just on the ground that when most of us talk of consciousness we find it a puzzle to know what we mean.

In this paper, I am going to explore Campbell's account of how conscious perceptual attention enables reference. However, I disagree with his theory of perception, and so, after outlining his account, I shall seek to reformulate what I think is acceptable in his account in my own terms. My account draws to an extent on Chisholm[2] and to an extent on Lewis[6]. What I offer is an account of the perceptual underpinnings of *de se* belief (i.e. belief concerning the self). My account draws on their accounts of such belief and of how our use of indexicals and demonstratives is based upon it. It draws on Chisholm's account of how reference is founded on *de se* belief as underpinned by perception. Nevertheless my account of perception is very different from that of Chisholm. In my discussion, I shall argue that some of the evidence on neural information processing to which Campbell refers supports my account rather than his. I shall also refer briefly to some work on perception by Wheeler[14] and Clark[3].

2 Campbell on Perceptual Demonstratives

Philosophically, Campbell seeks to distinguish, but, harmonise, three levels: that of conceptual thought about a person's surroundings (for example thought using demonstratives), that of perceptual experience (for example conscious attention as it enables use of a demonstrative), and the neural information-processing level (for example binding as the neural correlate of conscious attention).

Here is a succinct statement by Campbell of his account of the subject's knowledge of reference of a perceptual demonstrative and its role in verification of sentences that employ it (op.cit. p.34):

> Your knowledge of the reference of the demonstrative is constituted by your conscious attention to the object itself. And this conscious attention to the object is what causes, and justifies, the use of the particular information-processing procedures, reaching down to the feature maps, to verify propositions about that object.

He supports his account with examples, arguments about the underlying psychological processes and neurophysiological mechanisms and philosophical argument. I find enough of these examples intuitive and enough of the arguments cogent to be persuaded that there is something in his view that needs to be incorporated into any adequate view of perceptual demonstratives.

Consider the following example (op.cit. p.4):

> ...suppose that ... your neighbour whispers in your ear 'Who's that man there?' To understand the remark you need to know who he means. So you need to single out the right person visually. It really is conscious attention that matters here. If, as you listened to your neighbour, the neural circuitry underpinning visual awareness blinked out of operation, leaving your visuomotor circuitry intact, it could happen that your visuomotor system, remote from consciousness, managed to lock onto the right person, so that you could, to your surprise, point to the right person ...you would not know who was being referred to until normal service was resumed and you achieved experience of the person.

The idea *that your visuomotor system could lock onto the right person even though you were not conscious of it* is supported by reference to blindsight cases (op.cit. p. p.7-8)[13]. It is when your conscious attention to the right person confers on you the ability to respond to the question appropriately that it constitutes knowledge of the referent — when you respond, it causes and justifies your responses as Campbell says.

Campbell points out that conscious attention is not to be confused with (foveal) focus on an object of perception. There is evidence to show that humans can consciously attend to objects on which they are not focusing (op.cit. p.9)[12][8]. Intuitively, one is told, 'Don't look now, but that man just over there seems to recognize you', and you manage to shift your attention without focusing on him (ibid).

He provides a careful discussion of what he takes the neural correlate of conscious attention, in a normally functioning human, to be. There is a 'Binding Problem' (op.cit. pp.30-31) in our perceptual systems, because, even within one sensory system (e.g. the visual one) information about different properties of an object is initially processed separately and then needs to be recombined, if we are to single out an object to which properties jointly belong. This recombination is highly selective. It is only done for a restricted portion of visual input. This makes it plausible that it is the correlate of conscious attention. According to the Triesman[10] model it is because information about each property retains, throughout processing, association with location in the sensory field that binding is accomplished. Recombination of information about the properties can occur, because information about location makes it possible. Campbell favours the Crick Koch[4] hypothesis that synchronised firing of appropriate neurons indicates that binding has been accomplished. Thus it is the neural correlate of consciousness in a normal human being. There are complications here, because it is argued that the pure Triesman hypothesis cannot be sustained. Information about location is important, but information about other parameters may also be significant. Also we may succeed in identifying an object even though its apparent location is wrong (as when we look through a prism or into a mirror). However, Campbell recognises that at this stage of neural processing there is nothing corresponding to a singular term that refers to a particular object, thus if binding is not achieved through information about location, it is achieved by information about properties.

One might think, given his identification of binding as the neural correlate of conscious attention, that Campbell would conceive of conscious attention as representation of location and property clusters, but he does not. Indeed, he argues against a Representational view of Perception, and in favour of what he calls a Relational Theory of Perception. For him, the content of visual perception is a scene that consists of concrete particular objects and their (visually-available) properties. In my opinion, his particular theory of perceptual content neither coheres well with the neurophysiological account that he gives, nor is intuitive nor is explanatory of demonstratives. Nevertheless, I believe some of his aims in constructing his theory of perception are correct. In section 3, I shall articulate these aims, further delineate the theory of perception that he constructs as a means to furthering them, and then I shall list problems that I see in his theory. In section 4, I shall construct a theory of my own and argue that it is better suited to fulfilling Campbell's aims, and accords better with the neurophysiology, than his own theory.

3 Problems with Campbell on Perception

What are Campbell's aims in constructing his theory of conscious perception? They are much more ambitious than simply to explain Perceptual Demonstratives. He seeks (i) to provide an account that explains how knowledge of the external world is possible, (ii) to explain reference, (iii) to explain intentionality (iv) to provide an account of perception which incorporates conscious attention. These aims are to be achieved by constructing a theory in which the external world is constitutive of perceptual content (thus knowledge of the external world is built in), perception is not representational (thus it does not presuppose referential semantics, but can be used to explain it), and is not intentional (it does not consist in a relation to intentionally inexistent objects — objects that might be hallucinatory instead of real).

Thus he tells us that his Relational View (or Theory) of perception is to explain knowledge as follows (op.cit. p.114-5):

> ... only the [Relational View], on which experience of an object is a simple relation holding between perceiver and object, can characterize the kind of acquaintance with objects that provides knowledge of reference.

To understand how on his view both particular objects and their intrinsic properties are constitutive of the phenomenal character of experience, consider (op.cit. p. 116):

> On a Relational View, the phenomenal character of your experience, as you look around the room, is constituted by the actual layout of the room itself: which particular objects are there, their intrinsic properties, such as colour and shape, and how they are arranged in relation to each other and to you.

He insists that the content of conscious experience is not the same as the informational or representational content of sub-personal processes in perception as follows (op.cit. p.118):

> ... those familiar with scientific work on vision sometimes assume that consciousness of the world is what happens when at some stage in the cognitive processing, the contents being processed acquire the extra dimension of being 'subjectively available'. The tendency is to suppose that it is the very same contents that are cognitively processed as figure in the contents of consciousness (cf. Marr 1982 ...) ... On the Relational View of experience, we have to think of experience of objects as depending jointly on the cognitive processing and the environment. Experience, on this view, cannot be understood simply as a matter of cognitive contents becoming subjectively available.

Finally, he rejects the claim that conscious experience is intentional (op.cit. p.122):

> We are not to take the intentional character of experience as a given; rather experience of objects has to be what explains our ability to think about objects. That means that we cannot view experience of objects as a way of grasping thoughts about objects. Experience of objects has to be something more primitive than the ability to think about objects, in terms of which the ability to think about objects can be explained.

How is conscious attention to be incorporated into Campbell's account of perception? He seems to offer us different accounts of it, or at any rate he tells us about various properties of it (a) it correlates with binding (but this does not tell us what it is) (b) it has a causal role, which includes the role of determining how we verify propositions about the object to which we attend (c) it has a justifying role (d) it can be metaphorically described as experiential highlighting of the object to which we attend.

On my view, there is a variety of problems with Campbell's view. However, I think that the problem that is most obvious with respect to perceptual demonstratives is that it is not obvious how the metaphor of 'experiential highlighting' is to be understood on his theory, and, if that is not intelligible, then it is unclear how conscious attention justifies anything. In his metaphor, Campbell talks of experiential highlighting by using the metaphor of marking text with a marker pen. Consider his intuitive example (op.cit. p.4):

> ... the lecture fails to grip ... you look idly at the other people in the audience, your gaze resting now on this person, now on that. In effect

you highlight now one aspect of your experience, now another. In effect, you put a yellow highlighter now over one or another part of your visual experience ...

If the highlighting metaphor is close to the truth, then the apparent properties of perceived objects change when they become highlighted. They, as it were, turn yellow, that is, they have some different apparent property as a result of highlighting. This property is not meant to be, for example, that they are in focus, because, as we have seen in section 2, Campbell recognises that we may consciously attend to objects even though they are not in focus. Indeed, it seems a problem to identify anything that, on Campbell's view of perceptual content or phenomenal character, can play the role of highlighting, because these are constituted out of the objects themselves, their arrangements, properties and relations to yourself.

At least, one might say that what does the 'highlighting' in perception is attention itself, but then one needs to be clear about what attention itself is, how it can be a component of the content or character of perception, and what it is to be conscious of it. Is attention to an object to be treated as one of the relations to an object of perception that is constitutive of phenomenal character? We have a suggestion that it is in Campbell's treatment of joint attention in which two or more people jointly attend to the same object of perception and it is constitutive of their experience that this is so[1] (op.cit. p161):

> On a Relational View, joint attention is a primitive phenomenon of consciousness. Just as the object you see can be a constituent of your experience, so too it can be a constituent of your experience that the other person is with you, jointly attending to the object. This is not to say that in a case of joint attention, the other person will be an object of your attention. On the contrary, it is only the object that you are attending to. It is rather that, when there is another person with whom you are jointly attending to the thing, the existence of the other person enters into the individuation of your experience.

Thus it seems that attention, even someone else's attention, may be constitutive of the phenomenal character of an experience. Suppose one is jointly attending to an object with someone else, who leaves without your noticing it, because, after all, on Campbell's account of joint attention, you need not be attending to them, then on Campbell's account the content and character of your visual experience will have altered, but you will have failed to notice it, so you will 'take it' that the person is still there (op.cit. p.165). That is to say, your problem will be that you have the wrong belief that the person is still there. It is not that you will have a sensory illusion of the person still being there. Also, on Campbell's account you might, so far as I can tell, be having the experience of jointly attending with

---

[1] Joint attention is not essential to understanding a perceptual demonstrative — one person may overhear another and identify the object of their perception without jointly attending to it with them (op.cit. p.164), but Campbell argues that it is a phenomenon of special interest in coordination

someone else to an object without believing that you are. At least, you could fail to believe in joint attention and still fulfil the functional requirements on joint attention that he makes (op.cit. p.165), which are that the presence of the other person causes and sustains your perceptual experience.

Campbell handles the important cases of intra-personal tracking of an object over time and cross-modal cases of a person attending to an object with more than one sensory system in a comparable way (op.cit. p128-131). For him, the fact that it is one and the same object that is being tracked and/or perceived by more than one sense is enough to ensure that the phenomenal content and character of the perceptual experience is that of perceiving the same object. On this account, if you fail to believe in the success of your tracking, or in the cross-modal perception of one and the same object, then your experience must still have the phenomenal content and character that it would when you rightly believe that you are tracking or cross-modally perceiving. Thus, for example, there will be no difference in the phenomenal content and character of a case where you see yourself as yourself in a mirror and the case in which you see yourself as someone else. The difference will lie at the level of who you 'take' (or believe) the object of perception to be, not at the level of perceptual content. This is important for the present paper, because, as I have already announced, my strategy for explaining demonstratives is to draw on work by Lewis and Chisholm, which makes *de se* belief (i.e. belief concerning the self) fundamental to reference to particular objects and thus to the use of demonstratives.

Consider also Campbell's handling of hallucinations. If you perceive a dagger floating in the air in front of you and, in another case, you hallucinate such a dagger then the perceptual content and character of the two experiences must be different on his view, not because the hallucinatory dagger appears to shimmer or somesuch, but simply because in the first case there actually is a dagger and in the second case there is not (op.cit. p.116-117). Thus Campbell's theory is a form of the disjunctive theory of perception (ibid), in which the perceptual experience in veridical perception and hallucination necessarily differs, even though one fails to notice that it does. However, Campbell prefers to describe his theory as the Relational Theory or View of perception, because standard disjunctivists analyse veridical perception as having perceptual demonstratives that are constitutive of the very experience itself, whereas Campbell cannot accept this because he seeks to explain reference and intentionality in terms of experience. Thus he cannot have any reference or intentional inexistence that is constitutive of perceptual experience itself.

Is it possible to have a theory of perception that recognises intuitive distinctions and identities in appearance itself where Campbell's position only recognises distinctions and identities at the level of belief? That is, a theory which can be used to explain perceptual demonstratives and yet which can also recognise the appropriate identities and difference in cases of tracking, cross-modal perception of the same object as the same object, hallucination, joint attention and self-perception.

4 Toward a Better Analysis of Perception

The aim of this section is to provide a better theory of perception than Campbell's, one that is capable of analysing phenomena at the level of appearance, which Campbell has to handle at the level of belief, and that can also handle Campbell's conscious attention or 'experiential highlighting'. Before going into its details, let us illustrate the difference between the account of perception that I shall propose and Campbell's. Take the example of hallucinating a dagger in front of me, and suppose, as in Campbell's example, the experience of the hallucination and the experience of actually seeing a dagger are indistinguishable to the subject. I take it that there is something in common in the two cases, which is that, in both cases, *visually I appear to have a dagger in front of me*. The difference between the two cases is that, in veridical perception, there is a dagger before me and, in the hallucination, there is not.

It may help, in understanding my view, to flag how, in the example, I deal with content and character of perception, and also to note how I take my account to be relational.

Content: this is a property, in the example, it is the property of *having a dagger before me*

Relation: this is a relation to the content and it is the relation of [its] *appearing to* [me]

Character: this is expressed through adverbial modification, in this case the property appears to me *visually* - of course not all visual character is the same, so character arises from subject[2], relation, adverbial modifier, and property taken together

I shall call this theory of perception the *Property-Content Theory*.

On the *Property-Content Theory*, just what is a property? There are many views of what a property is, for example a function from possible worlds to extensions in those worlds, something intensional, but not to be understood in terms of possible worlds (e.g. Chisholm[2, pp.4-9]), or something extensional. It is possible to combine the basic features of the Property-Content Theory with any of these accounts of properties. However, I am inclined to conceive of the fundamental properties to which I relate in perception as extensional (thus *perceptible* properties[3] are identified with classes of objects in the actual environment[4]). In an hallucination, properties appear to me *whose presence and absence in the environment I am normally able to identify*. I do not have the

---
[2] Character may vary according to the state of the subject.
[3] An underlying motivation here is to develop a view of observables that is coherent with van Fraassen's Philosophy of Science[11]. Van Fraassen eschews metaphysical commitment to the existence of possible worlds, even though he employs the machinery of possible worlds in elaborating theoretical models. For him, the aim of science is not the literal truth of theories, but explanation of the observables, and the latter need to be treated extensionally.
[4] A way of working out the view that I propose would be through acquaintance. The idea would be that those properties that can appear to one at any given time are

space to defend this view in detail here, although I shall briefly return to the matter later in this section. Part of what follows from it is that in cases where there is hallucination of a kind of object which does not exist — e.g. a horned man — I need to treat the hallucination as a compound of the appearance of a horned being before me with the appearance of a man before me (located exactly where the horned being is), because there are no horned men. I have the ability visually to distinguish horns, and I have the ability to distinguish men, but I do not have the ability to distinguish horned men, because there are no horned men (I would be able to pick them our were there any, but that does not enable me to pick them out in the actual world). It is worth mentioning this aspect of my precise version of the Property-Content Theory here, because Campbell stresses that his *Relational* View treats perception as a relation to an external world. My view is externalist too, because of properties being treated extensionally.

It may also help to understand my view if we contrast it with a variety of other views:

Sense datum: the sense datum view introduces special objects of perception distinct from the things we would normally take to be objects of perception, but I introduce no special objects, just properties,

Disjunctive view and Campbell's view: these two views take both relation and content to differ between veridical perception and hallucination whereas I take it that the fundamental relation ('appearing to') is the same for both and so is the content,

Representational theories: on my view, the correct way to categorise conscious perceptual appearance is as the presentation[5] of a property, not as the representation of anything - thus, with Campbell, I think that we may be able to use perception to explain representation without any vicious circularity,

Perception as belief: in perception there is *appearance*, but there need not be belief, thus someone taking LSD and hallucinating need not be deluded into believing their hallucinations to be veridical,

---

either properties with which one is acquainted through a past or present instance or are compounds of such properties. However, this Hume-like view is not quite my view. On my view, what matters is that one's perceptual system enables one to identify classes of objects in the actual environment. This normally occurs in us through historical acquaintance with instances of those classes, but in principle a mode of perception might be enabled, and one might have hallucinations in it, even though one lacked actual acquaintance with instances of its classes.

[5] It is important here to understand the relation between the level at which there is a presentation of a property, and the underlying level of neural informational processing. The former level is a level to be described in a *specification* language, whereas the latter level is to be described at the level of algorithm and implementation. The former level is the level that Marr[7] (rather misleadingly I think) calls *'computational'*, Pylyshyn[9] *'semantic'* and Fagin et al[5]. the *'Knowledge Program'* level. Since my position does not accord exactly with Marr, Pylyshyn or Fagin, I would prefer to think simply of the *result-specification* level.

The adverbial theory: on Chisholm's adverbial theory of perception, special adverbs (modal modifiers), such as 'redly' are introduced (bringing with them logical problems about how each special adverb — 'redly' — relates to its predicate — '_ is red', thus if I have an hallucination of a bloody dagger, then I might be appeared to redly and daggerly[6], whereas my view needs no such proliferation of special adverbs, but only the standard 'visually', 'tactilely' and so on.

In what follows, I shall develop my account of perceptual demonstratives by reference to Lewis (op.cit.) and Chisholm (op. cit.) and their account of *de se* belief. I shall attempt to offer an account of the perceptual underpinnings of this. I have already indicated that the Property-Content theory of perception that I advocate does not analyse perception in terms of belief but in terms of appearance. Perceptual appearance has phenomenal character, which abstract belief does not and it only provides a reason to believe, it does not provide belief itself.

Now let us turn to the question of how *de se* attitudes in general may be handled, beginning with belief. Consider Lewis' paper[6, p.520-1]. He has a thought experiment about belief *de se* in which he thinks about gods. In that example, there are two gods who are omniscient with respect to propositions. They each know all propositions that are true of the world. They each live on top of a mountain and what one throws down is thunderbolts whilst what the other throws down is manna. Yet they each may have considerable ignorance, for it does not follow from the way we have described them that each god knows what it itself throws, or, more generally, which god it is. Thus a god could know that Thor throws thunderbolts, yet not know what it itself throws, because it does not know that it itself is Thor. Lewis' account of what these gods lack is that the god who throws thunderbolts fails to self-ascribe the property of throwing thunderbolts (even though he attributes that property to Thor), and the god who throws manna fails to self-ascribe throwing manna. Thus the knowledge they fail to have is a kind of non-propositional knowledge, which consists in self-ascriptions of properties. These self-ascriptions are beliefs, on Lewis' account of beliefs, they are beliefs *de se*. Indeed Lewis seeks to analyse all beliefs in terms of such *de se* beliefs[6, p.538]. In particular, if I have an object that I identify in a belief by a perceptual demonstrative, then it will be an object to which I self-ascribe some relation. An example that he gives[6, p.543], taken from Kaplan, is that a man may notice in a mirror that a certain man, 'That man', has pants which are on fire, and yet not appreciate the relevance of this to himself, because he does not realise that he himself is the man whose pants are on fire. Lewis describes the relation that the man has to 'that man' as the relation of 'watching', which I take it is different from self-ascription in the sense of belief. Thus, in Lewis' paper (op. cit.), we are presented with an embryonic theory of perception.

---

[6] Perhaps, this the wrong adverb to introduce, but I think it is problematic which adverbs to introduce on Chisholm's account.

Let us try to articulate Lewis' account using the Property-Content Theory of perception that I have proposed. Perhaps we can think 'watching' occurs when one has a certain visual appearance and there is an object to which one is related as one appears to be. Thus one might say that *visually one appears to have a man with burning pants reflected in a mirror before one*, and there might be such a man. There are three cases: (1) in which one does not appear to the man as oneself (although it is not an appearance of a man who is other than oneself), (2) in which one appears to oneself to be other than oneself, and (3) in which one does appear to oneself to be oneself:

1. $Visually(Appears(w, \lambda x \exists y \exists z (Man(y)$ & $Reflected\_in(y,z)$ & $Mirror(z)$ & $Before(x,z)$ & $Burning\_pants(y)))$
2. $Visually(Appears(w, \lambda x \exists y \exists z (Man(y)$ & $Reflected\_in(y,z)$ & $Mirror(z)$ & $Before(x,z)$ & $Burning\_pants(y))$ & $x \neq y))$
3. $Visually(Appears(w, \lambda x \exists z (Man(x)$ & $Reflected\_in(x,z)$ & $Mirror(z)$ & $Before(x,z)$ & $Burning\_pants(x)))$

As I understand Campbell's view, in the case that the man with burning pants is oneself, the perceptual content and character of one's visual experience would be just the same in cases (1) and (3). There would be no such case as (2), at the level of perceptual experience, because non-identity with oneself is not a property of the perceived object. Instead case (2) would be handled at the level of belief, although exactly how I am not sure. At any rate, the account that I propose is that we adopt a Lewis-like account at both the level of appearance and belief. However, the example requires refinement of our simple account of 'watching'. Watching does not require that there be an object to which one is related wholly as one appears to be — else one would not be watching in case (2) — all that is required is that there be an object to which one is related partly as one appears to be.

On my account, what is the difference between the level of appearance and the level of belief? I have already said something in answer to this question, but let me amplify. I take it that, in typical humans, a sensory appearance can provide a reason for a belief, desire, or action, but that having a sensory appearance does not commit one to a belief content. Perhaps, in an infant or animal, an appearance might provide a reason for action without being mediated by conceptual belief. In conceptual belief one's belief has a logical structure which is expressible in some logical language with a compositional semantics and for which there are rules of inference. In contrast, both appearance and Lewis' *de se* belief are relations one has to a property, and the property is not envisaged as having logical structure. Thus a cat or infant might have *de se* appearance and belief without having conceptual belief. However, for them one would need to understand the psychological (perhaps indeed the neuro-psychological) underpinnings of cognitive processing in them in order to understand the functional roles of their states. Whether or not their processes would process the realisers of *de se* belief concerning a complex property by processing neural realisers of *de se* belief concerning simpler constitutive properties is a matter to be settled by neuropsychological investigation. With respect to perceptual appearance, I take it that its primitive content is not to be treated as conceptual content, and so to understand its role we need to understand our psychological mechanisms.

As for attention, I would propose to treat relations of attention as relations that can play a role in the phenomenal content and character of sensory appearance. Thus I might visually appear to have a man on the left to which I am attending, whereas I am not attending to a man on my right:

$$Visually(Appears(w, \lambda x \exists y \exists z(Attends\_to(x, \lambda v(Man(y) \,\&\, Left\_of(y, v))) \,\&\, Man(z) \,\&\, Right\_of(z, x))))$$

I propose:

- to handle the tracking of objects by recognising that temporal relations that may be constitutive of perceptual appearance,
- to handle cross-modal perception by allowing that a property concerning perception in one sensory modality may be constitutive of the content of appearance in another sensory modality
- to handle joint attention by recognising that another person's attention to one's own object of attention may be constitutive of the content of perceptual appearance.

Campbell treats hallucinatory content as different from the content of veridical perception. On the account that I propose, there is no need to do this. Perceptual appearance is a relation between oneself and a property of having some object present to one. Thus someone with a visual hallucination of a dagger in front of them has a relation to the property of having a dagger before them. That person visually appears to have a dagger before them, but there need be no dagger there, yet the content need not differ from the veridical case, because it is a property, not a dagger, that needs to be in common.

Indeed, the theory of perception that I propose can be used to handle another case that Campbell discusses[1, p.124], which he takes from Martin Davies. In that case, two people are in exactly qualitatively identical prison cells with qualitatively identical contents except for the people themselves. Each can be seeing an empty cup before him. Some of us have an intuition that perceptual content will be the same in the two cases, but Campbell insists that it is not. On the account that I propose the content is a property, and this can be the same in both cases. It is the people who are different not the property that they each appear to have (and in this case happen to have) of having an empty cup before them.

I submit that my proposed account of perception can handle Campbell's cases as well as his does. It can even be argued that it meets his concerns. It is not representationalist, because it analyses the content of perception as a presented property, not as a representation. Thus it does not introduce representations (which have reference) into perceptual content and it does not propose the intentional inexistence of objects. Campbell might propose that it is not genuinely a Relational Account of experience, because it is not externalist. However properties may be thought to be external to the individual. Indeed, on the account of perception proposed by Wheeler and Clark[15,14,3], the external properties of objects play an important role in normal perception. This is because of the phenomenon that they call 'non-trivial causal spread', which obtains when processes previously thought to be internal to a system are recognised to be non-trivially dependent on causes in the environment. The cues that we have stored neurophysiologically which allow us to interact with an external scene may be nothing like as rich as the properties in the surrounding environment. In veridical perception, we can scan the actual scene, treating the instances of properties in it as a substitute for memory or for a massive visual buffer. In hallucination, there will not be richness, so one would expect an hallucination of a dagger to different from the genuine case. Nevertheless, if the hallucinator has enough information stored to enable recognition of genuine daggers, then the hallucinator may relate visually to *the property of a dagger being present* even when there is no such dagger. Wheeler and Clark themselves have a representationalist position on perception, but that does not prevent me from incorporating their view into mine, giving me an externalist account. Thus, for me, it matters that there are basic properties in the environment that an hallucinator would be able to recognise in a normal case. Obviously some hallucinations are fantastic, but fantastic properties are constituted out of actually instanced[7] properties; these latter are also properties which perceivers have the ability to recognise, when functioning normally, in their environment.

In his discussion of the Binding Problem, Campbell himself recognises that the Binding Problem is solved through information about property clusters, not through a neurophysiological representation of an individual object. This accords well with the account of perception that I propose.

With respect to perceptual demonstratives, I propose to employ Chisholm's account of these[2], but to substitute my own account of perception for his. With respect to indexicals, I propose to follow the Lewis account which, in its fully developed form, takes self-ascription to be self ascription by an ephemeral self[6, p.527], not self-ascription by a self that endures over time. Thus 'now' is in a sense 'my' time and the past and future belong to earlier and later selves. Does the fact that my proposed account of perception is externalist mean that we should not expect to have the problems in the use of perceptual demonstratives with which I began this paper? I think we can still have such problems. Artists are interesting because they attend to different properties of the visual environment from more mundane people, and the originality of artists illustrates how individuals may diverge[8] in the perceptual properties to which they attend.

5 Conclusion

Campbell's work on perceptual demonstratives has been explored, but his account of perception has been found unsatisfactory. A different account developed

---

[7] This is a departure from Lewis who holds that *de se* belief is in the head[6, p.526]
[8] This potential for divergence provides one basis for my account of *intrinsically intentional states* in an earlier paper[16]

from Lewis' and Chisholm's work on *de se* attitudes has been substituted. It is proposed that this account of perceptual demonstratives can also draw on Campbell's work on perceptual attention and its relation to the Binding Problem.

References

[1] J. Campbell. *Reference and Consciousness*. Oxford University Press, 2002.
[2] R. Chisholm. *The First Person*. Harvester Press, 1981.
[3] A. Clark. *Being There*. Bradford Books. MIT Press, Cambridge MA, 1998.
[4] F. Crick and C. Koch. Towards a neurobiological theory of consciousness. *Seminars in the Neurosciences*, 2:253–73, 1990.
[5] R. Fagin, J.Y. Halpern, Y. Moses, and M.Y. Vardi. *Reasoning about Knowledge*. MIT Press, 1995.
[6] D. Lewis. Attitudes de dicto and de se. *Philosophical Review*, pages 513–543, 1979.
[7] D. Marr. *Vision: A Computational Approach*. Freeman & Co., San Francisco, 1982.
[8] M.I. Posner. Structures and functions of selective attention. In T.J. Boll and B.K. Bryant, editors, *Clinical Neuropsychology and Brain Function: Research, Measurement and Practice*, volume 7 of *The Master Lecture*, pages 171–202. The American Psychological Association, 1988.
[9] Z. Pylyshyn. *Computation and Cognition*. M.I.T. Press, Cambridge, Mass., 1984.
[10] A. Treisman. Features and objects. *Quarterly Journal of Experimental Psychology*, 40A:201–37, 1988. The Fourteenth Bartlett Memorial Lecture.
[11] B.C. van Fraassen. *The Scientific Image*. Clarendon Press. Oxford University Press, 1981.
[12] H. von Helmholz. *Physiological Optics*. Dover, 1866/1925.
[13] L. Weiskrantz. *Blindsight*. OUP, 1986.
[14] M. Wheeler. Two threats to representation. *Synthese*, 129:211–231, 2001.
[15] M. Wheeler and A. Clark. Genic representation: Reconciling content and causal complexity. *British Journal for the Philosophy of Science*, pages 103–135, 1999.
[16] R.A. Young. The mentality of robots. *Proceedings of the Aristotelian Society*, Supp., 1994.

Perceiving Action from Static Images: The Role of Spatial Context

Elisabetta Zibetti[1] and Charles Tijus[2]

[1]Department of Psychology
Jordan Hall, Bldg. 420 - Stanford University
Stanford, CA 94305-2130
ezibetti@psych.stanford.edu

[2]Laboratoire CNRS- FRE 2627 Cognition et Usages,
Université Paris 8 - 2, rue de la Liberté,
93526 Saint-Denis Cedex 02 FRANCE
tijus@univ-paris8.fr

Abstract. In this paper we discuss the role of spatial context in interpreting and understanding actions from the perception of static images. Our proposal is that people use spatial context, which is to say the relations among objects, to infer actions and make predictions about future states even when no physical events are directly perceived. We begin by presenting an overview of relevant theories for action understanding. We then discuss the role of spatial context as a prerequisite for acting as well as for action goal attribution, by presenting and discussing an experimental study. The addressed question is how representation of motion in static images maps onto perception of an action goal. We tested the extent to which action representations attributed to a static stimulus can be changed when the physical relational properties of objects are manipulated. Results show that the goal attribution, considered as a highly informative component for the interpretation of the perceived action, is carried out by a contextual categorization process which takes into account both the objects' physical and relational properties, and the semantic relationships between them.

1 Introduction

As adaptive creatures, humans need not only to know their environment but also to appropriately anticipate changes in it. To anticipate means to "guess" the future evolution of an event that has not yet been produced. To do so, our brain combines information about objects and events from the retinal array with knowledge and expectations to make reasonable guesses about what is present in the scene and what could append. Yet the question remains: how do people do use the perception of their environment, attending selectively to some aspects, when anticipating possible events and action? Our proposal is that people use spatial context, which is to say the relations among objects[1]. Those relations can be topological (the object that is on, under, on the left of another), those derived form the surface properties (the darkest, the biggest object), but they can also be contextual. The latest are potential relational properties, extremely related to the semantic of the objects. For example, if the considered object is a pen and it is located on a desk office, it would easily be considered as an object useful to write. If the pen is located on the shelf of a bookstore, it would likely be considered as an object to buy, to sale or to offer. In our everyday lives, we hardly know anything for sure, but because the world is a fairly regular place, we usually surmise correctly. The situations with which one is faced are usually dynamic, and their comprehension is built ad-hoc, according to the situation's evolution. The processing of objects in their spatial context allows observers to make inferences about the future course of events and action. These inferences about future states can give coherence and establish relations between what one has just perceived and what one might expect based on prior knowledge.

We will first present and discuss some aspects of relevant theories for action understanding and then present our approach based on contextual categorization. The role of spatial context as prerequisite for understanding actions and making predictions will be examined through an experimental study using static images. Results will be discussed according to our approach based on contextual categorization.

2 Relevant Theories for Action Understanding

The constructivist theory and the ecological approach are two of the main theoretical positions about the process of perceiving events and actions. The former privileges the "top-down" processing of information, with the generation of preliminary assumptions around the stimulus, while the latter privileges a "bottom-up" processing directed by the stimulus. The top-down approach aims at consolidating assumptions by "going down" towards the source to check if what is awaited is realized. According to the bottom-up approach, successive interpretations of the physical stimulus lead to the final result. A third approach, contextual categorization, is a mixed bottom-up and top-down approach, which provides a dynamic interface matching external data with knowledge.

2.1 The Constructivist Theory

The constructivist theory [e. g., 12, 27] claims that our experience teaches us to draw general conclusions starting from rather limited sensory information. According to this theory, our perceptive inferences, called "unconscious inferences" by Von Helmholtz [27], are generated almost "automatically" through the intervention of memory. The human sensory system appears to be in charge of selecting the information from which the cognitive system will generate inferences. Perception is then reduced to a

---

[1] By „ objects " we mean inanimate objects such as bottles, baguettes, etc as well as animated entities such as human being.

"passive" process of information acquisition directed by goals and activation of related knowledge. Actions could be comprehended by understanding the goals and plans of the agents. This identification of the agents' goals is the crucial aspect of plans recognition.To the extend of this view, plan recognition theories were developed in psychology and AI in order to account for the understanding of actions being performed by an agent [15]. Plan recognition is a powerful technique, particularly when the observer is able to attribute a goal to an agent. Still, it is too limiting to be a complete model of action understanding. For instance, having to cross a street will activate a whole action plan that comprises taking care of cars and looking at red lights. In opposite, the constructivist approach does little to explain how we attribute the goal of crossing a road to someone who is paying attention to cars and red lights without having in mind the whole plan of every person we might see. In addition, little change in the situation (the person being a policeman, for instance), might need another plan.

This kind of rational default inferences about goal attribution seems to show that a system just concentrating on plan recognition will fail to be a complete understanding system.

2.2 The Ecological Approach

The ecological approach, supported by Gibson [7, 8], questions the constructivist theory. According to Gibson, traditional research on perception neglected the active interaction between an organism and its environment in the search for and the selection of information. Human sensory systems do not passively select the information from which the brain will emit inferences. They have evolved in such a way to be sufficiently sensitive to the complexity of the environment that they can generate immediate perceptive experiences without interposition of any inferential process, or any stored plan. For the defenders of this approach [e.g., 11,16, 24, 25], perception would be seen as a direct process where the relevant information for action is "discovered" rather than "built".

The principal goal of such an approach is to understand how an organism manages to perceive the physical events that take place in its environment. A central concept of this approach is "Affordance" [6, 7]. Affordance is a set of principal properties of an object captured by our perception. These properties are basically functional and indicate how the objects can be used. Thus Gibson claims that what we primarily perceive of an object are the possibilities of actions that it suggests to us rather than its conceptual or categorical identity: "to perceive an affordance is not to classify an object" [8, p.135]. However, the Gibsonian approach hardly explains why we do not systematically cross all roads even if invited to do so by a red light.

2.3 Contextual Categorization as a Constructivist and Ecological Approach

Whether one assumes a Gibsonian orientation or a constructivist one, it can hardly be disputed that perceptions are caused by cues present in the environment and that they depend both on the context in which they are processed, and on the prior knowledge of the observer.

Today the debate between these two approaches centers around the primacy of bottom-up processes over those which are top-down. It seems obvious to us that these two processes coexist in the interpretation of visually perceived actions Combining top-down and bottom-up processing is certainly not new, either in cognitive psychology [e.g., 26] or in artificial intelligence [e.g., 19]. However, there is a lack of knowledge about how those two processes interact. Thus in our opinion, it is more worthwhile to analyze how their respective roles are articulated. We propose that the process that allows making the link between perception and cognition during action comprehension is not classification, but a contextual categorization process [20, 28]. This assumption finds its roots in the general principles of categorization as a basic activity of human cognition, that particularly stresses the relational properties of objects in their context.

In our categorization based approach, context is central. Our definition of context is similar[2] to Turner's [21] definition of "the world state", which is to say, "the state of the agent's world at some particular time: i.e., all features of the world, including all objects in existence, their properties and internal states, and the relationships between them" (p.376). Our context definition is also closed to Pomerol & Brezillon [14] position. According to those authors "The notion of context offers an alternative view to knowing how to capture that part of knowledge which is related to decision making and action [...] the notion of context does certainly not explain know how[3], but it helps to understand how experienced people with a recognized know how adapt their behavior according to the circumstances" (p. 463).

Contextual categorization links two basic concepts of human cognition : (i) the *categorization*, a process by which environmental information is organized and received; and (ii) the *context* (environmental and temporal) in which information relating to events and objects is anchored. Contextual categorization is highly dependent on the physical properties of objects — which is to say that the perception of objects can not be dissociated from the perception of actions prerequisites (the parts of objects that allow acting) and lastly from action inferences based on the use of seen objects.

Contextual categorization is the grouping and differentiating of whole sets of present objects relative to each other while processing their properties. The construction of such contextual categories is made ad-hoc. The categories are progressively created or updated through a process the simultaneously integrates context, objects, and semantic knowledge. This integration is made possible by the creation of categories which activate the common semantic properties of the objects and, among them, the actions that can be done with and/or to these objects. Between alternatives, the meaning that contributes most specifically to the coherence of the situation is favored. For example, running cars and motorcycles would be grouped as "running vehicles" and differentiated from the red light only if they are both present and moving, while both "running vehicles" and a red light will be grouped as "things to be aware of". Notice that present unmoving cars, although they belong in memory to the class of running vehicles, will not be instances of the contextual category of vehicles. Such a process allows an observer to perform the goal of crossing the street as well as to attribute to a present person the goal of crossing the street.

---

[2] However, differently from Turner [22, 23], we considered that conetxt can not be represented only by " contextual schemas [c.f. 13, for a discussion on this subject].

[3] For those authors, the terms "know how" are referred to the knowledge that people use in order to act.

3 Spatial Context as Prerequisite for Action Interpretation

Denis [3], for instance, notes that "No cognitive theory could neglect to consider the way in which the individuals "experiment" the space, the way in which they memorize it, create "mental or material" representations and the way in which they use these representations to plan their displacements or to anticipate displacements of others" (p. IX). We would readily add that the observer must also process constraints imposed on the objects by the specific spatial environment.

When inferring the causes and/or consequences of a visual event, much weight is generally accorded to the temporal context and, in our opinion, not enough to the spatial context. For example, if someone sees a car moving down a street under construction, understanding the action entails anticipating that the car will soon stop because the street does. These environmental constraints steer the observer's interpretation in terms of action (in this case, moving vs. stopping).

Despite the fact that most actions do not appear in isolation, but rather as parts of a complex sequence of events, observers often recognize actions without first seeing all the pieces of a sequence. They are able to infer action and produce causal reasoning based simply on environmental context, without requiring the direct perception of movement. When an observer recognizes an action even with no explicit temporal information (e.g. in a painting or an icon), is it probably because he/she is processing the spatial information provided by the environmental context which is a prerequisite for potential action.

Let us take a everyday life example. Suppose someone knocks at the door while you are watching a soccer match on tape. You press the "pause" button, and the ball is suspended in mid-air in front of a goalkeeper with outstretched arms. A guest comes in and when he sees the screen, says, "the goalkeeper is going to catch the ball." What information has allowed him to interpret this static image in terms of action? As we view it, even if the guest does not know what has just happened, the information provided by the properties of the situation (i.e., the ball suspended in mid-air, the goalkeeper with outstretched arms) plus folk knowledge of the world's physical principles (e.g., the ball is going to come down, because of gravity) are enough for him to infer and to anticipate the event that is about to happen.

Thus, in our opinion, the processes by which actions are are based on (i) the information provided by object properties, including relational properties; and (ii) the folk knowledge that people possess about the physical principles of the world. The latter determines the possible transformations of objects (e.g., the ball falling down) and the activation of action representations (e.g., the goalkeeper catching the ball). Indeed, for an action to occur, the objects must necessarily possess the properties that make that action possible: the object's inherent properties as well as its properties in relation to other objects. For example, if you press the "pause" button when the ball is on the ground near a player such as Maradona, then your guest would probably say, "Maradona is going to kick the ball" instead of "catch the ball." Thus, even a static situation conveys the dynamic information on which anticipation is built.

Neither the event, nor the action, nor their temporal organization, can be determined if they do not satisfy a certain number of physical prerequisites which strongly condition the action interpretation. These prerequisites often take the form of contextual interactions (in the example above, the position of the ball in relation to the arm of the player), and they probably intervene at different levels of the data processing. Perceptive information represents the input that allows inferences about action [10] by providing "the entry" responsible for the process of understanding [2]. But it is probably at the semantic level that the environmental context intervenes in a crucial way, by activating actions concepts according to the objects' functional properties that become salient thanks to the contextual categorization process. Therefore, it is necessary to consider the information provided by the properties of the objects (among them relational properties), as well as the naive knowledge about physical principles of the world that are considered by various researchers to be an integral part of our perceptive and representational system [e.g., 9, 17; 18].

Contextual prerequisites are also fundamental to establishing causality, and the attribution of an action to a designated agent. For example, if a character located outside a house moves towards the door, it is probably because he/she is going to enter. The actual action carried out in this example (to move) must satisfy sufficient conditions (to move in the direction of the door) to involve the consequent event (to enter the house). In order to enter, one must satisfy not only the prerequisite "to move," but also all the prerequisites for the achievement of the goal: "to move towards the door", "to open the door", and "to cross the threshold".

4 Goal Attribution from Static Images

An image has an internal organization. When faced with an image, however, we "see" much more than what we perceive. When an observer perceives, for example, the image of a character, he/she builds a representation that is not only a physical entity but also a rich unit of information concerning the character: goals, roles, intentions, and even emotional states. Where does the attribution of roles, goals, and intentions come from? Wouldn't it be the result of a contextual and categorical encoding of this character starting from the processing of his/her physical and relational properties in context ?

In order to infer an action, it is necessary to attribute goals to the character. We can regard the goal two ways: as the representation of a possible or impossible state congruent with the constraints provided by the situation; or as a future state, based on the observer's anticipation of what might be necessary in order for a still unrealized phenomenon to be realized. We believe that the identification of categories that lead to the interpretation of action starting from static images is carried out by the contextual categorization, which processes objects' relational properties as action prerequisites [17, 20, 28, 29]. Whether in a static image, (e.g., a sitting man) or a dynamic one (e.g., a man in walking position), the action is not made clear by the movement itself, as in animations or films, but can be inferred from the salient properties of the objects in their context.

How does this attribution of goal take place? When, for example, one sees a motionless person on the train platform with his ticket and luggage, one can quickly anticipate that the person will probably take the train. This inference is the result of linking the properties of the four relevant objects: the person, the luggage, the station platform, and the train ticket. However, if one sees neither the ticket nor the luggage but a bunch of flowers, a more congruent goal with the perceived situation, based on the objects' relational properties, is that this person is waiting to welcome somebody.

Putting the various elements (the person, the station platform and the bunch of flowers) in relational context provides us clues about the goal.

Thus, when one observes an image, the selection of information can depend not only on the objects' inherent properties abut also on their relational properties to other objects of the context. "Ticket" and "train" represent cognitive categories related to the same semantic field. In the same way, "bread" and "bakery", or "bar" and "bottle," are attached semantically. Thus, if the image presents somebody with a baguette under his arm and very close to a bakery, one should make the inference that the person has just bought bread, which is saying that he/she has just achieved his/her goal. If the image places the person near two shops, a bar and a bakery, the relation between bread and bakery will be facilitated by the salient baguette. The semantic relation between the two objects is established despite the fact that if one already has bought bread it is unlikely that one will be on his way to the bakery.

We tested the extent to which action representations attributed to a static stimulus can be changed when the situation's physical properties are modified. Consequently, we hypothesized that action representations can affect the perception of events even when no event has occurred. Pictures should then be a good way to study how observers infer action from images — which is a common sense idea. By manipulating the' relational properties of the objects depicted, we can begin to see how observers infer actions and goals from static images.

5   Experiment

We start exploring this idea by carrying out an experiment to study the role played by physical context and by the relational properties of objects in the inference of action. The experiment reduces the richness of the visual context to few elements: a person, a destination, an object, and a path (figure 1). This experimental choice, which could appear "context reductive", allows us study the effects of the contextual environment in the process of action understanding by introducing a minimal variation in the stimulus.

The specific objective of this experiment is to study in what ways the processing of physical and contextual properties of a static image (figure 1) guides the goal attribution to an agent in the absence of any physical event. In other words, how does representation of motion in static images map onto the perception of an action goal?

Our specific assumption is that participants, asked to indicate a direction for the character on the path, will preferentially choose that one which is in the continuity of the principal path. Under the conditions 2A, 2B, 2C and 2D (figure 1), we should observe a higher number of participants who indicate an UPWARDS direction, whereas under the, 3A, 3B, 3C and 3D conditions (figure 1) we should observe a higher number of participants indicating a DOWNWARDS direction. If this prediction is true, it will indicate that the properties of the situation constrain possible anticipations of an observed behavior.

As for the possibility the participants attribute goals to the agents by taking into account the relational properties of the objects, let us assume that the semantic relations established between the objects are the results of the process of contextual categorization. Thus we would expect that when the character depicted is carrying a specific object (baguette or bottle), he is quickly associated with the corresponding shop (bakery or bar). This association should be represented in the choice of the direction that leads to the shop. In particular, for situations C1, C2, C3 (baguette-bakery), although it is incongruous to attribute to an agent the goal of moving towards the bakery when he/she already has bread, we expect that the relation established between these two objects is a stronger determinant of path direction than that one established between bread and bar. And finally because the semantic relation between bottle and bar is weaker than that one between bottle and wine-shop these two objects provide a weaker determinant of path direction than that one established between bread and bakery (D1, D2, D3).

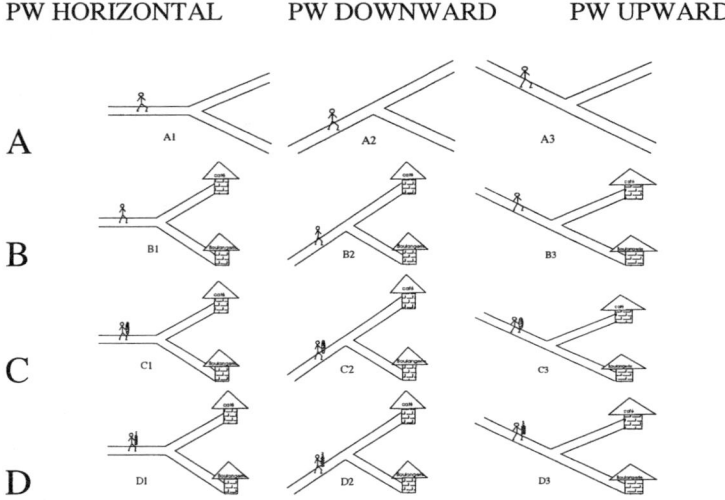

Fig. 1. The 12 experimental conditions

Participant. Sixty voluntary second-year students of the psychology department of Paris 8 University took part in the experiment: 48 girls and 12 boys ranging from 19 to 29 years old.

Material. Twelve different (15 x 20 cm) motionless images were presented to the participants. Each of the 12 images depicts a character on a path which leads from left to right to a fork. The left part of the path is called "the principal path". The orientation of this part of the path is manipulated — rotated 45° either upwards or downwards from horizontal. We obtain thus 3 types of forks (Figure 1: Images 1, 2, 3) in which only the first part of the path changes in its orientation.

The character is located halfway down the principal path before the fork. He can carry an object — a baguette (figure 1: images C) or a bottle (figure 1: images D) — or no object (figure 1: images A and B). At the end of each of the two paths after the fork, there can be a house or no house (which can be seen as a destination). The two houses (one located at the end of the upwards path and the other at the end of the downwards path) are graphically identical. On the roof a sign indicates if it is a bar or a bakery (figure 1: images B, C, D).

According to the presence or absence of the elements "OBJECT", "HOUSE", and according to the orientation of the PRINCIPAL PATH, we obtain 12 images corresponding to 12 experimental conditions. The 12 images are presented to the participants in a booklet (an image per page). In order to avoid effects of a temporal nature between the presentation of the various images, we inserted between the images a distracting task. Participants were asked to solve a mathematical operation of the type "17- 9=?".

Procedure. The experiment takes place collectively in a course room. Each participant has a booklet of 22 pages. On the first page he/she is given the following instruction: *In this booklet, there are several pages which present: an image which shows a character on a path; and a mathematical operation which you must answer. Even if these images are one following the other, each image MUST be treated independently of the others. (1) when you see the image with the character, you must as soon as possible indicate the path which would be indicated by the majority of the people if they were asked: "which path will the character take?". You have time limits but it will be sufficient to accomplish your task. Once you have carried out the task, wait for my signal before turning the page. The participants have 4 seconds to indicate the path that they considered for the character, and 6 seconds for the mathematical task.*

We coded the answers of the participants according to the path indicated by the arrow drawn by the participant: the DOWNWARDS path (B) =1; and the UPWARDS path (H) =O. We analyzed (i) the effect of changing of the orientation of the principal path and (ii) the effect of the factor context. The context is defined by the presence or absence of objects other than the character and the path. Four different contexts thus are obtained: In condition A, the context is defined by the fork and the character; in condition B, we added the presence of the two destinations, bar and bakery; in condition C, in addition to the other elements, the character is carrying a baguette; and in condition D the character is carrying a bottle.

6 Results

(i) The effects of the orientation change of the principal path.
The ANOVA was carried out on the average frequency of the DOWNWARDS answers by the participants for each of the three modalities (PW-HORIZONTAL, PW-UPWARDS and PW-DOWNWARDS) of the principal path (PW) orientation. The ANOVA carried out highlights an effect of the orientation change on the direction chosen by the participants ($F(3, 2832) = 22.78$, $p<.0001$) (figure 2).

The "post-hoc" analysis (PLSD Fisher) shows that when the orientation of the principal path aligns with the path which goes DOWNWARDS (images A3, B3, C3, D3), rather than HORIZONTAL, (images A1, B1, C1, D1), ($M=.49$; $p<0001$), or UPWARDS, (images A2, B2, C2, D2), ($M=.38$; $p<0001$), participants chose significantly more often a DOWNWARDS path for the character ($M=.75$). These results support the assumption that the properties of the path affect the direction to be taken.

(ii) The context effect.
The ANOVA on the DOWNWARDS answers in each of the four modalites of the context (A, B, C, D) highlights an effect due to the contextual properties ($F(3, 2832) = 2.89$, $p<.05$).

The post-hoc analysis (PLSD Fisher) shows that the rate of DOWNWARDS answers for the images D1, D2, and D3, (M=.23) is statistically different from that of the images B1, B2, B3, B4 (M = 34; p<.05) and from the images C1, C2, C3, C4 (M = 36; p<.01) (figure 3). This result indicates that the introduction of a destination and an object (baguette or bottle) has an effect on the choice of the goals attributed to the character by the observers.

(iii) Effects of the interactions between the factors "Orientation of Principal Path" and "Context".
The ANOVA does not show any global effect due to the interaction of the two principal factors: change of orientation of the "principal path" (1, 2, 3) and context (A, B, C, D).

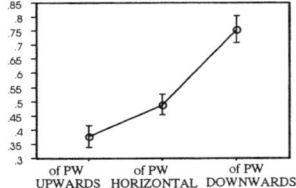
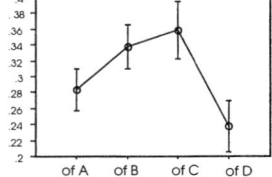
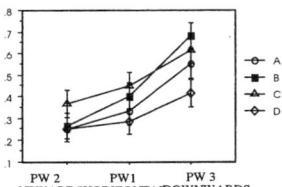

Fig. 2. Effect of the orientation of the "principal path" (1, 2, 3) on the rate of DOWN-WARDS answers

Fig. 3. The context effect (A, B, C, D) on the rate of DOWNWARDS answers

Fig. 4. Effects of the interactions between the factors "orientation principal path" and "Context"

However, as shown in figure 4, in the C2 condition, compared to the A2, B2, and D2 conditions, participants indicate significantly more often a downwards path for the character — i.e. towards the bakery. This result deserves to be underlined, because, although the path affords an UPWARDS direction, the baguette in the character's hand leads the participants to direct the character preferentially towards the bakery, which is on the DOWNWARD fork. On the contrary, no increase of the DOWNWARD answers is noticed for condition D3, ("mirroring to the C2 but with a bottle instead of the bread). Finally for condition B (no object), no effect of the interaction between CP (1 and 3) and B is noticed.

7 Discussion and Conclusions

The results of this experiment clearly highlight that the environmental context exerts a strong influence on the predictions of action and the attribution of goals to the agent. As the effects of changing path orientation show, participants tend to indicate that (after the fork) the character will take the path with the same orientation as the principal path. This result can be explained in term of "affordance"— the path that is the continuation of his path says to him, to some extent, "take me"—, but also in terms of physical relational properties. Some aspects of Gibson's theory[4] [6, 8] seem quite appropriate for explaining the type of "minimum" information needed for understanding a dynamic situation and interpreting action. In accordance with a Gibsonian ecological approach, we think that what humans perceive about the environment is mainly what it "affords" in terms of action. Thus, children who cannot yet walk perceive the characteristics of the surfaces, such as resilience or elasticity, which can afford crawling around on them [4, 5]. If the concept of affordance is relevant to an acting participant in a given environment, why shouldn't it be relevant when a subject is simply observing events as they take place? If adults and children are equally able to isolate this type of information from surface properties in order to perform their own actions, it is reasonable to think that they are also capable of inferring the ability of another agent to carry out the same action. Nevertheless, an alternative explanation would be that participants, on the basis of their past knowledge, prefer to continue along paths unless there's a reason to turn aside. But how to explain combining perceived information (the two roads) and knowledge (past experience) in terms of cognitive processes ? We advocate that participants establish a relation between the parts of the path. What is found in common is the same direction. This shared property allowed the creation of a common category. The cognitive economy provided by this organization of the spatial environment in contextual categories could explain why participant resist to attribute the agent a change in the direction along the path. The information — however static — provided by the environment may suggest possible actions depending on the environment and the properties of potential agents. Thus, processing contextual environmental properties[5] would enable an observer to anticipate events to come.

According to our approach, the understanding of the events that occur in a given environment is based on processing the properties provided by this environment. Thus, an object "affords" a given action because it possesses the specific characteristics which make that action possible. However, links between relational properties and actions are not inherent to objects but arise from cognitive attribution. We assert that this cognitive attribution comes under the heading of what we call the contextual categorization process.

---

[4] We wish to point out that Gibson's "direct realism" can hardly explain causal and intentional attribution phenomena, which are major components in understanding events in terms of actions However, our objective is to show that some of the phenomena reported by Gibson, phenomena pertaining to or derived from "affordance", may be explained in terms of categorization if action is considered as a non-perceived property. For instance, the property "can fly" will be attributed to an object if it is categorized as a "bird".

[5] One of the environment's relational properties that seem to be essential is the support surface of objects. According to Gibson [8], "terrestrian events" may be divided into three broad groups: (i) Changes of color and of surface texture, (ii) modifications of a surface and (iii) changes of layout, which include an object's displacement in space. The last group includes breaking and changing the direction or orientation, which characterize the support surface. Like Gibson, we think that the surface on which an object lies (whenever subject to modification) is very important in understanding changes. Indeed, given that space cannot be perceived [6], perception of the characteristics of the support surface play a major role in indicating how to perceive the modifications an object is subject to in space and over time.

Contextual categorization makes use of the properties that an object has in relation to other objects. Relational properties include location (to the right of, above, etc.), surface or structural properties, semantic and even actions arising from functional and procedural properties, and they relationships between objects. These relational properties are extrinsic to the object, and can change as the situation evolves. According to Anderson [1], categorization involves ascribing non-perceived properties to categorized objects. An action can be inferred as a non-perceived property in two ways: by ascribing to objects contextual categories which are actively created by processing the physical environment; and by accessing knowledge about the actions most likely to ensue from a given event. Emergent interpretation and anticipation can be seen as a process of structuring the prerequisites for action — prerequisites that reside in the relational properties of objects.

What is specific to this experiment is the presentation of semantically rich objects. Their relational properties allow the attribution of action goals to the characters. As the results of the context effect show, the goal is selected starting with the processing of the specific relational properties of each image. According to each image, the relations established between the two salient objects (baguette/bakery and bottle/bar) quickly activate the construction of a category that gathers the two semantically close elements.

The creation of this new category — a consequence of the semantic proximity between the object carried by the character, the destination of the character, and shared properties of these two variables — allows the attribution of a goal. The results are particularly striking for the context C. In this condition, in spite of any logic which would like that the bread possession makes it unnecessary to go to the bakery, participants preferentially indicate that the character should go in the direction of the bakery (downwards).

This type of result seems to indicate that the comparison of the physical properties of the context is carried out in a very early stage of data processing. The goal attribution, considered as a highly informative component for the interpretation of perceived action, is carried out by processing the physical and relational objects' properties, through the contextual categorization process. Creating new contextual categories covering certain type of properties allows the observer to infer and to anticipate the action-goal he/she considers most congruent with the situation.

Finally, at a completely speculative level, one could notice that if the goal activation was to be guided by a pre-stacked plan evocation, effects such as that observed under the condition C would be difficult to interpret. No schema-plan indicates that when one already has bread, he/she should go to a bakery, yet an approach based on contextual categorization helps to explain this type of errors. The results of this study should contribute to issues of the role of context and semantic bias in perception and anticipation of action.

Acknowledgments. The research reported herein was supported by the CNRS (France) and the Lavoisier Program (Grant ES.22/BFE/2002 340392L). The authors are grateful to Dylan Jhirad for reviewing of this paper.

References

1. Anderson, J.R. The adaptative nature of human categorization. *Psychological Review, 98*(3), (1991) 409–429.
2. Bonnet, C. Visual perception in context. *Cahiers de Psychologie, 6*(2), (1986) 137–155.
3. Denis, M. Avant-propos. In M. Denis (Ed.), *Langage et Cognition Spatiale* (pp.9–13). Paris: Masson. (1997).
4. Gibson, E. Introductory essay: What does infant perception tell us about theories of perception?. *Journal of Experimental Psychology: Human Perception and Performance, 13*(4), (1987) 515–523.
5. Gibson, E., & Schmuckler, M. Going somewhere: An ecological and experimental approach to development of mobility. *Ecological Psychology, 1*, (1989) 3–27.
6. Gibson, J.J. *The perception of the visual world.* Boston: Houghton-Mifflin (1950).
7. Gibson, J.J. The theory of affordance. In R.E. Shaw & J. Brandsford (Eds.) *Perceiving, Acting, and Knowing.* Hillsdale, New Jersey: Erlbaum (1977).
8. Gibson, J.J. *The ecological approach to visual perception.* Boston, Ma.: Houghton-Mifflin (1979).
9. Hubbard, T.L. How consequences of physical principles influence mental representation: The environmental invariants hypothesis. In P.R. Killeen & W.R. Uttal (Eds.). *Fechner Day 99: The end of 20th century psychophysics. Proceedings of the 15th Annual Meeting of the International Society for Psychophysics* Tempe, Az: The International Society for Psychophysics (1999) 274–279.
10. McArthur, L.Z., & Baron, R.M. Toward an ecological theory of social perception. *Psychological Review, 90*(3), (1983) 215–238.
11. Neisser, U. *Cognitive Psychology*, Englewood Cliffs, New Jersey: Prentice-Hall (1967).
12. Piaget, J. La psychogénèse des connaissances et sa signification épistémologique. In M. Piatelli-Palmarini (Ed.). *Théories du langage théories de l'apprentissage. Le débat entre Jean Piaget et Noam Chomsky* Paris: Seuil (1979) 53–64.
13. Pomerol, J.-Ch., & Brézillon, P. Dynamic between contextual knowledge and proceduralized context. In V. Akman, P. Bouquet, R.Thomason, R. Young (Eds.). *Lectures Notes in Artificial Intelligence, vol. 1688, Modeling and Using Context,* New-York: Springer (1999) 284–295 .
14. Pomerol, J-Ch, & Brézillon, P. About some relationships between knowledge and context. In P. Bouquet, L. Serafini, P. Brézillon, M. Benerecetti, F. Castellani (Eds.). *Lectures Notes in Artificial Intelligenec vol. 2116, Modelling and Using Context.* New-York: Springer. (2001) 461–464.
15. Schmidt, C. F., Sridharan, N. S, & Goodson, J. L. The plan recognition problem: An intersection of Psychology and Artificial Intelligence. *Artificial Intelligence: Special Issue on Applications to the Sciences and Medicine, 11,* (1978) 45–83.
16. Shaw R., & Bransford J. Approach to the problem of knowledge. In R. Shaw, & J. Bransford (Eds.), *Perceiving, acting and knowing.* Hillsdale, New Jersey: Erlbaum (1977).
17. Siskind, J. M., Axiomatic support for event perception. *Proceedings AAAI-94 Workshop on Integration of Natural Language and Vision Processing,* (1994) 153–160.
18. Siskind, J. M., Grounding language in perception. *Artificial Intelligence Review, 8,* (1995) 371–391.
19. Thorrisson, C. Machine Perception af Real-Time Mulimodal Natural Dialogues.in: P. McKevitt (Ed.), *Language, Vision & Music.* Amsterdam: John Benjamins (in press).
20. Tijus, C.A. Contextual categorization and cognitive phenomena. In V. Akman, P. Bouquet, R. Thomason, & R.A. Young (Eds.). *Lectures Notes in Artificial Intelligenec vol. 2116, Modelling and Using Context.* New-York: Springer. (2001) 316–329.

21. Turner, R. M. Model of explicit context representation and contextual knowledge for intelligent agents. In J.G. Carbonell, & J. Siekmann (Eds), *Lectures Notes in Artificial Intelligence, vol. 1688, Modeling and Using Context,* New-York: Springer. (1999b) 375–388.
22. Turner, R.M. Context-mediated behavior for intelligent agent. *International Journal of Human-Computer Studies, Special issue on Using Context in Applications, 48,* 3, (1998) 307–330.
23. Turner, R.M. Context-mediated behavior: An approach to explicitly representing contexts and contextual knowledge for AI applications. *Working Notes of the AAAI-99 Workshop on Modeling and Using Context in AI Applications, AAAI Technical Report* (1999a).
24. Turvey, M. T., & Shaw, R. E. The primacy of perceiving: An ecological reformulation of perception for understanding memory. In L. G. Nilsson (Ed.), *Perspectives on Memory Research.* Hillsdale, New Jersey: Erlbaum (1979).
25. Turvey, M. T., & Shaw, R.E. Memory (or, knowing) as a matter of specification non representation: Notes towards a different class of machines. In L. S. Cermak & F. I. M. (Eds.), *Levels of processing and human memory.* Hillsdale, New Jersey: Erlbaum (1978).
26. Tversky, B. Form and Function. In L. Carlson & E. van der Zee (Eds.) *Functional and Spatial Features in Language and Space.* Oxford: Oxford University Press (In press).
27. Von Helmholtz, H. *Treatise on physiological optics, Vol. III.* J. P. C. Southall (Ed.), New York: Dover (1925).
28. Zibetti, E. *Catégorisation Contextuelle et compréhension d'événements visuellement perçus et interprétés comme des actions.* Thèse de Doctorat, Université de Paris 8 (2001).
29. Zibetti, E., Hamilton, E., & Tijus C.A. The role of Context in Interpreting Perceived Events as Action. In V. Akman, P. Bouquet, R.Thomason, R. Young (Eds.). *Lectures Notes in Artificial Intelligence, vol. 1688, Modeling and Using Context,* New-York: Springer (1999). 431–441.

How to Define the Communication Situation: Determining Context Cues in Mobile Telephony

Louise Barkhuus

The IT University of Copenhagen
Department of Design and Use of IT
Glentevej 67, DK-2400 Copenhagen, Denmark*
barkhuus@it.edu

Abstract. Mobile telephony is increasing in use and is presently the most common mobile communication technology. Furthermore, the technological possibilities for context-aware features for mobile phones are increasing. The out-of-context communication that occurs from mobile telephony leads to a need for exploring how users deal with situational cues in order to evaluate proposed context-aware features. We present an exploratory case study of mobile telephone users where the communication situation is examined, in particular in relation to how users deal with context information. The study finds that users infer about context cues to a high degree and adapt accordingly. This indicates that context based features in mobile telephony should focus on the specific context measures when they are needed instead of focusing on general measures that are always available.

Keywords: Context-aware applications, human-computer interaction, mobile telephony.

1 Introduction

Mobile communication is a fast growing feature of people's lifestyle; the spread of mobile telephony has increased highly through the last decade in virtually all parts of the world due to recent improvements in wireless and processor technologies as well as common market forces. Although social behavior seems to adapt to the new way of communicating, mobile telephony is still an instant, out-of-context type of communication technology [8]. It is instant because the mobile phone can be kept on constantly and therefore is easily available. The communication is out-of-context because people initiate communication without knowing the other person's situation before placing or answering a call. In order to deal with this lack of context knowledge, several more aware and adaptable functions have been suggested by researchers [8,12]. These functions address the social problems arising when a person is available anywhere, anytime by

---

* Currently visiting researcher at UC Berkeley, The Department of Engineering and Computer Science, working with Assistant Adjunct Professor Anind Dey.

letting the user define his/her own context to the possible callers. Other aware functions include location tracking of previously specified individuals, a service that is already in use some places [14].

One issue that has not been considered when designing and evaluating these context-aware features is how the communication situation is defined according to the communicators themselves. It is an essential step towards evaluating the proposed technologies as well as for designing new context-sensitive features. Especially relevant is how users deal with the limited set of context cues that is available when communicating via mobile phones.

This paper presents an exploratory case study, focusing on mobile phone users. The aim is to determine how users deal with this out-of-context communication tool; we then propose to use this knowledge for evaluating context-aware services for mobile phones.

2 Context-Aware Computing

Originally considered to be a subset of ubiquitous computing, context-aware computing is increasingly viewed as essential to a truly ubiquitous computing environment [9]. Context-aware computing is described as "an application's ability to detect and react to environment variables autonomously" [1]. From the initial proposals for context-aware computing (e.g. [11]), to the latest advances (e.g. [3]), the development of context-aware features have been widespread across many aspects of technologies. Not limited to communication purposes, the technologies include, active badges [15], intelligent white boards [11], tour- and conference guides [4,2] as well as context-aware mailing list systems [3]. They all rely on one or more piece of context information such as location, identity of user or time to act accordingly.

2.1 Existing Context-Aware Applications for Mobile Communication

Unlike general context-aware applications, only a few applications focusing on mobile communication devices have been developed. Even fewer are actually employed in real consumer devices. An example of a context-enabled application is the context-call application by Schmidt et al. [12]. It enables the user to preset a status to for example, 'meeting', 'working' or 'at home'. The caller is then prompted with the status before the call is sent through the receiver. Another relevant context-aware application is the function that changes the time on a mobile phone depending on which GSM antenna it receives information from. When the user enters a new time zone, the time is adjusted automatically. The function is implemented in some mobile phones but works sporadically. Finally the localization of other mobile phone users is offered by some subscription providers. It requires that the user defines who should be able to locate him and that the friends have accepted to be localized as well [13,14].

2.2 Context as Real Life Cues

People rely on 'cues' in their everyday life in order to act according to a certain situation; they depend on personal observations in order to understand the situation's higher level. These cues can be compared to the context information that context-aware applications use to adapt according to the situation or environment.

Although the concept of context information is essential to context-aware applications, the notion of context is highly complex. When reviewing past literature, it appears that many researchers have attempted to define how situations are perceived differently by individuals. Examples include Greenberg, who claims that it is not possible to define a context in terms of sensor information because situations fluctuate and are unpredictable in nature [5]. The overall situation that a person is part of, is not solely made up of measurable context information, because the person's own perception of the situation is part of the context. What is possible, however, is to augment technology with functions that take certain aspects of the situation into account and thereby make the use more smooth and transparent for the user. Where early studies define context solely as context measures such as location and the user's task [11], more recent research realize that sensor information is only part of the definition of a certain context [3].

2.3 Research within the Use of Mobile Telephony

Research within mobile telephony includes numerous studies of user habits and behavior. The studies often focus on the impact that this type of communication has on the users [8,10]. One commonality of these studies is that situational context is rarely addressed as a separate research topic. Instead the studies cover actions and behavior of individuals or groups of users. Some look at privacy issues and find that the level of wanted privacy depends on the person, who is to receive the sensitive information, more than the overall situation [7,6]. Others look at the acquired practices of mobile communication from both social and technical perspectives and highlights that not knowing the receiver's situation, affects the person's willingness to call a mobile phone number [10]. Most of the studies mention that mobile communication differs from other communication in that the social context means a great deal to the way in which the communication takes place. None of the studies however, consider how users comprehend situational cues within conversation. Where most of the studies of mobile phone use focus on behavior *around* the communication situation, our study examines the actual communication context.

3 Case Study Design and Method

The present study is designed as an exploratory empirical study of situational context in mobile communicators. By interviewing users of mobile telephony, the users' contextual cues and habits of use are traced. The study consists of seven semi-structured qualitative interviews carried out in the participants' own environment (office, own home, school cafeteria, etc.).

In order to attempt to define the communication context, three context cues were selected as initial focus; they are likely to be part of the overall communication situation. The three pieces of information that are initially investigated are location, time[1] and identity of the person communicated with. Naturally these were not openly revealed to the participants, but by targeting these measures along with the exploration of others, the interviews become more focused than if very open questions are asked.

3.1 Participants

The participants were selected from one main criterion: they had to be fairly high level users of mobile telephony. This resulted in a causal restriction to age, since the heavy users are more commonly found in young age groups, with teenagers as well as adults under 30. The age of the participants ranges between 19 and 27, with an average age of 23. Their experience with mobile phones ranges from 3 years to 7 years with an average of 5.8 years and most of the participants were using their phone between 5 and 10 times a day (phone calls as well as text messaging).

3.2 Study Setup

The data was collected as a digital audio recording and transcribed shortly after. The transcription was performed in two ways, a word-by-word transcription and a summarized version describing the essence of each answer. This way, comparing the different answers is easier. Because it is exploratory in nature and because of the limited number of participants, none of the answers were quantitatively measured; the interviewees' expressions regarding the use and communication situation were merely compared and analyzed. The transcriptions were analyzed for general patterns of the use of mobile phones, in order to trace the situational context where the communication takes place. Since the study is qualitative and limited to a small number of participants, the findings merely provide *indication* of use as it relates to the communication situation.

4 The Out-of-Context Communication Situation

Most of the participants acknowledge that they are at times uneasy communicating in the uncertain environment that mobile telephony provides. Because they never know neither location or activity of the other person, it is evident

---

[1] The study distinguishes between time and relative time, that being the user's actual time and the user or receiver's perceived time-frame (e.g. "do you have time to talk?").

that communication depends on situational cues, both implicit and explicit. The findings support the notion that people rely on cues in their decision of how to communicate through their mobile phone and that these cues are complex. What will be explored here, however is how the users deal with the initial limited context knowledge.

4.1 Mobile Telephony Depends on Specific Situational Cues

All participants state, at one point or another, that their actions in many cases depend on the situation. One example of dependent use is a participant who, without hesitation, says that he greets virtually every caller differently except, of course, when the display shows no number or caller. Almost half of the participants have defined different ring tones for different people. That way they know if they should "run to the phone or can just relax", as one participant phrased it.

This finding is not a surprise of any kind, partly because of common sense but also because other research mentions this characteristic [12]. However, it does complicate the overall communication situation by creating a communication environment where much of the conversation evolves around context and less on content. The participants not only depend on cues such as location and present activity of the other person, but it is in certain cases a 'necessity' for the participants as well; some of them actually call their friends *just* to infer their friends' situation. The three participants who do this all state that it would be more efficient if technology facilitated features could give this kind of information. One states:"I often meet with my friends Saturday afternoon, but it is a casual thing, so I call them up to ask where they are; if they are at a cafe or something. Then I decide if I am close enough to go. If I knew where they were, I would just make plans to be in the neighborhood".

4.2 Inferring to Others' Context

While most of the participants express that their behavior depends on the overall situation, the way they deal with the situation is different. Where some of them tend to ask about all possible information, others are able to focus the conversation on the purpose and claim that context requests are omitted in highly goal oriented and short conversations. What appears in analyzing the results though, is that the situational cues that the participants request, are ranked in a specific order of importance. Besides the three measures initially studied, one extra measure was found to be important. The context information that the participants claim to make their communication situation rely on are, according to the participants' perceived importance: identity, location, relative time and social situation.

The most essential factor to the communication situation that the participants mention is the identity of the person that they are communicating with. It is technology facilitated by caller id, which all the participants use. Even though most of them will answer the phone if the number is unrecognized, they are more hesitant and suspicious. They have, in a sense, become reliant on the caller id function.

Besides identity, the participants verbally request other information in order to compensate for the out-of-context communication. The location is on the top of the list; even without suggesting it to them, most of them claim to ask where the other person is as the first question. They are then able to better understand the situation in which the other person is in. Another way of dealing with the unpredictability of the situation is to ask if the person who receives the call has time to talk. This is almost as common to infer about as location. Last, a rather complex piece of context information is requested fairly often. The participants phrase it as "what are you doing?", meaning what activity the other person is engaged in. One participant regarded this measure as essential and admitted that she sometimes sent a text message before calling her boyfriend: "He works in such a small cubicle with two desks, the person in front of him can hear everything. If I knew what he was doing, it would be much easier. I sometimes send him a message to ask if he can talk".

What is striking about the results is how the participants are willing to spend much time telling or asking the other person about their context. They voluntarily submit much of the context information in initial conversation. Some of the participants give an impression that most of their conversations evolve around where the communicators are and what they are doing. Only two participants say that they use the phone for short targeted conversations such as arranging meetings and giving messages. The others are willing (or curious enough) to share information about their situation in detail. In this sense the 'out-of-context communication' becomes 'inferring-of-context communication'.

4.3 People's Adaptability

A strong characteristic of the results is the participants' willingness to adapt their behavior to the available context in the communication situation. As mentioned before, they will most often ask the other's location and then adapt the conversation according to received information. As one of the participants expressed: "If I know that my friend is on the bus then I will not ask very personal questions". They also adapt to their own context by talking quietly in public places and not voluntarily providing very private information when they know others can hear.

The adaptive behavior is like second nature to the most of the participants. This is probably due to the time span in which they all have owned a mobile phone. Other research suggests that behavior change and that the user become more comfortable with mobile communication according to how accustomed the user is to dealing with it [7].

5 Discussion and Future Research

The present study shows that, to deal with out-of-context communication, the user collects separate situation information and adapts actions and behavior accordingly. The user's perception of the communication situation is highly influenced by context cues that are collected implicitly and explicitly. Inferring to context information is not just a mean to the goal but sometimes a goal in itself.

The ability and tendency for mobile phone users to adapt is not in any way surprising, since out-of-context communication requires flexibility, however it is highly relevant when evaluating context-aware features for mobile communication. The users request the information they need in order to adapt their behavior *when* they are communicating, therefore the technology should support the users' requested situational cues at the needed time. Instead of supporting location tracking that is always on, a service that displays the caller's name *and* the caller's present location when a call is received might be much more useful to the communicator than the proposed Friendfinder [14]. The adaptability also indicates that context-aware service should be careful when providing autonomous functions that leaves little choice to the user. For example, the user is usually aware that he has entered a new time zone and, although it may seem reasonable to change the time on mobile phone, the user might deliberately want to keep the old time setting for various reasons (e.g. to know what time it is 'at home'). A more flexible feature would be to prompt the user and ask if the time should be changed.

The ability as well as the desire to adapt according to situation suggest that implementing services with limited scope and facilitating one piece of context information at a time, is more likely to succeed, than complex multi-faceted context-aware features. The user will adapt behavior to the new service if it is simple and facilitates his actual need in the communication situation.

The fact that people request context information, supports the existence of context supporting services. By having more situational cues available before or during the conversation, the 'inferring-of-context communication' is minimized and the users can focus on the content of the conversation. When evaluating context-awareness in mobile communication, however, it is important to pay attention to the cases where the purpose of the communication situation is the context request in itself. Just because the only purpose is to get information that could be obtained by the aid of technology, it is not always desirable to receive the information this way.

5.1 Future Research

The case study is by no means exhaustive within context measures. It has not been the initial purpose to extract all possible situational cues that are present in mobile communication. Furthermore other research, e.g. observational studies, should support this study in order to be able to generalize on the behalf of other types of mobile computing. Future research should be conducted to find if facilitating the communication situation with situational cues are truly going to give users a better experience or if the cues are merely a need created out of technical possibilities.

Acknowledgement. I would like to thank all the participants for their time they took out of their schedule to be interviewed. I also thank Professor Anind Dey for his valuable comments on the poster abstract and layout.

References

1. L. Barkhuus. Context information vs. sensor information: A model for categorizing context in context-aware mobile computing. In W.W. Smari and W. McQuay, editors, *Fourth International Symposium on Collaborative Technologies and Systems*, pages 127–133, San Diego, CA, 2003. The Society for Modeling and Simulation International.
2. K. Cheverst, N. Davies K., Mitchell, and A. Friday. Experiences of developing and deploying a context-aware tourist guide: The GUIDE project. In *Proceedings of MOBICOM 2000*, pages 20–31, Boston, Massachussetts, 2000. ACM Press.
3. A. K. Dey, G. D. Abowd, and D. Salber. A conceptual framework and a toolkit for supporting the rapid prototyping of context-aware applications. *Human-Computer Interaction*, 16(2–4):97–166, 2001.
4. A. K. Dey, D. Salber, G. D. Abowd, and M. Futakawa. The conference assistant: Combining context-awareness with wearable computing. In *Proceedings of the 3rd International Symposium on Wearable Computers*, pages 21–28, Los Alamitos, CA: IEEE, 1999.
5. S. Greenberg. Context as a dynamic construct. *Human-Computer Interaction*, 16(2–4):257–269, 2001.
6. S. Lederer, J. Mankoff, and A. K. Dey. Who wants to know what when? privacy preference determinants in ubiquitous computing. In *Proceedings of CHI 2003*. ACM Press, 2003.
7. C. Licoppe and J. P. Heurtin. Managing one's availability to telephone communication through mobile phones: A french case study of the development dynamics of mobile phone use. *Personal and Ubiquitous Computing*, 5(1):99–108, 2001.
8. P. Ljungstrand. Context awareness and mobile phones. *Personal and Ubiquitous Computing*, 5(1):58–61, 2001.
9. T.P. Moran and P. Dourish. Introduction to this special issue on context-aware computing. *Human-Computer Interaction*, 16(2–4):87–95, 2001.
10. L. Palen, M. Salzman, and E. Youngs. Discovery and integration of mobile communications in everyday life. *Personal and Ubiquitous Computing*, 5(2):109–122, 2001.
11. B. Schilit, N. Adams, and R. Want. Context-aware computing applications. In *Proceedings of the 1st International Workshop on Mobile Computing Systems and Applications*, Los Alamitos, CA, 1994. IEEE.
12. A. Schmidt and H.W. Gellersen. Context-aware mobile telephony. *SIGGROUP Bulletin*, 22(1):19–21, April 2001.
13. AT & T. m Mode: Find friends. www.attws.com/mmode/features/findit, March 2003.
14. Telia. Friendfinder, hitta dina vanner med mobilen. www.teliamobile.se, October 2002.
15. R. Want, A. Hopper, V. Falcão, and J. Gibbons. The active badge location system. *ACM Transactions on Information Systems (TOIS)*, 10(1):91–102, 1992.

How to Use Enriched Browsing Context to Personalize Web Site Access

Cécile Bothorel[1] and Karine Chevalier[1,2]

[1]France Telecom R&D, DMI/GRI, 2 Av. Pierre Marzin, 22300 Lannion, France
[2]LIP6, Université Paris VI, Paris, France
{cecile.bothorel, karine.chevalier}@francetelecom.com

Abstract. Using a browsing context is one of the keys to web site access personalization under particular constraints. With poor user information modeling, which is a common situation, a web site cannot be adapted to the current user. Assuming the current clickstream is the only known information about a web site user (no profile, no past sessions, no identification, no content analysis of viewed pages), we propose here a method to enrich the browsing context and enhance the current user model. In a batch mode profile-based enriched navigation patterns are computed. In on-line mode, Navire, a personal agent and its matching rule engine continually re-adapts the browsing context with pre-calculated profiles. Based on the current up-to-date context, Navire personalizes the access to a web site.

Keywords: Representing context and contextual knowledge, Human-computer interaction, Context and the Web.

1 Introduction

How to facilitate site content access to users? How to make visiting a site natural and pleasant? These questions are important aspects for site designers. One answer is to personalize the site to the users. Personalization includes an adapted presentation, the use of an agent-based personal assistant or a recommender system.
Personalization requires personal data to fit one particular user. The first solution to acquire personal information is to ask it explicitly to users: users are asked to give very precise profiles, about what they need, about what they want and how they want it. With such rich information, recommendations and personalization could be performed and they are likely of good quality. Unfortunately, users are not very cooperative with the communication of their personal data.
Another solution is to use automatic learning techniques to acquire users profiles. For instance, some systems learn profile 'by example' [16], they propose items to a user, and analyze if she accepts, what she modifies, etc. The actions are reflected in the profile, which can lead to empirical advice.
Knowing the best and most complete profile about someone can't prevent any system to use the 'wrong' part of this profile. The recommendation may be relevant for a user, 'in general', but not for her current needs. Moreover the profile can turn out to be of

bad help when a user is browsing for someone else... Even if this extreme case is not frequent, the identification problem is not always an easy task: when the identification data is an IP number, how to deal with employees in large companies whose machine is hidden behind a proxy? When a cookie is used, a web site logs a browser instead of the person behind the browser.

Advising someone means to supply the right information, in the right manner, at the right moment, to the right person, and of course, if the need exists.

In this paper, we discuss a personalization method (1) which does not ask any profile information to browsing users, and (2) which has the aim to use the context to perform useful recommendations. We detail in a first part how to use a browsing context in order to personalize a web site for uncharacterised users. Then we present a browsing personal agent, called Navire (stands for Navigation Recommender).

2 Context and Personalization for Uncharacterised Users

We are addressing the personalization problematic in particular circumstances:
- No profile is required from users (no 'inconvenience' in filling long forms),
- No identification is required from users (no cookie, no login, no IP, ...),
- Users may be unique or occasional visitors (no browsing historical storage),
- The content of viewed pages is not analyzed (no topic discovery, no ontology...).

These constraints have been defined on account of experienced situations in which one or more had been unsolved problems or had biased results.

Our main challenge is to offer personalization to all users, to the ones who have provided profile information, and also to the ones who have chosen not to. Their sequences of clicks are our only 'identification' criteria. We choose to make a recommender system based only on current user navigation.

The question is now: how could we obtain 'intelligent information' during a browsing session without any identification consideration. What if we use the context as a media to convey rich information?

Lieberman wrote about the need to take into account the context [11]. He points out the importance of the context in which advice is given and the necessity to identify the relevant parts of the context (time, place, history of interaction, etc.). The argument is discussed in [10]. Considering that a browsing session is composed of several phases (several successive subjects tackled for example), if the advice provided by a recommender system concerns 'obsolete' subjects — or in the worst case, no interesting subject at all — users may disable the assistant or quit the site. [17] makes the distinction between 'good' and 'useful' recommendations. A recommendation may be 'good', but not perceived as 'useful'. 'The ultimate effectiveness of a recommender system is dependent on factors that go beyond the quality of the algorithm'. From a user's perspective, such a system must (among others) inspire trust, be somewhat transparent and explain recommendations. We are convinced that one way to be useful, is to make the recommendations occurring at the right moment. While browsing, the context of a click is a necessary condition to obtain an accurate link recommendation.

One can be interested in the practical use of context through (location and time) data aiming at making easier person mobility. Such applications are called context-aware applications [7]. But when context becomes knowledge, in relationships with the user, we speak of context-based systems. In Artificial Intelligence, according to Brézillon's survey about context [4], 'context does not intervene directly in the learning (or in the knowledge acquisition, or in human-machine problem solving) but is seen as constraints'.

Anderson & Horwitz [1] suggests a way to assist users with routine web browsing. They define an enriched informational context as time of the day, time since last access, and recent seen topics. The system detects user's habits from her last sessions and dispatches recommendations relatively to the current context.

Our constraints of the personalization problematic involves, first, that nothing is known about a user who starts a visit on a web site, i.e. we do not have any historical context, and second, we do not have any content information about the seen pages.

We study a way to enhance information about the navigation. We suppose that navigation may depend on personal characteristics. We define our browsing context as a collection of inferred information describing a user. We use a method that links navigation data and user characteristics. Our challenge is to enrich any navigation with personal data. Then, when visiting a web site, the context of one's navigation becomes 'who' he or she may be. Consequently, regarding on 'who' a user may be, we choose the 'what' she may want and recommend it.

3   Navire: A Contextual Browsing-Based Personal Agent

Navire is a personal agent using browsing context to predict navigation and then recommend links on a web site. We base our assistant knowledge on a log mining method called SurfMiner that we have developed.

The SurfMiner method [5] performs the 'web mining' process of combining user profiles and navigation patterns. Its originality is to link navigation patterns with features of personal data. This learning process is done with data from 'reference users'. They are particular users who accept to provide a list of personal characteristics (like age, job…) and some navigation sessions on the web site.

SurfMiner relies on two assumptions: first, *two users similar in a socio-demographic way have got similar navigation on a web site*; and second, *two users similar in a navigation way on a web site have got similar needs and preferences*.

Taking into account these principles, we can learn two kinds of knowledge:

- Navigation patterns enriched with users characteristics. This knowledge provides the following kind of rule: 'if a user visits this page then with a confidence of 77%, this user is about 20 years old.' This enriched navigation patterns are association rules in the form of 'navigation pattern → user description' (NP→UD).
- Navigation patterns specialised to users communities. This knowledge provides the following kind of rule: 'if a user is a woman, with a confidence of 79%, she will visit this page of the site, after this page.' This specialized navigation patterns are association rules in the form of 'user description → navigation pattern' (UD→NP).

The SurfMiner learning method is done in two ways: (1) Navigation patterns are extracted from all site navigations of reference users and then, personal data from the users who have followed the patterns characterize the navigation patterns; (2) Groups of reference users are created with regards to their personal data and then, for each reference users' group, navigation patterns are extracted from the site navigation of the reference users belonging to the group.

Of course, it is a rather heavy datamining process, but it needs to be done only once, in a bootstrap phase.

The SurfMiner rule database can be used 'in online mode' to generate some predictions about a current site user. The predictions can be page recommendations (What will the user probably do?) or knowledge about the person (Who is she?). These predictions are based on the user's current navigation and deduced from the rule database.

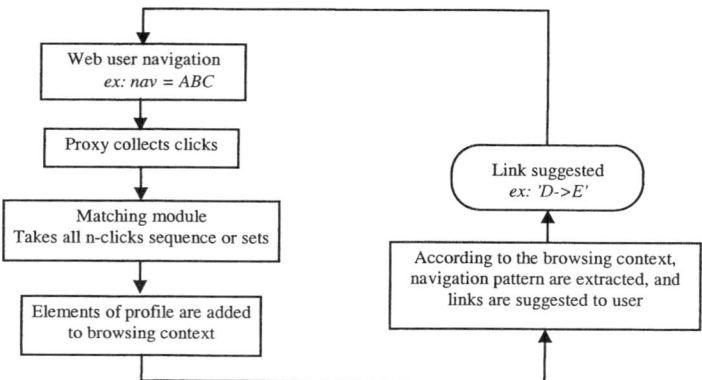

Fig. 1. Navire is a personal assistant that catches a web site user navigation via a proxy, stores knowledge deduced from the navigation. According to the up-to-date browsing context, the agent searches a navigation patterns database and determines links to recommend to the user.

Taking into account the current session clickstream, observed from a proxy in the case of Navire, the task is to match the current navigation sequence with the pre-defined database of profiles. Navire is constructing a browsing context with elements of profiles found in the database.

As a user comes and visits the site, Navire's matching engine calculates all navigational clicks combinations from the current clickstream. Navire matches each combination with the SurfMiner pre-calculated rules. As the user is browsing, pieces of profiles are learnt. Then, according to the profile learnt, our assistant tries to match it with the corresponding rules, and may recommend links deduced from the learnt profile. Profile construction is an iterative process. The agent automatically re-adapts the calculated session profile through each new browsing action.

The context can be seen as a dynamic, and up-to-date storage mechanism, in which useful knowledge constraints the recommendation engine.

Navire produces more precise recommendations than classical web mining method would do. Classical navigation patterns are discovered from the whole browsing population, and consequently underline the general site usage. The SurfMiner method reduces the searching space to a users community that reflect a specific user profile.

4 Evaluation of Click Prediction Constrained with Personal Data

An evaluation of the SurfMiner method is carried out on data of Internet usage. This data comes from Internet traffic log of a group of a thousand people extracted from a NetValue panel in 2000. They are used and enriched in the framework of a partnership between France Télécom R&D and NetValue about Internet usage [12].
35 attributes describe users: (1) some characteristics are directly provided by users, such as age, town size or date of their fist connexion to Internet; (2) others characteristics are deduced from their Internet use, such as the total number of Internet sessions done during the year 2000, the number of search engine requests done, or their type according to the Internet communication services which they used. The navigation data were previously processed in order to add to each HTTP request a session id and the user id. Each HTTP request is composed of a date, the asked url. A session is a sequence of url appearing in a chronological order. ([12] gives a complete description of tools and methods used to perform those treatments).
We construct 5 sets of data that correspond to sessions done in 2000 on 5 different web sites (anpe.fr, boursorama.com, liberation.fr, mp3.com and voila.fr) and descriptions of users. We randomly decompose data into two distinct sets:
– A training set used to learn navigation patterns enriched of users descriptions and navigation patterns specialized to a group of users. This set contains descriptions and navigation tracks of around 80% of users panel randomly selected.
– A test set that allows us to evaluate the learned navigation patterns. This set contains descriptions and navigation tracks of the remaining 20% of users panel.

We apply the SurfMiner method on the training data. The SurfMiner method generates two kinds of rules: the NP→UD rules and the UD→NP rules. To evaluate the prediction quality of the rules, we observe the percentage of good predictions on the test data (table 1 and 2).

Table 1. Percentage of correct predictions of NP→UD rules.

	Prediction with NP→UD rules		Global distribution-based prediction	
	% Correct predictions	Number of predictions	% Correct predictions	Number of predictions
Mp3.com	44.78%	5062	23,40%	5025
liberation.fr	42,61%	1217	27,32%	12036
voilà.fr	65,60%	44575	23,43%	44554
boursorama.fr	55,39%	533926	21,26%	528406
anpe.fr	48,63%	918657	27,20%	808293

Table 2. Percentage of users without good predictions of NP→UD rules.

	% users without correct predictions	Number of users for whom a prediction is possible	Number total of users
mp3.com	31,58%	38	38
liberation.fr	57,89%	19	35
voilà.fr	31%	100	163
boursorama.fr	16,00%	50	50
anpe.fr	5,56%	36	36

We perform a first evaluation on NP→UD rules. The test consists in using the NP→UD rules on each session (and each point of the session) to predict the user profile. A rule is activated when a navigation pattern (condition of the rule) is observed in the considered part of the session. If the user description (consequence of the rule) belongs to the user profile then the prediction is correct. We compare this result with a 'global distribution-based' prediction, that takes into account the users repartition according to personal characteristics. The global distribution-based prediction consists in drawing randomly one user characteristic, on the template of the global population repartition: if we have 33% of women in the whole population, we generate randomly twice more 'man' characteristic than 'woman' characteristic. When a SurfMiner rule is activated and predicts the value of a user characteristic then a global distribution-based prediction is generated on the same characteristics.

The table 1 shows the percentage of correct predictions with NP→DU rules comparing to the percentage of correct global distribution-based prediction. We could note (c.f. tab.1.) that rules generated by SurfMiner show better predictions than 'global distribution-based' ones. These good results could be a little revised because some NP→UD rules have a weak risk of bad prediction. In fact a few rules have quite evident conclusion, for instance one can predict the age bracket of a user is between 21 and 55 but the global population is concentrated between 21 and 55.

We perform a second evaluation on UD→NP rules. The test consists in predicting navigation for each user present in test data. A rule is activated when a user characteristic (condition of the rule) matches with the description of the current user. If the navigation pattern (consequence of the rule) is observed in the current user's sessions then the prediction is correct. The Tab.2 presents the percentage of users without correct predictions. This percentage is computed relatively to the number of users for whom a prediction is done. Tab.2 shows very uneven results. This difference may be due to the number of rules used to make a prediction. For the anpe.fr site, there is 3215 rules, and for the liberation.fr site, there's only 7 rules. The prediction ability of the set of UD→NP rules is not the same for the two web sites. These rules are interesting to produce NP→UD→NP rules.

The primary results presented here are encouraging. We are working on the rule discovery to increase the accuracy of predictions and then the prediction will become usable to enrich a context of web site navigation.

5 Related Work

The first way to make recommendations or personalization of a site visit is to implicate personally users: ask them to fill a form depicting them [see among many http://www.alapage.com], or give them tools to modify themselves some part of the site (the welcome page like MyYahoo [17]). This approach requires a notable effort from users who are not always ready to carry it out. Other systems use ratings of objects (pages, music, movies...) as a profile, they are based on collaborative filtering [8][14][15]. This method uses knowledge about liked and disliked objects to give some advices about a new object. The same problem of lack of participation arises.

On the contrary, our method is based on the observation of what users do on the site. Usage may lead to some preferences deduction and then allow the use of collaborative filtering based methods [2]. If a user views a page, the system rates it positively, as if the user had annotated it. Amazon.com is a good example of the use of this kind of method. But the past usage can bring wrong ideas about actual interests of a user. In fact user interests are diverse and fluctuating. Our approach takes into account only things that the user is doing at the 'recommending moment'. We consider that the current navigation is a key issue to personalize the access to a web site.

The current clickstream can be exploited in two manners: analyzing the content of the viewed pages, or using only the sequence of urls clicked in the same session.
The first case involves content analysis. The recommender system FAB [2] constructs a user profile with themes present in site pages visited by the user. FAB recommends pages with her favored themes or pages visited by users who have a similar thematic profile. In the same way, [9] constructs a thematic profile with the content of viewed pages and adjusts recommendations with user's interests. We choose not to base our method on any content analysis. A relationship between web pages can be more complex that the fact of having a similar content. [6] points out that two pages can be frequently visited together even if the two pages have no similar thematic.
For the second case, a navigation patterns extraction is done on the web site log. Navigation patterns are frequent page-sets and/or frequent page-sequences, independently of the associated content. The navigation patterns are used to recommend pages relatively to current session [3][13][6]. Pattern discovery is processed on the whole site activity. These techniques provide interesting results for usage studies, but for recommender systems, advices may be too general.
Our work focuses on an extension of the second case. In the same manner, we use the current clickstream, but we enrich it with complex knowledge that combines navigation pattern and user characteristics in order to construct a user profile.

6 Conclusion and Future Work

In this paper, we have discussed a method to personalize Internet browsing on a web site when a user is poorly described (no profile, no identification, no browsing historical storage, no content-analysis of viewed pages). We have seen that we can consider the clickstream as an identification criterion and create a rich context. Enriching the browsing context with knowledge acquired in a batch data-mining process provides clues to recommend relevant links to an unknown user. The context knowledge consists of elements of user profiles associated with navigation patterns. The context changes as the navigation goes on, and each navigation action triggers a profile update.
In our system, the principle is to use the very scarce personal data we may be able to collect —from users filling personal description, from experts describing the population visiting a site, or from panel audience specialists—to learn navigational profiles and apply those learned profiles to any user.
In future work, we will analyze the usage made of the recommendations by the users, and will consequently adapt the choice of rules stored into the browsing context. If we observe (in a long enough session) that recommendations are not used, we will tune the confidence threshold higher so that more relevant rules will be chosen. More generally, we will study which knowledge would be significantly add to the browsing context.

References

[1] Anderson C.R., Horvitz E. (2002). Web Montage: A Dynamic Personalized Start Page, WWW2002, May 7–11, 2002, Honolulu, Hawaii, USA.
[2] Balabanovic M., Shoham Y. (1997). Fab: Content-Based, Collaborative Recommendation. In Communications of the ACM, vol.40, n°3, March 1997, pp 66–72.
[3] Borges, J., Levene, M. (1999): Data Mining of User Navigation Patterns. In Proceedings of the Workshop on Web Usage Analysis and User Profiling, 1999, pages 31–36.
[4] Brézillon P. (1999) Context in Artificial Intelligence: I. A survey of the literature. Computer & Artificial Intelligence, 18(4): 321–340.
[5] Chevalier K., Corruble V., Bothorel C. (2002). SurfMiner: connaître les utilisateurs d'un site. DVP 2002, Juillet 10–11 2002, Brest, France.
[6] Cooley, R., Tan, P., Srivastava, J. (1999): WebSIFT: The Web Site Information Filter System. In Proceedings of the Web Usage Analysis and User Profiling Workshop, 1999.
[7] Dey, A., Abowd, G. (2000). 'Towards a Better Understanding of Context and Context-Awareness', Workshop on the what, who, where, when and how of context-awareness at CHI 2000, April 3rd 2000, The Hague, The Netherlands.
[8] Konstan J.A., Miller B.N., Maltz D., Herlocker J.L., Gordon L.L, and Riedl J. (1997). GroupLens : Applying Collaborative Filtering to Usenet News. In Communications of the ACM, vol.40 , No 3 (Mars 1997), pp. 77–87.
[9] Lieberman, H. (1995). Letizia: An agent that assist Web browsing. In proceedings of the International joint Conference on Artificial Intelligence in Montreal, 1995 (pp 924–929).
[10] Lieberman H., Selker T. (2000). Out of context: Computers systems that adapt to, and learn from, context. IBM Systems Journal.
[11] Lieberman H. (2001). Interfaces That Give and Take Advice, Chapter 21 of BOOK, p. 481. ISBN...
[12] Beaudouin V., Assadi H., Beauvisage T., Lelong B., Licoppe C., Ziemalicki C., Arbues L., Lendrevie J. (2002). Parcours sur Internet: analyse des traces d'usage. Rapport RP/FTR&D/7495, France Telecom R&D, Net Value, HEC.
[13] Mobasher, B., Dai, H., Luo, T., Nakagawa, S., Yuqing, S. (2000). Discovery of Aggregate Usage Profiles for Web Personalization. In Proceedings of the Web Mining for E-commerce Workshop (WebKDD), August 2000.
[14] Pennock D.M, Horvitz E. (1999). Collaborative filtering by Personality Diagnosis : a hybrid memory- and model-based approach'. IJCAI Workshop on Machine Learning for Information Filtering, International Joint Conference on Artificial Intelligence, 1999
[15] Resnick P., Iacovou N., Suchak M., Bergstrom P., Riedl J.(1994). GroupLens: An Open Architecture for Collaborative Filtering of Netnews'. In Proceedings of ACM 1994 Conference on Computer Supported Cooperative Work, Chapel Hill, NC: Pages 175–186.
[16] Shearin S., Lieberman H. (2001). Intelligent Profiling by Example, IUI'01, January 14–17, 2001, Santa Fe, New Mexico, USA.
[17] Mamber U., Patel A., and Robison J. (2000). Experience with Personalization on YAHOO!!! In Communications of the ACM, volume 43, number 8, august 2000.

Modular Partial Models: A Formalism for Context Representation

Harry Bunt

Dept. of Computational Linguistics and AI, Tilburg University
P.O. Box 90153, 5000 LE Tilburg, The Netherlands
harry.bunt@uvt.nl

Abstract. This paper motivates and outlines a logical formalism for the representation of a dialogue agent's beliefs and goals in a computationally attractive way, and is meant to be a basis for effective and efficient context representation in a dialogue system.

1 Introduction

For a dialogue system to understand natural language or multimodal dialogue contributions from a user, and to react intelligently and cooperatively, it must have access to and be able to reason with context information of a diverse nature. In particular, to generate an appropriate dialogue act, a dialogue system has to consider what (it believes that) the user wants, knows, and does not know. And typically, a dialogue system has very limited information of that kind: its knowledge of what the user wants and knows is highly incomplete. So one fundamental requirement on a context representation formalism is that it must deal with highly incomplete information.

This paper is concerned with the formal representation of that information, which forms part of the totality of the information that is necessary and sufficient for the interpretation and generation of the contributions in a dialogue, which we call the *dialogue context*.

Logical formalisms for describing 's knowledge or beliefs have been based on Hintikka's analysis in terms of possible worlds. On this approach, an A agent considers a multitude of logically possible alternative worlds, and the things that he knows are true in a subset of those worlds, the K_A-worlds. (The things that he *believes* to be true form another subset, the B_A-worlds, with $K_A \subset B_A$.) A direct implementation of this idea is computationally very unattractive for context representation, since knowledge growth in this approach corresponds to eliminating possible worlds from the K_A-worlds. To represent an agent who has little knowledge means representing a huge set K_A, and we already noticed that a dialogue system typically has little knowledge of what the user knows and wants. In contrast with this 'eliminative' view on knowledge, we will develop an *incremental* formalism, where an agent's representation structures grow as his knowledge grows.

By relating different types of context information to the interpretation of different types of communicative action, we have argued in Bunt (1999) that the various categories of context information have rather different logical properties, and bring different requirements on the complexity of a formalism for representing context information. In this analysis, we related the study of context modelling to the theory of dialogue acts developed in Dynamic Interpretation Theory (see Bunt, 2000). According to this theory, the interpretation of a dialogue act leads to assumed mutual beliefs about each other's knowledge of the domain. When we look at the preconditions for performing dialogue acts of the various types, however, we note that these conditions do not contain mutual beliefs but only *finite nestings* of belief. For instance, if agent A asks a question to agent B in order to obtain the information X, then from the assumed rationality of a dialogue agent it follows that A does not believe his question to be superfluous, i.e., A believes that B does not already know that A wants to know X. This condition is schematically of the form (1):

(1) A believes not B knows A wants A knows X

Upon interpreting A's utterance as intended, B will build up the belief that (1) is the case:

(2) B believes A believes not B knows A wants A knows X

If there's no evidence to the contrary, A will assume that B has understood his utterance correctly, so A will build up the belief that (2) is the case:

(3) A believes B believes A believes not B knows A wants A knows X

We see that complex nestings of belief/know and want/goal attitudes may easily arise; hence a formalism for representing context information in a way that is adequate for a dialogue system should be equipped to handle such nestings.

The approach that we propose for context representation is in the tradition of model-theoretic semantics, and uses *partial models*, i.e. models which make certain formulas neither true nor false but undecided, capturing the observation that a dialogue agent's knowledge of the context is incomplete.

Making knowledge representations incremental rather than eliminative is important for efficiency reasons. There is one more way in which representations can be designed to be efficient for modelling dialogue contexts. Context representations change as a result of processing dialogue inputs. Now the processing of a dialogue contribution typically affects only certain parts of the context, and leaves most of the context untouched. For instance, consider again the example of A asking a question to B. When B understands the utterance as the intended question, a belief of the form (2) is added to his context representation, i.e. a belief about A's beliefs about B's beliefs about A's goals. Other beliefs that the question may give rise to, concern the belief that A wants to know X and the belief that A believes that B knows X. These are all beliefs about As beliefs and goals and not, for instance, beliefs about the domain under consideration. Separating different kind of beliefs can help to *modularize* the processes updating context models and evaluating expressions against context models, allowing these processes to access only certain parts of the representation structure. We will design our models in such a way that they are highly structured, consisting of "modules" that are linked in various ways. Since our models have the properties of modularity and partiality, we call them *modular partial models* or MPMs. MPMs have an intuitively obvious implementation in the form of linked (mini-)databases.

We will describe modular partial models for representing the beliefs of two communicating agents. The communicative behaviour of an agent in a given context will depend on what he believes to be true, not on whether these beliefs are actually true. For modelling the understanding and generation of a dialogue agent, we will therefore not be concerned with what is actually true in the communicative situation, but with what each of the participants believes to be the case. Belief is usually distinguished from knowledge in that knowledge is justified true belief, but since an agent always thinks that his beliefs are true and may be assumed to have some justification for his beliefs (at least for the kind of factual beliefs intended here), an agent cannot clearly distinguish between what he *believes* and what he *knows*. We will therefore drop this distinction and use the terms 'believe' and 'knows' interchangeably.

2 Modular Partial Models

We will describe an agent's beliefs in first-order logic extended with a belief operator, and call this language DFOL: *Doxastic First-Order Language*. Modular partial models will be used to interpret DFOL.

An MPM is in essence a structured set of partial interpretation functions (or 'valuations'), somewhat akin to Fagin, Halpern and Vardi's 'knowledge structures' (Fagin, Halpern & Vardi, 1984). One of these functions plays a particular role, as its extension represents the elementary facts known to an agent. Using S to denote the agent in question, $F_s(p)$ is true if S believes that p; $F_s(p)$ is false if S believes that not p; and $F_s(p)$ is undefined if S doesn't know whether p;

In doxastic first-order logic, we can take the following forms of incomplete information into account:

1. Propositional incompleteness: knowing that $p \vee q$ without knowing whether p or whether q; known incompleteness in another agent's knowledge: S knows that U does not know that p.
2. Predicate-logical incompleteness concerns existentially quantified knowledge and partial knowledge about a predicate. For instance, you know that there were three Marx brothers, that Groucho and Harpo were two of them, and you don't know the third but you do know it was *not* Karl Marx.

To deal with incomplete knowledge, MPMs have the following provisions:

1. a) To capture that S knows that p or q, associated with the valuation F_s are two partial functions F_{s_1} and F_{s_2} one assigning true to p, the other assigning true to q. These functions are called '*alternative extensions* of F_s'; they can be thought of as representing alternative ways in which F_s can be extended when more information becomes available.

b) If S believes that it is not the case that U believes that p, then associated with F_s is a function $F_{s\sim u}$ that assigns true to p.
2. a) 'Anonymous referents': a kind of pseudo-objects for representing information about individuals whose existence is considered within the scope of a certain propositional attitude.
   b) To represent 'negative knowledge' about the extension of a predicate, we split every valuation F into two functions: $F = <F^+, F^->$, assigning 'positive' and 'negative' extensions to predicate terms. For instance, $F_u^+(Marx\text{-}brothers) = \{\text{Groucho}, \text{Harpo}\}$; $F_u^-(Marx\text{-}brothers) = \{\text{Karl}\}$.

To formally define MPMs, we first define the set of indices of valuations; their significance is explained below.

Definition 1. The set of indices $\mathcal{I}$ of a modular partial model M_i is the following collection of strings: $i \in \mathcal{I}$; if $j \in \mathcal{I}$ then js, ju, $j \sim s$, $j \sim u$ and $j\text{nex} \in \mathcal{I}$, and if k is a natural number, then $j_k \in \mathcal{I}$.

Definition 2. A modular partial model for DFOL is a five-tuple

$M_\alpha = <D, \mathcal{N}, \mathcal{F}, F_\alpha, \mathcal{A}>$, where:
- D is a domain of individuals;
- $\mathcal{N}$ is an indexed set of finite sets N_j of anonymous referents, with $N_j \cap D = \emptyset$ for every $j \in \mathcal{I}_s$;
- $\mathcal{F}$ is an indexed set of pairs $<F_i^+, F_i^->$ of partial functions assigning values to DFOL terms, satisfying the contraints mentioned below;
- $F_\alpha \in \mathcal{F}$;
- $\mathcal{A}$ is an indexed set of subsets of $\mathcal{F}$, specifying the (non-empty) alternative extensions present in the model.

The index α of the valuation F_α is called the *root index* of the model. The set of indices as defined here is infinite; in a given model, for most indices i it will be the case that $F_i = N_i = A_i = \emptyset$. We will use $\mathcal{I}_M$ to denote the subset of indices for which at least one of these sets is not empty. The various indices have the following significance:

1. F_s assigns to DFOL terms the denotations they have according to agent S.
2. F_{su} does the same according to S's beliefs about U's beliefs and so on.
3. Numerical indices designate alternative extensions.
4. $F_{s\sim u}$ assigns predicate terms denotations which S believes that U does not believe them to have. Similarly for $F_{su\sim s}$, etc.
5. Indices of the form i nex represent negative existential knowledge (see below).

The explicit representation of negative beliefs in MPMs is necessary because of the partial character of beliefs. This creates the danger that 'positive' and 'negative' parts of an MPM contain conflicting information. The set of valuations of an MPM is therefore required to meet certain consistency conditions, such as $F_i^+(P) \cap F_i^-(P) = \emptyset$ for every predicate constant P.

Anonymous referents are introduced by those sets N_j which are not empty. An anonymous referent a is thus introduced at some point in an MPM, namely at the index i where $a \in N_i$, which gives it something like a *scope*, comparable to that of discourse referents in DRT (Kamp and Reyle, 1993): when introduced at index i, it may be used (*"is available"*) at every index of the form $i\gamma$. We use 'Av_i' to denote the set of anonymous referents, available at index i.

As an illustration, consider the MPM M_s depicted in DRT-like form in Fig. 1.

$M_s = < \{\text{ann}, \text{eve}\}, \{N_s, N_{sunex}\}, \{F_s, F_{su}, F_{sunex},\}, F_s, \emptyset >$, where
$\quad N_s = \{\text{a}_1\}, N_{sunex} = \{\text{a}_2, \text{a}_3\},$
$\quad F_s^+ = \{< Q, \{\text{ann}, \text{eve}\} >, < R, \{< \text{a}_1, \text{ann} >\} >\},$
$\quad F_{su}^+ = \{< R, \{< \text{a}_1, \text{ann} >\} >\},$
$\quad F_{sunex}^+ = \{< P, \{\text{a}_2\} >\}.$
$\quad F_{sunex}^- = \{< Q, \{\text{a}_3\} >\}.$

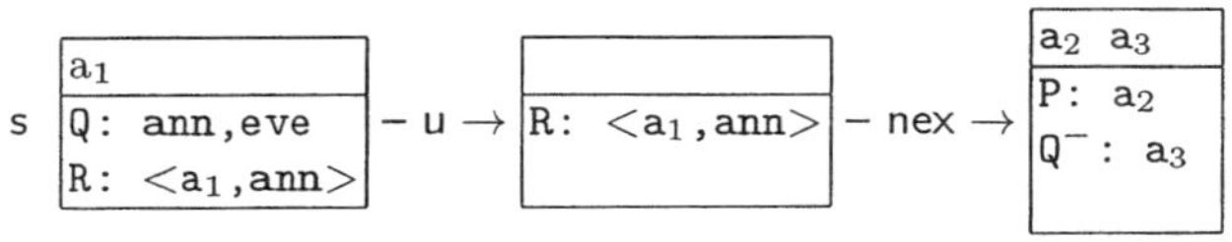

```
                                                            ┌──────────┐
                           ┌──────────────┐                 │ a₂ a₃    │
              ┌──────────┐ │              │                 │ P: a₂    │
          s   │ a₁       │ │              │                 │ Q⁻: a₃   │
              │ Q: ann,eve│ - u → │ R: <a₁,ann> │ - nex →    │          │
              │ R: <a₁,ann>│ │              │                │          │
              └──────────┘ └──────────────┘                 └──────────┘
```

S believes that $Q(eve)$ and that $Q(ann)$;
S believes that there is an a_1 such that $R(a_1, ann)$ and
U believes that $R(a_1, ann)$;
S believes that U believes that there is no a_2 such that $P(a_2)$.
S believes that U believes that there is no a_3 such that not $Q(a_3)$.

Fig. 1. *Example of a simple* MPM.

The definition of MPMs must be supplemented with certain *normalization constraints*, to avoid unintended and undesirable ways of using alternative extensions. Normalization constraints for instance rule out a model where S's belief that p and q is represented not by $F_s^+(p) = F_s^+(q) = \text{true}$, but by $F_s^+(p) = \text{true}$ and F_s having two alternative extensions F_{s1}, F_{s2} such that $F_{s1}^+(p) = \text{false}$ and $F_{s2}^+(q) = \text{true}$. This would amount to modelling that $p \wedge (\neg p \vee q)$, rather than $p \wedge q$. An MPM that satifies all normalizaton constraints is called *normal*.

Truth in a modular partial model
A modular partial model is intended to represent the beliefs of a dialogue agent. We will write $S \Vdash \phi$ to denote that S believes that ϕ. To define the truth conditions of $S \Vdash \phi$ we will use the relations of *verification*, denoted by $\models$, and *falsification*, denoted by $\dashv$. These relations are defined by simultaneous recursion. In the definition, we use the notation $M_i[a/x]$ to designate the submodel that differs from M_i at most in that $F(x) = a$ for all $F \in \mathcal{I}_{M_i}$.

Definition 3. A formula of the form $S \Vdash \phi$ is true in a modular partial model $M = < D, \mathcal{N}, \mathcal{F}, F_s, \mathcal{A} >$ iff $M_s \models \phi$.

The verification and falsification of a DFOL formula by a (normal) MPM M_i are defined as follows.

1 Let P be a k-ary predicate constant and t a sequence $(t_1,..,t_k)$ of individual constants or variables. Let F_i^* be the same as F_i except possibly for some arguments t_j, for which $F_i^*(t_j) \in \mathsf{Av}_i$ ($j = 1,..,k$). We write $F(t)$ to abbreviate $<F(t_1),..,F(t_k)>$.
  a. $M_i \models P(t) \iff F_i^+(t) \in F_i^+(P)$ or $F_i^+(t) \in F_{i\,\mathsf{inex}}^-(P)$ or $i \in \mathcal{A}_j$ for some $j \in \mathcal{I}_s$ and $M_j \models P(t)$
  b. $M_i \dashv P(t) \iff F_i^-(t) \in F_i^-(P)$ or $F_i^+(t) \in F_{i\,\mathsf{inex}}^+(P)$ or $M_j \dashv P(t)$ for some $i \in \mathcal{A}_j, j \in \mathcal{I}_s$.

The remaining clauses apply to any DFOL expressions ϕ, ψ.

2 a. $M_i \models \neg\phi \iff M_i \dashv \phi$
  b. $M_i \dashv \neg\phi \iff M_i \models \phi$

For $A = S$ and $\alpha = s$, or $A = U$ and $\alpha = u$:

3 a. $M_i \models \phi \vee \psi \iff M_i \models \phi$ or $M_i \models \psi$, or for every index $j \in \mathcal{A}_i$, $M_j \models \phi$ or $M_j \models \psi$, or the index i is of the form $\gamma * \alpha$ and $M_\gamma \models (A \Vdash \phi \vee A \Vdash \psi)$
  b. $M_i \dashv \phi \vee \psi \iff M_i \dashv \phi$ and $M_i \dashv \psi$

4 a. $M_i \models A \Vdash \phi \iff M_{i\alpha} \models \phi$ or $M_{i\sim\alpha} \dashv \phi$ or $i \in \mathcal{A}_j$ for some $j \in \mathcal{I}_s$ and $M_j \models A \Vdash \phi$.
  b. $M_i \dashv A \Vdash \phi \iff M_{i\alpha} \dashv \phi$ or $M_{i\sim\alpha} \models \phi$ or $i \in \mathcal{A}_j$ for some $j \in \mathcal{I}_s$ and $M_j \dashv A \Vdash \phi$.

5 a. $M_i \models \exists x: \phi \iff$ there is an $a \in D \cup \mathsf{Av}_i$ such that $M_i[a/x] \models \phi$ or for every index $j \in \mathcal{A}_i$, there is an $a \in D \cup \mathsf{Av}_j$ such that $M_j[a/x] \models \phi$, or the index i is of the form $\gamma * \alpha$ and $M_\gamma \models A \Vdash \exists x: \phi$
  b. $M_i \dashv \exists x: \phi \iff$ there is an $a \in N_{i\,\mathsf{nex}}$ such that $M_i[a/x] \models \phi$ and there is no $a \in D \cup \mathsf{Av}_i$ such that $M_i[a/x] \models \phi$.

3 Modular Partial Models as Context Representations

The definition of truth in a normal MPM allows us to prove that every *honest* set D of DFOL formulas, that is every set of formulas that characterizes a logically possible state of information, has a unique normal MPM which verifies exactly the formulas of D plus their logical consequences.

The notion of 'honesty' of knowledge has been introduced by Halpern & Moses (1984) in relation to knowledge bases, and concerns the logical constraints on the possibility of honestly claiming to *only know that* ψ for some formula ψ.[1] For instance, one cannot honestly claim to *only know whether* p, without knowing that p or that $\neg p$. Given the truth definition of DFOL formulas for normal MPMs,

---

[1] For an analysis of honesty in epistemic logic see Van der Hoek et al. (1996).

it can be proved that, if ψ is an honest DFOL formula, then there exists a unique normal MPM M^ψ in which only ψ is true and any formula logically entailed by ψ (like $\psi \wedge \psi$), while leaving the truth of all other formulas undefined. This easily generalizes to honest *sets* of formulas, i.e. sets of formulas whose conjunction is an honest formula.

To make MPMs useful for modelling an dialogue agent's context, we have to also take a goal attitude into account. The addition of goal attitudes to MPM is technically rather straightforward. We add indices of the form $\alpha!$ to refer to partial knowledge states that α has the goal to achieve, and we add to Definition 3 a clause for the truth conditions of S has the goal that ϕ, designated by $S \triangleleft \phi$:

6 a. $M_i \models A \triangleleft \phi \iff M_{i\alpha!} \models \phi$ or $M_{i \sim \alpha!} \dashv \phi$ or $i \in A_j$
for some $j \in \mathcal{I}_s$ and $M_j \models A \triangleleft \phi$.

b. $M_i \dashv A \triangleleft \phi \iff M_{i\alpha!} \dashv \phi$ or $M_{i \sim \alpha!} \models \phi$ or $i \in A_j$
for some $j \in \mathcal{I}_s$ and $M_j \dashv A \triangleleft \phi$.

where A stands for S or U (dialogue agents) and α for the corresponding index s or u, respectively). Additional consistency conditions make sure that goals are not incompatible with beliefs.

Context construction and update
An MPM representing the current dialogue context, as viewed by an agent who participates in the dialogue, can be constructed effectively by using the notion of *merging* two MPMs, which consists of taking the unions of the respective domains, of the sets of anonymous referents, of the valuations, and of the non-empty alternative extensions, and checking that the resulting structure satisfies the consistency conditions (because the participating MPMs might contain incompatible information).

Given an initial state M_s and a dialogue contribution that brings the information ψ, the agent performs the merge $M_s \otimes M^\psi$. If this merge fails, the failing consistency condition(s) will indicate in what respects ψ is incompatible with the goals and beliefs of the recipient agent. The resolution of such conflicts is not a matter of logic, but of pragmatic strategy taking into account the roles of the participating dialogue agents, each agent's authority on which information, and so on. The logic of MPMs provides the formal foundations on which to base pragmatically appropriate belief revision strategies. More generally, strategies concerning the persistence of information require pragmatic considerations for maintaining dialogue contexts, and are beyond the scope of this paper.

4 Concluding Remarks

In Bunt (1999), we argued that the various types of information that constitute an agent's dialogue context differ in logical complexity, such as in their time dependence and in the necessity to consider recursive nesting within belief attitudes. We argued that each type of context information should preferably be represented in the logically simplest formalism that would be adequate given the intrinsic complexity of the information. This brings up an issue of how to integrate different representation formalisms in context representations. We believe that the formalism of minimal partial models, as outlined here, is promising in this respect. For example, we have argued in Bunt (1999) that information concerning the dialogue history can be represented adequately by means of fairly simple, shallow feature structures. Since the language of feature structures is logically simpler than first-order predicate logic, feature representations are easily absorbed in DFOL and in MPMs. More generally, the ideas of modularity and partiality on which MPMs are based can evidently be applied not only to first-order logic, but to a wide range of languages for describing semantic content and pieces of knowledge.

Acknowledgements. I would like to thank Elias Thijsse for detailed comments on an earlier, more extended version of this paper. Thanks are also die to Jeroen Geertzen and Yann Girard for developing an implementation of MPMs in the context of the MATIS dialogue project.

References

Bunt, H. (1999). Context Representation for Dialogue Management. In P. Bouquet, L. Serafini, P. Brezillon, M. Beneceretti, and F. Castellani (eds.) *Modeling and Using Context.* Lecture Notes in Artificial Intelligence Vol. 1668, pp. 77–90. Berlin: Springer.

Bunt, H. (2000). Dialogue pragmatics and context specification. In Harry Bunt and William Black, editors, *Abduction, Belief and Context in Dialogue. Studies in Computational Pragmatics.* Amsterdam: Benjamins, 81 – 150.

Fagin, R., J. Halpern, and M. Vardi (1984) A model-theoretic analysis of knowledge: preliminary report. In *Proc. 25th IEEE Symposium Foundations of Computer Science.*

Halpern, J. and Y. Moses (1986). Towards a theory of knowledge and ignorance: a preliminary report. In *Proc. AAAI Workshop on Non-Monotonic Reasoning,* New Paltz, NY, pp. 125–143.

Hoek, W. v. d., J. Jaspars, and E. Thijsse (1996). Honesty in partial logic. *Studia Logica* 56(3), 326–360.

Kamp, H. and U. Reyle (1993). *From discourse to logic: introduction to model theoretic semantics of natural language.* Dordrecht: Kluwer Academic Press.

Contextual Modeling Using Context-Dependent Feedforward Neural Nets

Piotr Ciskowski

Institute of Engineering Cybernetics,
Wrocław University of Technology,
Wybrzeże Wyspiańskiego 27, 50 370 Wrocław, Poland
cis@polandmail.net *

Abstract. The paper addresses the problem of using contextual information by neural nets solving problems of contextual nature. The models of a context-dependent neuron and a multi-layer net are recalled and supplemented by the analysis of context-dependent and hybrid nets' architecture. The context-dependent nets' properties are discussed and compared with the properties of traditional nets considering the Vapnik-Chervonenkis dimension, contextual classification and solving tasks of contextual nature. The possibilities of applications to classification and control problems are also outlined.

1 Introduction

The phenomena and objects in real world can rarely be considered in isolation. The parameters describing them often change according to the environmental conditions, called the context. A learning machine, such as a neural net, dealing with such systems should take the information about the context into consideration in order to improve its performance when using only the primary features of the analyzed object. The idea of changing the net's functioning in reaction to changes of the environment arose in many applications, frequently under different names, e.g. context-sensitive [7] or even adaptive [6]. The possibility of improving systems' performance by identifying and managing the existence of context-sensitive features among the input data has recently been noticed by the researchers. The context-dependent neural net's model, presented in the paper (introduced in [4] and developed in [3]), being the generalization of the traditional neural net's model, may be very useful for applications of neural nets in problems showing contextual dependencies among data.

The idea of context-dependent neural nets, the way of using contextual information by them and the model of a context-dependent perceptron has already been presented in [2]. Here we only recall that the **context** is defined as all the factors that influence (improve) the decision making, or other kind of information processing, performed by the learning algorithm (here a neural net), but

---
* The work is supported by KBN grant in the years 2002-2005

other from the data on which the decision is taken or which are directly processed. The contextual data may be determined by the environmental conditions in which the reasoning takes place, or in which the primary data occurred.

The difference between the context-dependent neuron's model and the traditional one is the division of inputs into two groups: primary inputs (supplied with primary features of the problem, useful for solving the problem regardless of the knowledge about other features) together with context-sensitive inputs (supplied with context-sensitive features of the problem, useful only when considered together with other contextual features), and contextual inputs (also supplied with context-sensitive features - those describing the context, environment of the problem begin solved). The first two types of inputs are grouped in the primary inputs vector $\overline{\mathbf{X}}$, while the last one in the context vector $\overline{\mathbf{Z}}$. The neuron's weights to the primary inputs, as well as the offset, depend on the vector of contextual variables: $w_s = w_s\left(\overline{\mathbf{Z}}\right)$, $s = 0, 1, \ldots, S$. Thus we obtain a model of the context-dependent neuron in the form

$$y = \Phi\left[w_0\left(\overline{\mathbf{Z}}\right) + \sum_{s=1}^{S} w_s\left(\overline{\mathbf{Z}}\right) x_s\right] \quad (1)$$

where x_s denotes the s-th primary input, y is the neuron's output, $\overline{\mathbf{Z}}$ denotes the vector of contextual variables, while Φ is the activation function. The model is presented in fig. 1. The context-dependent weights are given by: $w_s\left(\overline{Z}\right) = \overline{A}_s^T \overline{V}\left(\overline{Z}\right) = [a_{s1}, a_{s2}, \ldots, a_{sM}]\left[v_1\left(\overline{Z}\right), v_2\left(\overline{Z}\right), \ldots, v_M\left(\overline{Z}\right)\right]^T$, where the coefficients $a_{s,m}$ are now the neuron's parameters, spanning the dependence of weights on the vector of contextual variables (for more details on the model, as well as training algorithms, please refer to [2,3]).

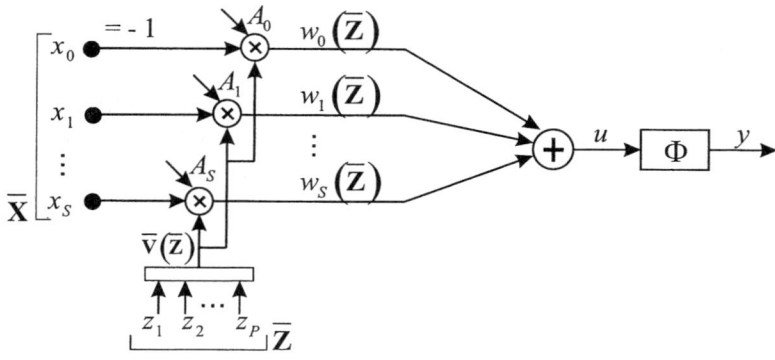

Fig. 1. Model of the context-dependent neuron

2 Context-Dependent Nets' Architectures

As in traditional nets, single context-dependent neurons presented in previous section may be used for building larger structures, appropriate for solving more complicated problems. Various nets' architectures may be developed by using traditional or context-dependent weights in different parts of the net. This choice is done according to the nature of the problem to be solved - the analysis of the features describing it and the way of applying them to the net.

A neural net may be used to solve a problem characterized both by primary and context-sensitive features. We assume that the features are already recognized and divided into these groups and supplied on the appropriate inputs, mentioned above. The primary features are useful for solving the task regardless of their context, that is should be treated by the net the same way in all the contexts. The context-sensitive features require adapting the way of processing them to the information about the context. Therefore one may think of a hybrid net, in which some weights (for connections to context-sensitive inputs) are context-dependent and others (for connections to primary inputs) are traditional - constant after training, that is the same in all the contexts. Different structures of hybrid nets are presented in [3]. Fig. 2 presents a general diagram of a hybrid traditional/context-dependent net. The net's first layer is divided into two parts: traditional net (using constant weights) operating on strictly primary inputs, and context-dependent net (using context-dependent weights) operating on context-sensitive inputs. The next layers of the net are presumed to be either traditional or context-dependent or hybrid. Using the same types of nets in all these parts results in two simplified models: only traditional and only context-dependent net.

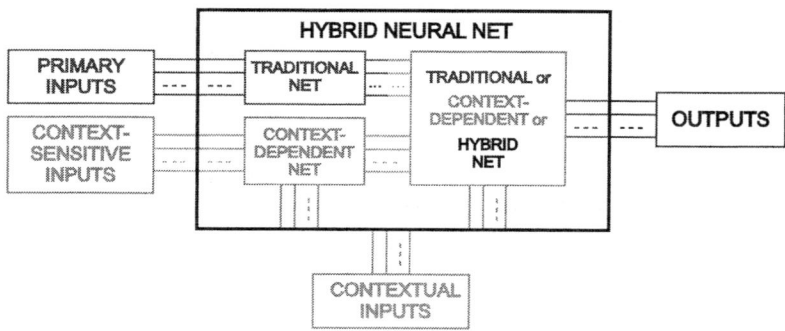

Fig. 2. General diagram of hybrid traditional/context-dependent net

For simplicity, in purely context-dependent nets the primary and context-sensitive inputs are grouped in one input vector $\overline{X}$ called the vector of primary features. All the net's weights are context-dependent. Apart from simplifying the net's architecture, this assumption allows using the effective training algorithms for context-dependent nets, introduced in [3].

If some additional knowledge of the problem being solved suggests using traditional weights for some features, the net's designer may simply set some of the basis functions or coefficients for these weights to zero. These weights will act as traditional in the hybrid net. We may also hope that the weights which do not need to be context-dependent should automatically converge to fixed values during training. Therefore it is also advisable to analyze the weights' change throughout the contexts when the learning process is completed in order to recognize the weights which should be constant. These two procedures allow combining the hybrid net with the pure context-dependent net model. Algorithms for contextual pruning of the net's structure are presented in [3].

Although not considered in this paper, a combination of the context-dependent net's model with the recurrent model of Ellman's net - that is an architecture in which the input layer recognizes the context and supplies it on the contextual inputs, may also be worth further investigation.

3 The Vapnik-Chervonenkis Dimension of CD Nets

The Vapnik-Chervonenkis dimension is one of the quantities used in machine learning theory for estimating their generalization abilities and the effectiveness of learning. The idea of the growth function and VC-dimension is well presented in [1]. A parallel theory of the VC-dimension for context-dependent neural nets is developed in [3]. Generally speaking, we may define the VC-dimension as the maximal number of points in the input space the learning machine is able to dichotomize in all possible ways.

The VC-dimension of a traditional perceptron using S primary inputs is equal to the number of its adjustable parameters, that is $S+1$. The appropriate theorem is given in [1].

The traditional perceptron may be supplied with only primary features of the data. Then a neuron with S inputs is able to dichotomize $S+1$ points in its S-dimensional input space. If we also supply the neuron with information about the context of these points (encoded on P inputs and added to the S primary inputs already being used), then the neuron is able to dichotomize $S+P+1$ points in the joint input space. In both cases the neuron's parameter space is of the same dimensionality as its input space.

The equivalent of the above mentioned theorem for the context-dependent perceptron, derived in [3], presented in [2], says that for a context-dependent perceptron with $S \in N$ real primary inputs, using the base function vector consisting of M independent base functions, the Vapnik-Chervonenkis dimension equals $M(S+1)$.

The separating abilities of traditional and context-dependent perceptrons are strictly related to their parameter space. The dimensionality of the traditional neuron's parameter space is equal to the number of its weights W. Analogically, for the context-dependent neuron it is equal to the number of its coefficients

$A = MW$. The traditional neuron transforms both primary and contextual data the same way, therefore a traditional neuron using S primary and P contextual inputs acts the same as a neuron using $S + P$ traditional (primary) inputs - it is able to produce an discriminating hyperplane in its W-dimensional, that is $(S + P + 1)$-dimensional, parameter space. Let us recall that the parameter space of the traditional neuron is of the same dimensionality as its input space (including the bias).

The context-dependent perceptron produces a discriminating hyperplane in its A-dimensional, that is $M(S + 1)$-dimensional *parameter* space. However, this hyperplane is in fact a hypersurface in the $S + P + 1$-dimensional *input* space. This leads to much more adjustable decision boundaries that context-dependent nets are able to produce in the joint input space. Therefore their discriminating abilities are more powerful in classification tasks which are complicated in the context domain.

4 Discriminating Hypersurfaces of CD Nets

Neural nets may be considered as tools for statistical modeling and prediction. Their aim is to produce statistical models of the processes generating the training data. These models allow making predictions on new data. They may involve decision boundaries in classification tasks. The input data may be represented as vectors (points) in the input space. It is known that the simple traditional perceptron is able to model a linear border between two classes. Such a decision boundary is suitable for simple (linearly separable) classification tasks, in which the class conditional densities of the input data belong to the exponential family of functions, e.g. the most widely assumed Gaussian density.

The model of context-dependent perceptron modifies the linear part of the neuron, responsible for producing the decision boundary. It may be now much more complicated than in the linear model. The densities, for which the decision borders may be modeled directly by a single neuron, should also belong to the exponential family, but the parameters of the densities may vary from one context to another. Our approach may be considered as a form of mixture distribution, where in general the conditional density for each class is composed of a set of M simpler densities, each one for a different context.

Thus the context-dependent neuron is still able to model only linear decision boundary in one context (in the input space consisting only of primary inputs and input data from the same context). However, the parameters of this linear boundary may change from one context to another - some nonlinearity of the generator of the data is already modeled by the single neuron.

Although a single traditional neuron is only able to model a linear decision boundary between two classes of input data, nonlinear decision borders may also be modeled by traditional neural nets. This may done by using multi-layer nets - the nonlinear decision boundary are composed of linear boundaries produced by the hidden layer's neurons. The same strategy is used in context-dependent nets for modeling a non-linear boundary in one context.

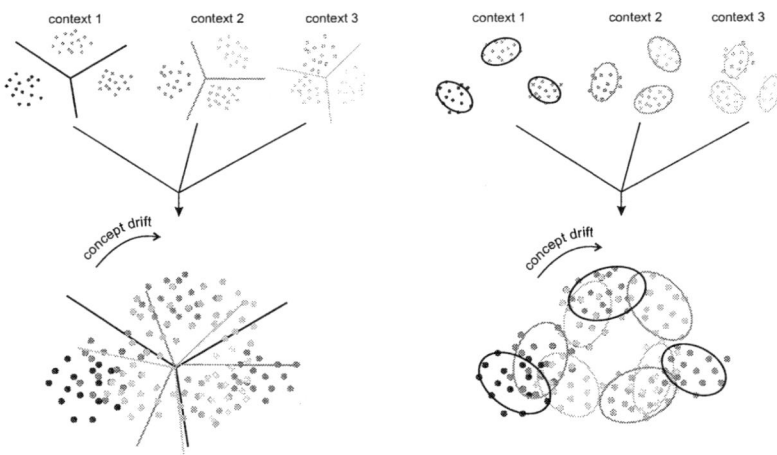

Fig. 3. Example of a concept drift in the decision boundaries produced by feedforward (left) and radial basis function nets (right)

However, the nonlinear decision boundaries may shift through the contexts not changing their complexity much - when the nature of the data generator is the same in all the contexts, changing only its parameters (see fig. 3). Then both tasks of modeling the decision boundaries and their change through the contexts need to be modeled by the same hidden neuron in the traditional net. This leads to adding more neurons to the hidden layer, or even adding more hidden layers.

As mentioned before, in context-dependent nets the task of modeling the nonlinear boundaries between the class conditional density distributions is separated from the task of modeling their parameters' drift across the contexts. The former is performed by the combination of decision boundaries produced by the hidden layer's neurons (as in traditional nets), while the latter is modeled by the choice of basis functions and estimation of the coefficients for the weights. The separation of these problems allows using simpler nets (using less hidden neurons), clarifying its structure and making it more adjustable separately in the traditional and contextual domain. Finally it simplifies the learning process of context-dependent nets and creates possibilities for new training algorithms.

5 Modeling a Magnetorheological Damper

In this section an example of using a context-dependent net for modeling the magnetoreological damper is presented - a device used for response reduction in structures subjected to strong earthquakes. The model of the MR damper was developed in [5]. The damper's response - force F, to the stimulating displacement x is given by

$$F = c_1 \dot{y} + k_1 (x - x_0) \qquad (2)$$

$$\text{where } \dot{y} = \frac{1}{c_0 + c_1} \left[\alpha z + c_0 \dot{x} + k_0 (x - y) \right] \quad (3)$$

$$\text{and } \dot{z} = -\gamma |\dot{x} - \dot{y}| z |z|^{n-1} - \beta (\dot{x} - \dot{y}) |z|^n + A (\dot{x} - \dot{y}) \quad (4)$$

The parameters $\alpha(u), c_1(u)$ and $c_0(u)$ are dependent on the voltage ν applied to the damper through the variable u:

$$\dot{u} = -\eta (u - \nu) \quad (5)$$

The parameters $k_0, k_1, x_0, \gamma, \beta, A, n, \eta$ are constant values determined experimentally (here the values given in [5] are used).

One may notice that the nonlinear response of the damper to the stimulating displacement signal is also non linearly dependent on the voltage supplied to the device. A traditional net treats both signals x and ν the same way - as the input signals to the modeling device, and learns the mapping of these two inputs to the output force signal F. A context-dependent net learns directly the mapping of the displacement x to the damper's response F, and separately the dependence of this mapping on the voltage ν.

The damper has been modeled in SIMULINK and the data obtained from simulations was then used as the training set for traditional and context-dependent neural nets. Different structures of nets have been tested. In order to simplify the learning of the input-output mapping, the primary input signal for the nets was not the history of displacement, but its velocity $\dot{x}$. The history of the voltage ν applied to the damper was supplied on the primary inputs of traditional nets, and on the contextual inputs in context-dependent nets. The nets were trained to learn the force F of the damper's response.

The performance - predicted response to sinusoidal displacement and random voltage - of a traditional net with 20 hidden neurons and of the context-dependent 10-1 net using 6 basis functions is presented in fig. 4. The nets have comparable number of adjustable parameters. The mean square error is 0.0648 and 0.0219, respectively.

Because of the necessity of modellng both the response to the displacement and its dependence on the applied voltage the traditional net is not able to learn the mapping for different voltages as smoothly as the context-dependent one, especially during the voltage change. The structure of the context-dependent net is more appropriate for the model of the device.

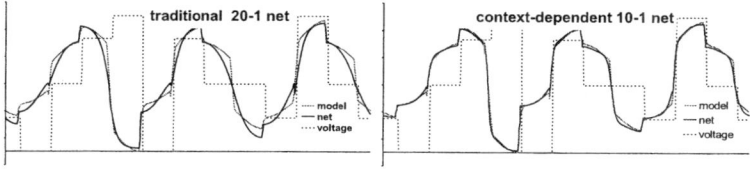

Fig. 4. Performance of traditional and context-dependent nets on damper data

6 Conclusions

The models of context-dependent neural nets have been considered in the paper. The idea of context-dependence seems to have a great potential. The basic learning algorithms for one layer and multilayer nets are developed and presented in [3], as well as efficient algorithms taking advantage of the contextual dependencies in the training data and the new phenomena occurring during context-dependent nets' training - strategies of using contextual training data, contextual overfitting and contextual pruning of the net's parameters. Future work may be concentrated on developing the context-dependent learning algorithms for other types of feedforward networks such as radial basis function (RBF) nets and Kohonen self organizing maps. The global minimization techniques are worth considering. Implementing context-dependence to recurrent networks seems to be very demanding but also interesting task. The examples presented in the paper illustrate the behavior of context dependent nets in classification tasks and their potential in modeling devices or phenomena sensitive to some context. Using context-dependent nets for real life problems, where the possibility of using contextual features in the data is clear and natural, is a promising field for further research.

References

1. M. Anthony and P. Bartlett. *Neural Network Learning: Theoretical Foundations.* Cambridge University Press.
2. P. Ciskowski. Vapnik-chervonenkis dimension of a context-dependent perceptron. In *Proc. Third International and Interdisciplinary Conference on Modeling and Using Context, CONTEXT 2001*, Dundee, UK, 2001.
3. P. Ciskowski. *Learning of context-dependent neural nets.* PhD thesis, Wroclaw University of Technology, 2002.
4. E. Rafajlowicz. Context-dependent neural nets – problem statement and examples. In *Proc. Third Conference Neural Networks and Their Applications*, Zakopane, Poland, May 1999.
5. B. Spencer Jr., S. Dyke, S. M., and J. Carlson. Phenomenological model for magnetorheological dampers. *Computer of Engineering mechanics, ASCE*, 123(3):230–238, 1997.
6. R. Watrous and G. Towell. A patient-adaptive neural network ECG patient monitoring algorithm. In *Computer in Cardiology*, Vienna, Austria, September 10–13 1995.
7. D. Yeung and G. Bekey. Using a context-sensitive learning for robot arm control. In *Proc. IEEE International Conference on Robotics and Automation*, pages 1441–1447, Scottsdale, Arizona, May 14–19 1989.

Context-Based Commonsense Reasoning in the DALI Logic Programming Language*

Stefania Costantini and Arianna Tocchio

Università degli Studi di L'Aquila
Dipartimento di Informatica
Via Vetoio, Loc. Coppito, I-67010 L'Aquila - Italy
{stefcost, tocchio}@di.univaq.it

Abstract. In this paper we will discuss the context management features of the new logic programming language DALI, aimed at defining agents and multi-agent systems. In particular, a DALI agent, which is capable of reactive and proactive behaviour, builds step-by-step her context. Context update is modelled by the novel concept of "evolutionary semantics", where each context manipulation is interpreted as a program transformation step. We show that this kind of context-based agent language is well-suited for representing many significant commonsense reasoning examples.

1 Introduction

The new logic programming language DALI [Co99], [CT02], [CGT02] has been designed for modelling Agents and Multi-Agent systems in computational logic. Syntactically, DALI is close to the Horn clause language and to Prolog. DALI programs however may contain a new kind of rules, reactive rules, aimed at interacting with an external environment. The environment is perceived in the form of external events, that can be exogenous events, observations, or messages from other agents. In response, a DALI agent can either perform actions or send messages. This is pretty usual in agent formalisms aimed at modelling reactive agents (see among the main approaches [KS96], [DST99], [Fi94] [Ra91], [Ra96]), [SPDEK00].

What is new in DALI is that the same external event can be considered under different points of view: the event is first perceived, and the agent may reason about this perception, then a reaction can take place, and finally the event and the (possible) actions that have been performed are recorded as past events and past actions. Another important novel feature is that internal conclusions can be seen as events: this means, a DALI agent can "think" about some topic, the conclusions she takes can determine a behaviour, and, finally, she is able to remember the conclusion, and what she did in reaction. Whatever

---

* Research partially funded by MIUR 40% project *Aggregate- and number-reasoning for computing: from decision algorithms to constraint programming with multisets, sets, and maps* and by the *Information Society Technologies programme of the European Commission, Future and Emerging Technologies* under the IST-2001-37004 WASP project. Many thanks to Stefano Gentile, who has joined the DALI project, has cooperated to the implementation of DALI, has designed the language web site, and has helped and supported the authors in many ways.

the agent remembers is kept or "forgotten" according to suitable conditions (that can be set by directives). Then, a DALI agent is not a purely reactive agent based on condition-action rules: rather, it is a reactive, proactive and rational agent that performs inference within an evolving context.

The *evolutionary semantics* of the language consists of a sequence of logic programs, resulting from subsequent transformations, together with the sequence of the Least Herbrand Models of these programs. This makes it possible to model an evolving agent incorporating an evolving context. In this way, it is possible to reason about the conclusions reached and the actions performed by the agent at a certain stage, or, better, in a certain context.

In this paper we want to demonstrate that the features of the DALI language allow many forms of commonsense reasoning to be gracefully represented.

A prototype implementation of the DALI language has been developed by the authors of this paper at the University of L'Aquila. The implementation, together with a set of examples, is available at the URL [CGT02].

2 Context-Dependent Reasoning in Everyday Situations

A DALI program is syntactically very close to a traditional Horn-clause program. In particular, a Prolog program is a special case of a DALI program. Specific syntactic features have been introduced to deal with the agent-oriented capabilities of the language, and in particular to deal with events.

Let us consider an event incoming into the agent from its "external world", like for instance $bell\_ringsE$ (postfix E standing for "external"). From the agent's perspective, this event can be seen in different ways.

Initially, the agent has perceived the event, but she still have not reacted to it. The event is now seen as a present event $bell\_ringsN$ (postfix N standing for "now"). She can at this point reason about the event: for instance, she concludes that a visitor has arrived, and from this she realizes to be happy.

visitor_arrived :- bell_ringsN.

happy :- visitor_arrived.

As she is happy, she feels like singing a song, which is an action (postfix A). This is obtained by means of the mechanism of internal events: this is a novel feature of the DALI language, that to the best of the authors' knowledge cannot be found in any other language. Conclusion $happy$, reinterpreted as an event (postfix I standing for "internal"), determines a reaction, specified by the following *reactive rule*, where new connective :> stands for *determines*:

happyI :> sing_a_songA.

In more detail, the mechanism is the following: goal $happy$ has been indicated to the interpreter as an internal event by means of a suitable directive. Then, from time to time the agent wonders whether she is happy, by trying the goal (the frequency can also be set in the directive). If the goal $happy$ succeeds, it is interpreted as an event, thus triggering the corresponding reaction. I.e., internal events are events that do not come from the environment. Rather, they are goals defined in some other part of the program.

For coping with unexpected unpleasant situations that might unfortunately happen to ruin a good day, one can add a directive of the form:

keep happyI unless ⟨ *terminating_condition* ⟩.

stating in which situations *happy* should not become an internal event. ⟨*terminating_condition*⟩ is any predicate, that must be explicitly defined in the program, and is attempted upon success of *happy*. This formulation is elaboration-tolerant, since it separates the general definition of happiness, from what (depending on the evolution of the context) might "prevent" happiness.

Finally, the actual reaction to the external event *bell_ringsE* can be that of opening the door:

bell_ringsE :> *open_the_doorA*.

After reaction, the agent is able to remember the event, thus enriching her reasoning context. An event (either external or internal) that has happened in the past will be called *past event,* and written *bell_ringsP*, *happyP*, postfix P standing for "past". External events and actions are used also for sending and receiving messages. Then, an event atom can be more precisely seen as a triple $Sender : Event\_Atom : Timestamp$. The $Sender$ and $Timestamp$ fields can be omitted whenever not needed.

The DALI interpreter is able to answer queries like the standard Prolog interpreter, but it is able to handle a disjunction of goals. In fact, from time to time it will add external and internal event as new disjuncts to the current goal, picking them from queues where they occur in the order they have been generated. An event is removed from the queue as soon as the corresponding reactive rule is applied.

3   Coordinating Actions Based on Context

A DALI agent builds her own context, where she keeps track of the events that have happened in the past, and of the actions that she has performed. As soon as an event (either internal or external) is reacted to, and whenever an action subgoal succeeds (and then the action is performed), the corresponding atom is recorded in the agent database. By means of directives, it is also possible to indicate other kinds of conclusions that should be remembered. Past events and past conclusions are indicated by the postfix P, and past actions by the postfix PA. The following rule for instance says that Susan is arriving, since we know her to have left home.

is_arriving(susan) :- *left_homeP(susan)*.

The following example illustrates how to exploit past actions. In particular, the action of opening (resp. closing) a door can be performed only if the door is closed (resp. open). The window is closed if the agent remembers to have closed it previously. The window is open if the agent remembers to have opened it previously.

open_the_doorA :- *door_is_closed*.

door_is_closed :- *close_the_doorPA*.

close_the_doorA :- *door_is_open*.

door_is_open :- *open_the_doorPA*.

It is possible to have a conjunction of events in the head of a reactive rule, like in the following example.

rainE, windE :> *close_windowA*.

In order to trigger the reactive rule, all the events in the head must happen within a certain amount of time. The length of the interval can be set by a directive, and is checked on the time stamps.

It is important to notice that an agent cannot keep track of *every* event and action for an unlimited period of time, and that, often, subsequent events/actions can make former ones no more valid. In the previous example, the agent will remember to have opened the door. However, as soon as she closes the door this record becomes no longer valid and should be removed: the agent in this case is interested to remember only the last action of a sequence. In the implementation, past events and actions are kept for a certain (customizable) amount of time, that can be modified by the user through a suitable directive. Also, the user can express the conditions exemplified below:

keep open_the_doorPA until close_the_doorA.

As soon as the *unti* condition (that can also be *forever*) is fulfilled, i.e., the corresponding subgoal has been proved, the past event/action is removed. In the implementation, events are time-stamped, and the order in which they are "consumed" corresponds to the arrival order. The time-stamp can be useful for introducing into the language some (limited) possibility of reasoning about time. Past events, past conclusions and past actions, which constitute the "memory" of the agent, are an important part of the (evolving) context of an agent. The other components are the queue of the present events, and the queue of the internal events. Memories make the agent aware of what has happened, and allow her to make predictions about the future.

The following example illustrates the use of actions with preconditions. The agent emits an order for a product P of which she needs a supply. The order can be done either by phone or by fax, in the latter case if a fax machine is available.

need_supplyE(P) :> *emit_oder(P)*.

emit_order(P) :- *phone_orderA*.

emit_order(P) :- *fax_orderA*.

fax_orderA :- *fax_machine_available*.

If we want to express that the order can be done either by phone or by fax, but not both, we do that by exploiting past actions, and say that an action cannot take place if the other one has already been performed. Here, *not* is understood as default negation.

need_supplyE(P) :> *emit_order(P)*.

emit_order(P) :- *phone_orderA, not fax_orderPA*.

emit_order(P) :- *fax_orderA, not phone_orderPA*.

4 Evolutionary Semantics

The declarative semantics of DALI is aimed at describing how an agent is affected by actual arrival of events, without explicitly introducing a concept of state which is incompatible with a purely logic programming language. Rather, we prefer the concept of context, where modifications to the context are modelled as program transformation steps. For a full definition of the semantics the reader may refer to [CT02]. We summarize the approach here, in order to make the reader understand how the examples actually work.

We define the semantics of a given DALI program P starting from the declarative semantics of a modified program P_s, obtained from P by means of syntactic transformations that specify how the different classes of events are coped with. For the declarative semantics of P_s we take the Well-founded Model, that coincides with the the Least Herbrand Model if there is no negation in the program (see [PP90] for a discussion). In the following, for short we will just say "Model". It is important to notice that P_s is aimed at modelling the declarative semantics, which is computed by some kind of immediate-consequence operator, and not represent the procedural behaviour of the interpreter.

For coping with external events, we have to specify that a reactive rule is allowed to be applied only if the corresponding event has happened. We assume that, as soon as an event has happened, it is recorded as a unit clause (this assumption will be formally assessed later). Then, we reach our aim by adding, for each event atom $p(Args)E$, the event atom itself in the body of its own reactive rule. The meaning is that this rule can be applied by the immediate-consequence operator only if $p(Args)E$ is available as a fact. Precisely, we transform each reactive rule for external events:

$p(Args)E \; :> \; R_1, \ldots, R_q.$

into the standard rule:

$p(Args)E \; :\text{-} \; p(Args)E, R_1, \ldots, R_q.$

Similarly, we have to specify that the reactive rule corresponding to an internal event $q(Args)I$ is allowed to be applied only if the subgoal $q(Args)$ has been proved.

Now, we have to declaratively model actions, without or with an action rule. Procedurally, an action A is performed by the agent as soon as A is executed as a subgoal in a rule of the form

$B \; :\text{-} \; D_1, \ldots, D_h, A_1, \ldots, A_k. \quad h \geq 1, k \geq 1$

where the A_i's are actions and $A \in \{A_1, \ldots, A_k\}$. Declaratively, whenever the conditions $D_1, \ldots, D_h$ of the above rule are true, the action atoms should become true as well (given their preconditions, if any). Thus, the rule can be applied by the immediate-consequence operator. To this aim, for every action atom A, with action rule

$A \; :\text{-} \; C_1, \ldots, C_s. \quad s \geq 1$

we modify this rule into:

$A \; :\text{-} \; D_1, \ldots, D_h, C_1, \ldots, C_s.$

If A has no defining clause, we add clause:

$A \; :\text{-} \; D_1, \ldots, D_h.$

In order to obtain the *evolutionary* declarative semantics of P, as a first step we explicitly associate to P_s the list of the events that we assume to have arrived up to a certain point, in the order in which they are supposed to have been received. We let $P_0 = \langle P_s, [] \rangle$ to indicate that initially no event has happened.

Later on, we have $P_n = \langle Prog_n, Event\_list_n \rangle$, where $Event\_list_n$ is the list of the n events that have happened, and $Prog_n$ is the current program, that has been obtained from P_s step by step by means of a *transition function* Σ. In particular, Σ specifies that, at the n-th step, the current external event E_n (the first one in the event list) is added to the program as a fact. E_n is also added as a present event. Instead, the previous event E_{n-1} is removed as an external and present event, and is added as a past event.

Then, given P_s and list $L = [E_n, \ldots, E_1]$ of events, each event E_i *determines* the transition from P_{i-1} to P_i according to Σ. The list $\mathcal{P}(P_s, L) = [P_0, \ldots, P_n]$ is called the *program evolution* of P_s with respect to L.

Notice that $P_i = \langle Prog_i, [E_i, \ldots, E_1] \rangle$, where $Prog_i$ is the program as it has been transformed after the ith application of Σ. Then, the sequence $\mathcal{M}(P_s, L) = [M_0, \ldots, M_n]$ where M_i is the model of $Prog_i$ is the *model evolution* of P_s with respect to L, and M_i the *instant model at step i*.

Finally, the *evolutionary semantics* $\mathcal{E}_{P_s}$ of P_s with respect to L is the couple $\langle \mathcal{P}(P_s, L), \mathcal{M}(P_s, L) \rangle$.

The DALI interpreter at each stage basically performs standard SLD-Resolution on $Prog_i$, while however it manages a disjunction of goals, each of them being a query, or the processing of an event.

5 A Complete Example: Barman and Customer

Below we show the DALI code for two agents: *Barman*, who is the shopkeeper of a cafeteria, and *Gino*, who is a customer coming in to drink a beer.

The barman waits for events of the form $C : requestE(P)$ where C is the name of the customer agent, and P is the product he would like to get. For instance, we may have $C = Gino$ and $P = beer$. The barman examines the request, and if Pr is available at a cost A, he asks the customer for payment (in this cafeteria you pay in advance!). Otherwise, he tells the customer that there Pr is not available. The action $messageA(C, M)$ consists in sending message M to agent C.

      Barman

    C:requestE(Pr) :> *examine_request(C,Pr).*

 examine_request(C,Pr) :- *not finished(Pr), cost(Pr,A),*

          messageA(C,ok(Pr)), messageA(C,please_pay(Pr,A)).

 examine_request(C,Pr) :- *finished(Pr), messageA(C,no(Pr)).*

The barman concludes that Pr is finished if the quantity left in store is zero. This conclusion is an internal event, and thus (via the next rule) triggers a reaction, that consists in ordering a supply of the product, but *only if the order has not been issued already*: in fact, in the body of the rule there is a check that there is not in the memory of the agent past action $order\_productPA(Pr, Q1)$.

If the payment arrives (event $paidE(C, Pr, A1)$), then the barman makes some checks. First, if he *remembers* that Pr is finished (in fact, $finishedP$ is a past event), he tells again the customer that Pr is finished, and that he should take the money back. Otherwise, if the customer has paid an amount $A1$ which is different from the cost A, he will be required again to pay. Finally, if everything is ok, Pr will actually be served to the customer. Then, $serve(C, Pr)$ is interpreted as an internal event, and will cause the available quantity of Pr to be updated.

Barman (continued)
  finished(Pr) :- *quantity(Pr,0)*.
  finishedI(Pr) :> *not order_productPA(Pr,Q1), order_productA(Pr,Q)*.
  C:paymentE(Pr,A) :> *check_payment(C,Pr,A)*.
  check_payment(C,Pr,A1) :- *finishedP(Pr), messageA(C,no(Pr)), messageA(refund(A1))*.
  check_payment(C,Pr,A) :- *cost(Pr,A), A =/= A1, messageA(C,please_pay(Pr,A))*.
  check_payment(C,Pr,A) :- *serveA(C,Pr)*.
  serveI(C,Pr) :> *update_quantity(Pr)*.

We have now to explain one more reason why it is useful to use internal events also form a procedural point of view. In fact, one may wonder why not write a rule such as:

  check_payment(C,Pr,A) :- *serveA(C,Pr), update_quantity(Pr)*.

Consider however that the Barman might receive several concurrent requests by several customers. Therefore, these requests are to be "contextualized", i.e., they have to be considered in a sequence, keeping in mind the information about the available quantity of each product. Procedurally, a purely reactive rule would produce concurrent attempts to update the same quantity. The use of internal events prevents any problem of "dirty update": in fact, the internal events to be reacted to are put in a FIFO queue. Then, the different updates to the quantity of Pr are performed one at a time, and cannot interfere with each other.

Moreover, the mechanism of internal events is more elaboration-tolerant since it separates the phase where the agent becomes aware of something, and the phase where the agent decides what to do in consequence. Rules for updating the quantity are straightforward, and those for making the order have been reported in a previous example.

The code for the customer agent might look for instance like the following. Agent Gino is thirsty whenever he has played tennis. Then, as a reaction ($thirsty$ is an internal event) he asks the barman for a beer. If he is told by the Barman that the beer is finished, as a reaction he asks for a coke. He pays when requested by the external event $please\_payE(beer, amount)$ coming from Barman. The rule for payment is general, and can be used for either beer or coke. Notice that Gino is disappointed whenever what he asked for is not available. This conclusion is drawn from the present event $finishedN(Pr)$ coming from Barman.

  Gino
  thirsty :- *play_tennisPA*.
  thirstyI :> *messageA(Barman,request(beer))*.
  Barman:please_payE(Pr,A) :> *messageA(Barman,paymentE(Pr,A))*.
  Barman:finishedE(beer) :> *messageA(Barman,request(coke))*.
  disappointed :- *Barman:finishedN(Pr)*.

6 Concluding Remarks

We have presented some examples of context-based commonsense reasoning in the formalism of DALI logical agents. Their ability to behave in a "sensible" way comes from the fact that DALI agents are not just reactive, but have several classes of events, that are coped with and recorded in suitable ways, so as to form a context in which the agent performs her reasoning. A simple form of knowledge update and "belief revision" is provided by the conditional storing of past events and actions. In the future, more sophisticated belief revision strategies as well as full planning capabilities and a real agent communication language will be integrated into the formalism.

References

[Co99] S. Costantini. Towards active logic programming. In A. Brogi and P. Hill, (eds.), *Proc. of 2nd International Works. on Component-based Software Development in Computational Logic (COCL'99)*, PLI'99, Paris, France, September 1999. http://www.di.unipi.it/ brogi/ ResearchActivity/COCL99/ proceedings/index.html.

[CGT02] S. Costantini, S. Gentile, A. Tocchio. DALI home page: http://gentile.dm.univaq.it/dali/dali.htm.

[CT02] S. Costantini, A. Tocchio. A Logic Programming Language for Multi-agent Systems. In S. Flesca, S. Greco, N. Leone, G. Ianni (eds.), *Logics in Artificial Intelligence, Proc. of the 8th Europ. Conf., JELIA 2002*, Cosenza, Italy, September 2002, LNAI 2424, Springer-Verlag, Berlin, 2002

[DST99] P. Dell'Acqua, F. Sadri, and F. Toni. Communicating agents. In *Proc. International Works. on Multi-Agent Systems in Logic Progr., in conjunction with ICLP'99*, Las Cruces, New Mexico, 1999.

[Fi94] M. Fisher. A survey of concurrent METATEM – the language and its applications. In *Proc. of First International Conf. on Temporal Logic (ICTL)*, LNCS 827, Berlin, 1994. Springer Verlag.

[KS96] R. A. Kowalski and F. Sadri. Towards a unified agent architecture that combines rationality with reactivity. In *Proc. International Works. on Logic in Databases*, LNCS 1154, Berlin, 1996. Springer-Verlag.

[PP90] Przymusinska, H., and Przymusinski, T. C., *Semantic Issues in Deductive Databases and Logic Programs*. R.B. Banerji (ed.) *Formal Techniques in Artificial Intelligence, a Sourcebook*, Elsevier Sc. Publ. B.V. (North Holland), 1990.

[Ra96] A. S. Rao. AgentSpeak(L): BDI Agents speak out in a logical computable language. In W. Van De Velde and J. W. Perram, editors, *Agents Breaking Away: Proc. of the Seventh European Works. on Modelling Autonomous Agents in a Multi-Agent World*, LNAI, pages 42–55, Berlin, 1996. Springer Verlag.

[Ra91] A. S. Rao and M. P. Georgeff. Modeling rational agents within a BDI-architecture. In R. Fikes and E. Sandewall, editors, *Proc. of Knowledge Representation and Reasoning (KR&R-91)*, pages 473–484. Morgan Kaufmann Publishers: San Mateo, CA, April 1991.

[SPDEK00] V.S. Subrahmanian, Piero Bonatti, Jürgen Dix, Thomas Eiter, Sarit Kraus, Fatma Özcan, and Robert Ross. *Heterogenous Active Agents*. MIT-Press, 2000.

An Ontology for Mobile Device Sensor-Based Context Awareness

Panu Korpipää and Jani Mäntyjärvi

VTT Technical Research Centre of Finland, VTT Electronics, Kaitoväylä 1,
P.O. Box 1100, FIN-90571 Oulu, Finland
{Panu.Korpipaa, Jani.Mantyjarvi}@vtt.fi

Abstract. In mobile computing, the efficient utilisation of the information gained from the sensors embedded in the devices is difficult. Instead of using raw measurement data application specifically, as currently is customary, higher abstraction level semantic descriptions of the situation, context, can be used to develop mobile applications that are more usable. This article introduces an ontology of context constituents, which are derived from a set of sensors embedded in a mobile device. In other words, a semantic interface to the sensor data is provided. The ontology promotes the rapid development of mobile applications, more efficient use of resources, as well as reuse and sharing of information between communicating entities. A few mobile applications are presented to illustrate the possibilities of using the ontology.

1 Introduction

The ongoing rapid development in mobile and ubiquitous computing has led to an increasing interest on sensors embedded into mobile devices. Sensors can be used to acquire information from the environment and the usage situation, context, in order to increase the usability of the device. Currently however the use of sensors is highly application specific, and sensor information is mostly utilised as raw numerical data. The information value and the usefulness of raw measurement data is very low for the application developer. Development efficiency is impaired and reuse is difficult, not to mention the inefficiency and variability of the resulting implementations. A semantic interface to the sensor data is needed. Low level sensor data can be abstracted for general use. This article introduces an ontology designed for utilising context abstractions, derived from a set of sensors embedded in a mobile device.

In this paper we refer to context as semantic symbolic expressions abstracted from numerical sensor data. Usually, the processing of context information from sensors is carried out using signal processing methods to extract suitable features [1,2]. The suitability of extracted features should reflect the concepts from the real world and they should be important for applications [2].

The ontology and syntax for representing context has been fairly superficially treated in the context awareness literature. Dey and Abowd [3] give a rough categorization of context into four classes, location, time, activity and identity. Otherwise, the vocabulary they use is strongly application dependent and actually consists of a list of widgets, fixed sources of context. Crowley et. al. [4] give a process-based approach of representing conceptual components. They bind the meaning to processes, and by that choice their system is in fact very similar to the widget-based approach, but differs in focusing on the transformation of information from measurements or contexts to other contexts. Winograd [5] emphasises the importance of creating ontologies for distributed environments that provide the application writer with a representation of the aspects of context that are relevant to program execution. Furthermore, the goal is seen in finding the right level of description, which abstracts away from implementation details, but is still specific enough to enable the inferring of actions based on context.

Context usually consists at each time instant of multiple partial descriptions of a situation, which we refer to as context atoms. Context atoms may already be useful without any further processing. An application may be either interested in a single part of the whole context, or many parts of it. Sometimes many context atoms, at a certain time instant, together form a single description of an event, a higher level context. A higher level context may also be formed by a causal sequence. Flanagan et. al. [6] give a formal definition for generating higher level contexts.

The ontology we describe in this article facilitates vector representation of context for inferring composite contexts. It has been used as an underlying representation for explorative analysis of the structure and dynamics of higher level contexts by segmenting a time series of atoms, and for unsupervised clustering in an attempt to raise the abstraction level of context data [6,7]. Mäntyjärvi et. al. [8] use the representation as a basis for experiments in utilising information from multiple mobile devices in recognizing context of a group of mobile terminals and their users collaboratively.

In summary, the contribution of this paper is introducing a mobile device sensor-based context ontology, a semantic interface, to data that is acquired from a set of sensors embedded in a mobile device. The use of the common ontology increases the information value and usability of the data for the application developer, as well as facilitating reuse and information sharing between communicating parties. To validate the feasibility of the approach, a few real-world applications that utilise the ontology are presented.

2 Sensor-Based Context Ontology

2.1 From Raw Measurement Data to Semantic Context

The objective in context atom extraction is to extract features describing a concept from the real world to facilitate the versatile exploitation of context information. Let's take for example, illumination. The shift from dark to normal illumination conditions and to bright is fuzzy rather than crisp. Quantizing the dynamic range of a feature with fuzzy quantization results in a more expressive representation of context information [9]. Various signal processing and feature extraction methods have been utilised to extract a set of features from sensor signals that describe concepts from the real world [7]. The approach provides a simple way to add semantic meaning to extracted features. However, it is required that features are already scaled to a certain range and the dynamics of features is examined with correspondence to real world situations. The quantization allows us to present context information in a vector format that opens possibilities to process context information, and enables us to connect low level context information to applications [2].

2.2 Goals and Design Rationale for the Ontology

The principles on which the design was based on, in order of importance, are:

1. Domain. The ontology is developed for the purpose of utilising mobile device sensor-based context information.
2. Simplicity. The ontology, relations that are used, and expressions should be simple enough to be easily utilised by application developers. The completely expressive and detailed ontology is not very useful if it is too complex compared to the necessary level of detail required by most applications.
3. Practical access. The ontology should enable practical, meaningful, intuitive, and simple queries and subscriptions to the context information.
4. Flexibility, expandability. The representation should support the addition of new context elements to the ontology, as well as new or complementary relations.
5. Facilitate inference. The representation should enable efficient inference by the recognition engines as well as application control. It should not restrict the inference to any single method, since no single inference method today is optimal for every type of problem.
6. Genericity. The ontology should support different types of context information.
7. Efficiency. The representation should be memory efficient as well as support time efficient inference methods.
8. Expressiveness. The possible amount of detail in describing any single context and the versatility of the expressions should be high.

2.3 Structure of the Ontology

Each context is described using seven properties (below). Each context expression is required to contain at least *Context type* and *Context* or *Value*, in order to facilitate the practical manipulation, storing, and usage of context information.

- *Context type*: Category of the context. All subscriptions and queries must have context type as the primary parameter. Each context type is allowed to have one context at a certain moment in time, thus the context type is the unique identifier of context. Context type concepts form a tree structure.
- *Context*: The symbolic "value" of context type. Is usually used together with context type, forming a verbal description (table 1).
- *Value*: "Raw", numerical optional value or feature describing context.
- *Confidence*: An optional property of context describing the uncertainty of context. Typically a probability or a fuzzy membership of context, depending on the source.
- *Source*: An optional property which can be used to describe the semantic source of context. It can be used by a client interested only in contexts from a specific source.
- *Timestamp*: The latest time when a context occurred.
- *Free attributes*: Can be used to specify the context expression freely. May contain any additional properties about details that are not included in the other properties.

Resource Description Framework (RDF) (www.w3c.org) [10] was used as the formal syntax for describing both structure and vocabulary of the ontology in order to enable information sharing.

2.4 Vocabulary

The vocabulary of sensor-based context constituents is presented in a list form for compactness, table 1. The hierarchy which is formed based on the constituents of each context type, separated by a colon, can be used for semantically querying and subscribing to information concerning branches of the context tree, instead of handling only individual context atoms described using the full context type path.

There are two categories of context types based on their values. The first category always has a quantisizable, valid value in all situations. Such contexts types are for example temperature, stability, light and sound intensity, etc. All possible values at any moment can be labeled as one of the contexts defined for the type, unless the sensor output is erroneous. The contexts in the other category only occasionally have a meaningful, defined value, and for those context types there must exist a NotAvailable context. The purpose of NotAvailable context is to convey the information that none of the contexts defined for the concerned context type is valid, and the current state is something other than the last value that was detected. Table 2 shows a few examples of context descriptions which are processed from raw sensor measurement signals and represented as instances of the context ontology.

Table 1. The vocabulary of sensor-based contexts in a list form

Context type (category of context)	Context
Environment:Sound:Intensity	Silent
Environment:Sound:Intensity	Moderate
Environment:Sound:Intensity	Loud
Environment:Light:Intensity	Dark
Environment:Light:Intensity	Normal
Environment:Light:Intensity	Bright
Environment:Light:Type	Artificial
Environment:Light:Type	Natural
Environment:Light:SourceFrequency	50Hz
Environment:Light:SourceFrequency	60Hz
Environment:Light:SourceFrequency	NotAvailable
Environment:Temperature	Cold
Environment:Temperature	Normal
Environment:Temperature	Hot
Environment:Humidity	Dry
Environment:Humidity	Normal
Environment:Humidity	Humid
User:Activity:PeriodicMovement	FrequencyOfWalking
User:Activity:PeriodicMovement	FrequencyOfRunning
User:Activity:PeriodicMovement	NotAvailable
Device:Activity:Stability	Unstable
Device:Activity:Stability	Stable
Device:Activity:Placement	AtHand
Device:Activity:Placement	NotAtHand
Device:Activity:Position	DisplayDown
Device:Activity:Position	DisplayUp
Device:Activity:Position	AntennaDown
Device:Activity:Position	AntennaUp
Device:Activity:Position	SidewaysRight
Device:Activity:Position	SidewaysLeft

Table 2. Examples of context instances based on the ontology

Example context	Context type	Context	Value	Confidence	Source	Free attributes
1	Environment: Light:Type	Natural	-	1	Device Sensor	Confidence = Fuzzy membership
2	Environment: Temperature	Warm	21	0,8	Device Sensor	Confidence = Fuzzy membership ValueUnit = Celsius
3	Device: Activity: Placement	AtHand	-	1	Device Sensor	Confidence = Crisp

Depending on the case, a confidence field may contain, e.g., fuzzy membership, probability, or crisp zero or one. A statement that directly names the accurate meaning of the confidence can be included in the free attributes field if it is necessary.

2.5 Subjective and Objective Contexts

Conceptually, contexts in the ontology of the table 1 can be divided into two groups, subjective and objective. Contexts derived from the measurements such as for instance light level, temperature and humidity, can have different interpretation in different situations or by people from different cultures. Clearly the fuzzy quantization is not fully satisfactory, at least in the case of temperature. A partial solution to avoid the subjectivity problem is to offer the applications the value of context in addition to the semantic interpretation. The use of direct value is especially useful in the case of temperature. The application can use either the symbolic (subjective) context or the value. Concerning other subjective contexts in the ontology than temperature, it is not as practical to use the value, at least if the value is known to be rapidly changing.

2.6 Naming Policy

To avoid naming conflicts between context information providers, a policy is defined. Java style conventions are adopted for naming (Java.sun.com/docs/codeconv). The context type description prefix is hence of following form:

domain_suffix.domain_base.[domain_prefix,
...]:device_internal_/_external:context_type

The last part separated by a dot in the prefix defines whether the context is device internal or provided by a third party. If a context is device internal, this field is empty, and if the context is provided by any third party, this field must contain an *x*. There may be many domain prefixes. Table 3 contains a few examples of naming.

Table 3. Examples of naming of context types from different providers

Origin of context	Context type
Device internal context	com.device_manufacturer:environment:light:intensity
Third party context	fi.vtt.ele:x:environment:temperature
Third party context	fi.vtt.ele.ais.tie:x:device:activity:stability
Third party context	fi.oulu:x:environment:humidity

2.7 Inference

The ontology is designed to facilitate any inference framework, or alternatively none. Uncertainty of context is described by the confidence attribute, which can be utilised in several inference frameworks, such as probabilistic networks, clustering, or case-based reasoning [11]. Hence, there is no central inference mechanism nor framework, nor is the inference mechanism restricted in any way. The idea is to distribute the inference (context recognition or application control), which is made invisible for the client by using a central blackboard manager. The approach is commonly known from blackboard systems [12].

3 Applications of Context Ontology

Context dependent information representation in user interface applications and browsing have been proven to gain from manipulating and processing context information [2]. The approach utilises a part of the ontology. Fuzzy controllers are used to guide the adaptation according to context. Environment audio guides the adaptation of volume of operating tunes. Fig. 1 presents screenshots of context based information representation of a bus timetable service, illustrating how fontsize, illumination of a display, and service content are controlled using context information.

Fig. 1. Adaptive information representation on the display of a mobile terminal during service browsing. Current contexts in screenshots are; 1a) Environment:Light:Intensity Normal, User:Activity:PeriodicMovement NotAvailable; 1b) User:Activity:PeriodicMovement RunningFrequency; 1c) Environment:Light:Intensity Bright, User:Activity:PeriodicMovement WalkingFrequency

4 Discussion

The representation seems to satisfy the needs of the domain of sensor-based context in a mobile device, concerning the applications that were discussed. The model is straightforward, clear, and enables practical usage. The ontology can be expanded when new contexts are discovered, and the inference framework is not bound to any single method. The efficiency of the representation is dependent on the chosen inference method. First order logic inference is not always applicable. Many efficient reasoning frameworks utilise a vector form of input data. Each context can form an element in the context vector, which represents a generic description of a more complex higher level context, and can be bound to meaning for understandability [13].

5 Conclusions

Semantic context descriptions can be used to develop sensor data-utilising mobile applications. This article introduced an ontology of context constituents, which are derived from a set of sensors embedded in a mobile device. The ontology promotes simplified development of sensor-based mobile applications, more efficient use of development and computing resources, as well as reuse and sharing of information between communicating entities. A proof-of-concept prototype application, adaptive information display, was discussed to illustrate the possible ways of using the ontology. In the future, the ontology will gradually be expanded to cover wider areas of context within the mobile computing domain. RDF is examined as a promising syntax for sharing the context information among devices, providers and users.

Acknowledgements. The authors would like to acknowledge the funding provided by NOKIA.

References

1. Schmidt, A., Aidoo, K.A., Takaluoma, A, Tuomela, U., Van Laerhoven, K., Van de Velde, W.: Advanced Interaction In Context. Lecture Notes in Computer Science, Hand Held and Ubiquitous Computing (1999) 89–101
2. Mäntyjärvi, J., Seppänen, T.: Adapting Applications in Handheld Devices Using Fuzzy Context Information. To be published in Interacting with Computers Journal, Vol. 15(3), Elsevier (2003)
3. Dey, A.K., Abowd, G.D.: Towards a Better Understanding of Context and Context-Awareness. CHI 2000 Workshop on The What, Who, Where, When, Why and How of Context-Awareness (2000)
4. Crowley, J., Coutaz, J., Rey, G., Reignier, P.: Perceptual components for context awareness. Proceedings of the International conference on ubiquitous computing, Springer-Verlag (2002) 117–134
5. Winograd, T.: Architectures for context. Human-Computer Interaction, Vol. 16 (2001) 401–419
6. Flanagan, J.A., Mäntyjärvi, J., Himberg, J.: Unsupervised Clustering of Symbol Strings and Context Recognition. IEEE Conference on Data Mining (2002) 171–178
7. Mäntyjärvi, J., Himberg, J., Korpipää, P., Mannila, H.: Extracting the Context of a Mobile Device User. Proceedings of the International Symposium on Human-Machine Systems, Kassel, Germany (2001) 445–450
8. Mäntyjärvi, J., Huuskonen, P., Himberg, J.: Collaborative Context Determination to Support Mobile Terminal Applications. IEEE Wireless Communications, Vol. 9(5) (2002) 39–45
9. Cox, E.: The Fuzzy Systems Handbook; A Practinioner's Guide to Building, Using and Maintaining Fuzzy Systems, 2nd edition. AP Professional, NewYork (1998)
10. Ahmed, K., et. al.: XML Meta Data. Wrox Press (2001)
11. Mitchell, T.: Machine learning. McGraw-Hill (1997)
12. Engelmore, R., Morgan, T. (Eds.): Blackboard systems. Addison-Wesley (1988)
13. Korpipää, P., Koskinen, M., Peltola, J., Makelä, S.M., Seppänen, T.: Bayesian Approach to Sensor-Based Context Awareness. To be published in Personal and Ubiquitous Computing, Springer-Verlag (2003)

The Use of Contextual Information in a Proactivity Model for Conversational Agents

Marcello L'Abbate and Ulrich Thiel

Fraunhofer IPSI, Dolivostr. 15
64293 Darmstadt, Germany

{labbate, thiel}@ipsi.fraunhofer.de

Abstract. For conversational agents engaging in a natural language-based interaction with web site users in service exchange applications, simple entertaining "chatting" is not sufficient. Instead, the agent needs to be cooperative by trying to provide relevant information about products and/or the conditions of purchasing. In this paper we analyze how the proactive behaviour of conversational agents can be used to increase the general user satisfaction. The results of two case studies allow us to outline a generic proactivity model for information agents based on retrieval mechanisms and search heuristics.

Keywords. Human-computer interaction, Autonomous agents and multiagent systems.

1 Introduction

The Internet has changed the way people communicate, do business, and even play games. Many people now use the Internet for banking, retails and business-to-business document and service exchanges. The growth of the internet (and e-commerce applications) is changing the way companies interact with customers. With the new era underway, companies must prepare for the web based communication and develop a new competitive landscape. According to the technology research company Forrester Research, today's Web is "DUMB, BORING, and DUSTY"[1] in terms of customer contact. There is an increasing need to upgrade the efficiency of application specific conversations, improve new-word-detection/learning capability during the conversation, and the agent's ability to identify the customer's intention clearly in order to provide accurate information and respond quickly with user-friendly interaction. This paper describes a way of improving the utility of natural language based interfaces, towards a web which is less dumb, less boring, and not "dusty" anymore. Our starting point was the utilization of text-based interfaces ("chatterbots"), usually only capable of amusing their users with some chatting functionality and the improvement of their capability by adding functions exploiting contextual information.

---

[1] http://www.forrester.com

2 From Chatterbots to Conversational Agents

A "chatterbot" (i.e. a "chatting robot") is a software system that attempts to simulate the conversation or "chatter" of a human being, often entertaining the user with some "smalltalk". "Eliza" and "Parry" were well-known early attempts at creating programs that could at least temporarily fool a real human being into thinking they were talking to another person. A chatterbot's knowledge base consists of a collection of dialogue management rules, which use different techniques for processing the user's input. These techniques may range from simple text parsing to more complex logic-oriented methodologies based on inference mechanisms. Applying a rule may have mainly the effect of determining the output text by taking into account some keywords extracted from the preceding user's input sentence. The output is sometimes presented to the user in combination with a graphical visualization, e.g. a cartoon, of the related emotion [7]. On the Internet, chatterbots are commonly used as interactive guides during a web site tour [5]. While showing the different pages of the web site, the system tells the user about the functionality and the main controlling mechanisms of the currently displayed page. This kind of help service may be requested anytime by new users. Such a facility implies a need to make chatterbots continuously available and accessible. Indeed, human intervention is needed only for the design and implementation process of the knowledge base, while the control of the dialogue is carried out autonomously by the system.

Recently, chatterbot technology has been improved in order to allow for a more effective management of a dialogue. For instance, a supplementary feature of chatterbots is the treatment of a user as an individual having specific needs, preferences, etc. Dialogue files of user sessions are decomposed to extract facts about information needs, attitudes towards items (e.g. desires), known items, etc, thus producing a model of the user representing her/his interests and background in the dialogues. On the basis of the generated user profile, the chatterbot can better support users during the interaction, providing personal recommendations and incentives and helping users in problematic situations during the search for specific information.

The features identified so far are commonly used by the literature for characterizing software agents, which are indeed defined as personalized, continuously running and semi-autonomous software systems [9]. Therefore, we can claim that enhanced chatterbots are software agents based on the natural-language interaction paradigm.

From our point of view, the most engaging and challenging improvement for conversational agents is proactivity: conversational agents which do not simply react in response to their environment, but are instead able to exhibit a goal-directed behavior, can have a more widespread range of use. Agents' applications are more efficient and powerful if they are able to take the initiative in both problematic and doubtful situations, in which persuasive skills are required. These considerations gain even more in importance if we apply the technology to a specific domain. For instance, within the e-commerce domain, advertisements and product offers can have a more positive impact if carried out proactively [4]. Providing an offer of the most suitable product at the most suitable time and with the most convenient conditions has always been a high priority within economic contexts. But this can only be fully achieved if the application is capable of identifying and reaching a complex goal by its own initiative, i.e. not as a consequence of an explicit request.

3 Identifying a Complex Goal

There are two different kinds of dialogue goals to be reached by conversational agents. First, it is necessary to provide a suitable response to the user's utterance. Mainly, this is achieved by the application of a dialogue management rule, and is strictly based on the last input sentence. Most of the existing chatterbot systems are designed with the aim of reaching only this kind of goal. Their success is measured upon the ability to interpret the last user input and their power relies on the size of the rule base: the more keywords can be recognized and processed, the more appropriate answers can be generated. In case of no rule to apply, standard sentences are output, such as "I don't understand you" or "please rephrase your last sentence".

The second class of goals has a more complex nature and is tied to underlying application problems to be solved by the software system employing the agent as user interface. A goal of this kind does not originate from the user's input sentence, but is already determined at the system's implementation stage. As regards the point of time at which problems are solved, another classification can be given:

- **Problems subsisting through the whole dialogue session.** The solving of these problems is the main objective of the entire system. The dialogue management rule base is conceived in a way that the relevant information to be used is elicited from the user during the dialogue, combined and processed by an expert subsystem and then presented as a result at the end of the session. These agents concentrate on a specific domain and are usually able to understand a limited vocabulary of input terms. The TRIPS system [3], for instance, concentrates on the transportation/logistics domain with the aim of assisting a human manager in finding the most appropriate evacuation plan for areas affected by natural disasters.

- **Problems arisen with the last user input.** Particular user requests can only be satisfied by accessing external data sources, such as catalogues of products or continuously updated information boards. A dialogue management rule in this case manages to activate parallel processes searching for a set of possible results. A weighting mechanism provides a ranking, in order to find out the most relevant record to deliver to the user. As we will see in the following a proactive behavior can be considered for increasing the quality of the retrieval task, by exploiting any available contextual information to include in the information search.

- **Problems not yet arisen, but most likely to occur in the future.** Users of conversational systems are sometimes clustered into categories according to classes of profiles and patterns of usage. By examining a structured recording of the already known dialogue, the agent tries to assign a new user to one of the predefined classes and predict his behavior by considering past dialogue sessions of other users pertaining to the same category. In this way, possible problems can be anticipated and tried to be solved in advance. These problems may range from requests for clarifications to more elaborated configuration tasks. A framework for the proactive recognition and solution of these anticipated problems is given in section 4.2.

4 Reaching a Complex Goal

In order to reach a complex goal, dialogues carried out by conversational agents need to be designed and planned in advance automatically by the system. Sequences of rules have to be arranged in plans, providing a strategic path to reach the desired goals. Sometimes it will be necessary to substitute the current active plan with another one, more suitable and efficient to solve the current or upcoming problem. The decision about what plan to choose to replace the currently obsolete one has to be taken proactively, as it cannot be made by the user.

By examining the following two case studies, we will show how a proactive behavior can be used for increasing a conversational agent's ability to solve product search problems (section 4.1), and anticipate clarification questions (section 4.2).

4.1 Product Search with COGITO

The main aim of the EU-funded project COGITO (IST-1999-13347) was the integration of a rule-based conversational agent within the Internet e-commerce site of Bertelsmann Online (BOL), an international bookseller that has online shops in several countries [8]. An additional component called "prompter" implements an intelligent search functionality which the agent needs to act as a shop assistant by applying a set of search heuristics. Depending on the current dialogue and the information stored in the user profile, the prompter checks whether one of the heuristics is suitable for generating a query to the product database. An explicit user request starting with "I am searching for…" or "I would like to buy…" may cause the agent switching to a specific internal state, in which rules for the generation of a query can be triggered. The set of attributes of the query takes contextual data into account, such as keywords from previous utterances. General search strategies can be employed with the aim of transparently collecting information to be used for recommending a potentially interesting product [6]. Depending on the chosen search strategy, the query to generate will consist of a generic template, updated with the values of the variables coming from the interpreted dialog sentences. The generated query may lead to unsatisfactory results, e.g. because too few or too many items are retrieved. Therefore, once the system has evaluated the result set, it may be necessary to refine the query in order to enhance the quality of the set of results.

The following example dialogue shows a situation in which collected contextual information can be used proactively for retrieving more relevant results:

Susanna:	Nice to meet you again, Peter. I am pleased that you have found your way back to BOL.DE. What can I do for you?
User:	I'm looking for a book.
Susanna:	Then this is the right place to be! We have hundreds of thousands of deliverable books. Are you looking for a book by a specific author?
User:	Yes
Susanna:	What is the name of the author?
User:	King
Susanna:	I shall check whether we have a book by the author King….

Out of this short piece of dialogue the system recognizes the task of generating a query to the product database, consisting of the search for books written by an author named "King". Without further processing the query, the first ranked books of the search result are by the author of horror literature "Stephen King". Books by other authors are found further down the list, which means that the user should scroll down a long list if he was not looking for a book by Stephen King. A proactive consideration of the available contextual information leads to a refinement and improvement of the relevance of the retrieved set of results. If we assume that the profile of the user "Peter" implies constraining the query to books pertaining to the category "Computers and Internet" the result set may change significantly (see Figure 1).

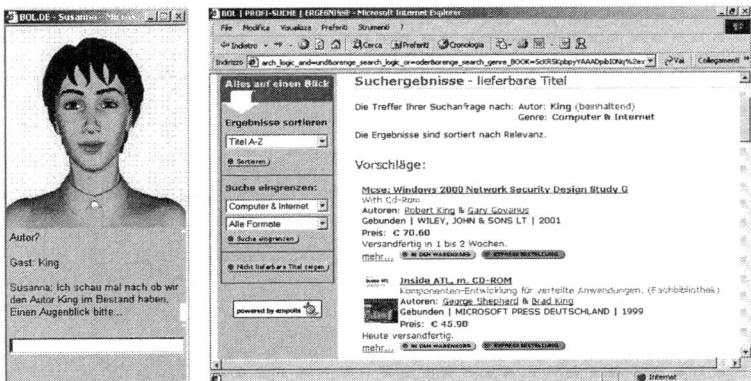

Fig. 1. A search result considering the category "Computers and Internet"

The evaluation of the COGITO prototype showed that the proactive expansion mechanisms introduced by the prompter increased the overall effectiveness of the system. More details about the evaluation results can be found in [1].

4.2 Anticipating Problems with VIP-Advisor

The key objective of the ongoing EU-funded Project VIP-ADVISOR (IST-2001-32440) is to develop a virtual personal insurance and finance assistant capable of natural language interaction. This personal assistant is specialized in risk management counseling for Small and Medium Enterprises (SMEs) but could be extended towards general insurance counseling for private persons. The interface supports speech recognition and synthesis in order to make the advisor easier and, thus, more convenient to use. Through online translation mechanisms it will be possible to use the advisor in different languages. The project builds upon an existing static tool (the Risk Manager Online) provided by the user organisation (Winterthur insurances). The existing tool takes the user through a Q&A session with predefined questions before producing a risk analysis matrix. VIP-Advisor supports the Risk Manager's functionality by taking so-called 0-level support into new dimensions for capturing user requirements about identifying risks as well as providing help and expert advice on risks. An Interaction Manager component is responsible for coordinating and synchronizing the multimodal interaction with the system, as the user can simultaneously talk via a microphone and make selections directly on the Risk Manager Forms via the pointing device and keyboard. The dialogue management is based on predefined dialogue plans monitored by the Interaction Manager. A plan consists of the realization of a strategy for guiding the interaction through a predefined sequence of dialogue steps. Whenever a user's intervention causes a deviation from the current plan, a new one has to be selected and substituted for the no longer valid plan. As a key technique for the selection of a new plan to apply, VIP-Advisor uses Case Based Reasoning: solutions that were used to solve old problems are adapted for solving new ones. The system relies on a continuously updated set of successful dialogue plans, whose abstractions are indexed by a retrieval engine in order to identify dialogues (i.e. *cases*) similar to the current situation. The measuring of the similarity is carried out by considering available contextual information, such as the user profile and a structured dialogue history.

The chances of achieving a successful dialogue can also be increased by proactively suggesting alternative actions to the user. In this way problems that may potentially arise are forecast and a solution offered. For this purpose, the ongoing dialogue has to be monitored in real time by examining its extended log file and considering alternative dialogue plans. These plans are selected according to the dialogue history and processed in the background by using the actual user input. By examining the dialogue context, an appropriate point of time may be identified for exchanging the current plan with the new one following an alternative strategy. This process can be reversed whenever an utterance expressing a rejection to the proposed alternative is identified. In this case the original plan is applied again. Figure 2 sketches the procedure described so far. The thick line in the figure represents the flow of the current dialogue. Plan 1 guides the dialogue and after every step *i* the context is determined. It contains both static profile information about the user (P_n) as well as dynamic information extracted from the dialogue topics (T_n). The updated context is continuously compared with the ones from the case base. In case of a correspondence the control of the dialogue switches from plan 1 to the plan having the corresponding context. Step n+1 of plan 2 may contain the solution to a problem which may be raised by the user in some further steps. After step n+1 of plan 2 the control of the dialogue is passed back to plan 1.

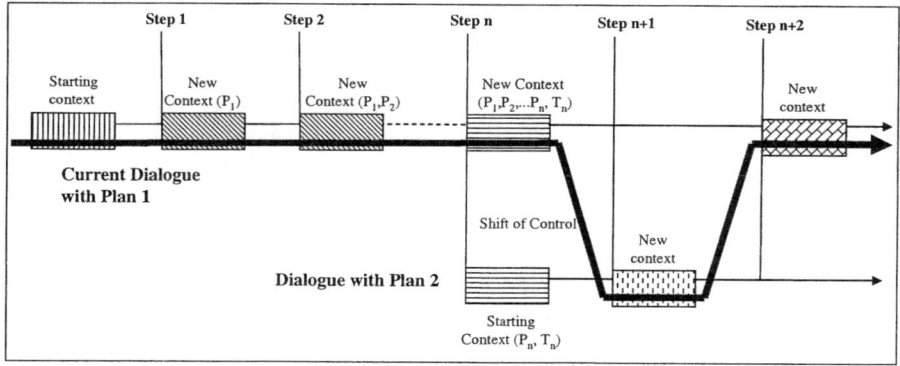

Fig. 2. Proactive switching of plans

5 Modeling Proactivity

Both COGITO and VIP-Advisor offer a system user interface, based on the natural language interaction paradigm. As they meet the requirements of being personalized, continuously running and semi-autonomous, we classify them as conversational agents. In addition to the capability of being reactive to the users' requests, by handling their input appropriately and providing suitable responses, both systems adopt proactive behaviors for improving the efficiency and therefore the user satisfaction in the interaction process. The general effectiveness of their proactive approach is achieved by use of the same underlying technology. The area of Information Retrieval provides a robust theoretical framework to rely on as well as widely accepted criteria for the quality assessment of search results. Performance issues can be overlooked, as the field's maturity already provides enough experience and solutions. The agents need only to identify the correct point of time at which a retrieval step is necessary and accurately specify the query to be posted by including available contextual information. The underlying retrieval engine as well as its weighting mechanisms will do the rest of the job, extracting the most relevant record out of the available information corpus, even if the contained data is not carefully prepared or formatted.

The two applications discussed here adopt a different dialogue management approach. The agent used in COGITO relies on a chatterbot-like organization of the rule base, allowing for a broader choice of the topics of discussion. The dialogue planning task is implemented by constraining the application of some rules to others previously applied. On the other hand, dialogues in VIP-Advisor follow predefined plans, restricting the user's freedom to change the topic of discourse, but allowing for a more efficient steering of the conversation. This diversity is required because of the different nature of the support offered by the agents in their respective systems: the cooperation achieved in VIP-Advisor refers to the usage of the Risk Manager tool, which encompasses a predefined sequence of interaction steps. The initiative has to remain on the system's side, as the user may not deliberatively change the order of the tasks to perform. In COGITO, the user has to elaborate his own strategy in terms of dialogue requests for achieving his goals and needs therefore to take the own initiative in formulating his desires. As a result, the kind of proactivity offered by a conversational agent has to take into account which side of the interaction process carries the burden of a dialogue leader: if it stays on the system's side, proactive planning may be helpful, but if it stays on the user's side, proactivity can only enhance the quality of the system's responses in short-term objectives.

In order to finally arrive at a proactivity model based on the considerations made so far, we move from an extension given to the Stimulus-Response model: proactivity is the ability and freedom to choose a response to a stimulus [2]. The stimulus in conversational agents is determined by contextual information: the dialogue history, the user profile and the specification of the underlying problem to be solved are used as input data for the generation of a query for an information retrieval-based process, performing the choice. Its outcome, (i.e. the product suggestion, the clarification text or even the new plan to apply) represents the generated response, presented to the user in view of his acknowledgement. Formally, these thoughts may be expressed by the following function:

$$Re_{n+1} = R(UI_n) +_e ProC(IR(Context(P_1, P_2, ..., P_n, T_n)))$$

The Response *Re* for the dialogue step *n+1* is determined by the normal system's output generation function *R* applied to the user input *UI* provided at step *n*. By using the operator $+_e$ the response is extended with the result of a retrieval process *IR* which takes into consideration the *Context* determined up to step *n*. It is based on the profile information $P_1, P_2, ..., P_n$ and a specification of the dialogue topic *T* of step *n*. A retrieval process will only be executed if there is a need for proactivity. This is decided by the function *ProC*.

In future activities we will concentrate on enhancing and finalizing the model, in order to take into account further variants of goal-directed behaviours. The conversational agents considered so far engage in cooperation dialogues, improved with useful recommendations and interesting suggestions. A different scenario is the consideration of dialogue partners, with the aim of being influential and convincing. A product or service is not only recommended but presented as *the* solution to the user's request, persuading him that there are no available alternatives. In order to reach that level of self-confidence and hence allow for a more appropriate tailoring of suggestions, it will be necessary at one side to increase the user's trust in the system and on the other side to interpret and evaluate available contextual information.

References

1. Andersen, V. and H.H.K. Andersen. „Evaluation of the COGITO system". Deliverable 7.2, IST-1999-13347, Risoe National Laboratory, DK, 2002.
2. Covey, Stehpen R. "The 7 Habits of Highly Effective People" Publisher: Franklin Covey Co. (1989)
3. Ferguson, George and Allen, James "TRIPS: An Intelligent Integrated Problem-Solving Assistant," in Proceedings of the Fifteenth National Conference on Artificial Intelligence (AAAI-98), Madison, WI, 26–30 July 1998, pp. 567–573.
4. Guttman, R., Moukas, A., and Maes, P., "Agent-mediated Electronic Commerce: A Survey", Knowledge Engineering Review, June 1998.
5. Hayes-Roth, B., Johnson, V., Van Gent, R., and Wescourt, K., "Staffing the Web with Interactive Characters", Communications of the ACM, 42(3), pp 103–105, 1999
6. L'Abbate, Marcello; Thiel, Ulrich "Helping Conversational Agents to find Informative Responses: Query Expansion Methods for Chatterbots" in Proceedings of the Autonomous Agents and Multi Agents System Conference July 2002, Bologna, Italy
7. Paradiso, Aldo; L'Abbate, Marcello, "A Model for the Generation and Combination of Emotional Expressions". In: C. Pelachaud, I. Poggi (Eds.) "Multimodal Communication and Context in Embodied Agents". Workshop Proceedings of the fifth International Conference on Autonomous Agents (AA'01), May 2001, Montreal, Canada (pp. 65–70) Montreal: ACM Press, 2001
8. Thiel, Ulrich; L'Abbate, Marcello; Paradiso, Aldo; Stein, Adelheid; Semeraro, Giovanni; Abbattista, Fabio; Lops, Pasquale "The COGITO Project: Intelligent E-Commerce with Guiding Agents based on Personalized Interaction Tools" In: J. Gasos and K.-D. Thoben (Eds.) "e-Business applications: results of applied research on e-Commerce, Supply Chain Management and Extended Enterprises" Section 2: eCommerce, Springer-Verlag, 2002
9. Wooldridge, M., J. and Jennings, N. R. (1995) "Intelligent Agents: Theory and Practice" The Knowledge Engineering Review 10 (2) 115–15

GloBuddy, a Dynamic Broad Context Phrase Book

Rami Musa[1], Madleina Scheidegger[2], Andrea Kulas[3], and Yoan Anguilet[4]

[1] MIT-EECS, Cambridge, MA 02139, USA
rmusa@mit.edu
[2] MIT-CMS, Cambridge, MA 02139, USA
mscheid@mit.edu
[3] MIT-CMS, Cambridge, MA02139, USA
kulas@mit.edu
[4] MIT-EECS, Cambridge, MA 02139, USA
anguilet@mit.edu

Abstract. GloBuddy is a dynamic broad-context phrase book that assists English speaking travelers in finding words and phrases, in a foreign language, that are relevant to their current situation or context. Unlike traditional phrase books, such as Berlitz, GloBuddy achieves a broad range of contexts by using common sense knowledge from the OpenMind database and the OMCSnet, in conjunction with a translation program. When GloBuddy is launched on a hand-held device travelers will be able to turn to it for help in any context or situation.

1 Introduction

Travelers in foreign countries often find themselves in situations where they have to communicate in a language that, at best, they barely speak. Therefore, many travelers take along dictionaries as well as phrase books, such as Berlitz[2], when they travel. Phrase books contain a limited set of phrases and words and their translations, categorized into a few very common situations in which travelers might find themselves. However, often travelers encounter situations which are not covered in phrase books and using a dictionary to look up all necessary words often is not a time efficient alternative.

For example, if one gets involved in a car accident, one may have to deal with the police, the insurance company or the hospital. Even worse, one might get arrested and go to jail. Phrase books might help in asking for assistance or in making a phone call but, typically, they will not prove useful when explaining the circumstances of an accident or even when communicating with a policeman or a lawyer.

What travelers need is a new solution that can help them communicate in any context or situation. A good new solution would be interactive, easy to use and faster than a phrase book or a dictionary. The authors, at the Massachusetts Institute of Technology Media Laboratory, have found a solution that uses commonsense knowledge to address the proposed broad context feature of a new better solution.

GloBuddy is an interactive broad-context application that uses common sense knowledge together with a translation capability to provide English-speaking travelers with words and phrases relevant to any possible situation. It can easily be launched on a hand-held device making it convenient for the traveler to take it along.

GloBuddy borrows its translation ability from an independent service. To provide words and phrases relevant to a certain situation, GloBuddy uses common sense knowledge from the OpenMind database[1] at the Media lab as well as the OMCSnet, a semantic network representation of OpenMind. For example, you can type in your situation ("I have been arrested") and GloBuddy retrieves common sense surrounding that situation and serves it to a translation service. GloBuddy might return "If you are arrested you should call a lawyer" or "Bail is a payment that allows an accused person to get out of jail until a trial". GloBuddy can also return useful words such as "prison, bail, policeman, lawyer, etc" together with their translations.

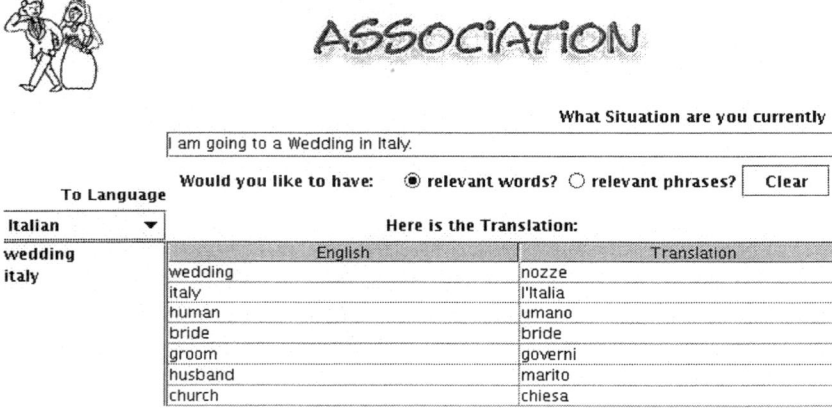

Fig. 1. GloBuddy: Word Association

With variable performance levels, the current version of GloBuddy fulfills all of the criteria listed above. GloBuddy can be used in a broad range of contexts. It is fast, interactive and easy to use. Furthermore, it is easy to expand and add new features to GloBuddy, or for it to be incorporated into a different application.

---

[1] The OpenMind database is a repository of knowledge that constitutes "commonsense", all those aspects of the world that humans understand so well and take for granted. OpenMind currently contains more than 500,000 natural language assertions built from the contribution of more than 10,000 people. For more information visit the OpenMind Common sense Project webpage at http://openmind.media.mit.edu or read [1]

2 Implementation

2.1 Translation

All translations in the program are done by querying Altavista Babelfish, an online translation service. The user can set both the language of the text to be translated, and the language to which the text should be translated. The text is then passed to Babelfish, and the returned translation is displayed. Translation can currently be performed between English, French, German, Italian, Portuguese and Spanish.

2.2 Association of Words

This functionality takes in a description of a situation in English. It then filters the description for keywords, using a predefined list of words to be excluded. GloBuddy then uses the OMCSnet through an http interface.

GloBuddy queries OMCSnet with each keyword to extract all nodes that are directly connected to the keyword. For each output word two scores are calculated. The first is a relevance score, which equals the number of keywords to which it is directly connected. The second is a frequency score, which equals the number of times the word appears in the total output across all keywords.

The output words are then sorted according to their relevance score. This ensures that outputs directly linked to a higher number of input keywords appear first. In other words, more relevant output words appear first. If multiple words have the same relevance score, they are then sorted according to their frequency scores. This deals with the situation where many of the output words are linked to the same number of keywords. In such a situation, the more often an output word appears in the total list of outputs, the more likely it is to be useful, and so it would be placed ahead of other words which appear less often.

The output is then translated to the language the user chose. The English output and the translated equivalent are then displayed, twenty at a time, to the user. The user can browse the output forward and back at will. In case GloBuddy fails to provide satisfactory results, the user can perform a more specific search. The user would choose a word, either a keyword or an output word, and an OMCSnet relation to perform a specific query.

2.3 Association of Phrases

This functionality currently uses two sources of phrases. The first is a collection of sentences and questions that are common in everyday conversation. Those are stored in GloBuddy under seven categories: "Food and Dining", "Accommodations", "Travel and Transportation", "Leisure", "Shopping", "Services" and "Health and Safety". The second source of phrases for this functionality is the OpenMind common sense database.

When using this functionality, GloBuddy starts by trying to guess the context of a situation that the user describes. More specifically, GloBuddy tries to guess whether the context of the user's situation fits in any of the seven categories mentioned above. If it does, GloBuddy queries the collection of phrases stored under the identified category for relevant phrases. If the user's situation does not fit any of the defined categories, GloBuddy turns to OpenMind to extract phrases.

Guessing the Category. For the purpose of making a guess, GloBuddy maintains a list of relevant words and concepts for each of the seven categories mentioned above. Those lists are compared against a set of words relevant to the situation described by the user. A score for each category is calculated. The seven scores are then used to decide whether there is a fit between the user's situation and any particular category.

To generate a list of words relevant to the user's situation, GloBuddy first uses the filtering method described in section 2.2 to extract keywords from the user's description of the situation. GloBuddy then queries the OMCSnet with each keyword and extracts the words directly connected to it. The output for each keyword is then compared with the list of predefined words and concepts for each category. A percentage score is calculated for every keyword-category combination by dividing the number of words from the keyword's output which are present in the list of relevant words and concepts for the category, by the total number of output words. A final score for each category is then calculated by summing across the keywords. The category whose score is at least 15% larger than any other score is taken as the relevant category. If there is no score that satisfies this requirement the user is asked for feedback.

When GloBuddy is unable to identify a match between the user's situation and a predefined category or if GloBuddy mis-guesses the relevant category, the user can choose to manually pick the category that GloBuddy should query. The user can also choose to proceed with the "Other" category.

Selecting Phrases. After making the guess, GloBuddy proceeds to querying for phrases. If a category other than "Other" was identified, GloBuddy looks for phrases, stored under the identified category, that contain any of the keywords from the user's situation description or any of the words that are directly connected to all of the keywords in OMCSnet. Phrases that contain more of the search words are provided to the user first together with their translations.

If GloBuddy could not identify a relevant category for the user's input, it defaults to "Other" and uses OpenMind. GloBuddy uses the filtering method described in section 2.2 to extract keywords from the user input. For each of the keywords, GloBuddy extracts fifty phrases from OpenMind and constructs a temporary phrase collection. At the same time, GloBuddy queries OMCSnet with each keyword and extracts the words directly connected to it. GloBuddy then queries the temporary phrase collection with the keywords and OMCSnet output and serves the phrases and translations in a similar approach to that used in querying the predefined categories.

The use of a small temporary phrase repository when in the "Other" mode was due to a current constraint that limits queries to OpenMind to one keyword at a time. Work is currently underway to allow for direct multiple-word queries to the OpenMind database. The limited temporary repository solution will then no longer be necessary. This should allow for better output in the "Other" mode through wider access to the OpenMind knowledge.

3 Evaluation of the Performance

The performance evaluation of GloBuddy was based only on the performance of the association functionalities. The translation functionality was borrowed from an outside source and therefore its evaluation is of little relevance to the performance evaluation of GloBuddy. There are three aspects of the association functionality that needed to be evaluated: a) association of words, b) identifying the context and c) retrieving phrases from OpenMind.

In order to evaluate how well the program performs in the word association mode, GloBuddy was queried on about fifty travel situations. The output was then examined and the number of relevant words in the top twenty and the top forty output words was counted. It was found that there were some examples for which GloBuddy was unable to find any relevant words. This occurred in less than ten percent of the cases. In the cases where GloBuddy did find relevant words, on average, more than fifty percent of the top twenty output words were relevant. Within the top forty output words the relevant words were usually less than half of them, but more than a third of them.

In order to evaluate how well the program performs when guessing the context of a situation, GloBuddy again was queried on about fifty travel situations. In about seventy percent of the cases, GloBuddy correctly matched the situation with one of the seven pre-defined context categories. In most of the cases when it did not, it was unable to determine a category match and turned to the user for feedback. In only one or two cases did GloBuddy determine a wrong match.

In the current implementation, GloBuddy's performance in the "Other" mode is low. In most cases, there are a few phrases that are relevant to the input situation. However, many of the sentences are too general to be of any practical use. Some relevant concepts are often present in the output sentences, and hence useful in travel situations, although the complete sentences may not be relevant.

As an example of a good output in the "Other" mode, providing GloBuddy with the situation "I need to go to the library to work on my research paper", returns useful sentences such as "a library can lend a book" and "a research paper would make you want to find information." In contrast to these applicable outputs, given the situation "I got into a fight with a drunk man at the bar," GloBuddy retrieves irrelevant sentences such as "People can fight each other" and "if you want to fight inflation then you should print less money."

4 Discussion

GloBuddy achieves the goals proposed earlier for a new solution to the language problem that travelers face. GloBuddy is interactive, fast and easy to use. GloBuddy is not limited to any particular context or language. Not only does GloBuddy make use of the known useful phrases found in phrase books, but it also uses common sense knowledge from OpenMind to provide applicable phrases in any situation. Furthermore, GloBuddy uses a semantic representation of OpenMind, OMCSnet, to provide words useful to any given context.

GloBuddy is a good example of the benefits of using common sense knowledge in interactive applications. Common sense knowledge can be used to develop solutions for problems and to improve current solutions. It is hard and impractical to think of all possible scenarios of situations that travelers could face, let alone to compile all relevant phrases. It is equally hard and impractical to compile all relevant words related to all possible situations. However, with common sense knowledge, the association and categorization can happen at run time and travelers can receive the necessary help they need.

GloBuddy performed better than expected in both word association, with fifty percent of the top twenty outputs consisting of relevant words, and context guessing, with more than seventy percent of the test cases matched properly. However, it performed worse than would be hoped for when retrieving phrases out of OpenMind. This low performance is due to the current limitations in the interface to OpenMind, as was discussed at the end of section 2.3.

When evaluating the performance of GloBuddy there were a few travel situations in which it consistently performed poorly. The reason for this is that the knowledge in OpenMind, and thus OMCSnet, is not consistently distributed across all topics. There are some areas for which OpenMind has more information, and thus gets a better correlation, than others. For example, if the word flight appears in the input, the output is going to have many words about birds and few of them to do with planes. This has to do with the fact that there are many more facts in OMCSnet about birds and flying than planes and flying.

5 Future Work

The basic functionality intended for GloBuddy has been created in this version, but many improvements and extensions can be introduced.

One important issue is to improve the quality of word associations taken out of OMCSnet since they directly and indirectly affect the performance of the program. There are still too many nonsensical associations. There are a number of ways to improve this such as, for example, tweaking the current scoring mechanism or using a new context identification approach.

Another important enhancement to GloBuddy is to increase the sources of potential relevant phrases. For instance, GloBuddy might have the ability to create new sentences from templates using natural language processing. Alternatively, GloBuddy might be able to mutate sentences from OpenMind to take

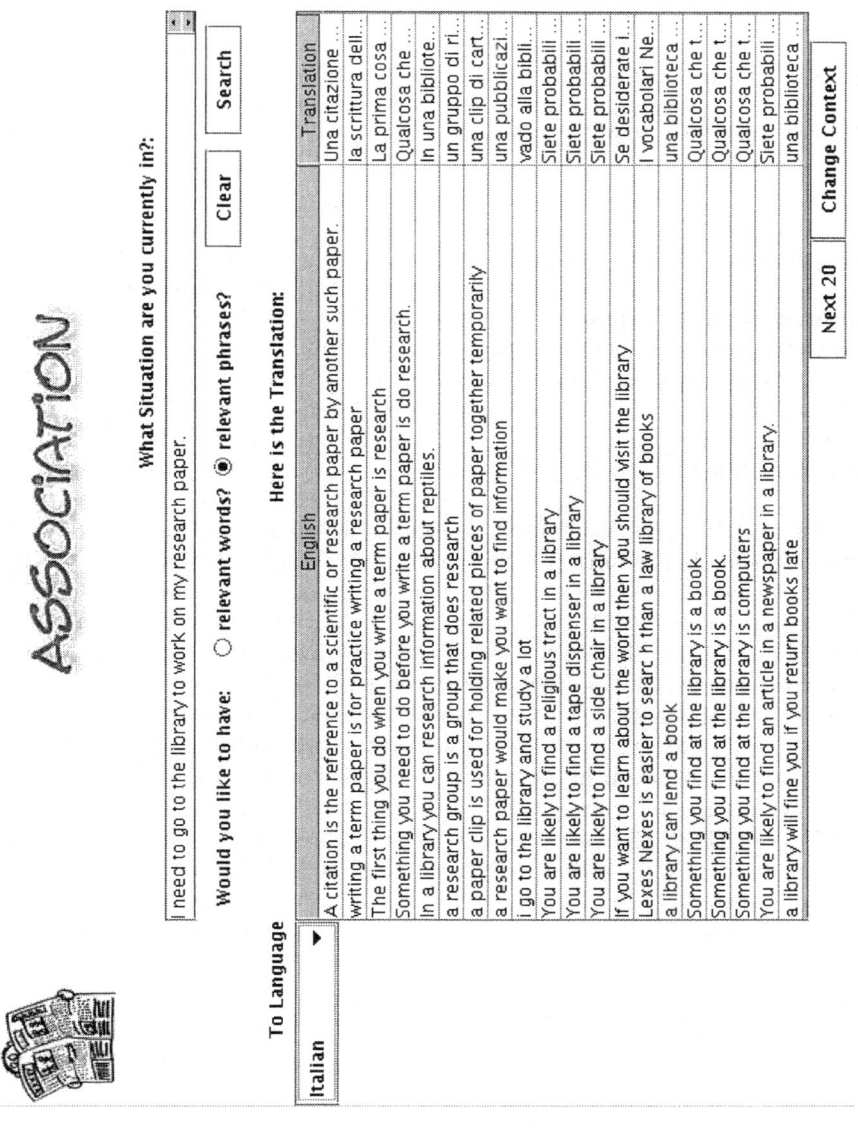

Fig. 2. GloBuddy: Phrase Output from OpenMind

out irrelevant parts, or to rephrase them as questions. A third alternative is to construct a smaller OpenMind-like database, that contains common sense questions and sentences that are common to everyday conversation.

GloBuddy can also be improved by learning from user input when it makes wrong guesses about context or when it mistakenly filters out important keywords from the input. Another learning opportunity would be for GloBuddy to track which words, users expand using the specific query functionality, and put those higher up on the list in future queries for the same situation.

In the current version of GloBuddy, all interfaces to external programs are done through http. This interface was selected since it existed for all programs, and would allow the use of a free translation program. This will be changed to an interface which does not require the internet, in the next version of GloBuddy.

To move GloBuddy onto hand-held devices, various small changes need to be introduced. For one, the user interface might need to be adapted to a smaller screen space. A smaller screen will also mean that fewer words and phrases can be displayed at the same time, making it important that the relevant phrases are presented first. This may require a better algorithm for determining which phrases are more relevant than others.

6 Related Works

To the authors' best knowledge, GloBuddy is the first research attempt to address the travelers' language difficulty problem by searching for an efficient, easy to use, broad-context solution. In the field of computer translation, it is now possible to translate words and phrases among different languages, a functionality that GloBuddy borrows from an existing solution, Altavista's Babelfish. However, beyond simple translation GloBuddy also provides collections of useful words and phrases that would enhance the user's conversational ability much faster than reverting to simple translation repeatedly. This seems not to have been done in any other application.

Acknowledgments. The authors which to thank Professor Henry Liebermann, Hugo Lin and Push Singh for their help and support in developing GloBuddy. The authors also which to thank the Common Sense Reasoning for Interactive Application class at the MIT Media Laboratory for providing feedback.

References

1. Push Singh, et, al.: Open Mind Common Sense: Knowledge Acquisition from the General Public. AAAI-2002
2. Berlitz: Spanish Phrase Book. Berlitz Publishing Company, 2001

Exploiting Dynamicity for the Definition and Parsing of Context Sensitive Grammars

Emanuele Panizzi

Università di Roma La Sapienza, Dipartimento di Informatica,
Via Salaria, 113, 00198 Roma, Italy
panizzi@dsi.uniroma1.it

Abstract. Classical Context Sensitive languages can be parsed by dynamically adding or removing production rules from the rules set. The grammar is modified according to the context, and evolves during parsing, allowing to take the context into consideration at the syntactical level and not in a separate semantic analysis. This approach has proven greater flexibility in the definition of programming languages, improving the readability and maintainability of program codes and the general usability of the language, thus improving the human-computer interaction. Examples of implementation of classical context sensitive languages as well as examples derived from practical applications are provided.

1 Introduction

It is well known from literature that it is not easy to parse context sensitive (CS) grammars. In fact it is not possible to recognize a context sensitive language using traditional parsers, that are based on context free (CF) languages. Many languages, however, cannot be defined by CF grammars. For example many programming languages have constructs that are inherently CS (eg. declarations of variables before their use, checking of agreement between formal and actual parameters in subprograms), so there is no CF grammar able to completely define these languages. For this reason, the analysis of common programming languages is traditionally performed in two different phases [1]: a subset of the language is described by a CF grammar, while all the CS part of the language is checked during the so called "semantic analysis".

This work shows how it is possible to parse classical CS languages using a parser able to modify the grammar according to the context. A parser with this capability, the theory of which is described in [3], is called dynamic parser and can add, hide or definitively delete production rules. The grammar defined can evolve during parsing, as opposed to traditional parsers that are statically defined by the grammar (think to the parser generators, e.g. yacc [2], which get a grammar definition in input and return in output the source code for the parser).

The parser used for this research is called Zz and has been developed in the frame of the APE supercomputer project [4,5]. A main programming language and many other tools (like compilers, application libraries, interfaces, debuggers and translators [6,7,8]) have been developed using this parser. All this languages have a certain degree of context sensitiveness, as will be shown below.

The structure of this article is the following: next section describes the dynamic parser Zz and its features; the third section shows an example of classical CS grammar and analyzes its implementation using the dynamic parser; section four shows two examples from practical applications.

2 Description of the Dynamic Parser Zz and Its Usage

The main feature of the Zz parser is that it is able to extend the syntax by adding production rules during parsing. In this section we explain this feature while giving some description on the usage of the parser. For a deeper detail, the reader is redirected to [3] and [10]. Zz is an open source project, and it can be downloaded from SourceForge [9].

2.1 Dynamical Creation of Production Rules

The initial grammar which Zz is based on is composed of a very tiny set of rules, terminals and nonterminals (the *kernel*) that allow basic operations. The input files to be parsed contain new rules that will be added to the rules set. New terminals and nonterminals are automatically defined as they occur in a new rule.

Each rule is in a context-free form: with one nonterminal in the left side and a number of space-separated terminals and nonterminals in the right side. Terminals and nonterminals are strings of characters, but nonterminals are followed by a caret (^) and a parameter name. A slash (/) indicates the beginning of a rule and an arrow (->) separates the left side from the right one.

The following example is a good Zz input file. It shows a definition of a new rule and a sentence that uses it. The rule uses two kernel nonterminals (stat and int) and defines four terminals (I, am, years, old). The parameter age is associated to the nonterminal int.

```
/stat->I am int^age years old
I am 37 years old
```

The nonterminal stat may be considered as the starting nonterminal of each Zz grammar, while int is able to parse a positive integer number. The second row of this example can be parsed by Zz using the newly defined rule.

Rules may have an attached action, enclosed in braces. The action content will be parsed when the rule is matched. Syntax used in an action must be defined when the action is parsed, but not necessarily before. The parametes associated to nonterminals can be used within the action and it is also possible to use other variables, without declaring them, as in the input file below (note that print writes to the output file and the & sign connects two strings):

```
/stat->I am int^age years old {
  /y = 2003 - age;
  /print "you were born in "&y;
}
I am 37 years old
```

The output file for this example will thus be: `you were born in 1966`.

An action can return a value, as in the example below:

```
/stat->Tell me the code of colour^rgb {/print "The code is "&rgb}
/colour->red {/return "255,0,0"}
/color->gray {/return "128,128,128"}
Tell me the code of gray
```

The output file of this example is: `The code is 128,128,128`

Actions can contain the definition of new production rules. The new rules will be added to the rules set when the action is executed, i.e. parsed. The example below shows how to define new colour names during parsing (note that the `ident` nonterminal is a *kernel* nonterminal able to parse identifiers – alphanumeric strings starting with an alphabetic character).

```
/stat->Define ident^newcolour as int^r int^g int^b {
        /colour->newcolour {/return r&","&g&","&b }
}
Define white as 255 255 255
```

2.2 Dynamical Deletion of Production Rules

Zz provides a mechanism to hide or definitively delete production rules. One or more rules can be defined inside a scope, and this scope can be popped out from the rules set, pushed in back again or deleted, allowing the defined language to be extended or reduced dynamically. Consider the following extension to the colour input file:

```
/push scope specialcolours
/colour->cyan { /return "0,255,255"; }
/colour->fuchsia { /return "255,0,255"; }
Tell me the colour of cyan
/pop scope
Tell me the colour of cyan
```

The output for this example is:

```
The code is 0,255,255
Syntax error
```

In fact, after the `/pop scope` statement, there is no rule in the rules set able to accept the colour cyan. To re-insert the two rules above, it is sufficient the statement `/push scope specialcolours`, without need of specifying the rules again. To delete permanently the rules, we can use the following Zz statement: `/delete scope specialcolours`.

3 Definition of a Typical Context Sensitive Language Using the Dynamic Parser

It is well known [1] that CF grammars can keep count of two items but not three. For example, the language $L_1 = \{a^n b^n | n \geq 1\}$ is context free (and not definable by any regular expression) and can be expressed by the Zz grammar:

```
/stat->S^$ {/print ok}
/S->a S^$ b
/S->a b
```

($ is a reusable dummy parameter). After these rules Zz will accept the following two input lines, but not the third one (in which the count of a's is different from the count of b's):

```
a a a b b b
a b
a a b b b
```

The language $L_2 = \{wcw | w \in (a|b)*\}$ consists of two equal strings, each one containing a's and b's in the same order and quantity, separated by a c; L_2 represents the declaration and use of identifiers in programming languages and it can be demonstrated that it is not a CF language.

In fact the second string (the w after the c) depends on the first string (in this simple case they must be identical). It is possible to accept every string as w, and only when the c is found it is known what string is expected after it. The second w string of wcw depends on the first, thus on the context. So it is not possible to parse L_2 with a traditional parser, because it is not known *a priori* what is to be expected after c.

On the other hand we can implement such a language using Zz because it is possible to define an incomplete grammar at the beginning and to complete the definition when the context is known. We will gradually illustrate this technique in three steps.

3.1 First Step: Dynamical Definition Based on Context

As a first approximation we add the necessary grammar to parse the second w only after parsing the first w. The following Zz syntax defines the language L_2:

```
/stat->first^$ c second^$ {/print ok}
/first->l2string^w { /second->w }
/l2string->{/return ""}
/l2string->l2string^s l2char^t {/return s&" "&t}
/l2char->"a" {/return "a"}
/l2char->"b" {/return "b"}
```

The first rule has two non terminals (`first` and `second`) and the terminal c between them. We want to stress here that the `second` nonterminal has not yet been defined at this point. But Zz only requires that it be defined when needed, i.e. before being used. In fact we will define it after the first w is matched because only at that point we know how to do it.

When the c terminal is matched, the rule /second->w is ready and can be reduced, provided the input string belongs to L_2. Note that w has been replaced by its value, for example if the input were a a b c ... then the rule would have been /second->"a a b".

In case the input string be of type $wcw'|w \neq w'$, Zz will report a syntax error, because w' cannot be reduced to `second` (there is no other rule /second->...): Zz expected a w, the syntax is not correct.

The grammar to accept the first w string is a normal CF grammar. the grammar for the second w string is a CF grammar as well. But the L_2 language that we are parsing is a Context Sensitive language.

3.2 Second Step: Exiting from Context

We consider now another input to the parser, a b b c a b b, following the previous one. This would cause the enhancement of the grammar with the rule /second->"a b b". Note that, because this rule is defined as soon as the first part of the string (w) is matched, we must previously delete the old rule /second->"a a b", if we want to to prevent that the string a b b c a a b (that does not belong to L_2) be accepted. To do this, we put the definition of the rule /second->... in the scope L2 which is pushed just before the definition of the rule and is deleted when the full string wcw is matched (in the action attached to the rule /stat->first^$ c second^$, see below). In other words: first we increment the grammar according to the context, and then, after parsing the whole string, we remove the production rules introduced, in order not to accept the old w in a further string. In fact we remove the rules when we exit the context, i.e. when we finish parsing the string wcw.

Below is the grammar after the introduction of the scope:

```
/stat->first^$ c second^$ {/delete scope L2; /print ok}
/first->l2string^w {
  /push scope L2
  /second->w
}
/l2string->{/return ""}
/l2string->l2string^s l2char^t {/return s&" "&t}
/l2char->"a" {/return "a"}
/l2char->"b" {/return "b"}
```

3.3 Third Step: Cleaning

Finally, if we consider the case in which the string does not belong to the language, we notice that it is necessary to delete the L2 scope also if the rule

/stat->first^$ c second^$ is not matched. We need to clean the rules set when the input sentence does not belong to the language. For this reason we introduce in the definition of the L_2 grammar the rule

/stat->first^$ c any^$ {/delete scope L2; /print nok}

that will be matched only if it is not possible to match the other one. Note that any is a *kernel* nonterminal that can match any sentence.

So, each time we parse a text to decide if it belongs to L_2, we first increment and then decrement the grammar. At each moment the grammar defined is in a context free form, but, thanks to its adaptive nature, it globally defines a context sensitive language.

4 Application Examples

Several applications have benefited of this approach and are based on evolving grammars and the Zz parser: the Tao language and its compilers for the APE parallel computers, the APE symbolic debugger, some application libraries for compute intensive codes, the Seismic Migration Language, etc.

In this section, two of these applications will be described in their core aspects, in order to show how the context dependence in programming languages can be exploited at the syntactical level. This approach allows greater flexibility in the definition of programming languages, improving the readability and maintenability of program codes and the general usability of the language and thus improving the human-computer interaction.

4.1 Permutations

The first example is the grammar defined by the author for the Seismic Migration Language [6], that can be exemplified as a language to accept permutations.

Suppose we want to accept permutations according to patterns like:

```
x y z -> x z y
x y z -> z y x
```

where x, y and z are identifiers; suppose also that we can't accept all of the six permutations, but only the two shown above.

Based on these patterns, and in a manner similar to the definition of the L_2 grammar in the previous section, we write the following syntax that evolves according to the context and accepts permutations like those reported above.

```
/stat->first_part^$ "->" second_part^$ {
  /delete scope permutations;
  /print ok
}
/first_part->ident^x ident^y ident^z {
  /push scope permutations
```

```
  /second_part->x z y
  /second_part->z y x
}
```

The possible second parts are defined only when the first part is parsed. The identifiers x, y and z will be substituted by the actual indices used in the permutation, and the rules second_part will be defined according to the context. We show a possible continuation for the input file and the corresponding output:

```
alpha beta gamma -> alpha gamma beta
alpha beta gamma -> gamma beta alpha
alpha beta gamma -> gamma alpha beta
ok
ok
Syntax Error
```

In the Sesismic Migration Language this approach has been successfully used to describe in a compact and readable form the data distribution and the FFT and Migration operations on large numerical matrices. The actions attached to the rules produced object code interacting with the compiler back-end.

4.2 Types and Variables

Another example of the implementation of the above approach in a real compiler is due to the Zz authors [10] and consists in the syntax for the definition of variables in the Tao language for the APE parallel computer [11].

Tao integer expressions are parsed by the following (simplified) CF grammar:

```
/integer_expr->integer_expr^e "+" integer_term^t
/integer_expr->integer_term^t
/integer_term->integer_term^t "*" integer_fact^f
/integer_term->integer_fact^f
/integer_fact->"(" integer_expr^e ")"
/integer_fact->integer_var^v
```

The actions, whose purpose is to produce the assembly related to the parsed expression, are omitted for simplicity.

In this grammar, like in the L_2 grammar, we use a nonterminal (namely integer_var) that is not defined; till now, in fact, no production rule with that nonterminal on the left has been defined. The corresponding nonterminal in L_2 was second.

Like in L_2, we provide a syntax that automatically adds a new rule to the grammar when a variable is declared, as shown in the following (we remind that in Tao an integer variable can be declared using the word integer followed by an identifier):

```
/stat->integer ident^x { /integer_var->x }
```

When a variable is declared, e.g. integer foo, a new rule is created:

```
/integer_var->foo
```

This rule extends the grammar of the integer expressions and will be matched if the variable foo is used inside an integer expression. The context was created when the foo variable was declared and the integer expression can be parsed if it is compatible with the context.

Note that, if a non declared variable (say bar) is used inside an integer expression, it is not accepted by the parser (there is no rule /integer_var->bar) and the parsing stops with a syntax error. bar is unknown in this context.

5 Conclusion

The approach described in this paper is based on the dynamic creation and deletion of production rules during parsing. Using Zz, it is possible to define an incomplete grammar and extend it when needed according to the context. This approach takes into consideration the context at the syntactical level and not in a separate semantic analysis. The grammar defined at any given moment during parsing is in context-free form, but the ability to create and delete production rules allows for parsing of Context Sensitive languages.

This approach has proven flexibility in the definition of programming languages [11], compilers and other tools [8], improving the readability and maintenability of program codes and shortening them, so enhancing the human-computer interaction in this kind of applications.

References

1. Aho A.V., Sethi R., Ullman J.D., "Compilers: Principles, Techniques, and Tools", Addison-Wesley, 1986
2. "Yacc: Yet Another Compiler-Compiler", http://www.combo.org/lex_yacc_page/yacc.html
3. Cabasino S., Paolucci P.S., Todesco G.M., "Dynamic Parsers and Evolving Grammars", ACM SIGPLAN Notices 27, 1992
4. Panizzi E.,"APEmille: a parallel processor in the teraflop range", Nucl. Phys. Proc. Suppl. 53:1014-1016, 1997
5. Panizzi E., Sacco G.,"The APEmille project", Lecture Notes in Computer Science, High Performance Computing and Networking, HPCN2000
6. Cabasino S., Paolucci P.S., Panizzi E., Todesco G.M., "A Parallel Digital Signal Processing Language, Specialized for Seismic Migration, Built with Evolving Grammars", 1999
7. Bartoloni A. et al., "The Software of the Ape100 Processor", International Journal of Modern Physics C, 4 (1993) 969-976
8. "DBQ: Quadrics Symbolic Debugger", Alenia Spazio – Quadrics, http://www.casaccia.enea.it/APE100/documents/Dbq/dbq_man.ps
9. OpenZz Project Homepage on SourceForge, http://openzz.sourceforge.net
10. "Zz: The Tao Engine", http://cvs.sourceforge.net/cgi-bin/viewcvs.cgi/ *checkout*/openzz/openzz/doc/zzdoc.html?rev=1.3&content-type=text/html
11. Cabasino S., Dautilia R., Paolucci P.S., Todesco G.M., "The TAO Language", Alenia Spazio – Quadrics, http://www.casaccia.enea.it/APE100/documents/Tao/taolng.ps

Co-text Loss in Textual Chat Tools

Mariano Gomes Pimentel, Hugo Fuks, and Carlos José Pereira de Lucena

Pontifícia Universidade Católica do Rio de Janeiro
Rua Marquês de São Vicente, 225
22453-900 Rio de Janeiro, Brazil
{mariano, hugo, lucena}@inf.puc-rio.br

Abstract. The research presented in this article investigates a problem related to the lack of understanding of messages exchanged during chat sessions. There is confusion in the majority of the textual chat tools when various people converse at the same time. Sometimes a participant does not identify the relationship of a new message with a previous one, and is unable to establish a conversation thread—within this research this phenomenon is denominated "co-text loss." The causes, consequences and frequency of this phenomenon are discussed in this article. Two textual chat tools that have been developed to try to reduce the occurrence of co-text loss are presented in this article, together with the results obtained through the use of these tools during synchronous debates among undergraduate and postgraduate students in distance learning courses.

1 Introduction

The research presented in this article began with an analysis of the chat sessions that took place during the Information Technology Applied to Education (ITAE) course [1], which is an on-line course taught by the Computer Science Department of the Catholic University of Rio de Janeiro using the AulaNet learningware [2]. Although enthusiastic about the "different and interesting" activity, the participants of the debates in the ITAE course frequently believe the conversation is confusing: "It is not easy to communicate through such a chaotic tool" (Humberto); "Really chaotic!" (Geraldo); "I liked this debate…however I couldn't follow linearly what was being discussed" (Marcelo) [1].

In a chat with some participants talking at the same time, the result is a tangle of messages where, in many situations, it is difficult to identify who is talking to whom about what—this problem is being denominated here as "co-text loss." and is detailed in Section 2. Two textual chat tools that were developed to try to reduce co-text loss are presented In Section 3. The conclusion is presented in Section 4.

---

[1] In the chat transcript fragments published in this article, the real names of the participants were substituted for pseudonyms. The texts were originally in Portuguese and then translated into English. The original transcripts are available at [3, 4].

2 Co-text Loss

The objective of this section is to define the "co-text loss" phenomenon (Section 2.1) and also to present the investigations about causes (Section 2.2), consequences (Section 2.3) and its frequency (Section 2.4).

2.1 Definition of Co-text Loss

The initial inspiration for the identification of co-text loss was the perception that the text that resulted from a chat session displayed some features that are similar to text in hypertext: both are non-linear. This similarity spurred an investigation into whether in a chat session a problem similar to the classic hypertext problem also occurred: *disorientation* or *lost in hyperspace* [5]. And, in fact, some participants felt "lost" during chat sessions, as shown in the chat transcript in Text 1.

Text 1. Stating co-text loss. Source: Debate 1 of the ITAE 2000.1 edition (first semester 2000). In this debate a total of 289 messages were produced and sent by 9 participants

```
    24     <Liane>   Director, as far as I know it is a piece of
                     authoring software and not Groupware
    26     <Pablo>   in my understanding, authoring software contributes
                     to groupware
    30     <Liane>   I believe that it is just the contrary, that
                     groupware can help in the authoring process since
                     it can facilitate the communication process
                     between members of a team
 ▶  31     <Humberto> Contrary to what, Liane, I'm lost
    36     <Liane>   When I said the contrary, I didn't mean that
                     authoring defines groupware, but that groupware
                     makes authoring possible
    37     <Humberto> OK
    38     <Pablo>   How about both ways, Liane?
```

"Co-text[2] loss" was the term used in this study to designate the phenomenon that occurs in a chat session when a participant does not establish a conversation thread. Co-text loss occurs each time the reader is unable to identify which of the previous messages provides the elements that are necessary to understand the message that is being read. For example, in the transcript fragment in Text 1, it was necessary to identify that message 30 (of Liane) was counter-arguing message 26. Humberto was not able to make this association and expressed co-text loss in message 31.

Co-text loss can be verified through statements such as "what are you talking about?" or "I didn't understand." Such statements are called here *textual*

---

[2] "Co-text" designates surrounding text that has been written before or after an enunciation and that provides elements for understanding it. This term is used in Linguistics as an effort to solve the ambiguity of the word context, which has a wider meaning [6].

manifestations of co-text loss. It is necessary to emphasize that those statement should not always be considered as a manifestation of co-text loss. Upon declaring "I didn't understand," the participant may have identified the co-text of the message but not have understood it for another reason—for example; the participant might be manifesting the inconsistency or non relevance of the argument presented in the message. It is also necessary to emphasize that co-text loss is a cognitive phenomenon—the *textual manifestations of co-text loss* is only one of the consequences of co-text loss, and not the phenomenon itself.

2.2 Causes of Co-text Loss

From text that is "linear" and "well organized," as generally is the case in books, articles and magazine texts, one expects threading, concatenation, sequence of information. Although a given text may not be merely a chain of enunciations, it is this chain that provides a more legible text. Different than linear and well-organized text, text from a chat session is non-linear. The *linearity* of a chat text is defined here as the percentage of the messages that establish linearity; that is, the percent of the messages associated with the message immediately before them. The *non-linearity* of the chat session is the complement of this percentage.

The predominant non-linearity of a chat session[3] implies features that make it more likely for co-text loss to occur. The greater distance between associated messages[4] makes it more difficult to locate the referenced message, understand the cohesion mechanisms [7] and make the inference of the association between the messages. The non-linearity of chats also make it likely for a confluence of topics to occur: different topics are discussed in parallel, alternately. Although other factors could also be regarded as possible causes of co-text loss, for this research project, it was assumed that the non-linearity of chat sessions is the main cause of co-text loss.

2.3 Consequences of Co-text Loss

Figure 1 presents a simplified scheme of actions that participants could carry out after detecting co-text loss.

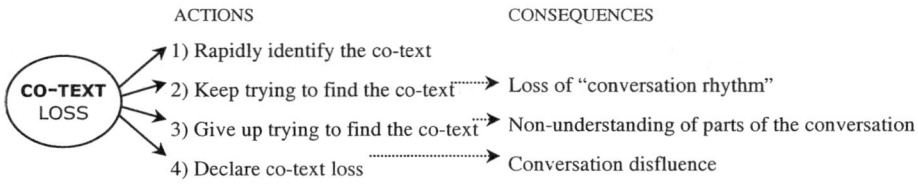

Fig. 1. Possible actions after co-text loss detection

---

[3] The debate sessions analyzed on this research project were 85% non-linear [4].
[4] On average, the messages were associated with the 6th previous message [4].

Faced with co-text loss, the participant seeks the co-text in the previous messages. If the co-text is quickly identified (action 1 of Figure 1), the conversation continues as if nothing had happened. If the participant does not quickly identify the co-text and continues to look for it in the previous messages (action 2 of Figure 1), this search will take up time and effort, and cause the loss of conversation rhythm—while looking for the co-text, the other participants will be continuing the conversation. If the participant stops looking for the co-text, and does not declare its loss (action 3 of Figure 1), she might not understand part of the conversation. If the participant states her co-text loss (action 4 of Figure 1, message 31 of Text 1), another participant may attempt to outline the unidentified co-text (message 36 of Text 1) and, eventually, the participant who lost the co-text may declare her understanding of it (message 37 of Text 1), and the conversation can continue. All of these messages—while necessary for co-text loss repair—cause conversational disfluency [8]: they do not contribute to the development of the topic of the conversation, interrupting the information flow.

The loss of conversational rhythm, the non-understanding of parts of the conversation and the disfluency of the conversation are potential consequences of co-text loss and characterize the phenomenon as a problem.

2.4 Frequency of Co-text Loss

The frequency of co-text loss situations that occurred during the debates of two ITAE editions was investigated for this research project.

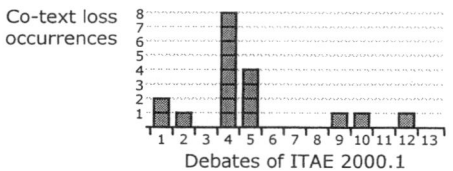
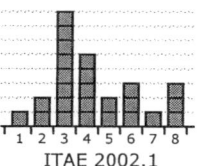

Fig. 2. Co-text loss occurrences observed during the ITAE 2000.1 edition. On average 7 participants were present and 336 messages were sent per debate. In the ITAE 2002.1 edition, on average 19 participants were present and 622 messages were sent per debate.

What can be seen, based upon the data presented in Figure 2, is the low average of situations in which co-text loss is declared: 2 situations per debate, 1 situation every 217 messages. However, based upon this data one should not conclude that co-text loss is a sporadic phenomenon. Not all co-text loss is manifested textually.

Another finding of the study was the reduction in the number of situations in which co-text loss was declared. There are two interpretations for this. The former is that over the course of chat sessions the group acquires experience, learns how to better interact and converse. The latter is that the participants learn to better tolerate the phenomenon, reducing their declarations of co-text loss. Even if co-text loss is reduced or becomes more tolerable over the course of the debates, it does not disappear completely, even after a number of chat sessions have been held.

3 Development of Chat Text Tools for Diminishing Co-text Loss

Two chat tools were developed in an attempt to reduce co-text loss and to help understanding the phenomenon: HyperDialog (Section 3.1) and Mediated Chat 2.0 (Section 3.2).

3.1 HyperDialog and the Explicit Threading of Conversation

While the majority of chat text tools organize messages in chronological order, the HyperDialog tool [4] was developed to structure messages by threads. Using this organization, the text sequences are evident. In an isolated thread, the conversation remains totally linear: each message is associated with the message that comes immediately before it. The hypothesis is that the threads mechanism would reduce co-text loss because it imposes structure to the non-linearity nature of the chat, which was identified as one of the main causes of co-text loss (see Section 2.2).

In order to evaluate whether the HyperDialog tool reduces co-text loss, it was put to use in the INformation Technology in Education (INTE) course, a subject taught by the Computer Science Department of the Federal University of Rio de Janeiro. In this course, the HyperDialog tool and a typical chat tool were used in different and intertwined sessions. In this evaluation, 5 debates sessions were held, each one lasting approximately 50 minutes. On average 11 participants were present and sent 173 messages per debate. The debate dynamics of this course was similar to the one of the ITAE course. Contrary to what was expected, there were still instances of co-text loss in the debates in which the HyperDialog tool was used. The data is presented below.

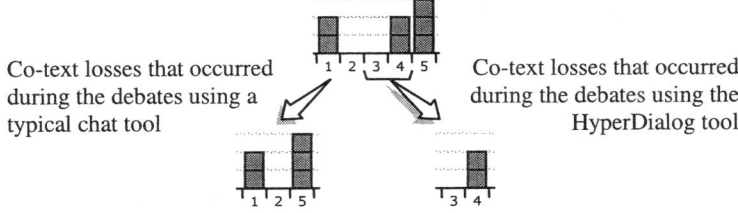

Co-text losses that occurred during the debates using a typical chat tool

Co-text losses that occurred during the debates using the HyperDialog tool

Fig. 3. Co-text loss occurrences observed during the INTE 2001.1 edition

Investigating the co-text losses manifested through the use of the HyperDialog tool, it was clear that these occurred to messages that lacked a proper association with the message it referred to. The omission of or error in making this association means supplying the reader with incorrect information—this can disorient the reader regarding the interpretation of the message or generate indisposition towards trying to understand it. When messages are incorrectly associated, the conversation threads become useless.

Analyzing the errors committed by the participants while establishing message associations using the HyperDialog tool, it was seen that approximately 92% of the messages were correctly associated. This percentage indicates that most of time the participants were able to converse while associating their messages during the chat session[5]. On the other hand, the 8% of the messages that were not properly associated indicated that the participants were having difficulty using the HyperDialog tool—improvements must be made to lower this percentage.

The evaluation of the use of the HyperDialog tool makes it clear that threads help to reduce co-text loss as long as the associations between the messages are correctly established. However, it is also evident that conversation threads make the chats more formal and degrade conversation fluency. This structuring is useful for activities in which understanding of the conversation is highly desirable, as is supposed for debates involving course subject matter. However, the use of threads perhaps is not appropriate for activities involving socialization and recreation, where conversational informality and fluency are highly desirable.

3.2 Mediated Chat 2.0 and the Group Conversation Techniques

The "Mediated Chat 2.0" tool [10] implements the following group conversation techniques: *free contribution*, where any participant can send a message at any time; *circular contribution*, where the participants are organized in a circular queue and, one by one, the first one of the queue can send a message; *single contribution*, where each participant must send a single message at any time; and *mediated contribution*, where only the selected participant can send messages. To take advantage of the conversation techniques, the following debate dynamics was proposed: the moderator was supposed to present the topic (mediated contribution); next, each participant was to send a message commenting on the topic (circular contribution); and then, they were to choose (single contribution) which of the commentary should be discussed in a free way (free contribution). This cycle—topic, comments, vote and free discussion—should be repeated 3 times. The hypothesis is that the use of conversation techniques would reduce co-text loss, as a result of the global organization of the conversation into well defined stages and not by the local organization of the messages that was made possible by the message thread mechanism aforementioned.

In order to evaluate whether the use of the group conversation techniques would reduce co-text loss, the Mediated Chat 2.0 tool was used during the ITAE 2002.2 edition (second semester 2002). In this evaluation, 8 chat sessions were held, with each one lasting approximately 50 minutes. On average, 10 participants were present and sent 364 messages per debate. A typical chat tool was used during the first 4 debates, and the Mediated Chat 2.0 tool was used in the last 4 ones. Co-text lost occurred during this debates according to the data presented in Figure 4.

---

[5] At the beginning of this research project, there was no available data that indicated the feasibility of using threading in chat tools. Currently, it is possible to find some studies about the use of threaded chats [9]. However, these studies show no data regarding the co-text loss phenomenon.

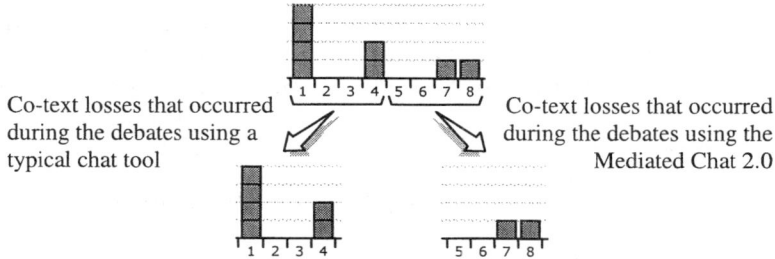

Fig. 4. Co-text loss occurrences observed during the ITAE 2002.2 edition.

Investigating the co-text losses that occurred during ITAE 2002.2, it was clear that they occurred only during the free conversation stage. It is precisely during the branching out phase of the free conversation, that other topics started to be discussed in parallel—a trait that makes co-text loss more likely. In the other stages, the conversation thread is easier to be followed.

Although co-text losses still occurred in the ITAE 2002.2 debates, on average only half of the co-text loss situations that had occurred during the debates of the other editions took place. One possibility is that the debate dynamics envisaged to make use of the conversation techniques forces the reduction of co-text loss occurrences. Although the conversation techniques could be enforced without using the Mediated Chat 2.0 tool—by resorting to authority and social protocol—its computational support stimulated the application of the new dynamics to the debate.

4 Conclusion

Textual chat tools have achieved widespread popularity and increasingly people want to use these tools in activities that go beyond socialization and recreation. In this research project the use of chat tools for running synchronous debates in on-line courses was investigated. In these debates, participants frequently complained about confusion during chat conversations. Among the symptoms that provided evidence of such conversational confusion was the recurrence of co-text loss.

The research presented in this article sought to investigate mechanisms that could render better organized chat conversations. In order to investigate these mechanisms, the research sought to look into whether these mechanisms reduced the manifestations of co-text loss. The results that were obtained indicate that the use of threads and the use of conversation techniques help to reduce the occurrence of co-text loss, increasing the understanding of the conversation. However, these mechanisms only partially solve the problem of co-text loss. In future work, other mechanisms such as the use of multiple pages of text to organize the chat conversation, and the imposition of limitations regarding the quantity of messages sent over a given period of time are going to be investigated.

Acknowledgments. The AulaNet project is partially financed by the Fundação Padre Leonel Franca, by the Ministry of Science and Technology through its Program of Excellence Nuclei (PRONEX) grant nº 76.97.1029.00 (3366), and also through its Multi-Agent Systems for Software Engineering Project (ESSMA) grant nº 552068/2002-0 and by the 1º article of decree 3.800, of the 20th of April of 2001. It is also financed by individual grants awarded by the National Research Council to: Carlos José Pereira de Lucena nº 300031/92-0 and Hugo Fuks nº 303055/02-2. Mariano Gomes Pimentel received an individual grant from the Council for the Improvement of Higher Teaching of the Ministry of Education. We would also like to thank Juliana Lucas de Rezende for the development of the Mediated Chat 2.0 tool.

References

1. Fuks, H., Gerosa, M.A. & Lucena, C. J. P.: The Development and Application of Distance Learning on the Internet. Open Learning – The Journal of Open and Distance Learning, v. 17, n. 1 (Feb 2002) p 23–38
2. Fuks, H.: Groupware Technologies for Education in AulaNet. Computer Applications in Engineering Education, NY, v 8 issue 3 & 4. Wiley-InterScience (December 2000) 170–177
3. Pimentel, M. G., Fuks, H. & Lucena, C. J. P.: Perda de Co-texto em sistemas de Bate-papo Textual. Technical Report, n. 01/03. Computer Science Department of the of the Catholic University of Rio de Janeiro, Brazil (January 2003)
4. Pimentel, M. G.: HiperDiálogo: ferramenta de bate-papo para diminuir a perda de co-texto. M. Sc. Dissertation. Computer Science Department of the Federal University of Rio de Janeiro, Brazil (April 2002)
5. Conklin, J.: Hypertext: An Introduction and Survey. IEEE Computer (Sept 1987)
6. Crystal, D.: A Dictionary of Linguistics and Phonetics. Blackwell, Oxford (1985)
7. Halliday, M. A. K., Hasan, R.: Cohesion in English. London: Longman (1976)
8. Shriberg, E. E.: Preliminaries to a Theory of Speech Disfluencies. Ph. D. Thesis of the University of California at Berkeley, USA (1994)
9. Smith, M, Cadiz, J. J., Burkhalter, B.: Conversation trees and threaded chats. Technical Report (June 2000) On-line document: ftp://ftp.research.microsoft.com/pub/tr/tr-2000-43.doc [Consultation date: 22nd January 2003]
10. Rezende, J. L.: Aplicando Técnicas de Comunicação para a Facilitação de Debates no Ambiente AulaNet. Forthcoming M. Sc. Dissertation. Computer Science Department of the Catholic University of Rio de Janeiro, Brazil (March 2003)

Context Proceduralization in Decision Making

Jean-Charles Pomerol and Patrick Brézillon

LIP6, Université Pierre et Marie Curie
4, Place Jussieu – 75252 Paris Cedex 05
{Jean-Charles.Pomerol, Patrick.Brezillon}@lip6.fr

Abstract. Although it seems obvious that decision making is a contextual task, papers dealing with decision making tackle rarely the problem of contextual information management. After a brief presentation of our view on context, we examine the contextual dimension of decision making. Then we explain our views about the acquisition of contextual data and the construction of a reasoning framework appropriate for decision making. We call this process proceduralization and we refer to a rational construction for action (rca).

Keywords: decision making, contextual knowledge, proceduralized context.

1 Introduction

Starting from the view that contextual knowledge exists as the background knowledge describing a decision situation, we argue that effective action requires the extraction of a subjective part of this contextual knowledge called proceduralized context. In other words a decision problem is interpreted as a case (in the artificial intelligence sense [14, 10, 11]) and the decision maker builds a reasoning for action using some the contextual clues contained in the case. This construction is called proceduralization. In this paper, we discuss of the process of proceduralization under the light of some psychological views and experiments about decision making.

2 Some Reminding about Context

To understand more accurately the relationships between decision making and context we have to be more specific about what we call context. From an engineering point of view, we can start from a definition of the context as the collection of relevant conditions and surrounding influences that make a situation unique and comprehensible [12, 1]. The difficulty with this definition is that there are "numerous interacting factors that people do not even pay attention to on a conscious level, and many of which are outside the ability of machine input devices to capture" [9]. In the above definitions the authors have clearly in mind the fact that the *context is not under the control of the observer*.

Let us take an example, in the control of a subway line [7]: a large amount of knowledge about trains, electricity, people's reactions, etc. contributes to make the situation unique, while some more particular conditions about the time, the day, the weather and so on, influence decision making more specifically. In other words, there is beforehand a common background context which is then specified by some conjectural and contingent influences. For example, the general context is subway control which differs from train or bicycle control although they share some mechanical laws and the particular context is specific to a line, a day, an hour, etc. These considerations explain why Tiberghien [25] defines context as the whole set of secondary characteristics of a situation or secondary properties of a cognitive or motivational state of an individual that can modify the effect of an effective stimulation (stimulus) or an oriented activity.

Thus, it would probably be wise to talk of primary and secondary contexts to distinguish between the general, relatively fixed primary characteristics of a situation, and the secondary characteristics. If we think about primary context, we must confess that it is difficult to avoid the word knowledge about this general background used by the operators to carry out their task. These are some of the reasons why in a previous study [5], we defined three types of context. First, the context which is shared by those involved in the problem and is directly but tacitly used for the problem solving. Second, the context that is not explicitly used but influences the problem solving. Third, the context that has nothing to do with the current decision making but is known by many of those involved. We call these three types of context respectively: proceduralized context, contextual knowledge and external knowledge.

Contextual knowledge is more or less similar to what people generally have in mind about the term 'context'. It contains some general information about the situation and the environment of the problem. Contextual knowledge implicitly delimits the resolution space (this idea is also evoked in [2]). It is always evoked by a task or, an event, but does not focus on a task or on the achievement of a goal but is mobilized according to a set of tasks, even though it has not yet been proceduralized for use. Contextual knowledge is on the back-stage, whereas the proceduralized context is on the front-stage under the spotlights. It is noteworthy that, as far as engineering is concerned, only the proceduralized context matters, but contextual knowledge is necessary because this is the raw material from which proceduralized context is made. One can say that contextual knowledge is proceduralized, not necessarily explicitly, to become the proceduralized context. In a sense, the proceduralized context is the contextual knowledge activated and structured to make diagnoses, decisions and actions. This aspect is discussed in a concrete way in the framework of the contextual graphs [4].

The proceduralized context is a part of the contextual knowledge that is invoked, structured and situated according to a given focus. The proceduralized context is the part of the contextual knowledge which is proceduralized before decision but having the decision making problem in mind. According to the current decision making step, a piece of the contextual knowledge either enters the proceduralized context or becomes external knowledge. Thus, the content of the context evolves continuously all through the decision making. Once the first pieces of contextual knowledge are mobilized, some other pieces of contextual knowledge, such as the position of the incident on the line, also enter the focus of attention and are proceduralized. Finally, whereas the contextual knowledge is rather independent of the subject, the proceduralized context is a personal construct for action.

3 Proceduralization

We argued that a crucial step in dealing with context as regards decision making is the proceduralization step [5, 6]. At this step, the decision maker picks into the back-stage context the data and information which he feels are necessary to make a decision in the case at hand. According to the previous discussion (section 2), the proceduralization acts as a filter upon the contextual knowledge. Actually, the process is twofold. On the one hand it consists of selecting relevant information. Thus, the proceduralized context appears as a sub-part of the whole contextual knowledge. Note that in the selection of the contextual elements for the proceduralized context, there is a part of arbitrary, and, as a consequence, a proceduralized context can be better than another one. On the other hand, the decision makers try to organize the knowledge in such way that the proceduralized context can be used for decision. The problem for the decision maker is then to build the rationales of the observed facts and, if possible, to anticipate the consequences of the possible actions. Let us call rationale construction for action (*rca* hereafter) the second part of the process.

Thus, the context is not only a set of objective characteristics describing a situation as is often claimed (see [24] for some usual views on context and knowledge), but the mental representation (as quoted in Cognitive Ergonomics for representing and interpreting external events) generated by the words of the description as well as the risk attitude also matter as regards decision. In other words, the proceduralization of identical facts depends on the mental representation of the decision maker that constitute a context evoked by the presentation. Presented in a context of loss or mortality, the same facts do not trigger the same decision as presented in a context of gains Tversky and Kahneman [26, 27, 28]. The reader could object that everybody knows that decision maker's mood clearly influences decision making. One can say that an optimistic person does not evaluate the consequences of an action and/or the probabilities of the events in same manner as a pessimistic one. The frame effect is seemingly different, the probabilities as well as the consequences are the same but the evaluation is changed because different visions are associated to the different representations. The difference does not come from the person's mood but from the words that describe the situation and of the brain images associated to the situation. This does not discard the necessity and the possibility to start from an objective description of the facts and an accurate diagnosis, but rises the question of *what is an objective representation in face of representation biases* ? Obviously, there is an interference with the image generated and/or recorded in the mind of a subject in the representation of a context. The sensible representation of the context depends on the experience of the subject.

The acquisition of the contextual facts is selective and is obviously sensitive to some availability and representativeness biases [15, p. 82]. It is likely that more recent or more striking events will be more easily recalled. This view is also common in cognitive science [1]. When the subject can easily get a representation of the situation because he is familiar with, the probability of occurrence and the prominence of the phenomenon have a chance to be increased (representativeness effect, see [27]). Using some prior knowledge, the subject captures a set of contextual facts that are extracted form what we called the contextual knowledge. When these facts are available, the proceduralization consists of structuring this facts in order to make them useful for decisions. The first step is to understand why the situation is that which is observed in order to anticipate the effects of possible actions. During the next step, what is important is to identify the causal and consequential links between the facts: this is the *rca* phase as introduced at the beginning of this section.

The *rca* process is even more prone to biases than selection. First, the subject can interpret positive correlation and contingency as causal rules. Moreover, it has been observed a phenomenon of search for dominance [17,18]. This search for dominance tends to justify *a posteriori* the choice by proving that the chosen action dominates the others. According to our view, this search for dominance is nothing else than a rationalization process either before decision or more generally after the decision. As such, when it occurs before action it is a kind of *rca*. Actually, the process of search for dominance is already carried out during the proceduralization because people privilege the causal and consequential links which reinforce their prior beliefs. To some extent the contextual knowledge is extracted, according to some choice which is already made, in order to justify this choice. As mentioned by Weinberger [29], to make a decision amounts to build a story that makes sense and whose the denouement is precisely the action commanded by the decision. Thus the *rca* process organizes the observed facts, records the ones that fits with the diagnosis on which the subject is anchored [19]. Then, continuing the story, the diagnosis opens the door to some anticipation and various scenarios. Once more, there is no decision without the capability of writing scenarios [21, 22]. This entails that the *rca* process is tale dependent! People very easily adhere to "good stories" even to explain purely contingent events. This is why so frequent are the beliefs of plot and/or purposeful action of powerful leaders when an accident or a catastrophe occurs (see [15, p. 83]; [20, p. 164]). In some sense, magic thought and illusory correlations [8] are frequent, this is a sign of a perverse proceduralization: people proceduralize without any evidence because they need these links to organize the "small world" structuring the decision at hand [3].

In a study about the management of the incidents on an underground line [7] we observed that the contextual knowledge is one of the main components of diagnosis construction. In other words, this means that the operators of the line try to gather as much as possible of contextual elements to know what is the context of the incident, because this context determines their diagnosis and the subsequent actions. We also observed that the proceduralization consists of building a diagnosis consistent with the incoming information. Moreover, the *rca* process also encompasses the design of some scenarios to allow an anticipation of the forthcoming events and of the results of the possible actions. In this process, the uncertainty is reduced by action postponement [22] and the prior gathering of the maximum of information. When one operator chooses to undertake an action, the uncertainty is rather about the current context than about future. In other words, it could be said that if the operator could know the exact context of his action, he could accurately anticipate its result. Thus, in the operator's decision making process, we think that the most important bias are those that are related to knowledge acquisition, namely anchoring, representativeness and availability. They can impede the proceduralization of the contextual knowledge both during the selection and the *rca*.

To sum up, we can say that there is a "construct" more or less rational before decision making occuring during the proceduralization of the contextual knowledge. This process is very important, it must be thoroughly checked and system supported because human beings tends to find "rational causes" even to purely contingent events. *One can say that human are contingency-averse, especially for unfavorable events*! It is thus necessary to bring some rationality for building scenarios and for proceduralizing the contextual knowledge. This is what we tried to do for subway control by using contextual graphs in which the user must specify the context value at each contextual node.

4 Conclusion

The contextual knowledge influences decision making, but data and information that are more or less shared by everybody do not influence *per se*. To understand the role of the contextual knowledge it is necessary to understand the cognitive process that occurs between context, apprehension and decision making. The cognitive biases shed some light on this process.

An essential characteristic of context is its dynamics. First, the understanding of external event depends of the current backstage knowledge available, but that is a knowledge that evolves. Second the decision making process (diagnosis) has also a contextual nature that depends on the decision-maker background, background that itself evolves. This is, in our viewpoint, a reason to make explicit the proceduralization process. As a side effect, there is a need for an incremental knowledge acquisition for improving the continuous building and rebuilding of the mental representation through the proceduralization process.

The first lesson is that the process of context management is twofold: contextual knowledge acquisition and rationale construction for action (*rca*). While the first step of contextual knowledge acquisition and interpretation is subject to the cognitive biases of anchoring and availability the second step is more difficult to describe and analyze.

Context acquisition results into a list of contextual elements which are instantiated when used by somebody. In the two models we recalled in the introduction, this knowledge is easily represented by context nodes in context graphs and by criteria in multicriterion decision. However, let us observe that this last representation is poorly adapted to context use because the criteria generally denote attributes controlled or to be controlled by the decision maker whereas the most significant contextual element which are not under control and are, as such, difficult to apprehend in a multicriterion analysis framework where it turns out that actions and criteria are thought as controlled, the former by definition and the latter by intention.

Actually, the proceduralization step is difficult to model. Let us recall that during this decisive step the subject picks out data and facts in the contextual knowledge in order to build his proceduralized context. This construction is vaguely similar to the conception of a story linking facts and consequences. At this step a structuration of the knowledge occurs resulting in a meaningful organization of the world preparing action (denoted *rca* in the paper). This rationalization separates diagnosis and the contextual elements for the diagnosis of the current state from the anticipated consequences. This is a temporal process for which scenario representation is adapted and, up to now, without no other competitive representation. The result of the proceduralization step is the basis for decision making.

Unfortunately the *rca* process can be impeded by many biases which are described in the literature, see [23] for a survey. Among these biases are illusory correlation [8], illusion of control [13, 16] and reinforcement, all consolidating false inferences, neglecting small probabilities and ignoring unfrequent events and consequently eliminating them from scenarios [19]. There are few possibilities to obviate these cognitive biases but validating scenarios and accumulating knowledge via graphical representations. This is what we tried to promote in the underground control case. It remains that identical contextual knowledge leads to different decisions because the proceduralization process is subjective. Efforts must be done by system designers to facilitate and decrease the subjectivity of the proceduralization process.

References

1. Anderson J.R., 1995, *Cognitive Psychology and its Implications*, Freeman, New York.
2. Bainbridge L., 1997, The change in concepts needed to account for human behavior in complex dynamic tasks. *IEEE transactions on Systems, Man and Cybernetics, 27*, 351–359.
3. Berkeley D. and Humphreys P., 1982, Structuring Decision Problems and the "Bias Heuristic", *Acta Psychologica 50*, 201–252.
4. Brézillon P. (2003) Contextual graphs: A context-based formalism for knowledge and reasoning in representation. To appear in a Research Report, LIP6, University Paris 6, France.
5. Brézillon P. and Pomerol J-Ch., 1999, Contextual Knowledge sharing and cooperation in intelligent assistant systems, *Le Travail Humain 62* (3), PUF, Paris, 223–246.
6. Brézillon P. and Pomerol J-Ch., 2001, Modeling and Using Context for System Development : Lessons from Experience, *Journal of Decision Systems 10*, 265–288.
7. Brézillon P., Pomerol J-Ch. and Saker I., 1998, Contextual and contextualized knowledege, an application in subway control, Special Issue on *Using Context in Applications, International Journal on Human-Computer Studies 48 (3)*, 357–373.
8. Chapman L.J. and Chapman J.P., 1969, Illusory Correlation as an obstacle to the use of valid psychodiagnostic signs, *Journal of Abnormal Psychology 74*, 271–280.
9. Degler D. and Battle L., 2001, Knowledge Management in Pursuit of Performance : the Challenge of Context, http://www.pcd.innovations.com/kminpursuit/id3-m.htm.
10. Gilboa I. and Schmeidler D., 1995, Case-Based Decision Theory, *Quaterly Journal of Economics 110*, 605–639.
11. Gilboa I. and Schmeidler D., 2000, Case-Based knowledge and Induction, *IEEE Transactions on Systems, Man and Cybernetics 30*, 85–95.
12. Hasher L. and Zack R.T. (1984) Automatic processing of fundamental information : the case of frequency of occurrence, *American Psychologist 39*, 1372–1388.
13. Kahneman D. and Lovallo D., 1993, Timid choices and Bold Forecasts: A cognitive perspective on Risk Taking. *Management Science 39*, 17–31.
14. Kolodner J., 1993, *Case-based Reasoning*, Morgan Kaufmann, San-Francisco.
15. Leake, D B, 1991. "Goal-based explanation evaluation" *Cognitive Science* 15(4).
16. McKenna F.P., 1993, It won't happen to me: unrealistic optimism or illusion of control, *British Journal of Psychology 84*, 39–50.
17. Montgomery H., 1983, Decision rules and the search for a dominance structure : towards a process model of decision making. In P.C. Humphreys, O. Svenson and A. Vari (Eds), *Analysing and aiding Decision Processes*, North Holland, 343–369.
18. Montgomery H., 1987, Image theory and dominance search theory : how is decision making actually done ? *Acta Psychologica 66*, 221–224.
19. Morel C., 2002, *Les décisions absurdes*, Galimard, Paris.
20. Piattelli-Palmarini M., 1995, *La réforme du jugement ou comment ne plus se tromper*, Odile Jacob, Paris.
21. Pomerol J-Ch., 1997, Artificial intelligence and human decision making, *European Journal of Operational Research 99*, 3–25.
22. Pomerol J-Ch., 2001, Scenario Development and Practical Decision Making under uncertainty, *Decision Support Systems 31*, 197–204.
23. Pomerol J-Ch., 2002, Decision Making Biases and Context, Brussels DSS Conference, *Journal of Decision Systems* to appear.
24. Pomerol J-Ch. and Brézillon P., 2001, About some relationships between Knowledge and context, Submitted.
25. Tiberghien G., 1986, Context and Cognition: Introduction. *Cahier de Psychologie Cognitive 6(2)*, 105–119.

26. Tversky A. and Kahneman D., 1982a, Judgment under uncertainty: Heuristics and biases, in *Judgment under uncertainty: Heuristics and biases*, Kahneman D., Slovic P. and Tversky A. (Eds.), Cambridge University Press, Cambridge, U.K., 3–20.
27. Tversky A. and Kahneman D., 1982b, Subjective probability: A judgment of representativeness, in *Judgment under uncertainty: Heuristics and biases*, Kahneman D., Slovic P. and Tversky A. (Eds.), Cambridge University Press, Cambridge, U.K., 32–47.
28. Tversky A. and Kahneman D., 1988, Rational Choice and the Framing of Decisions, in *Decision Making*, Bell D.E. *et al.* (Eds.), 167–192.
29. Weinberger D., 2001, Garbage in, Great Staff Out, *Harvard Business review 79 n°8*, 30–32.

GRAVA: An Architecture Supporting Automatic Context Transitions and Its Application to Robust Computer Vision

Paul Robertson[1] and Robert Laddaga[2]

[1] Dynamic Object Language Labs, Inc.,
9 Bartlet St #334, Andover MA 01810, USA,
probertson@doll.com
[2] Massachusetts Institute of Technology,
Artificial Intelligence Laboratory,
NE43-804, Cambridge MA, USA,
rladdaga@ai.mit.edu

Abstract. Conventional approaches to most image understanding problems suffer from fragility when applied to natural environments. Complexity in Intelligent Systems can be managed by breaking the world into manageable contexts. GRAVA supports robust performance by treating changes in the program's environment as context changes. Automatically tracking changes in the environment and making corresponding changes in the running program allows the program to operate robustly.
We describe the architecture and explain how it achieves robustness. GRAVA is a reflective architecture that supports self-adaptation and has been successfully applied to a number of visual interpretation domains.

1 Introduction

Image understanding programs have tended to be very brittle and perform poorly in situations where the environment cannot be carefully constrained. Natural vision systems in humans and other animals are remarkably robust. The applications for robust vision are myriad. Robust vision is essential for many applications such as mobile robots, where the environment changes continually as the robots moves, and robustness is essential for safe and reliable operation of the robot.

Although the complexity of the real world is overwhelming the complexity does not assert itself at the same time. At any instant a program is operating within a *context* in which the complexity is bounded. In AI we have been fairly successful at building systems that perform robustly within environments of restricted complexity. If we can divide the complexity of the world up into a collection of contexts, each of a bounded and manageable size, we can in principle consider the hard problem of making a robust embedded system as an easier problem formulated as the composition of a collection of manageable parts.

Even if we could know all the different states that the environment could be in, we wouldn't know *a priori* what state the environment would be in at any particular time. Consequently, in order to achieve robust image understanding, programs should determine the state of the environment at runtime and adapt to the environment that is found. In practice it is likely that the set of possible contexts cannot be explicitly enumerated *a priori*.

A premise of the self-adaptive approach is that it should be possible, at runtime, to synthesize context specific systems, to determine the need to change context and to self-adapt the program so that the program's context matches the state of the environment and operates robustly because each of its components is operating well within their optimal range.

The idea of self-adaptation is to adapt the program to a particular "context". In order to achieve this adaptation we build structural descriptions that facilitate dividing the model space into contexts and provide a mechanism for determining when a context is a good fit to the environment.

Many of the ideas prevalent in natural language and speech understanding have direct counterparts in computer vision. The first application of the GRAVA architecture [1] was to the interpretation of satellite aerial images. In that program satellite images were segmented into regions of homogeneous content and the regions were parsed, much as words are in a sentence to form a structural understanding of the image. Different image types are comprised of different kinds of regions, different colors and textures, and different parse rules. Rather than making one huge grammar that includes all textures and region types, it is better to have grammars, and optical models tailored to the context because tailored contexts provide greater accuracy and constraint. In that program the contexts as well as the grammars and region content models were learned from a corpus of images annotated by a human photo interpreter.

In this paper we describe GRAVA, an architecture for building self-adaptive programs, and describe its theory of operation.

2   An Overview of the GRAVA Architecture

Vision (and Robotics) systems lack robustness. They don't know what they are doing, especially when things change appreciably (i.e. in situations where technologies such as neural nets are ineffective).

Reflective architectures—an idea from AI—offer an approach to building programs that can reason about their own computational processes and make changes to them.

The reflective architecture [2,3] allows the program to be aware of its own computational state and to make changes to it as necessary in order to achieve its goal.

Much of the work on reflective architectures has been supportive of human programmer adaptation of languages and architectures rather than self-adaptation of the program by itself.

Our use of reflection allows the self-adaptive architecture to reason about its own structure and to change that structure.

2.1 Interpretation Problems

The problem of self-adaptive software is to respond to changing situations by re-synthesizing the program that is running.

Each component of the system "knows" what it is doing to the extent that it knows what part of the level above it implements (interprets).

The purpose of the reflective architecture is to allow the image interpretation program to be aware of its own computational state and to make changes to it as necessary in order to achieve its goal. The steps below provide a schematic introduction to the GRAVA architecture.

1. The desired *behavior* is specified in the form of statistical models by constructing a corpus.
2. The behavior, which covers several different imaging scenarios, is broken down into contexts. Contexts exist for different levels of the interpretation problem. Each context defines an expectation for the computational stage that it covers. Contexts are like frames but because the contexts are gathered from the data automatically it is not necessary to define them by hand.
3. Given a context a program to interpret the image can be generated from that context. This is done by *compiling* the context into a program by selecting the appropriate agents.
4. The program that results from compiling a context can easily know the following things:
   a) What part of the specification gave rise to its components.
   b) Which agents were involved in the creation of its components.
   c) Which models were applied by those agents in creating its components.
   d) How well suited the current program is to dealing with the current input.
5. The division of knowledge into agents that perform basic image interpretation tasks and agents that construct programs from specifications is represented by different reflective levels.

2.2 Reflective Interpreter for Self-Adaptation

Unlike traditional implementations, which have largely been supportive of human programmer adaptation of languages and architectures, we use reflection as a way of supporting self-adaptation of the program *by itself*. There are two principal differences in our use of reflection:

1. We open up the program to itself so that by knowing what it knows it can use what it knows to alter itself in order to respond to changes in the real world.
2. We do not wish to change the semantics of the program/language, we wish to change the program itself.

A reflective layer is an object that contains one or more "interpreter". Reflective layers are stacked up such that each layer is the meta-level computation of the layer beneath it. In particular each layer is generated by the layer above it.

Each layer can reflect up to the layer above it in order to self-adapt. The prototype GRAVA implementation is written in Yolambda [4] a dialect of Scheme [5]

```
(defineClass ReflectiveLayer
  ((description) ;; the (input) description for this layer
   (interpreter) ;; the interpreter for this layers description
   (knowledge)   ;; a representation of world knowledge
   (higherlayer) ;; the meta-level above this
   (lowerlayer)));; the subordinate layer
```

A reflective layer is an object that contains the following objects.

1. *description:* the description that is to be interpreted.
2. *interpreter:* a system consisting of one or more cascaded interpreters that can interpret the description.
3. *knowledge:* a problem dependent representation of what is known about the world as it pertains to the interpretation of the subordinate layer. For the face identification application knowledge consists of evidence accumulated from agents supporting each of the contexts (age, race, sex, lighting, and pose).
4. *higherlayer:* the superior layer. The layer that produced the interpreter for this layer.
5. *lowerlayer:* the subordinate layer.

The semantics for a layer are determined by the *interpret, elaborate, adapt* and *execute* methods which we describe in turn below.

Figure 1 shows the relationship between reflective layers of the GRAVA architecture.

Reflective Layer "n" contains a description that is to be interpreted as the description for layer "n+1". A program has been synthesized either by the layer "n-1" or by hand if it is the top layer. The program is the interpreter for the description. The result of running the interpreter is the most probable interpretation of the description—which forms the new description of the layer "n+1". All the layers (including "n") also contain a compiler. Unless the layer definition is overridden by specialization, the compiler in each layer is identical and provides the implementation with a theorem prover that compiles an interpreter from a description. The compiler runs at the meta level in layer "n" and uses the knowledge of the world at layer "n+1" which resides in level "n". It compiles the description from level "n+1" taking in to account what is known at the time about level "n+1" in the *knowledge* part of layer "n". The compilation of the description is a new interpreter at layer "n+1".

Below we describe the meta-interpreter for layers in GRAVA.

The interpret method is the primary driver of computation in the reflective architecture. The reflective levels are determined by the program designer. In order for the self-adaptive program to "understand" its own computational

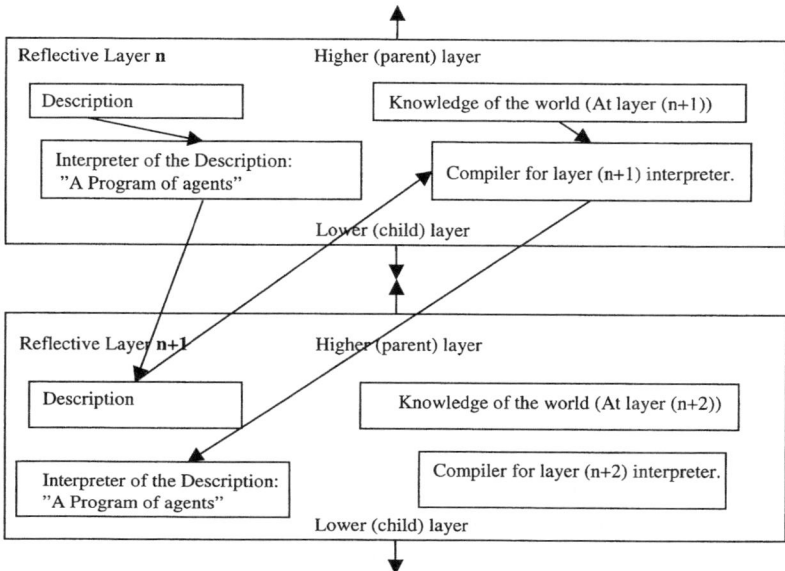

Fig. 1. Meta-knowledge and Compilation

structure, each layer describes the layer beneath it. In self-adapting, the architecture essentially searches a tree of meta-levels. This is best understood by working through the details of the architecture.

In the simplest of situations the top level application of "interpret" to the top layer results in the recursive descent of "interpret" through the reflective layers finally yielding a result in the form of an interpretation. Along the way however unexpected situations may arise that cause the program to need to adapt. Adaptation is handled by taking the following steps:

1. Reflect up to the next higher layer (parent level) with an object that describes the reason for reflecting up. It is necessary to reflect up because the higher level is the level that "understands" what the program was doing. Each level "understands" what the level directly beneath it is doing.
2. The world model (knowledge) that is maintained by the parent level is updated to account for what has been learned about the state of the world from running the lower level to this point.
3. Given updated knowledge about the state of the world the lower level is re-synthesized. The lower level is then re-invoked.

Armed with that conceptual overview of the interpret procedure we now explain the default interpret method.

```
1:(define (interpret ReflectiveLayer|layer)
2:   (withSlots (interpreter description lowerlayer) layer
3:     (if (null? interpreter)
4:        description
5:        (begin (elaborate layer))
6:              (reflectProtect (interpret lowerlayer)
7:                 (lambda (layer gripe) (adapt layer gripe))))))

8:(define (reflectionHandler ReflectiveLayer|layer gripe)
9:   (adapt layer gripe))
```

Line 3 checks to see if the layer contains an interpreter. If it does not the result of evaluation is simply the description which is returned in line 4. This occurs when the lowest level has been reached.

If there is an interpreter, the elaborate method is invoked (line 5). "elaborate" (described below) constructs the next lower reflective layer.

"reflectProtect" in line 6 is a macro that hides some of the mechanism involved with handling reflection operations.

(reflectProtect *form handler*) evaluates *form* and returns the result of that evaluation. If during the evaluation of *form* a reflection operation occurs the *handler* is applied to the layer and the gripe object provided by the call to reflectUp. If the handler is not specified in the reflectProtect macro the generic procedure reflectionHandler is used. The invocation of the reflection handler is not within the scope of the reflectProtect so if it calls (reflectUp ...) the reflection operation will be caught at the next higher level. If reflectUp is called and there is no extant reflectProtect the debugger is entered. Therefore if the top layer invokes reflectUp the program lands in the debugger.

When the reflection handler has been evaluated the reflectProtect re-evaluates the *form* thereby making a loop. Line 7 is included here to aid in description. It is omitted in the real code allowing the reflectionHandler method to be invoked. The handler takes care of updating the world model based on the information in *gripe* and then adapts the lower layer. The handler therefore attempts to self adapt to accommodate the new knowledge about the state of the world until success is achieved. If the attempt to adapt is finally unable to produce a viable lower level interpreter it invokes reflectUp and causes the meta level interpretation level to attend to the situation.

```
1:(define (elaborate ReflectiveLayer|layer)
2:   (withSlots (lowerlayer) layer
3:     (let ((interpretation (execute layer))
4:           (llint (compile layer interpretation)))
5:        (set! lowerlayer ((newLayerConstructor layer)
6:                          higherlayer: layer
7:                          description: interpretation
8:                          interpreter: llint)))))
```

The purpose of the elaborate layer is to build the initial version of the subordinate layer. It does this in three steps:

1. Evaluate the interpreter of the layer in order to "interpret" the layer's description. The interpretation of $layer_n$ is the description of $layer_{n+1}$.
   Line 3 invokes the interpreter for *layer* with *(execute layer)*. This simply runs the MDL agent interpreter function defined for this layer. The result of executing the interpreter is an interpretation in the form of a description.
2. Compile the layer. This involves the collection of appropriate agents to interpret the description of the lower layer.
   Line 4 compiles the new layer's interpreter. Layer n contains knowledge of the agents that can be used to interpret the description of layer $n + 1$. The description generated in line 3 is compiled into an interpreter program using knowledge of agents that can interpret that description.
3. A new layer object is instantiated with the interpretation resulting from (1) as the description and the interpreter resulting from *compile* in step (2) as the interpreter. The new layer is wired in to the structure with the bi-directional pointers (lowerlayer and higherlayer).
   In line 5, (newLayerConstructor layer) returns the constructor procedure for the subordinate layer.

The adapt method updates the world state knowledge and then recompiles the interpreter for the lower layer.

```
1:(define (adapt ReflectiveLayer|layer gripe)
2:   (withSlots (updateKnowledge) gripe
3:     (updateKnowledge layer))   ;; update the belief state.
4:   (withSlots (lowerlayer) layer
5:     (withSlots (interpreter) lowerlayer
6:       (set! interpreter (compile layer)))))
```

The representation of world state is problem dependent and is not governed by the reflective architecture. In each layer the world state at the corresponding meta level is maintained in the variable "knowledge". When an interpreter causes adaptation with a reflectUp operation an update procedure is loaded into the "gripe" object. Line 3 invokes the update procedure on the layer to cause the world state representation to be updated.

Line 6 recompiles the interpreter for the lower layer. Because the world state has changed the affected interpreter should be compiled differently than when the interpreter was first elaborated.

```
1:(define (execute ReflectiveLayer|layer)
2:   (withSlots (description interpreter knowledge) layer
3:     (run interpreter description knowledge)))
```

3 Conclusion

The GRAVA architecture has been successfully applied to a number of problems including the aerial image interpretation problem and a person/face recognition and tracking project (ongoing) [6,7]. Although the architecture doesn't depend upon corpus based methods the problem of generating the large number of models required for such a system to operate robustly makes corpus based systems particularly attractive.

We have developed methods for automatically inducing contexts from annotated corpora [8].

Since contexts are not random but are structurally related, transitions between contexts can be modeled as hidden Markov models (HMM) [9,10]. We are currently extending the architecture described in the paper to use HMM reasoning to optimize the context switching mechanism. We will present the results of this research at a later date.

Although GRAVA was developed as an architecture for building robust vision programs and so far has only been applied to vision problems there is no reason why, in principle, that GRAVA could not be applied to other interpretations including the interpretation of speech and natural language.

References

1. Robertson, P.: A Self-Adaptive Architecture for Image Understanding. PhD thesis, University of Oxford (2001)
2. Maes, P., Nardi, D.: Meta-Level Architectures and Reflection. North-Holland (1988)
3. Giroux, S.: Open reflective agents. In M. Wooldridge, J.P.M., Tambe, M., eds.: Intelligent Agents II Agent Theories, Architectures, and Languages. Springer (1995) 315–330
4. Laddaga, R., Robertson, P.: Yolambda Reference Manual. Dynamic Object Language Labs, Inc. (1996)
5. IEEE: Ieee standard for the scheme programming language. IEEE Standard 1178-1990, IEEE Pisscaataway (1991)
6. Robertson, P., Laddaga, R.: A self-adaptive architecture and its aplication to robust face identification. In: PRICAI 2002, Trends in Artificial Intelligence, Springer Verlag, LNAI 2417 (2002) 542–551
7. Robertson, P., Laddaga, R.: An agent architecture for information fusion and its application to robust face identification. In: Proceedings of the 21st International Conference on Applied Informatics, Innsbruck Austria. (2003) 132–139
8. Robertson, P., Laddaga, R.: Principle component decomposition for automatic context induction. In: Proceedings Artificial and Computational Intelligence, Tokyo 2002, ACI (2002)
9. Viterbi, A.: Error bounds for convolution codes and an asymptotically optimal decoding algorithm. IEEE Transactions on Information Theory **13** (1967) 260–269
10. Baum, L.: An inequality and associated maximization technique in statistical estimation for probabilistic functions of a markov process. Inequalities **3** (1972) 1–8

Speaking One's Mind

Alice G.B. ter Meulen

Center for Language and Cognition, University of Groningen,
Groningen, The Netherlands
atm@let.rug.nl

Abstract. Aspectual adverbs serve to create temporally coherent contexts. When prosodically marked by high pitch accents, they modify factual information while making direct reference to epistemic states of the agent regarding the actual or expected flow of events. Dynamic inferences background the agent's epistemic state, while adding new temporal information about events. In dialogue or multi agent settings this triangulation process serves to create common ground, to adjust plans after change in context and constrain the accommodation of presuppositions.

1 Introduction

In ordinary English aspectual adverbs such as *still, already, finally* and *still not* are used to create temporal coherence in local contexts, as well as to carry prosodic features that directly indicate the speaker's epistemic attitudes. Factual content may be questioned, accepted as true or rejected as false by the receiver. Information about his attitudes a speaker issues with first person authority, hence it is guaranteed to be veridical and directly referential, as it is caused by privileged access. Aspectual adverbs constitute hence a good case study of how temporal information is shared in the common ground by triangulation between two agents in the world (Davidson, 2001). For instance, when I say to you *"John is still asleep"*, you have reason to attribute to me the factual belief that John fell asleep sometime ago and has not woken up since. If the adverb *still* was prosodically marked in my utterance with a high pitch accent, you additionally have ground to attribute to me an epistemic state, i.e. the counterfactual expectation that John should have been awake by now. Anyone capable of communicating in English is able to distinguish the factual content of my utterance, which I invite you to share or question, from the subjective information about my mental state, privately owned and issued with first person authority.

As core indicators of such epistemic states, prosodically marked aspectual adverbs have an important role to play in contexts of planning, negotiating and reasoning in time, where information inferred at some time may require later adjustment to be preserved. Attribution of an epistemic state to another agent creates a context in which further communication is interpreted. Plans may have to be adjusted accordingly, when events happen earlier or later than expected. For example, if I tell you *"John is already asleep. Let's have dinner at 8."* you understand that, since we don't expect John to be at dinner and he fell asleep early, I propose to adjust our plan to have dinner to an earlier time, i.e. at 8. My subjective expectation that John would fall asleep later than he actually did, indicated by *already*, creates a context in which our common dinner plan is proposed to be advanced to an earlier time.

In addition, coherence of context is created by constraining the accommodation of presuppositions of the aspectual adverbs, requiring the presupposition to be verifiable in the given context. If the presupposition conflicts with asserted information introduced in the most recent update, accommodation cannot resolve it by changing the reference time to a later one,. This will lead us to conclude that a presupposition is not an as-if assertion, and only asserted content may induce context change.

The objectives of this paper are threefold:
(i) present a dynamic semantics of aspectual adverbs,
(ii) analyse how temporally coherence is created in context, preserving certain information while modifying other,
(iii) model the triangulation process of updating the common ground and the epistemic states of two agents in the world by extending the tools of Discourse Representation Theory ([2], [3]).

The paper offers an account of reasoning in time which exploits the aspectual adverb semantics, contributing factual content as well as information about agents' attitudes. Inferences remain valid when the facts change, if updates add locally consistent information and shift the reference time to a later one, whenever new facts are added which are incompatible with the given context. This temporal logic avoids any appeal to default assumptions or normal possible worlds in modelling our reasoning in time about time, regaining a sharp division between what may be inferred as valid conclusion in dynamic temporal logic and what may be inferred based on common knowledge of the world and causal relations.

2 The Dynamic Semantics of Aspectual Adverbs

This paper addresses how aspectual adverbs such as *still*, *not yet*, *already*, and *finally* are used in temporal reasoning, where the context may change during the interpretation of the premises. Syntactically, aspectual adverbs occur within INFL in VP clauses describing states, as in (1), but semantically they also contribute information about the events which initiate or terminate the described state.

1) John is not yet (*ne pas encore*)/already (*déja*) /still (*encore*)/no longer (*ne plus*)asleep.

What exacty is the information the aspectual adverbs contribute to the descriptive content? We seem able to infer from premises without any aspectual adverb, a conclusion which does contain one, as in (2).

2) a. When Mary arrived, John was asleep
b. John woke up
c. Bill left
d. |= When Bill left, John was *no longer* asleep

Eventhough there are no aspectual adverbs in premises in (2a-c), *no longer* in (2d) supports a valid conclusion. When *still* is added to the first premise of (2), this conclusion remains valid, as in (3).

3) a. When Mary arrived, John was *still* asleep
   b. John woke up
   c. Bill left
   d. |= When Bill left, John was *no longer* asleep

The occurrence of *still* in (3a) does not affect the conclusion in (3d), so at least in this inference *still* seems not to contribute any information at all. But it would be hard to maintain that *still* never expresses meaningful information, for in English it is used with high pitch intonation, ordinarily indicative of new and important content.

4) When Mary arrived, John was *STILL (H*H%)* asleep

As everyone competent in English understands, the marked prosody in (6) indicates that the speaker had counterfactually expected/hoped/feared that John would have woken up before Mary arrived. We use (6) to express some form of dissatisfaction or even irritation with the actual course of events. What the speaker is dissatisfied with systematically depends on the propositional content of the clause modified by the aspectual adverb. It is a matter of rhetorics or general pragmatics to determine in each context which attitude the speaker means to convey by the marked prosody, ranging from hope, fear, to expectation or even trust. In this paper the speaker's attitude will be left variable, using ATT as a generic relation of a speaker's attitude towards the content.

For the semantics of aspectual adverbs it is important to determine the presuppositions which any clause shares with its corresponding question and internally negated form. Proper answers to polarity questions share the presuppositions of the question. When the presupposition of the question is not accepted as common ground, another, stronger form of negation. i.e. denial (5d), must be used. In (5ab) *still* and *no longer* are seen to share a presupposition, not shared by *already* in (5c), and denied by *not yet* in (5d).

5) a. Was John still asleep when Mary arrived?
   b. No, he was no longer asleep.
   c. * No, he was already asleep.
   d. No, he had not even fallen asleep yet.

The interaction in (5) with internal and external negation shows that the four basic aspectual adverbs are clearly related by polarity in their temporal meaning. This relation is clarified below in the DRT-style of semantic representation.

In a similar vein, presuppositions of aspectual adverbs within clauses in discourse cannot be accommodated when the context created by a prior clause contains information that conflicts with them. Accommodating a presupposition in a context which does not already entail it is hence a much more constrained process, if it is considered a general repair strategy (cf. [1]). If (6) is assumed to constitute coherent discourse, it is clear that the presuppositions of falling or being asleep, i.e., being awake, cannot be accommodated when the context contains at the current reference time the contradictory information that the referent is asleep.

6) a. ?* John was already asleep. John fell asleep.
   b. ?* John was no longer asleep. John was asleep.

When the initial context containing the incompatible information is updated by asserting the information, instead of presupposing it, the reference time is shifted, so there is no problem whatsoever in coherently incorporating the content of the subsequent clauses in the resulting context, as in (7).

7) a. John was already asleep. He woke up and then he fell asleep again.
   b. John was no longer asleep. He fell asleep again, so then he was asleep.

Assuming an overall constraint of coherence of information, adding information by presupposition accommodation must hence be distinguished from asserting information. (6a/7a). Asserting information that is entailed by preceding clause creates incoherence, (6b/7b). Asserting information that is inconsistent with content of preceding clause cannot be repaired by first accommodating its presuppositions.

Aspectual adverbs modify the factual content contained in the clause in their scope. E. g. *John was already asleep* entails that John fell asleep at some point in the past, and *John is still not asleep* entails that John is not asleep yet, but falling asleep. It would be not just odd, but really misleading to state that John is still not asleep, if he is involved in actions, like jogging or cooking dinner, that are obviously incompatible with his falling asleep.

An informal model for the relative position of the aspectual adverbs within an interval modelling the onset and end of an event, is given in (8). We see that *not yet* and *no longer* relate to the negative phases, before its start and after the end, and *already* and *still* both relate to the positive phase of the event.

8) informal model:
   -NOT YET---------- |start| ----ALREADY------------STILL--|end|----NO LONGER----
   ---------------   + + + + + + + + + + + ++ + +   - - - - - - - - - -

To make this informal model more specific, the DRT techniques of declaring reference markers and relating these in conditions with descriptive predicates produces the following representations for the four basic aspectual adverbs.[1]

9) *John is already asleep*
   [r_0, r_1, e, j | PROG (sleep($e, j, +$)) & $e \supseteq r_0$ & r_0 = current &
   $r_1 \supseteq$ START(sleep($e, j, +$)) & $r_1 < r_0$ & SINCE(r_1, (sleep ($e, j, +$)))])]

In (9) John's sleeping is anchored to the current reference time and its presupposition that he fell asleep earlier is added by representing the aspectual adverb. Ordinarily presuppositions are not automatically included in the representation of a clause, although they can be added by presupposition accommodation. The event of falling asleep is telic, i.e. it does not contain subevents which themselves are events of falling asleep. It is included hence in a preceding reference time, and this reference time itself must introduced. The last condition specifies that the onset of John's sleeping was not just any past event of his falling asleep, but the one after which he remained asleep, i.e. the last time John fell asleep. The temporal adverb *since* serves to create this binding of John's falling asleep to his current state of being asleep.[2]

The other three aspectual adverbs are represented in (10)-(12), systematically using SINCE/UNTIL and the precedence order to reflect their polarity relations.

---

[1] The reader unfamiliar with DRT semantics is referred to [3] or [9] for an introduction.
[2] Cf. [3]. 628-635 for a discussion of the semantics of *since* and *until* in temporal contexts.

10) *John is still asleep*
   [r_0, r_1, e, j | PROG (sleep(e, j, +)) & e $\supseteq r_0$ & r_0 = current &
   $r_1 \supseteq$ END(sleep(e, j, +)) & $r_0 < r_1$ & UNTIL (r_1, (sleep (e, j, +)))])]

11) *John is not yet asleep*
   [r_0, r_1, e, j | PROG (sleep(e, j, -)) & e $\supseteq r_0$ & r_0 = current &
   $r_1 \supseteq$ START(sleep(e, j, +)) & $r_0 < r_1$ & UNTIL (r_1, (sleep (e, j, -)))])]

12) *John is not asleep anymore*
   [r_0, r_1, e, j | PROG (sleep(e, j, -)) & e $\supseteq r_0$ & r_0 = current &
   $r_1 \supseteq$ END(sleep(e, j, +)) & $r_1 < r_0$ & SINCE(r_1, (sleep (e, j, -)))])]

The DRT-construction rules for these adverbs are given in (13) in a simplied format.

13) a. [$_{IP}$ x [$_{INFL}$ *already* [$_{VP}$ λ y P (y)]]] => [r_0, r_1, e, x |
       PROG (P(e, x, +)) & e $\supseteq r_0$ & $r_1 \supseteq$ START(P(e, x, +)) & $r_1 < r_0$ &
       SINCE(r_1, (P (e, x, +)))])]

   b. [$_{IP}$ x [$_{INFL}$ *still* [$_{VP}$ λ y P (y)]]] => [r_0, r_1, e, x |
       PROG (P(e, x, +)) & e $\supseteq r_0$ & $r_1 \supseteq$ END (P(e, x, +)) & $r_0 < r_1$ &
       UNTIL(r_1, (P (e, x, +)))])]

   c. [$_{IP}$ x [$_{INFL}$ *not yet* [$_{VP}$ λ y P (y)]]] => [r_0, r_1, e, x |
       PROG (P(e, x, -)) & e $\supseteq r_0$ & $r_1 \supseteq$ START (P(e, x, +)) & $r_0 < r_1$ &
       UNTIL(r_1, (P (e, x, -)))])]

   d. [$_{IP}$ x [$_{INFL}$ *not* [$_{VP}$ λ y P (y)] *anymore*]] => [r_0, r_1, e, x |
       PROG (P(e, x, -)) & e $\supseteq r_0$ & $r_1 \supseteq$ END (P(e, x, +)) & $r_1 < r_0$ &
       SINCE(r_1, (P (e, x, -)))])]

The semantics of aspectual adverbs makes it possible to give information describing a state, but relating it in context to its future or past polarity transition in a systematic way. This constitutes an essentially indexical account of the aspectual adverbs and forms the basis for the semantics of the prosodically marked usage of aspectual adverbs presented in the next section.

3 Prosodically Marked Aspectual Adverbs

If aspectual adverbs are prosodically marked by a high pitch accent followed by a brief boundary tone, this indicates that the current course of events is different from what the speaker had subjectively thought the course of events would be like. The exact nature of the attitude of the speaker may vary greatly from one context to another, and is apt to be misunderstood by the recipient. To abstract from such intricacies of rhetorics, we use here the generic attitude ATT, relating the speaker (sp) to the onset or termination point of the described event.

Marked prosody is not naturally expressed on *not yet*, but English has an extensionally equivalent lexicalization, which does accept the prosody in *STILL not*. There may be phonological reasons why the first does not provide a suitable structure to carry such prosody. An answer to this issue would lead us much beyond the scope of the current paper. Accordingly in (14) the prosodically marked *STILL not* creates a contrast between the actual course of events, and what the speaker subjectively had thought it would be. It indicates that the actual course of events is slow in the eyes of the speaker, i.e. in his subjective state John would actually be asleep. His falling asleep should have been earlier, switching UNTIL to its counterpart SINCE to create the desired temporal binding.

14) *John is STILL NOT asleep*
   $[r_0, r_1, e, j \mid$ PROG (sleep(e, j, -)) & e $\supseteq r_0$ & r_0 = current &
   $r_1 \supseteq$ START(sleep(e, j, +)) & $r_0 < r_1$ & UNTIL (r_1, (sleep (e, j, -)))])] &
   ATT (sp, [- $\mid r_1 < r_0$ & SINCE(r_1, (sleep (e, j, +)))])]

Now it is easy to see what the other prosodically marked forms of the aspectual adverbs should be. Again, English has no prosodic marking for *not anymore*, as it uses *no longer* to express the contrastive presuppositions. The positive phase adverbs *already* and *still* are easily used with marked prosody.

15) *John is STILL asleep*
   $[r_0, r_1, e, j \mid$ PROG (sleep(e, j, +)) & e $\supseteq r_0$ & r_0 = current &
   $r_1 \supseteq$ END(sleep(e, j, +)) & $r_0 < r_1$ & UNTIL (r_1, (sleep (e, j, +)))])] &
   ATT (sp, [- $\mid r_1 < r_0$ & SINCE(r_1, (sleep (e, j, -)))])]

16) *John is no LONGER asleep*
   $[r_0, r_1, e, j \mid$ PROG (sleep(e, j, -)) & e $\supseteq r_0$ & r_0 = current &
   $r_1 \supseteq$ END(sleep(e, j, +)) & $r_1 < r_0$ & SINCE (r_1, (sleep (e, j, -)))])]
   ATT (sp, [- $\mid r_0 < r_1$ & UNTIL(r_1, (sleep (e, j, +)))])] &

17) *John is ALREADY asleep*
   $[r_0, r_1, e, j \mid$ PROG (sleep(e, j, +)) & e $\supseteq r_0$ & r_0 = current &
   $r_1 \supseteq$ START(sleep(e, j, +)) & $r_1 < r_0$ & SINCE (r_1, (sleep (e, j, +)))])] &
   ATT (sp, [- $\mid r_0 < r_1$ & UNTIL(r_1, (sleep (e, j, -)))])]

The contrasts induced by the prosodically marked aspectual adverbs always concern the speed with which the current course of events develops. Using *ALREADY* and *NO LONGER* the speaker registers her surprise at how early the transition took place, whereas with *STILL* and *STILL NOT* she indicates the transition should have taken place and hence its being late disappoints her. It is remarkable how much information is added to the meaning of the original basic four aspectual advers by the effective and efficient way of prosodically marking the corresponding expressions,

4 Dynamic Inference, Presuppositions, and Logical Entailment

In DRT the notion of logical consequence relates two DRSs, as stated in (17), requiring that a DRS K' is a logical consequence of DRS K representing the premises iff. any verifying embedding into the underlying models of the conditions in K can be extended to a verifying embedding of the conditions in K' (i.e. there is no model in which to make the premises true and the conclusion false). A DRS is pure just in case all reference markers used in its conditions are declared at that level or at a higher one. The conclusion cannot add any new reference markers in other words, doing justice to the static nature of drawing a conclusion from given information.

   DRT Definition of logical consequence
   Let K, K' be pure (...) DRSs. Thus K' is a logical consequence of K (K $\models$ K') iff the following condition holds: Suppose M is a model and f is a function from UK $\cup$ Fr(K) $\cup$ Fr(K') into UM, such that M $\models_f$ K. Then there is a function g $\supseteq$ UK' f such that M $\models_g$ K'. ([3], 305.)

The DRT account of aspectual adverbs declares at the top level the reference markers for reference times, which may also occur in K', when the conclusion is added to the given DRS K for the premises. Inferences based on aspectual adverbs should preserve the information about the subjective epistemic states of agents, while adding factual information about changes in the world by introducing new reference markers. For instance, in (18) it is first expressed in (18a) that John's falling asleep is late according to the speaker, but then the factual information is added in (18b) that it did happen, so the conclusion in (18c) is that he is now asleep but the speaker still found it late to happen. Attitudes are typically not affected when the facts of the world change. But after the factual change, the speaker must express his attitude with a ddiferent adverb which characterizes the polarity switch, e.g. in (18c) *finally*.

18) a. John was STILL not asleep.
    b. John fell asleep.
    c. So John was finally asleep.

18 a) $[r_0, r_1, e, j \mid \text{PROG (sleep}(e, j, -)) \& e \supseteq r_0 \& r_0 = \text{current} \&$
    $r_1 \supseteq \text{START(sleep}(e, j, +)) \& r_0 < r_1 \& \text{UNTIL } (r_1, (\text{sleep } (e, j, -)))])] \&$
    $\text{ATT (sp, [- } \mid r_1 < r_0 \& \text{SINCE}(r_1, (\text{sleep } (e, j, +)))])]$

18b) $[..., r_2 \mid r_2 \supseteq \text{START(sleep}(e, j, +)) \& r_0 < r_2 \& r_2 = \text{current}$

18c) $[r_0, r_1, r_2, e, j \mid \text{PROG (sleep}(e, j, +)) \& e \supseteq r2 \&$
    $\text{ATT (sp, [- } \mid r_1 < r_0 \& \text{SINCE}(r_1, (\text{sleep } (e, j, +)))])]$

The conclusion is verified at r_2 representing the attitude of the speaker that John should have fallen asleep before the first reference time r_0, which precedes r_2, by transitivity of the temporal precedence relation.

As illustration of an invalid dynamic inference with aspectual adverbs, consider (19). As in (18), the DRS for the premises (19a) introduces the speakers attitude that John should have fallen asleep earlier, i.e. he is late in falling asleep. This expectation is not affected when (19b) adds, as in (19b) above, the factual information that he fell asleep subsequently. Now we cannot consistently add (19c), because it would create a conflict in the speakers attitude, as she suddenly would be surprised by that fact that John has fallen asleep earlier than expected.

19) a. John was STILL not asleep.
    b. John fell asleep.
    c. $\not\models$ John was ALREADY asleep.

19 a) $[r_0, r_1, e, j \mid \text{PROG (sleep}(e, j, -)) \& e \supseteq r_0 \& r_0 = \text{current} \&$
    $r_1 \supseteq \text{START(sleep}(e, j, +)) \& r_0 < r_1 \& \text{UNTIL } (r_1, (\text{sleep } (e, j, -)))])$
    $\& \text{ATT (sp, [- } \mid r_1 < r_0 \& \text{SINCE}(r_1, (\text{sleep } (e, j, +)))])]$

19b) $[..., r_2 \mid r_2 \supseteq \text{START(sleep}(e, j, +)) \& r_0 < r_2 \& r_2 = \text{current}$

19c) $[r_0, r_1, r_2, e, j \mid \text{PROG (sleep}(e, j, +)) \& e \supseteq r_0 \& r_0 = \text{current}$
    $r_1 \supseteq \text{START(sleep}(e, j, +)) \& r_1 < r_0 \& \text{SINCE } (r_1, (\text{sleep } (e, j, +)))])] \&$
    $\text{ATT (sp, [-} \mid r_0 < r_1 \& \text{UNTIL}(r_1, (\text{sleep } (e, j, -)))])]$

Obviously, the speaker cannot reasonably consider John to fall asleep before and after r_0 while maintaining that she also expects him to be awake until r_3, hence at r_0. The verifying embeddings of the DRS for (19 a, b) cannot all be extended to verify the entire DRS for (19a, b, c). Attitudes or subjective information states are in general preserved when the facts change, which is why they ordinarily persist in dynamic factual updates. Of course, agents do change their attitudes, and sometimes also adjust them on the basis of factual evidence, if they are or would like to be considered rational. Such belief revision is of a different nature, perhaps probabilistic or inductive, rather than deductive, hence it has no intrinsic relation to the semantics of aspectual adverbs addressed in this paper.

All aspectual adverbs, regardless of their prosodic properties, trigger additional presuppositions about the polarity reversals of the event described by the predicate at the given reference time. For instance, *still P* presupposes that at an earlier reference time *P* was the case and ensures that this is preserved at the current reference time. This presupposition of *still* ensures that (20b) is a valid conclusion given (20a),

20) a. When Mary came home, John was still asleep.
   b. So John had fallen asleep before Mary came home and had been asleep since.

$[r_0, r_1, e_1, j, m, e_2 \mid$ come home $(e_2, m, +)$ & $r_0 \supseteq e_2$ & $r_0 =$ current &
   PROG (sleep($e_1, j, +$)) & $e_1 \supseteq r_0$ & $r_1 < r_0$ & $e_1 \supseteq r_1$ &
   $r_1 \supseteq$ START(sleep($e_1, j, +$)) & SINCE (r_1, (sleep ($e_1 j, +$)))

To make the paraphrase between (20a) and (20b) fully specific, an inference rule is needed which converts the simple past state in (20a) to the past perfect in (20b) of the descriptive content that John was sleeping. Although it would take more footwork to develop such a temporal logic with inference rules in the style of natural deduction systems, it is easy to formulate the rule in a preliminary form as in (21).[13]

21) PERF introduction rule:

   For all $e \supset r_n$ and $r_n < r_m \Rightarrow$ PERF (e) $\supset r_m$

   e.g. if John fell asleep at r_n, he had fallen asleep at r_m

Similar considerations apply to the inference (22), where *no longer* introduces the past moment of waking up, which triggers itself the presupposition that John must have been asleep until he woke up. The past perfect tense ensures that at the current reference time r_0, at which Mary came home, John's waking up had already taken place.

22) a. When Mary came home, John was no longer asleep.
   b. $\models$ John had been asleep until he woke up before Mary came home.

$[\, r_0, r_1, e_1, j, m, \mid e_2$ come home $(e_2, m, +)$ & $r_0 \supseteq e_2$ & $r_0 =$ current &
   PROG (sleep($e_1, j, -$)) & $e_1 \supseteq r_0$ & $r_1 < r_0$ & $e_1 \supseteq r_1$ & $r_1 \supseteq$ END(sleep($e_1, j, +$))
   & SINCE (r_1, (sleep ($e_1 j, -$))) & PROG (sleep($e_1, j, +$)) & UNTIL (r_1, (sleep ($e_1 j, +$)))

The accommodation of presuppositions in (non-empty) contexts is substantially constrained by the content of preceding clauses (cf. [1]). Violation of local consistency between the presupposition and prior asserted content may lead to unintelligibility or a breakdown of understanding between speaker and hearer. From (23a) we quickly conclude that John has woken up. But when the premises are presented in reverse order as in (23b), we cannot conclude that John has fallen asleep again, eventhough this is the presupposition of *still* in the second clause in (23b). However, if this presupposed information in (23b) is instead asserted, as in (23c), there is no problem in interpreting the discourse as coherent.

23) a. John is still asleep. John is no longer asleep. $\models$ John has woken up
   b. John is no longer asleep. John is still asleep. $\not\models$ John has fallen asleep again
   c. John is no longer asleep. John fell asleep (again). John is still asleep

The accommodation problem in (23b) is caused by a conflict of the presuppositions of *still*, which require that at an earlier reference time John had fallen asleep and had

---

[3] See [4], [5], [6], [7].

been asleep since. But this cannot be true in the context updated by asserting the stative clause that John is no longer asleep, which requires that he is not asleep at that given reference time. In acccommodating a presupposition in a non-empty context, a conflict between the presupposition of new content and the conditions which are true at the given reference time cannot be resolved by introducing another, intervening reference time at which the presupposed information is true. Accommodation is, like the DRT notion of logical consequence, defined in (17) above, static, i.e. unable to adjust the context and to introduce new reference times. Asserting information does introduce new reference markers and effectuates context change, which resets the current temporal reference.

As final consideration of how aspectual adverbs serve in adjusting context in a multiagent setting, let's briefly look at the way the counterfactual epistemic states are used in planning contexts. Suppose (24a) is uttered in a situation where agents share the information that they are to have dinner at 9, and that John is supposed to be asleep before dinner, hence he will not participate in the dinner.

24) a. John is ALREADY asleep, so let's have dinner at 8.
    b. Let's have dinner at 8. John is already asleep.

In (24a) John fell asleep earlier than the speaker had expected, as indicated by *ALREADY*. Now that he fell asleep before 8, the plan is adjusted to have dinner earlier. From (24b), reversing the order of the two clauses, in the context containing the plan to have dinner at 8, as asserted by the first clause, *already* may lose its subjective counterfactual temporal meaning. Instead, (24b) indicates that one of the first conditions necessary to fulfill the plan to have dinner at 8, i.e. that John should be asleep, has been satisfied. Elaborating the DRT account with subjective planning information and a shared common ground to which agents all have access would be a first enrichment of the semantic representations required for (24),

5 Semantics or Pragmatics?

Stalnaker ([10] 153-5) discussed two different ways to demarcate semantics from pragmatics, reflecting a difference in the role the notion of context plays in the explanation of the linguistic facts.

(1) a fact is pragmatic if it is independent from the truth conditional content and appeals to principles, maxims and inference rules other than logical deduction.

(2) meaning determines certain aspects of the interpretation of a speech act, and the context determiners other aspects of its interpretation. Information is characterized as semantic if it is based on rules which any competent speaker of the language must know to communicate effectively. Information is pragmatic when it relies on knowing certain factual circumstances under which the speech act was performed or knowledge of the world that may be used in determining what was said.

It should be evident that the DRT account of aspectual adverbs offered in this paper is semantic on both counts, for aspectual adverbs determines truth conditional content regarding epistemic attitudes of the speaker, and it is semantic in that it determines temporal content, relative to contextual information about reference times, which does not depend on matters of fact or common sense knowledge, but on linguistic competence. What remains for genuine pragmatics is to determine just which epistemic attitude the speaker wants to express by using marked prosody on an aspectual adverb. Perhaps a more detailed account of such issues relating to rhetorical relations arising in discourse needs to rely on a phonologically sophisticated analysis of the nature of the intonational contour used.

6 Conclusion

In this account of the dynamic semantics of aspectual adverbs a story, assumed to constitute coherent discourse, constitutes the premises from which the conclusion is drawn. The interpretation of the premises is itself modelled as a dynamic process in which the reference time is updated to later ones when dynamic information requires it. The construction rules for the DRSs are semantic in nature and the standard logical notion of entailment in DRT serves to characterize validity without any appeal to hairy notions such as a 'normal' course of events or 'normal possible world' or to common sense about what the world is like or how causal connections arise, as is needed in default logics (cf. [2]).

References

[1] Beaver, D., Presupposition, in [8] (1997) 939–1008.
[2] Lascarides, A. and N. Asher Temporal Interpretation, Discourse Relations and Commonsense Entailment, *Linguistics and Philosophy* **16.5** (1993) 437–493
[3] Kamp, H. and U. Reyle *From discourse to log(ic.* Kluwer, Dordrecht, (1993).
[4] ter Meulen, A. G. B. Dynamic Aspect Trees, with J. Seligman, *in*: L. Polos et al. (eds.), *Applied Logic: How, what and why. Logical approaches to natural language.* Synthese Library vol, 247. Kluwer. Dordrecht. (1994) 287–320.
[5] ter Meulen, A. G. B. *Representing Time in Natural Language. The dynamic interpretation of tense and aspect.* Bradford Books, MIT Press, Cambridge, (1995),
[6] ter Meulen, A. G. B. Chronoscopes: the dynamic representation of facts and events, *in* J. Higginbotham et al. (eds.) *Speaking about events.* Oxford University Press, NY and Oxford, (2000).
[7] ter Meulen, A. G. B. Situated reasoning in time about time, B. Löwe, et al. (eds.), *Foundations of the Formal Sciences II, Applications of Mathematical Logic in Philosophy and Linguistics*, in series Trends in Logic, Vol. 17. Kluwer, Dordrecht, (2003).
[8] van Benthem, J.F. A. K. and A.G.B. ter Meulen (eds.), *Handbook of Logic and Language*. Elsevier Science, Amsterdam, & MIT Press, Cambridge, (1997).
[9] van Eijck, J. and H. Kamp. Representing discourse in context. in [8] 179–237.
[10] Stalnaker, R. *Context and content.* Oxford Cognitive Science series. Oxford University Press, New York and Oxford, (1999).

Connecting Route Segments Given in Route Descriptions[1]

Ladina Tschander

University of Hamburg, Department for Informatics
Knowledge and Language Processing (WSV)
Vogt-Kölln-Str. 30, 22527 Hamburg, Germany
tschander@informatik.uni-hamburg.de

Abstract. This investigation focuses on specific cue phrases given in route descriptions. Since verbal route descriptions portray the route by conveying its route segments, these segments have to be connected via semantic and pragmatic analysis for constituting a route representation. I exemplify how sequential and descriptive cue phrases given in route descriptions assist the combination of spatial information used for building up internal representations. They trigger pragmatic inferences that are used for constituting the spatial relations of route segments and for constituting expectations about the environment. These representations establish the context for (virtual) navigation.

1 Introduction

Natural language descriptions are used to solve the problem how to come from A to B. To navigate successfully, two aspects of information are given in route descriptions: on the one hand, the spatial information about the environment and, on the other hand, the information about actions that have to be performed. In the model of a *geometric agent*[2] (given in [11]), we represent these information aspects separately during interpretation. The spatial information of the route description is represented as a net-like structure that abstracts from linguistic details of the route description. The action plan constitutes a sequence of commands, which employ a small set of imperative operators that point to the given route representation.

Main topics of this contribution are the questions how described route segments are connected in route representations, and how phrases as *und dann [and then]*, *am Ende [at the end]*, or *dahinter [behind it]* assist the interpretation of descriptions of unknown routes. Therefore, the representation of spatial information constituting the relevant context for navigation is focussed on.

---

[1] The research reported in this paper was conducted in the project ConcEv (Conceptualizing Events), which is supported by the DFG in the priority program 'Language Production' under grant Ha-1237/10 to Christopher Habel.
[2] The geometric agent navigates on routes in a virtual planar environment according to natural language instructions presented in advance. The goal of this investigation is to build a formal framework that demonstrates the performance of specific theories of natural language interpretation in the presence of sensing.

I consider the function of statements in route descriptions as to add their content to the representation (see [10]). However, context is not seen as a set of assumptions that are taken as shared in advance. Instead, the relevant context is constituted during interpretation using common knowledge of spatial concepts and knowledge about how lexical items refer to spatial concepts, as well as pragmatic rules. Thus, the established context is a representation of spatial conditions that gives rise for expectations about the environment.

So, in order to understand route descriptions (as given in hiking guides), enriched formal representations have to be gained from route descriptions. An important enrichment is that the information given by the statements has to be connected to each other. Before presenting the enrichments done by pragmatic inferences in section 3, the representation of route segments is given in section 2.

2 Route Segments

Route segments correspond to the spatial concept of paths. Paths are linear, directed, and bounded entities. They have two distinguished points (the starting point precedes every other point of the path, the final point is preceded by every other point of the path, see [3]). Segments are mentioned by phrases containing spatial prepositions or adverbs, or verbs of motion, since these expressions refer to the concept of paths by their meanings.

For example, the first statement (1a) specifies the first route segment completely by explicitly mentioning the starting position with the SOURCE prepositional phrase *von der Bushaltestelle Vahrenwinkelweg [from the bus stop Vahrenwinkelweg]* that is combined with the verb *führen [lead]* specified by the GOAL prepositional phrase *zur Autobahnunterführung [to the highway underpass]*. The second statement implicitly communicates a segment by mentioning the direction (1b), and gives a further segment (1c) with the verbal phrase *zum Kaiserstuhlweg kommen [come to the Kaiserstuhlweg]*.

(1) (a) Von der Bushaltestelle Vahrenwinkelweg führt uns der Weg bergab zur Autobahnunterführung.
[From the bus stop Vahrenwinkelweg the way leads us downhill to the highway underpass.]
(b) Wir behalten die Richtung bei
[We keep the direction (particle)]
(c) und kommen über einen Hang zum Kaiserstuhlweg.
[and come across a slope to the Kaiserstuhlweg.]

The interpretation process subdivides route descriptions into route segments by lexical items containing a path in their meaning. If a route segment is introduced by a verb of motion, its starting point or final point can be specified explicitly via prepositional phrases. Route segments given by directional prepositional phrases can be modified by another prepositional phrase, as well. Thus, syntactic and semantic information of verbal and prepositional phrases determine route segments.

I represent conceptualizations of route segments and routes using *referential nets* ([5], [6]). The referential net approach, which is kindred to *discourse representation theory* ([7]), was developed to model cognitively motivated linguistic processes, especially representations changing over time. In referential nets, all information about entities are associated with *referential objects* (refOs). RefOs are specified by their sort and by their descriptions. Since all information about an entity is bundled by its refO, refOs can be rearranged and the representation is independent of the linear order of the linguistic input. In the syntactic-semantic analysis, verbs of motion bring in *situation refOs* as well as *path refOs*. Path refOs are also introduced by directional prepositions, whereas locational prepositions give locations connected to regions. Nominal phrases and pronouns normally introduce *object refOs*.

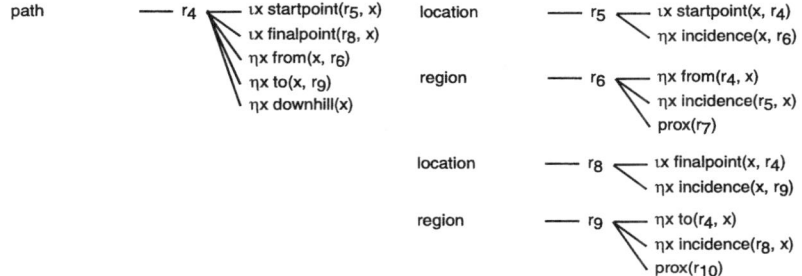

Fig. 1. Representation of the first route segment (1a)

Fig. 1 gives the representation of the first route segment gained by the semantic analysis. The refO r_4 is specified by the sort *path* and by five descriptions (as *ιx startpoint(r_5, x)* or *ηx from(x, r_6)*). The sort specifies the entity given by the refO, e.g. the sort of r_4 indicates that the refO refers to a path, which is a spatial entity.[3] According to the spatial knowledge, paths have two distinguished points that can be specified by regions mentioned by directional prepositional phrases. The description *ιx startpoint(r_5, x)*[4] means that r_5 is the location of the starting point. The beginning of the path from the region specified by r_6 is given by the description *ηx from(x, r_6)*. Therefore, the location of the starting point is in the region given by the description *ηx incidence(x, r_6)*. Entities of the environment[5] induce regions in their proximity (given as functional description *prox(r_7)*).

Since route descriptions depict the segments of a route, the connection of these segments is important for constituting representations of complete routes. The representation of the route is constituted by the combination of the path information given in the route description, this means refOs containing a *path* attribute have to be connected.[6] Therefore, the focussed question is how the connections of route segments are established during the interpretation of a route description. The required pragmatic inferences are given in the next section.

---

[3] Note that the path is not visually persistent, i.e. a path here is not a 'real-world' track.
[4] *ιx startpoint(r_5, x)* holds the information that r_4 is *the x which contains the starting point given by r_5*. *ηx* is the indefinite counterpart to the *ι*-operator (see [5]).
[5] These entities are given by refOs with the sort *immobile objects, mobile objects, tracks* or *buildings*. They are depicted in Fig. 3.
[6] Note that *path refOs* are not connected by the analysis given so far.

3 Representation of Routes

One view of the main task of pragmatic principles is that they play a crucial role in mapping semantic representations onto meanings of utterances (see [9]). They support the constitution of representations by establishing a relation between the semantic representation and a specific part of knowledge. In the case of route descriptions, the background knowledge is constituted by the knowledge about a general discourse structure (see [2]) and the spatial knowledge about paths and constellations of paths.

The semantic structure is constituted based on the syntactic information and the lexicon during the interpretation. Route segments are determined by the semantic structure. The connection of segments depends on the semantic structure, the knowledge of the pragmatic principles, as well as the background knowledge. The internal representation of a described route is the result of a semantic and pragmatic analysis in that the possible combinations of described route segments are minimized. Cue phrases and pragmatic inferences play a particular role in this minimization.

3.1 Cue Phrases as Connecting Elements in Route Descriptions

Cue phrases[7] are used for the combination of information given in different statements establishing coherence. In the case of route descriptions, they associate information units belonging to route segments. They do not only link information units (as logical operators) but also provide closer relations between the combined information units.

Two classes of cue phrases are observed in route descriptions. One class contains conjunctions as *und [and]*[8] or *dann [then]* or adverbs as *zunächst [first]*. The other one is constituted by phrases picking up mentioned information as pronominal adverbs, e.g. *dahinter [behind it]*, deictic adverbs as *hier [here]* or phrases like *Am Waldende [at the end of the forest]* or *in unveränderter Richtung [in unchanged direction]*. In the former class, the sequence is relevant (see [4] and [1] for a detailed analysis) and the sequence constitutes a closer relation between the given information units. The sequence is not only determined by the temporal course but also by the spatial ordering of objects. Due to the sequential order of route segments, a continuous movement is described, i.e. a continuous sequence of events is given. Therefore, I call cue phrases sensitive to the sequence of information **sequential cue phrases**. In the latter class, all cue phrases have in common that they spatially relate a new entity to an already mentioned one. Normally, these phrases describe a region or entity mentioned before used for (re-)orienting the bearer of motion. Furthermore, they interrupt the sequence of motion events. Cue phrases of this type are called **descriptive cue phrases**. Generally, cue phrases do not belong to the same syntactic category, and since they are kindred to linguistic material establishing coherence, it is to be expected that they appear frequently in descriptions of route segments.

---

[7] The conception of *cue phrases* is inspired by [8].
[8] However, *und [and]* does not count in every use as a sequential cue phrase. For instance, in *Schöner Tiefblick auf Bever und in das gleichnamige Tal [Beautiful view on Bever and into the valley of the same name]* the conjunction does not combine a sequence but two events in which one is part of the other.

Additionally, cue phrases do not only combine information but also trigger inferences used for the constitution of route representations.

The cue phrases in (2) belong to the sequential one.[9] As mentioned before, the sequence of route segments is given by the sequence of descriptions, and the order has not to be marked explicitly. Therefore, according to Grice's maxim of quantity and Levinson's Q-principle[10], cue phrases trigger inferences. Here, it can be inferred that there are no crossings before the *Bäckerstrasse* giving an expectation of a spatial condition of the environment.

(2) Von der Bushaltestelle Schenefeld Mitte gehen wir in die Nedderstrasse, auf der wir ZUNÄCHST die Bäckerstrasse UND ALS NÄCHSTES hinter dem Hotel Klövensteen die Hauptstrasse überqueren, um in den Uetersener Weg einzubiegen. [...]
[From the bus stop Schenefeld Mitte we go into the Nedderstrasse on which we cross FIRST the Bäckerstrasse AND NEXT behind the hotel
Klövensteen the main street in order to turn into the Uetersener Weg.]

The phrase *die Richtung beibehalten [to keep the direction (particle)]* in the second segment of (1) belongs to the descriptive cue phrases. According to its meaning, something for comparison has to be given to interpret the phrase. Assuming that cue phrases associate refOs with the same type of sort, and since the corresponding refO contains the sort *path*, the thing for comparison is the path mentioned by the preceding route segment. The meaning of the phrase specifies furthermore that the direction of the second path has to be specified. This is given by the description $\eta x\ direction(x, P, L)$ meaning that the path x has the direction of the path P at the location L (see Fig. 3). Thus, pragmatic principles specify which mentioned path and which given direction are selected for fulfilling the semantic demands of the phrase.

In comparison to sequential cue phrases that normally trigger inferences resulting in expectations about the environment, descriptive cue phrases trigger inferences about the spatial relations of paths.

---

[9] Some cue phrases can only appear in specific linguistic contexts, as *zunächst [first]*. It demands that the sequence is marked explicitly. But, only *und [and]* with its pragmatic reading is not as acceptable as if *als nächstes [next]* is given explicitly (note that an interruption fulfills the same function). There are two possibilities: either *zunächst [first]* blocks the pragmatic reading of *und [and]* or *zunächst [first]* demands a stronger sequential cue phrase; however, ordering cue phrases is beyond of the scope of this paper.

[10] Levinson ([9]) gives three principles: the Q-, the I-, and the M-principle. The Q-principle corresponds to Grice's first maxim of quantity ("Make your contribution as informative as is required"). The second maxim of quantity ("do not make your contribution more informative than is required") is reflected in the I-principle. This principle is responsible for a lot of enrichments. Enrichments has to be done, because a communicative strategy is that one does not provide unnecessary information, especially one would not say anything obvious. The last principle (M-principle) corresponds to the maxim of manner ("be perspicuous") saying that what is given in an abnormal way indicates an abnormal situation, or marked messages indicate marked situations.

3.2 Pragmatic Principles Constituting Route Representations

For constituting a route representation, it is (at least) necessary to combine the given paths (i.e. to connect r_4, r_{13} and r_{15} given in Fig. 3). This combination is supported by pragmatic principles establishing an enriched representation. Connections of route segments are only one kind of enrichment. As mentioned before, route segments are enriched by expectations on the spatial conditions of the environment, as well.

Although each aspect of a path can be specified, the goal is most frequently indicated. The source of a segment is frequently left implicit in route descriptions. Nevertheless, the connection of segments is unproblematic. According to the I-principle saying that paratactic adjunctions suggest a sequential occurrence of events, it can be additionally assumed that the sequence of descriptions of segments corresponds to the sequence of paths. Therefore, the mentioned goal of one segment is identical to the source of the following segment.

So, the first and second route segment described in (1) meet at the region given by the highway underpass. The second segment is given by *die Richtung beibehalten [keep the direction]*. The semantics of the phrase demands a path for comparison. As mentioned before, the direction of path given by the phrase has to be specified. However, due to the spatial knowledge there are two candidates given by the first segment: a global direction *downhill* and a local direction at the highway underpass.

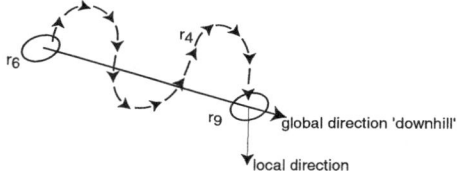

Fig. 2. Local and global directions

The direction *downhill* is valid for the whole described path. However, the way can contain several curves. Since paths are directed, they contain local directions at every position on the path. If a path contains curves, the local direction is not congruent with the global direction on every position (see Fig. 2). Which direction is selected, is determined by pragmatic inferences. During interpretation, the second route segment is first connected to the first segment. Then the I-principle selects the local direction at the end of the first path as a comparing direction. However, only the actual navigation can determine whether this selection is the right one.

Looking at different route descriptions one observes that changes of direction are given systematically. In contrast, the instruction to keep the direction is rarely given. Therefore, according to the M-principle, the instruction not to change the direction contains another information about the route. Since specifications of directions are normally given in a branching situation, more than one possible way to walk on can be expected at the highway underpass.

In the subsequent segment, the path is expressed by *über einen Hang [across a slope]* and the goal is specified by *zum Kaiserstuhlweg [to the Kaiserstuhlweg]*. Since *die Richtung beibehalten [keep going in the same direction]* connects the first and third route segment, it specifies the source of the third segment, as well.

These inferences are reflected in the referential net. It is enriched by the inferred refOs, as well as by the inferred designations. These designations are added to already existing refOs. The result of a pragmatic analysis is presented in Fig. 3.[11]

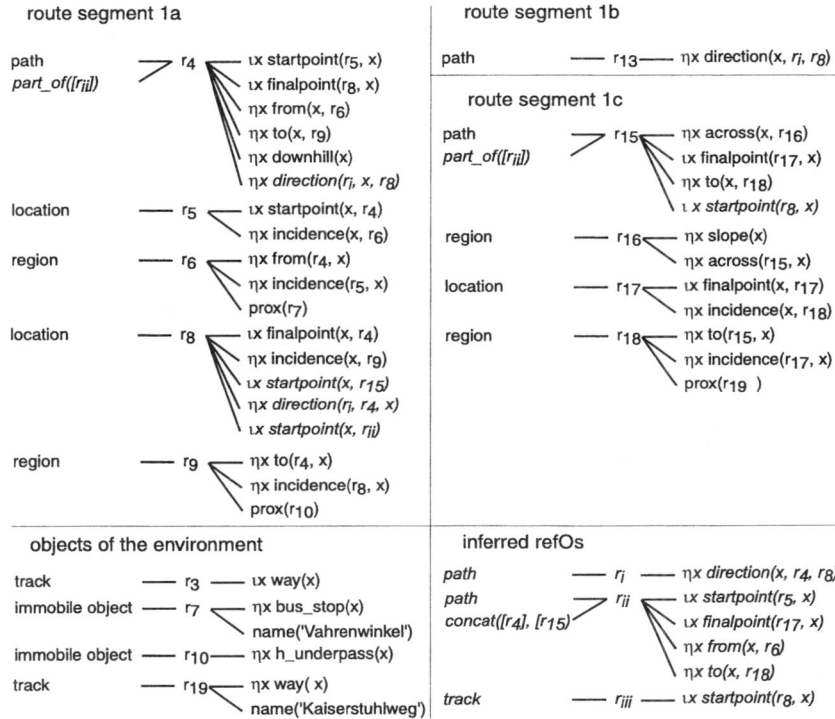

Fig. 3. Route representation of example (1): route segments (1a-c), inferred refOs, and objects of the environment constitute the route representation. Additionally, route segments (1a-c) are enriched by referred designations. Every inferred information is given in italics.

The refO r_{13} contains variables in its description $\eta x\ direction(x, P, L)$ that are identified by pragmatic inferences described above. According to the connection of the first and the second segment by the I-principle, the location L is given by the final point of the first path (r_8). Since the local direction of the path (r_4) at its final point is not yet given, a new refO is generated. It is a path refO r_i with a description $\eta x\ direction(x, r_4, r_8)$ meaning that r_i gives the direction of r_4 at r_8. Since the I-principle selects the local direction, the variable P is identified with r_i. Note that this connection is defeasible because of its pragmatic character.

The next connection is also a defeasible one. The following route segment is connected to the previous one by the sequential cue phrase *und [and]*. The cue phrase

---

[11] Note that Fig. 3 depicts only refOs which constitute the route representation. The whole representation contains additionally *situations refOs* (as situation —— r_1 —— $\eta x\ lead(x, r_2, r_3)$) that gives the temporal aspect of route descriptions as well as refOs (as mobile object —— r_2 —— $\eta x\ pron(x)$ / $\eta x\ human(x)$) representing the mobile objects given as pronouns in (1).

assists the combination of the path described by this route segment with a path described before. The last given path is r_{13} that is connected to r_4 (at the location given by r_8). Now, according to the pragmatics, the starting point of r_{15} is identified with the final point of r_4. Therefore, the paths r_4 and r_{15} can be concatenated constituting a new refO (r_{ii}). The generated refO r_{iii} grasps the expectation of another way at the highway underpass. This kind of route representations constitutes the context either for virtual navigation or navigation through the environment.

4 Conclusion

In this paper I presented the pragmatic function of cue phrases that are divided up into sequential cue phrases and descriptive cue phrases. They assist the establishment of route representations gained from verbal route descriptions given in advance. The semantic representation of the route contains gaps between segments that can be closed by the given cue phrases or via inferences. Since such combinations are marked as defeasible, they can be overridden during navigation. Furthermore, pragmatic principles and the spatial knowledge about paths enrich the representation of the route. These representations give the context for extracting action plans representing the actions to be performed.

References

1. Blakemore, D. & Carston, R. (1999). The Pragmatics of *and*-conjunctions: The non-narrative cases. *UCL Working Papers in Linguistics 11*.
2. Denis, M. (1997). The description of routes: A cognitive approach to the production of spatial discourse. *Cahiers de Psychologie Cognitive 16*. 409–458.
3. Eschenbach, C., L. Tschander, C. Habel & L. Kulik (2000). Lexical specifications of paths. In C. Freksa, W. Brauer, C. Habel & K.F. Wender (eds.), *Spatial Cognition II* (pp. 127–144). Berlin: Springer.
4. Grice, H.P. (1975). Logic and conversation. In P. Cole & J. Morgan (eds.), *Syntax & Semantics Vol.3* (pp. 41–58). New York: Academic Press.
5. Habel, C. (1986). *Prinzipien der Referentialität*. Berlin: Springer.
6. Habel, C. (1987). Cognitive linguistics: The processing of spatial concepts. In *T.A. Informations, Bulletin semestriel de l'ATALA, 28*, pp. 21–56.
7. Kamp, H. & Reyle, U. (1993). *From Discourse to Logic*, Dodrecht: Kluwer.
8. Knott, A. (1996). *A Data-Driven Methodology for Motivating a Set of Coherence Relations*. PhD thesis, Departement of Artificial Intelligence, University of Edinburgh.
9. Levinson, S.C. (2000). Presumptive Meanings: the Theory of Generalized Conversational Implicature. Cambridge: MIT-Press.
10. Stalnaker, R.C. (1999). *Context and Content*. Oxford: Oxford University Press.
11. Tschander, L., H.R. Schmidtke, C. Eschenbach & C. Habel (to appear). A Geometric Agent Following Route Instructions. In In C. Freksa, W. Brauer, C. Habel & K.F. Wender (eds.), *Spatial Cognition III*. Berlin: Springer.

Author Index

Ahn, David 1
Anguilet, Yoan 467
Arló Costa, Horacio 15
Arritt, Robert P. 29

Barkhuus, Louise 411
Bell, John 40
Bernstein, Philip 286
Bianchi, Claudia 54
Bothorel, Cécile 419
Bouquet, Paolo 66, 80
Brézillon, Patrick 94, 142, 204, 491
Bunt, Harry 427
Buvač, Saša 107

Chevalier, Karine 419
Ciskowski, Piotr 435
Costantini, Stefania 443

Fetzer, Anita 130
Fuks, Hugo 483

Ganet, Leslie 142
Giboreau, Agnès 154
Giunchiglia, Fausto 286
Gomes Pimentel, Mariano 483
Guha, R. 164
Gurevych, Iryna 272

Jang, Seiie 178

Komagata, Nobo 190
Korpipää, Panu 451
Kouadri Mostéfaoui, Ghita 204
Kulas, Andrea 467

L'Abbate, Marcello 459
Laddaga, Robert 499
Liu, Hugo 218

Mäntyjärvi, Jani 451
Magnini, Bernardo 66

McCarthy, John 164
Meulen, Alice G.B. ter 507
Musa, Rami 467
Mylopoulos, John 286

Nossum, Rolf 233

Paiva, Valeria de 116
Paletta, Lucas 245
Panizzi, Emanuele 475
Percus, Orin 259
Pereira de Lucena, Carlos José 483
Pomerol, Jean-Charles 491
Porzel, Robert 272

Richard, Jean-François 154
Robertson, Paul 499

Scheidegger, Madleina 467
Serafini, Luciano 66, 80, 286
Stojanovic, Isidora 300

Thiel, Ulrich 459
Thomas, Kavita E. 314
Thomason, Richmond H. 328
Tijus, Charles 142, 397
Tocchio, Arianna 443
Trautwein, Martin 342
Tschander, Ladina 517
Turner, Roy M. 29

Urdapilleta, Isabel 154

Whitsey, Mark 356
Widdows, Dominic 369
Woo, Woontack 178

Young, R.A. 383

Zanobini, Stefano 66
Zibetti, Elisabetta 397

Lecture Notes in Artificial Intelligence (LNAI)

Vol. 2448: P. Sojka, I. Kopecÿek, K. Pala (Eds.), Text, Speech and Dialogue. Proceedings, 2002. XII, 481 pages. 2002.

Vol. 2464: M. O'Neill, R.F.E. Sutcliffe, C. Ryan, M. Eaton, N. Griffith (Eds.), Artificial Intelligence and Cognitive Science. Proceedings, 2002. XI, 247 pages. 2002.

Vol. 2466: M. Beetz, J. Hertzberg, M. Ghallab, M.E. Pollack (Eds.), Advances in Plan-Based Control of Robotic Agents. Proceedings, 2001. VIII, 291 pages. 2002.

Vol. 2473: A. Gómez-Pérez, V.R. Benjamins, Knowledge Engineering and Knowledge Management. Proceedings, 2002. XI, 402 pages. 2002.

Vol. 2475: J.J. Alpigini, J.F. Peters, A. Skowron, N. Zhong (Eds.), Rough Sets and Current Trends in Computing. Proceedings, 2002. XV, 640 pages. 2002.

Vol. 2479: M. Jarke, J. Koehler, G. Lakemeyer (Eds.), KI 2002: Advances in Artificial Intelligence. Proceedings, 2002. XIII, 327 pages. 2002.

Vol. 2484: P. Adriaans, H. Fernau, M. van Zaanen (Eds.), Grammatical Inference: Algorithms and Applications. Proceedings, 2002. IX, 315 pages. 2002.

Vol. 2499: S.D. Richardson (Ed.), Machine Translation: From Research to Real Users. Proceedings, 2002. XXI, 254 pages. 2002.

Vol. 2504: M.T. Escrig, F. Toledo, E. Golobardes (Eds.), Topics in Artificial Intelligence. Proceedings 2002. XI, 432 pages. 2002.

Vol. 2507: G. Bittencourt, G.L. Ramalho (Eds.), Advances in Artificial Intelligence. Proceedings, 2002. XIII, 418 pages. 2002.

Vol. 2514: M. Baaz, A. Voronkov (Eds.), Logic for Programming, Artificial Intelligence, and Reasoning. Proceedings 2002. XIII, 465 pages. 2002.

Vol. 2522: T. Andreasen, A. Motro, H. Christiansen, H. Legind Larsen (Eds.), Flexible Query Answering. Proceedings 2002. XI, 386 pages. 2002.

Vol. 2527: F.J. Garijo, J.C. Riquelme, M. Toro (Eds.), Advances in Artificial Intelligence – IBERAMIA 2002. Proceedings 2002. XVIII, 955 pages. 2002.

Vol. 2531: J. Padget, O. Shehory, D. Parkes, N. Sadeh, W.E. Walsh (Eds.), Agent-Mediated Electronic Commerce IV. Proceedings, 2002. XVII, 341 pages. 2002.

Vol. 2533: N. Cesa-Bianchi, M. Numao, R. Reischuk (Eds.), Algorithmic Learning Theory. Proceedings 2002. XI, 415 pages. 2002.

Vol. 2541: T. Barkowsky, Mental Representation and Processing of Geographic Knowledge. X, 174 pages. 2002.

Vol. 2543: O. Bartenstein, U. Geske, M. Hannebauer, O. Yoshie (Eds.), Web Knowledge Management and Decision Support. Proceedings, 2001. X, 307 pages. 2003.

Vol. 2554: M. Beetz, Plan-Based Control of Robotic Agents. XI, 191 pages. 2002.

Vol. 2557: B. McKay, J. Slaney (Eds.), AI 2002: Advances in Artificial Intelligence. Proceedings 2002. XV, 730 pages. 2002.

Vol. 2560: S. Goronzy, Robust Adaptation to Non-Native Accents in Automatic Speech Recognition. Proceedings, 2002. XI, 144 pages. 2002.

Vol. 2569: D. Gollmann, G. Karjoth, M. Waidner (Eds.), Computer Security – ESORICS 2002. Proceedings, 2002. XIII, 648 pages. 2002.

Vol. 2577: P. Petta, R. Tolksdorf, F. Zambonelli (Eds.), Engineering Societies in the Agents World III. Proceedings, 2002. X, 285 pages. 2003.

Vol. 2581: J.S. Sichman, F. Bousquet, P. Davidsson (Eds.), Multi-Agent-Based Simulation II. Proceedings, 2002. X, 195 pages. 2003.

Vol. 2583: S. Matwin, C. Sammut (Eds.), Inductive Logic Programming. Proceedings, 2002. X, 351 pages. 2003.

Vol. 2586: M. Klusch, S. Bergamaschi, P. Edwards, P. Petta (Eds.), Intelligent Information Agents. VI, 275 pages. 2003.

Vol. 2592: R. Kowalczyk, J.P. Müller, H. Tianfield, R. Unland (Eds.), Agent Technologies, Infrastructures, Tools, and Applications for E-Services. Proceedings, 2002. XVII, 371 pages. 2003.

Vol. 2600: S. Mendelson, A.J. Smola, Advanced Lectures on Machine Learning. Proceedings, 2002. IX, 259 pages. 2003.

Vol. 2627: B. O'Sullivan (Ed.), Recent Advances in Constraints. Proceedings, 2002. X, 201 pages. 2003.

Vol. 2631: R. Falcone, S. Barber, L. Korba, M. Singh (Eds.), Trust, Reputation, and Security: Theories and Practice. Proceedings, 2002. X, 235 pages. 2003.

Vol. 2636: E. Alonso, D, Kudenko, D. Kazakov (Eds.), Adaptive Agents and Multi-Agent Systems. XIV, 323 pages. 2003.

Vol. 2637: K.-Y. Whang, J. Jeon, K. Shim, J. Srivastava (Eds.), Advances in Knowledge Discovery and Data Mining. Proceedings, 2003. XVIII, 610 pages. 2003.

Vol. 2639: G. Wang, Q. Liu, Y. Yao, A. Skowron (Eds.), Rough Sets, Fuzzy Sets, Data Mining, and Granular Computing. Proceedings, 2003. XVII, 741 pages. 2003.

Vol. 2645: M.A. Wimmer (Ed.), Knowledge Management in Electronic Government. Proceedings, 2003. XI, 320 pages. 2003.

Vol. 2663: E. Menasalvas, J. Segovia, P.S. Szczepaniak (Eds.), Advances in Web Intelligence. Proceedings, 2003. XII, 350 pages. 2003.

Vol. 2671: Y. Xiang, B. Chaib-draa (Eds.), Advances in Artificial Intelligence. Proceedings, 2003. XIV, 642 pages. 2003.

Vol. 2680: P. Blackburn, C. Ghidini, R.M. Turner, F. Giunchiglia (Eds.), Modeling and Using Context. Proceedings, 2003. XII, 525 pages. 2003.

Vol. 2689: K.D. Ashley, D.G. Bridge (Eds.), Case-Based Reasoning Research and Development. Proceedings, 2003. XV, 734 pages. 2003.

Vol. 2702: P. Brusilovsky, A. Corbett, F. de Rosis (Eds.), User Modeling 2003. Proceedings, 2003. XIV, 436 pages. 2003.

Lecture Notes in Computer Science

Vol. 2655: J.-P. Rosen, A. Strohmeier (Eds.), Reliable Software Technologies – Ada-Europe 2003. Proceedings, 2003. XIII, 489 pages. 2003.

Vol. 2656: E. Biham (Ed.), Advances in Cryptology – EUROCRPYT 2003. Proceedings, 2003. XIV, 429 pages. 2003.

Vol. 2657: P.M.A. Sloot, D. Abramson, A.V. Bogdanov, J.J. Dongarra, A.Y. Zomaya, Y.E. Gorbachev (Eds.), Computational Science – ICCS 2003. Proceedings, Part I. 2003. LV, 1095 pages. 2003.

Vol. 2658: P.M.A. Sloot, D. Abramson, A.V. Bogdanov, J.J. Dongarra, A.Y. Zomaya, Y.E. Gorbachev (Eds.), Computational Science – ICCS 2003. Proceedings, Part II. 2003. LV, 1129 pages. 2003.

Vol. 2659: P.M.A. Sloot, D. Abramson, A.V. Bogdanov, J.J. Dongarra, A.Y. Zomaya, Y.E. Gorbachev (Eds.), Computational Science – ICCS 2003. Proceedings, Part III. 2003. LV, 1165 pages. 2003.

Vol. 2660: P.M.A. Sloot, D. Abramson, A.V. Bogdanov, J.J. Dongarra, A.Y. Zomaya, Y.E. Gorbachev (Eds.), Computational Science – ICCS 2003. Proceedings, Part IV. 2003. LVI, 1161 pages. 2003.

Vol. 2663: E. Menasalvas, J. Segovia, P.S. Szczepaniak (Eds.), Advances in Web Intelligence. Proceedings, 2003. XII, 350 pages. 2003. (Subseries LNAI).

Vol. 2665: H. Chen, R. Miranda, D.D. Zeng, C. Demchak, J. Schroeder, T. Madhusudan (Eds.), Intelligence and Security Informatics. Proceedings, 2003. XIV, 392 pages. 2003.

Vol. 2667: V. Kumar, M.L. Gavrilova, C.J.K. Tan, P. L'Ecuyer (Eds.), Computational Science and Its Applications – ICCSA 2003. Proceedings, Part I. 2003. XXXIV, 1060 pages. 2003.

Vol. 2668: V. Kumar, M.L. Gavrilova, C.J.K. Tan, P. L'Ecuyer (Eds.), Computational Science and Its Applications – ICCSA 2003. Proceedings, Part II. 2003. XXXIV, 942 pages. 2003.

Vol. 2669: V. Kumar, M.L. Gavrilova, C.J.K. Tan, P. L'Ecuyer (Eds.), Computational Science and Its Applications – ICCSA 2003. Proceedings, Part III. 2003. XXXIV, 948 pages. 2003.

Vol. 2670: R. Peña, T. Arts (Eds.), Implementation of Functional Languages. Proceedings, 2002. X, 249 pages. 2003.

Vol. 2671: Y. Xiang, B. Chaib-draa (Eds.), Advances in Artificial Intelligence. Proceedings, 2003. XIV, 642 pages. 2003. (Subseries LNAI).

Vol. 2672: M. Endler, D. Schmidt (Eds.), Middleware 2003. Proceedings, 2003. XIII, 513 pages. 2003.

Vol. 2673: N. Ayache, H. Delingette (Eds.), Surgery Simulation and Soft Tissue Modeling. Proceedings, 2003. XII, 386 pages. 2003.

Vol. 2674: I.E. Magnin, J. Montagnat, P. Clarysse, J. Nenonen, T. Katila (Eds.), Functional Imaging and Modeling of the Heart. Proceedings, 2003. XI, 308 pages. 2003.

Vol. 2675: M. Marchesi, G. Succi (Eds.), Extreme Programming and Agile Processes in Software Engineering. Proceedings, 2003. XV, 464 pages. 2003.

Vol. 2676: R. Baeza-Yates, E. Chávez, M. Crochemore (Eds.), Combinatorial Pattern Matching. Proceedings, 2003. XI, 403 pages. 2003.

Vol. 2678: W. van der Aalst, A. ter Hofstede, M. Weske (Eds.), Business Process Management. Proceedings, 2003. XI, 391 pages. 2003.

Vol. 2679: W. van der Aalst, E. Best (Eds.), Applications and Theory of Petri Nets 2003. Proceedings, 2003. XI, 508 pages. 2003.

Vol. 2680: P. Blackburn, C. Ghidini, R.M. Turner, F. Giunchiglia (Eds.), Modeling and Using Context. Proceedings, 2003. XII, 525 pages. 2003. (Subseries LNAI).

Vol. 2681: J. Eder, M. Missikoff (Eds.), Advanced Information Systems Engineering. Proceedings, 2003. XV, 740 pages. 2003.

Vol. 2686: J. Mira, J.R. Álvarez (Eds.), Computational Methods in Neural Modeling. Proceedings, Part I. 2003. XXVII, 764 pages. 2003.

Vol. 2687: J. Mira, J.R. Álvarez (Eds.), Artificial Neural Nets Problem Solving Methods. Proceedings, Part II. 2003. XXVII, 820 pages. 2003.

Vol. 2688: J. Kittler, M.S. Nixon (Eds.), Audio- and Video-Based Biometric Person Authentication. Proceedings, 2003. XVII, 978 pages. 2003.

Vol. 2689: K.D. Ashley, D.G. Bridge (Eds.), Case-Based Reasoning Research and Development. Proceedings, 2003. XV, 734 pages. 2003. (Subseries LNAI).

Vol. 2692: P. Nixon, S. Terzis (Eds.), Trust Management. Proceedings, 2003. X, 349 pages. 2003.

Vol. 2694: R. Cousot (Ed.), Static Analysis. Proceedings, 2003. XIV, 505 pages. 2003.

Vol. 2695: L.D. Griffin, M. Lillholm (Eds.), Scale Space Methods in Computer Vision. Proceedings, 2003. XII, 816 pages. 2003.

Vol. 2701: M. Hofmann (Ed.), Typed Lambda Calculi and Applications. Proceedings, 2003. VIII, 317 pages. 2003.

Vol. 2702: P. Brusilovsky, A. Corbett, F. de Rosis (Eds.), User Modeling 2003. Proceedings, 2003. XIV, 436 pages. 2003. (Subseries LNAI).

Vol. 2704: S.-T. Huang, T. Herman (Eds.), Self-Stabilizing Systems. Proceedings, 2003. X, 215 pages. 2003.

Vol. 2706: R. Nieuwenhuis (Ed.), Rewriting Techniques and Applications. Proceedings, 2003. XI, 515 pages. 2003.

Vol. 2707: K. Jeffay, I. Stoica, K. Wehrle (Eds.), Quality of Service – IWQoS 2003. Proceedings, 2003. XI, 517 pages. 2003.

Vol. 2709: T. Windeatt, F. Roli (Eds.), Multiple Classifier Systems. Proceedings, 2003. X, 406 pages. 2003.

Vol. 2714: O. Kaynak, E. Alpaydin, E. Oja, L. Xu (Eds.), Artificial Neural Networks and Neural Information Processing – ICANN/ICONIP 2003. Proceedings, 2003. XXII, 1188 pages. 2003.

Vol. 2716: M.J. Voss (Ed.), OpenMP Shared Memory Parallel Programming. Proceedings, 2003. VIII, 271 pages. 2003.